BASIC MECHANICS

BASIC MECHANICS

BASUDEB BHATTACHARYYA
Associate Professor
Department of Aerospace Engineering & Applied Mechanics
Bengal Engineering and Science University, Shibpur
Howrah

SUBAL CHANDRA BERA
Professor of Mechanical Engineering
Narula Institute of Technology
Kolkata
and
Former Faculty, Department of Applied Mechanics
Bengal Engineering College
Howrah

Publishing Globally

NEW AGE INTERNATIONAL (P) LIMITED, PUBLISHERS
LONDON • NEW DELHI • NAIROBI
Bangalore • Chennai • Cochin • Guwahati • Hyderabad • Kolkata • Lucknow • Mumbai
Visit us at www.newagepublishers.com

Published by New Age International (P) Ltd., Publishers
First Edition: 2009
Reprint: 2018

GLOBAL OFFICES

- **New Delhi** **NEW AGE INTERNATIONAL (P) LIMITED, PUBLISHERS**
  7/30 A, Daryaganj, New Delhi-110002, (INDIA)
  Tel.: (011) 23253771, 23253472, **Telefax:** 23267437, 43551305
  E-mail: contactus@newagepublishers.com • Visit us at www.newagepublishers.com
- **London** **NEW AGE INTERNATIONAL (UK) LTD.**
  27 Old Gloucester Street, London, WC1N 3AX, UK
  E-mail: info@newacademicscience.co.uk • Visit us at www.newacademicscience.co.uk
- **Nairobi** **NEW AGE GOLDEN (EAST AFRICA) LTD.**
  Ground Floor, Westlands Arcade, Chiromo Road (Next to Naivas Supermarket)
  Westlands, Nairobi, KENYA, **Tel.:** 00-254-713848772, 00-254-725700286
  E-mail: kenya@newagepublishers.com

BRANCHES

- **Bangalore** 37/10, 8th Cross (Near Hanuman Temple), Azad Nagar, Chamarajpet, Bangalore- 560 018
  Tel.: (080) 26756823, **Telefax:** 26756820, **E-mail: bangalore@newagepublishers.com**
- **Chennai** 26, Damodaran Street, T. Nagar, Chennai-600 017, **Tel.:** (044) 24353401
  Telefax: 24351463, **E-mail: chennai@newagepublishers.com**
- **Cochin** CC-39/1016, Carrier Station Road, Ernakulam South, Cochin-682 016
  Tel.: (0484) 2377303, **Telefax:** 4051304, **E-mail: cochin@newagepublishers.com**
- **Guwahati** Hemsen Complex, Mohd. Shah Road, Paltan Bazar, Near Starline Hotel
  Guwahati-781 008, **Tel.:** (0361) 2513881, **Telefax:** 2543669
  E-mail: guwahati@newagepublishers.com
- **Hyderabad** 105, 1st Floor, Madhiray Kaveri Tower, 3-2-19, Azam Jahi Road, Near Kumar Theater
  Nimboliadda Kachiguda, Hyderabad-500 027, **Tel.:** (040) 24652456, **Telefax:** 24652457
  E-mail: hyderabad@newagepublishers.com
- **Kolkata** RDB Chambers (Formerly Lotus Cinema) 106A, 1st Floor, S N Banerjee Road
  Kolkata-700 014, **Tel.:** (033) 22273773, **Telefax:** 22275247
  E-mail: kolkata@newagepublishers.com
- **Lucknow** 16-A, Jopling Road, Lucknow-226 001, **Tel.:** (0522) 2209578, 4045297, **Telefax:** 2204098
  E-mail: lucknow@newagepublishers.com
- **Mumbai** 142C, Victor House, Ground Floor, N.M. Joshi Marg, Lower Parel, Mumbai-400 013
  Tel.: (022) 24927869, **Telefax:** 24915415, **E-mail: mumbai@newagepublishers.com**
- **New Delhi** 22, Golden House, Daryaganj, New Delhi-110 002, **Tel.:** (011) 23262368, 23262370
  Telefax: 43551305, **E-mail: sales@newagepublishers.com**

ISBN: 978-81-224-2694-6

C-17-09-10779

Printed in India at Print-O-Pack, Delhi.
Typeset at In-house, Delhi.

NEW AGE INTERNATIONAL (P) LIMITED, PUBLISHERS
7/30 A, Daryaganj, New Delhi-110002
Visit us at **www.newagepublishers.com**
(CIN: U74899DL1966PTC004618)

Dedicated

to

the student community who will use this book

PREFACE

Basic Mechanics, within the purview of WBUT syllabus, deals mainly the fundamental aspects of statics and dynamics of rigid body analysis and the basic preview to the study of deformable bodies mechanics. To achieve an in-depth knowledge of this basic subject, it needs very lucid elaboration of the text and also presentation of numerous worked out examples for complete understanding of the topic. Keeping these basic requirements in mind, we have tried to exert our best to come up with a most useful book.

In this newly revised edition, we have tried to rectify the errors and omissions detected earlier. The present version mainly features

- Thorough discussion of the topic in basic level, avoiding much detailing and critical discourse.
- Inclusion of multiple choice questions relevant to the topic.
- Provision of number of useful numerical examples for holistic understanding of the topic.

The forbearance as well as the active support of our family members and the useful suggestions by our colleagues and the commendable support from the editorial team of the publishing house consorted successfully to bloom this project. Though proper care has been adopted to make the book error free, any unintended mistake or error or omission noticed may be directly communicated to the publisher or the author(s).

Dr. Basudeb Bhattacharyya

Prof. Subal Chandra Bera

SYLLABUS

West Bengal University of Technology
New Syllabus for Basic Mechanics

1. Importance of mechanics in engineering; Introduction to Statics; Concept of particle and rigid body; Type of forces: collinear, concurrent, parallel; concentrated, distributed; Vector and scalar quantities; Force is a vector: Transmissibility of a force (sliding vector).
2. Introduction to Vector Algebra: Parallelogram law: Addition and subtraction of vectors; Lami's theorem; Free vector; Bound vector; Representation of forces in terms of *i.j.k.*; Cross product and Dot product and their applications.
3. Two dimensional force system; Resolution of forces; Moment; Varignon's theorem; Couple; Resolution of a coplanar force by its equivalent force-couple system; Resultant of forces.
4. Concept and equilibrium of forces in two dimensions; Free body concept and diagram; Equations of equilibrium.
5. Concept of Friction; Laws of Coulomb friction; Angle of repose; Coefficient of friction.
6. Distributed Force; Centroid and Centre of Gravity; Centroids of a triangle, circular sector, quadrilateral, composite areas consisting of above figures.
7. Moments of inertia; MI of plane figure with respect to an axis in its plane, MI of plane figure with respect to an axis perpendicular to the plane of the figure; Parallel axis theorem: Mass moment of inertia of symmetrical bodies, e.g. cylinder, sphere, cone.
8. Concept of simple stresses and strains; Normal stress Shear stress, Bearing stress, Normal strain; Shearing strain; Hooke's law; Poisson's ratio; Stress-strain diagram of ductile and brittle materials; Elastic limit; Ultimate stress; Yielding; Modulus of elasticity; Factor of safety.

9. Introduction to dynamics; Kinematics and Kinetics; Newton's laws of motion; Law of gravitation & acceleration due to gravity; Rectilinear motion of particles; determination of position, velocity and acceleration under uniform and non-uniformly accelerated rectilinear motion; construction of *x-t*, *v-t* and *a-t* graphs.

10. Plane curvilinear motion of particles; Rectangular components (Projectile motion); Normal and tangential components (circular motion).

11. Kinetics of particles; Newton's second law; Equation of motion; D' Alembert's principle and free body diagram; Principle of work and energy applied to particle and rigid bodies; Principle of conservation of energy; Power and efficiency.

CONTENTS

CHAPTER

1

INTRODUCTION TO STATICS

1.1 INTRODUCTION

Mechanics is a specialized branch of applied physics that treats the phenomena caused by the action of forces on material bodies, at rest or in motion. Mechanics is one of the oldest branches of science. Broadly mechanics can be classified as following:

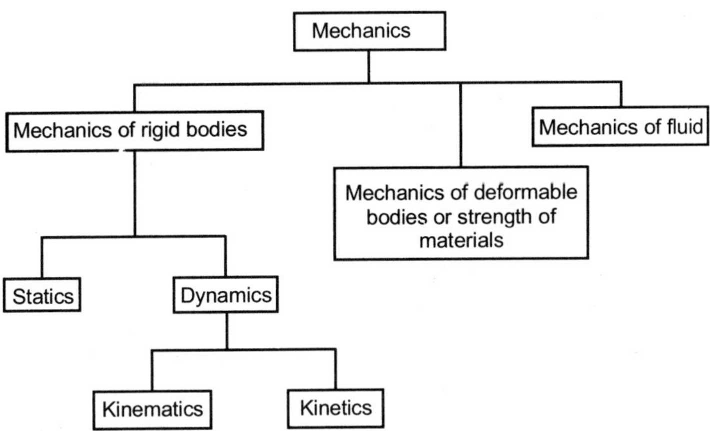

Statics deals with the conditions of equilibrium of bodies acted upon by forces *at rest*. Applications of some of the fundamental principles are found in the construction of pyramids in Egyptian era and of temples in Babylonian period. Archimedes was the pioneer who documented the works on statics in written form. In later days, Newton, Varignon, Stevinus, Pappus, St. Venant, Einstein etc., appeared as vanguard of development of statics to make it compatible to satisfy the ultimate requirements of modern day problems.

In *dynamics*, equilibrium of particles and bodies are studied, when those are *in-motion*. This specialization of mechanics is not too old in age as statics is. Dynamics is further classified into *kinematics*, which relates to study of *geometry of motion* without any reference to force and *kinetics*, which deals with motion of bodies with reference to a *system of forces*.

Fundamental concepts and relevant definitions, in context of mechanics are summarised as following:

1.2 SPACE

Space is the geometric region occupied by a body whose position is described by linear and angular measurements, with respect to the origin of a chosen co-ordinate system. In order to specify the position of a body in space, three independent co-ordinates are required. System of co-ordinates may vary, like, rectangular co-ordinate system (x, y, z) polar co-ordinate system (r, θ, z) or spherical co-ordinate system (R, θ, ϕ).

1.3 PARTICLE

A body with negligible dimensions is called a particle. From the view point of mathematics, a particle may be defined as a body whose dimensions approach to zero reducing it to a *point mass.* When the dimensions of a body are irrelevant to the description of its location (or to the action of forces applied to it) the body itself may be treated as a particle, in the analysis.

1.4 RIGID BODY

A body is said to be rigid, when the relative movements among its different parts are negligible i.e., the deformation of the body is considered to be negligible under the action of various forces on it.

Let's consider the lever AB which rests on fulcrum C as shown in Fig. 1.1. Under the action of two equal forces (F each at either end), if the lever AB bends slightly to its new position $A'B'$, the horizontal distance of either end of lever (A or B) decreases slightly, i.e., $A'C < AC$ or $B'C < BC$. This small variation of the length of the lever can be neglected safely, if we are interested to find out the reaction at the fulcrum on the lever, due to such loading, considering the static equilibrium condition. Although the body (under consideration) may not be rigid in the strict sense, we assume this to be rigid for the purpose of analysis, as statics deals generally with the calculation of external forces which act on the rigid bodies in static equilibrium.

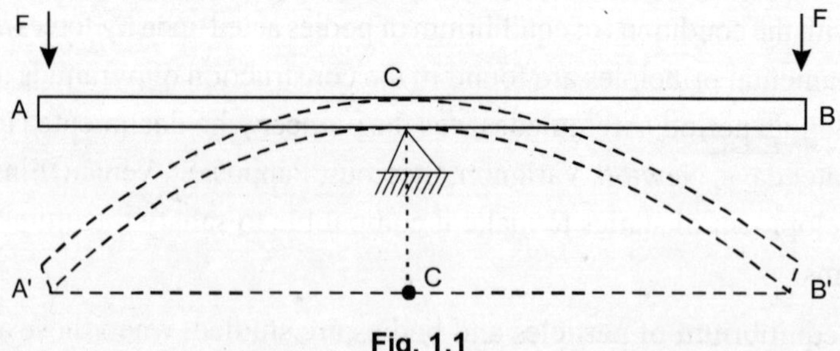

Fig. 1.1

1.5 FORCE

Force may be defined as any action that tends to change the state of a body (at rest or in steady motion), in the direction of its application. A force is characterized by its magnitude, the direction of its action and its point of application on the body. Force is a *vector quantity* and its properties are discussed in detail in the subsequent articles.

All the forces are basically seggregated in two types namely, *tensile force* and *compressive force*. Depending on nature and system of application, force system can be classified as shown in the tree diagram.

Brief description of these forces are provided as following.

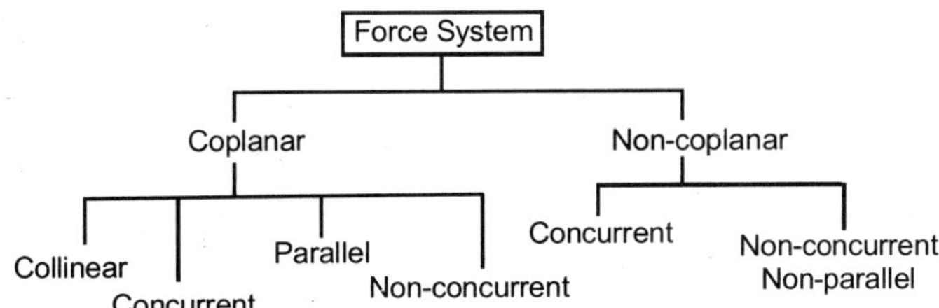

1.5.1 Tensile and Compressive Forces

With reference to Fig. 1.2 (*a*), a rigid bar *AB* is in equilibrium under the action of two collinear equal and opposite forces. If the bar is imagined to be cut into two pieces and they are isolated, each part must be in equilibrium as shown in Figs. 1.2 (*b*) and (*c*), which implies, the internal force *S'* induced in the bar *AB* must be equal to *S* in magnitude, opposite in direction and collinear in action. Thus we may conclude that the rigid bar *AB* is in tension and tensile force induced is *S*.

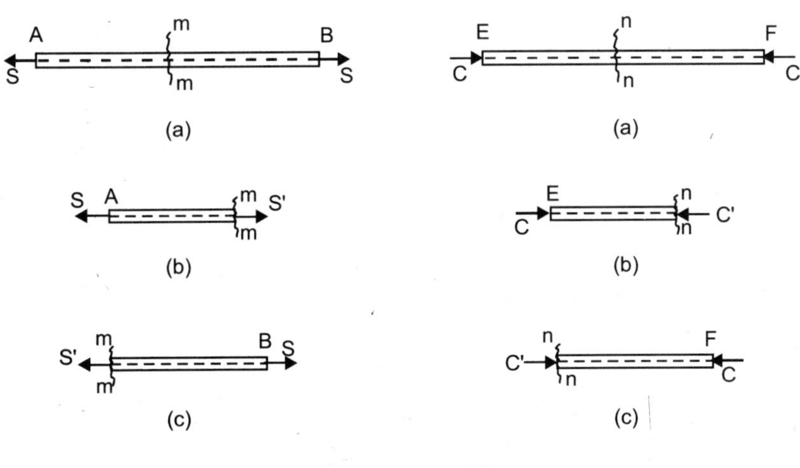

Fig. 1.2 **Fig. 1.3**

Similarly, with reference to Figures 1.3 (*a*), (*b*) and (*c*) we can explain that the rigid bar *EF* is under compression and the compressive force induced is *C*.

1.5.2 Coplanar and Non-coplanar Forces

When all the forces in any force-system remain in the *same plane*, those are called *coplanar forces*.

If those forces lie in *different plane* i.e., *in space*, those are called *non-coplanar* forces.

1.5.3 Concurrent and Non-concurrent Forces

If the lines of action of a number of non parallel forces acting on a body meet at or emerge from a common point the forces are said to be concurrent. And if the forces do not emerge from a point or converge to a point, they are called *non-concurrent* forces. Point of concurrency may be either in plane or in space.

Combination of different types of force systems and their respective equilibrium are discussed in the articles later.

1.6 DIMENSIONS AND UNITS OF SOME PARAMETERS

The dimensions and S.I units of some quantities, frequently encountered in the study of mechanics are listed below.

Table 1.1

Quantity	Dimensional Symbol	S.I. Units	
		Unit	Symbol
Mass	M	Kilogram	kg
Length	L	Meter	m
Time	T	Second	s
Force	MLT^{-2}	Newton	N
Moment of a Force or Torque	ML^2T^{-2}	Newton-meter	N·m
Momentum	MLT^{-1}	Kilogram-meter/sec	kgm/s

1.7 SCALARS AND VECTORS

Mechanics deals with two kinds of quantities, Scalars and Vectors. *Scalar* quantities are those with which a *magnitude alone is associated* and examples are, volume, mass density, speed, time, mass, energy etc. On the other hand, *vector* quantities are those which have *magnitude as well as direction*. Displacement, velocity, acceleration, force, moment, momentum etc., are examples of vector quantities.

1.7.1 Classification of Vectors

Physical quantities represented by vectors fall into one of the following three categories:

(i) **Free Vector:** *A free vector is one whose action is not confined to or associated with a unique line in space*. Thus, a force as a free vector having a definite magnitude and a fixed direction may have different lines of action.

(ii) **Fixed Vector (or Bound Vector):** *A fixed or bound vector is one for which a unique point of application is specified, apart from having definite magnitude and definite direction.* Therefore, a fixed vector occupies a particular position in the space.

(iii) **Sliding Vector (Transmissibility of a Force):** A sliding vector is one for which a unique line in the space must be maintained along which the vector acts. This implies that the point of application of a force may be transmitted along its line of action without changing the effect of the force on any rigid body to which it is applied. This is sometimes called "*transmissibility of a force or vector*". With reference to Fig. 1.4, the block *A* moves along the floor leftward direction, under the action of horizontal force *P*, whose point of application shifts continuously leftward direction along its line of action (dashed line), and hence force *P* may be called as a sliding vector.

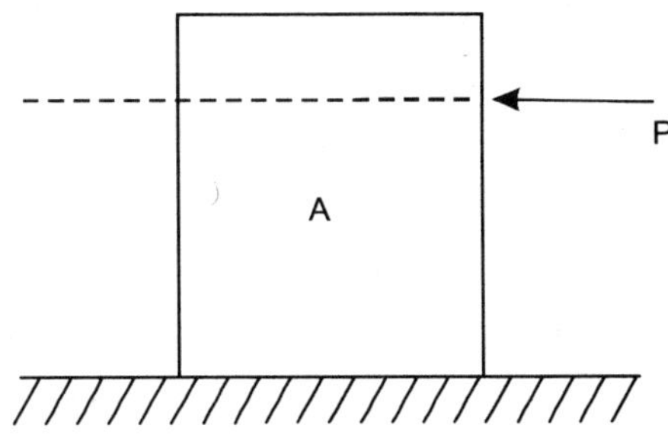

Fig. 1.4

However, irrespective of the categories as mentioned above, a vector having magnitude *unity* is called *Unit Vector*, whereas a vector having *zero* magnitude is called *Null Vector*.

1.7.2 Magnitude and Multiplication of Vector by a Scalar

If we consider a vectoı $\vec{P}$, then its magnitude = $|P|$ and will be presented with suitable unit. Thus $|P|$ is a positive scalar quantity.

The vector $-\vec{P}$ may be considered as the product of –1 and vector $\vec{P}$. Here $\vec{P}$ differs from $-\vec{P}$ only in direction opposite to each other having same numerical value.

1.8 COPLANAR COLLINEAR FORCES

If two forces act *on the same plane*, they are said to be *coplanar*, and if two forces have *same line of action* (their points of application on a body may be different), they are said to be collinear; however, the number of forces under considerations, may be more than two. *All the forces which are collinear, must be coplanar.*

1.8.1 Equilibrium of Collinear Forces

Any two forces can remain in equilibrium, only if they are equal in magnitude, opposite in direction and collinear in action.

In statics, the term *equilibrium* has got special implication. If we say, "*a rigid body is in equilibrium under the action of a number of forces*" we mean that the rigid body, under consideration, has neither translation nor any rotation in any direction, under the action of a number of forces on it.

1.8.2 Resultant of Coplanar Concurrent Forces

The law of parallelogram of forces is one of the basic principle of statics. According to this law, if two coplanar concurrent forces are said to act at any point A on a body [Fig. 1.5 (*a*)], the force vectors $\vec{P}$ and $\vec{Q}$ [Fig. 1.5 (*b*)] can be replaced by their equivalent $\vec{R}$ (called resultant force), which is the diagonal of the parallelogram formed by $\vec{P}$ and $\vec{Q}$ as its two sides, as shown in Fig. 1.5 (*c*).

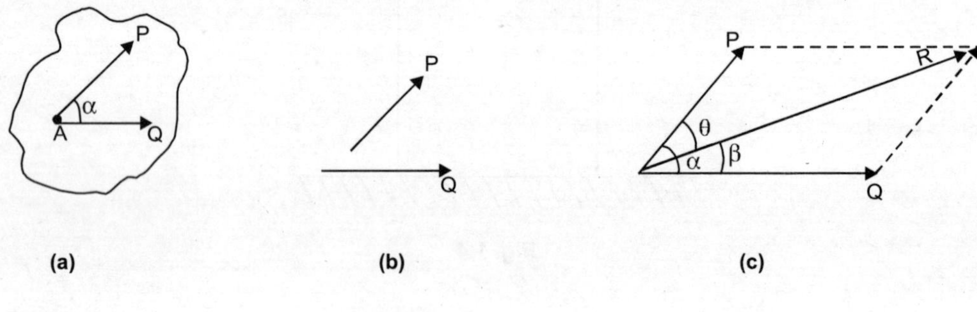

Fig. 1.5

On the other hand, according to law of triangle of forces, two free vectors $\vec{Q}$ and $\vec{P}$ may be added *head-to-tail* to obtain the same resultant $\vec{R}$, as shown in Fig. 1.5 (*d*). In any case this addition of vectors is represented by

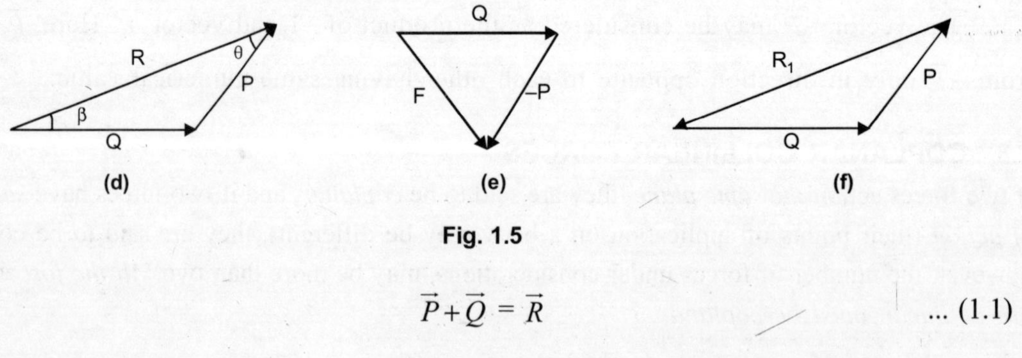

Fig. 1.5

$$\vec{P}+\vec{Q}=\vec{R} \qquad ... (1.1)$$

Magnitude of $\vec{R}$ can be computed using parallelogram law as following:

$$\left|\vec{R}\right| = \sqrt{\left|\vec{P}\right|^2 + \left|\vec{Q}\right|^2 + 2\left|\vec{P}\right|\left|\vec{Q}\right|\cos\alpha}$$

and
$$\sin\beta = \frac{\left|\vec{P}\right|}{\left|\vec{R}\right|}\sin\alpha,$$

and
$$\sin\theta = \frac{\left|\vec{Q}\right|}{\left|\vec{R}\right|}\sin\alpha$$

Vector addition is *commutative*, i.e., it doesn't matter, whether $\vec{P}$ is added with $\vec{Q}$ or vice versa, such that

$$\vec{P} + \vec{Q} = \vec{R} = \vec{Q} + \vec{P} \quad \text{... (1.2)}$$

Also vector addition is *associative*, i.e., in case of three free vectors $\vec{P}_1$, $\vec{P}_2$, and $\vec{P}_3$ we can write

$$\vec{P}_1 + \left(\vec{P}_2 + \vec{P}_3\right) = \left(\vec{P}_1 + \vec{P}_2\right) + \vec{P}_3 \quad \text{... (1.3)}$$

The difference $\left(\vec{Q} - \vec{P}\right)$ between two vectors can be easily derived by adding $-\vec{P}$ to $\vec{Q}$, as shown in Fig. 1.5 (*e*), in which case either triangle or parallelogram procedure may be used. This subtraction of vectors is represented by

$$\vec{Q} + \left(-\vec{P}\right) = \vec{Q} - \vec{P} = \vec{F} \quad \text{... (1.4)}$$

1.8.3 Equilibrium of Coplanar Concurrent Forces

With reference to Fig. 1.5 (*a*), the body under the action of forces $\vec{P}$ and $\vec{Q}$ is not in static equilibrium, if we assume that the resultant force $\vec{R}$ has some definite magnitude and rather the body will experience an unbalanced force in the direction of vector $\vec{R}$.

Under such circumstances, if an external force $\vec{R_1}$ is applied on the body at point A, which has some magnitude as $\vec{R}$, but direction opposite to that of $\vec{R}$, then we can imagine the body will be in static equilibrium, under the action of all three forces, $\vec{P}$, $\vec{Q}$ and $\vec{R_1}$. Addition of vectors $\vec{P}$, $\vec{Q}$ and $\vec{R_1}$ will result a *null vector*, having zero magnitude and also the free vectors of these forces will form a closed triangle, as shown in Fig. 1.5 (*f*).

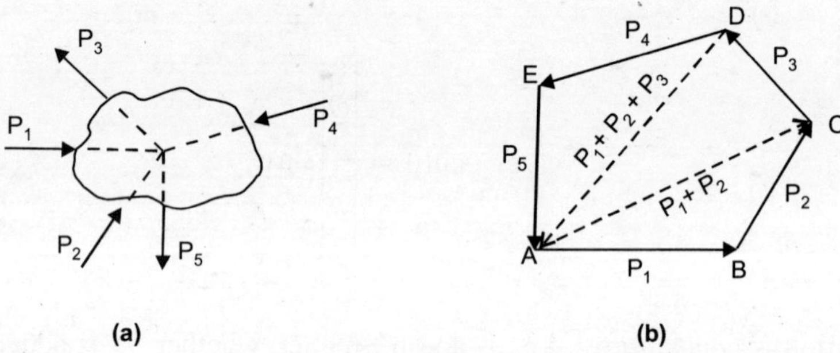

Fig. 1.6

Precisely speaking, three non-parallel forces can be in equilibrium, only when they are coplanar, concurrent and their free vectors build a closed triangle. Thus, we can readily draw the general conclusion as follows:

If a rigid body is in equilibrium, being acted upon by several concurrent, coplanar forces, then the free vectors of those forces, when geometrically added must form a closed polygon Figs. 1.6 (a) and (b).

1.9 LAMI'S THEOREM

It states that, *if three coplanar forces acting at a point on a body are in equilibrium, then each force is proportional to the sine of the angle between the other two.*

With reference to Fig.1.7(*a*), F_1, F_2, and F_3 are three coplanar concurrent forces. According to Lami's theorem, for a coplanar concurrent three force system remaining in equilibrium, the ratio of magnitude of each force and the sine of the angle included by other two forces is constant. So

$$\frac{|\vec{F_1}|}{\sin\alpha} = \frac{|\vec{F_2}|}{\sin\beta} = \frac{|\vec{F_3}|}{\sin\gamma}$$

This theorem can be proved very easily. As these three forces maintaining equilibrium are coplanar and concurrent, they can be expressed in a vector triangle ΔABC as shown in Fig. 1.7(*b*). Included acute angles are

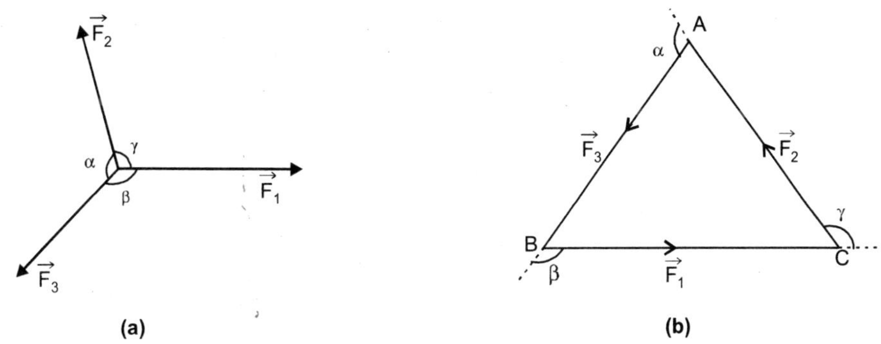

Fig. 1.7

$\angle ABC = (\pi - \beta)$, $\angle BCA = (\pi - \gamma)$ and $\angle CAB = (\pi - \alpha)$. Sides of the triangle AB, BC and CA represents forces $\vec{F_3}$, $\vec{F_1}$, and $\vec{F_2}$ respectively. Now, applying rule of sine in ΔABC, we obtain,

$$\frac{BC}{\sin\angle BAC} = \frac{CA}{\sin\angle ABC} = \frac{AB}{\sin\angle ACB}$$

$\Rightarrow$
$$\frac{|\vec{F_1}|}{\sin(\pi - \alpha)} = \frac{|\vec{F_2}|}{\sin(\pi - \beta)} = \frac{|\vec{F_3}|}{\sin(\pi - \gamma)}$$

$\Rightarrow$
$$\frac{|\vec{F_1}|}{\sin\alpha} = \frac{|\vec{F_2}|}{\sin\beta} = \frac{|\vec{F_3}|}{\sin\gamma}$$

1.10 REPRESENTATION OF FORCE USING UNIT VECTORS

It is sometimes convenient to express a vector $\vec{P}$ as the product of its magnitude and a vector $\vec{a}$ of unit magnitude and having direction corresponding to the vector $\vec{P}$. This vector $\vec{a}$ is called a unit vector which has no dimension.

$\therefore$
$$\vec{a} \text{ (unit vector in direction of } \vec{P}) = \frac{\vec{P}}{|\vec{P}|}$$

$\Rightarrow$
$$P = |\vec{P}|a \qquad \text{... (1.5)}$$

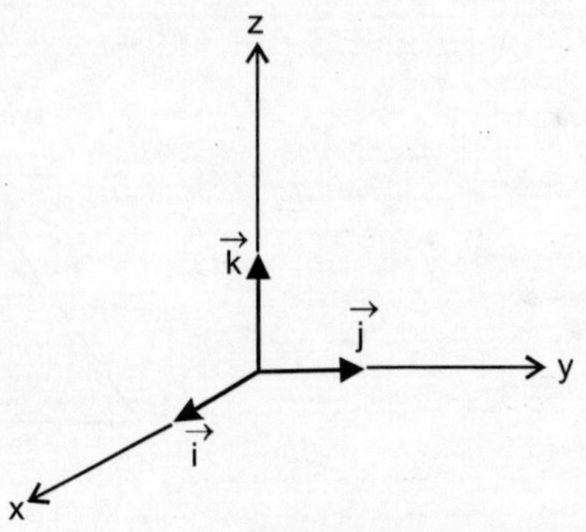

Fig. 1.8

Once the unit vector is established it can be used for any vector $\vec{Q}$ which is parallel to $\vec{P}$. Now when the unit vectors are represented along axes of three-dimensional co-ordinate system, these are expressed by $\vec{i}$, $\vec{j}$, and $\vec{k}$ along *x*, *y* and *z* axis respectively, as shown in the Figure 1.8. As the sum of a set of concurrent vectors is equivalent to the original vector, we can replace the vector $\vec{P}$ by its rectangular scalar components in the following manner:

$$\vec{P} = P_x \vec{i} + P_y \vec{j} + P_z \vec{k} \qquad \text{... (1.6)}$$

Here P_x, P_y, and P_z are components of vector $\vec{P}$ along *x*, *y* and *z* axis respectively.

If another vector $\vec{Q}$ becomes same in magnitude and direction with $\vec{P}$, we can write

$$P_x \vec{i} + P_y \vec{j} + P_z \vec{k} = Q_x \vec{i} + Q_y \vec{j} + Q_z \vec{k} \qquad \text{... (1.7)}$$

where from $P_x = Q_x, P_y = Q_y, P_z = Q_z$. ... (1.8)

1.11 RESOLUTION OF A FORCE

In general, three-dimensional analysis of problems in mechanics requires to resolve a force into its three mutually perpendicular components (called rectangular components), along directions of three mutually perpendicular cartesian co-ordinate axes *x*, *y*, and *z* respectively.

1.11.1 Three Dimensional Systems (Rectangular Components)

The force *F* acting at point *O* (arbitrarily chosen origin), as shown in Fig. 1.9, has the rectangular components given by

$$\left.\begin{aligned} F_x &= F\cos\theta_x \\ F_y &= F\cos\theta_y \\ F_z &= F\cos\theta_z \end{aligned}\right\} \qquad \text{... (1.9)}$$

Designating unit vectors as $\vec{i}$, $\vec{j}$, $\vec{k}$ along x, y, and z directions respectively, the force vector can be expressed as,

$$\vec{F} = F_x \vec{i} + F_y \vec{j} + F_z \vec{k} \qquad \text{... (1.10)}$$

and in terms of equation (1.9),

$$\vec{F} = F\left(\cos\theta_x \vec{i} + \cos\theta_y \vec{j} + \cos\theta_z \vec{k}\right) \qquad \text{... (1.11)}$$

whereas

$$F = \left|\vec{F}\right| \sqrt{F_x^2 + F_y^2 + F_z^2} \qquad \text{... (1.12)}$$

Now, introducing direction cosines of $\vec{F}$, which are

$$l = \cos\theta_x, \quad m = \cos\theta_y, \text{ and } n = \cos\theta_z$$

We may write from equation (1.11)

$$\vec{F} = F\left(l\,\vec{i} + m\,\vec{j} + n\,\vec{k}\right) \qquad \text{... (1.13)}$$

and also

$$l^2 + m^2 + n^2 = 1. \qquad \text{... (1.14)}$$

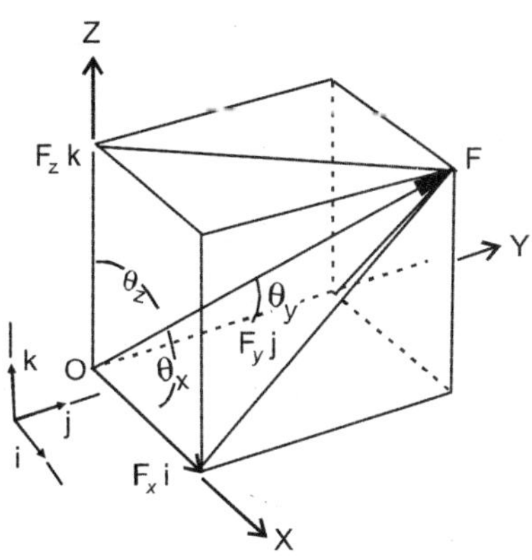

Fig. 1.9

Again the right hand side of the equation (1.13) may be regarded as the force magnitute F times a unit vector $\vec{n_F}$ whose scalar component are the direction cosines of the line of action of F, i.e.,

$$\vec{F} = F \cdot \vec{n_F} \qquad \text{... (1.15)}$$

such that,

$$\vec{n_F} = l\,\vec{i} + m\,\vec{j} + n\,\vec{k}\,.$$

In most of the cases of three-dimensional problems, the direction of a force is described by the following.

1.11.1 (a) Two Points on the Line of Action of the Force

If the co-ordinates of two points A and B (on the line of action of F) are as shown in Figure 1.10, then

$$\vec{F} = F\,\vec{n}_F = F\frac{\vec{AB}}{\left|\vec{AB}\right|} = F\frac{(x_2-x_1)\vec{i}+(y_2-y_1)\vec{j}+(z_2-z_1)\vec{k}}{\sqrt{(x_2-x_1)^2+(y_2-y_1)^2+(z_2-z_1)^2}}$$

Where, $$F = \left|\vec{F}\right|$$

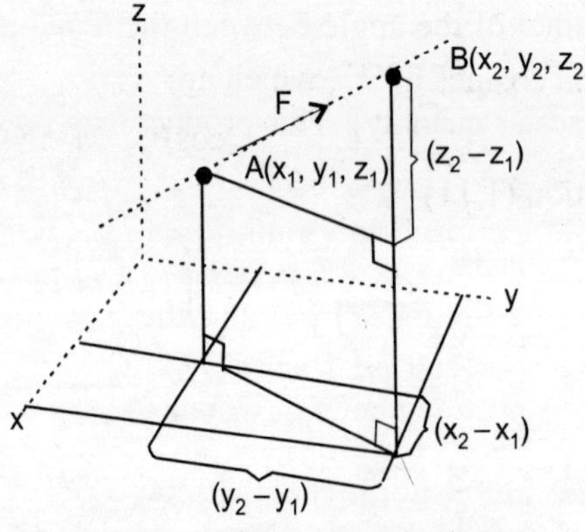

Fig. 1.10

1.11.1 (b) Two Angles which Orient the Line of Action

With reference to Fig. 1.11, let force $\vec{F}$ makes an angle ϕ with horizontal plane (X–Y) and component of $\vec{F}$ on horizontal plane makes an angle θ with X-axis. $\vec{F}$ can be resolved first into vertical and horizontal components as

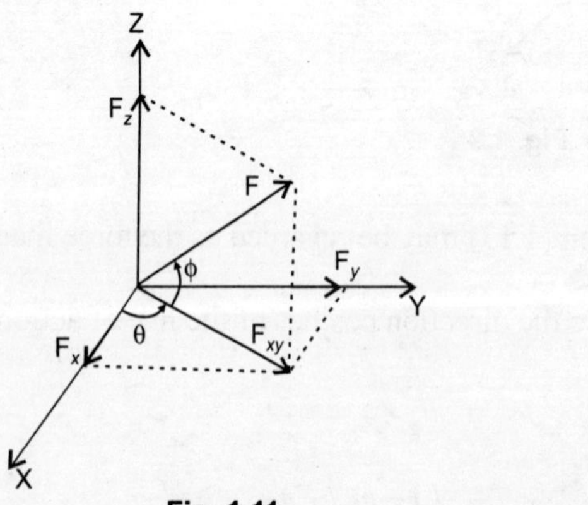

Fig. 1.11

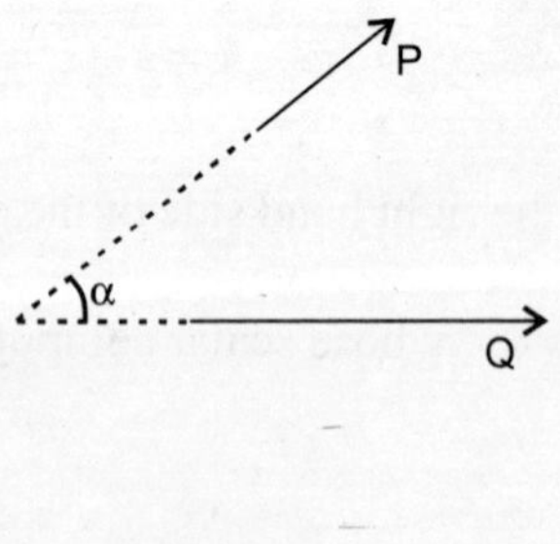

Fig. 1.12(a)

$$F_z = \vec{F}\sin\phi, F_{xy} = \vec{F}\cos\phi \qquad \text{... (1.17)}$$

The horizontal component F_{xy} is further resolved into *X* and *Y* components. Thus,

$$F_x = F_{xy}\cos\theta = \vec{F}\cos\phi\cos\theta$$

$$F_y = F_{xy}\sin\theta = \vec{F}\cos\phi\sin\theta \qquad \dots (1.18)$$

Hence, the quantities F_x, F_y, and F_z are desired scalar component of $\vec{F}$.

1.12 DOT (SCALAR) PRODUCT OF TWO VECTORS

By definition, the dot product of two vectors $\vec{P}$ and $\vec{Q}$ as shown in Fig. 1.12 (*a*), is the product of their magnitudes times the 'cosine' of the angle between them and is expressed as:

$\vec{P}\cdot\vec{Q} = |\vec{P}||\vec{Q}|\cos\alpha$ [which is a scalar quantity]. This product may be viewed either as the orthogonal projection $P\cos\alpha$ of $\vec{P}$ in the direction of $\vec{Q}$ multiplied by *Q*, or as the orthogonal projection $Q\cos\alpha$ of $\vec{Q}$ in the direction of $\vec{P}$ multiplied by *P*. In either case, the dot product is a scalar quantity.

Thus, for example, the scalar component $F_x = F\cos\theta_x$ of the force $\vec{F}$ as in Fig. 1.9, may be written as $F_x = \vec{F}\cdot\vec{i}$; $\vec{i}$ being the unit vector in the *X*-direction. Similarly $F_y = \vec{F}\cdot\vec{j}$ and $F_z = \vec{F}\cdot\vec{k}$

Properties of dot product to be remembered:

(*a*) It follows Commutative Law, $\vec{P}\cdot\vec{Q} = \vec{Q}\cdot\vec{P}$

(*b*) It follows Distributive Law, $\vec{P}\cdot\left(\vec{Q}+\vec{R}\right) = \vec{P}\cdot\vec{Q}+\vec{P}\cdot\vec{R}$

(*c*) From definition of dot product, considering unit vectors $\vec{i}, \vec{j}, \vec{k}$ along three mutually perpendicular axes *x*, *y* and *z*, we can write

(*i*) $\vec{i}\cdot\vec{i} = \vec{j}\cdot\vec{j} = \vec{k}\cdot\vec{k} = 1$ ($\because$ cos 0° = 1)

(*ii*) $\vec{i}\cdot\vec{j} = \vec{j}\cdot\vec{i} = \vec{i}\cdot\vec{k} = \vec{k}\cdot\vec{i} = \vec{j}\cdot\vec{k} = \vec{k}\cdot\vec{j} = 0$ ($\because$ cos 90° = 0)

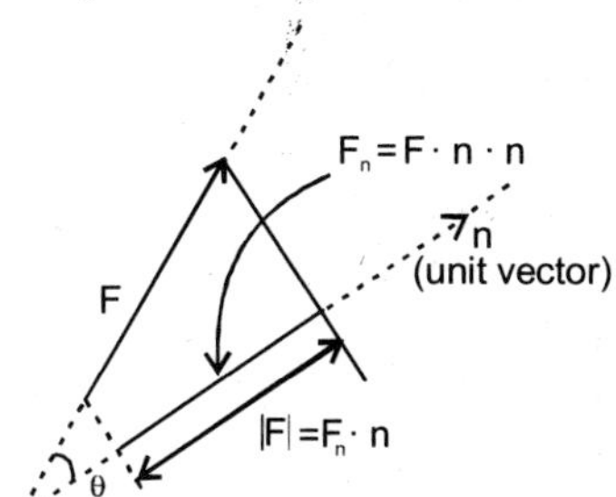

Fig. 1.12 (b)

It follows from which, that the dot product of two mutually perpendicular vectors, $\vec{P_1}$ and $\vec{P_2}$, always vanishes, $\vec{P_1} \cdot \vec{P_2} = 0$.

Now with reference to Fig. 1.12 (*b*), if $\vec{n}$ is a unit vector in a specified direction (other than directions of $\vec{i}$, $\vec{j}$, $\vec{k}$), the projection of $\vec{F}$ in the direction of $\vec{n}$, (i.e., if $\vec{F}$ is resolved along $\vec{n}$) has the magnitude $F_n = \vec{F} \cdot \vec{n}$ which is a scalar quantity and the same projection can be expressed as a vector in the form $\vec{F_n} = \left(\vec{F} \cdot \vec{n}\right)\vec{n}$, such that $F_n = \left|\vec{F_n}\right|$ and $\left|\vec{n}\right| = 1$

Let $\vec{F} = F\left(l\,\vec{i} + m\,\vec{j} + n\,\vec{k}\right)$ and $\vec{n} = a\,\vec{i} + b\,\vec{j} + c\,\vec{k}$ and then using properties of dot product $F_n = \vec{F}.\,\vec{n} = F\left(la + mb + nc\right)$ (which is a scalar quantity)

Again, $$\vec{F} \cdot \vec{n} = F \cdot n\cos\theta \text{ and } n = \left|\vec{n}\right| = 1 \qquad \dots (1.20)$$

Hence $$\theta = \cos^{-1}\left(\frac{\vec{F} \cdot \vec{n}}{F}\right) \qquad \dots (1.21)$$

In general, the angle between two vectors

$$\vec{P} = \left(P_x\vec{i} + P_y\vec{j} + P_z\vec{k}\right)$$

and $$\vec{Q} = \left(Q_x\vec{i} + Q_y\vec{j} + Q_z\vec{k}\right)$$

is given by $$\alpha = \cos^{-1}\left(\frac{\vec{P} \cdot \vec{Q}}{PQ}\right) = \cos^{-1}\left[\frac{P_xQ_x + P_yQ_y + P_zQ_z}{\sqrt{\left(P_x^2 + P_y^2 + P_z^2\right)\left(Q_x^2 + Q_y^2 + Q_z^2\right)}}\right] \qquad \dots (1.22)$$

Introducing expressions for direction cosines $P_x = Pl_1, P_y = Pm_1, P_z = Pn_1$ and $Q_x = Ql_2$, $Q_y = Qm_2$, $Q_z = Qn_2$ we rewrite $\vec{P} = P\left(l_1\,\vec{i} + m_1\,\vec{j} + n_1\,\vec{k}\right)$ and $\vec{Q} = Q\left(l_2\,\vec{i} + m_2\,\vec{j} + n_2\,\vec{k}\right)$ and

$$\alpha = \cos^{-1}\left[\frac{PQ\left(l_1l_2 + m_1m_2 + n_1n_2\right)}{PQ}\right] = \cos^{-1}\left(l_1l_2 + m_1m_2 + n_1n_2\right) \qquad \dots (1.23)$$

also $l_1^2 + m_1^2 + n_1^2 = 1$ and $l_2^2 + m_2^2 + n_2^2 = 1$.

It is to be noted that the dot product relationship may be applied in case of *non-intersecting* vectors also. With reference to Fig. 1.13, the dot product of two non-intersecting forces $\vec{P}$ and $\vec{Q}$ are given as $\vec{P} \cdot \vec{Q} = \vec{P_1} \cdot \vec{Q} = P_1 Q \cos\alpha = PQ\cos\alpha$, since P_1 and P are the same free vectors.

1.13 CROSS PRODUCT OF TWO VECTORS (VECTOR PRODUCT)

It is an interaction of two vectors which result in a vector quantity. Moment of a force is such an example of cross product.

If two vectors $\vec{P}$ and $\vec{Q}$ make an angle α, their cross product $\vec{R}$ will be :

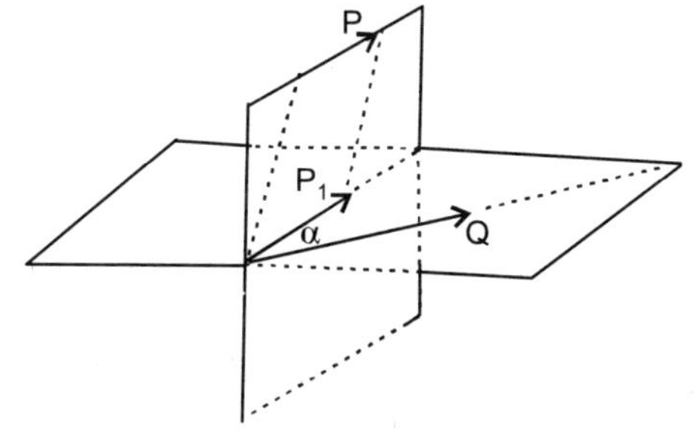

Fig. 1.13

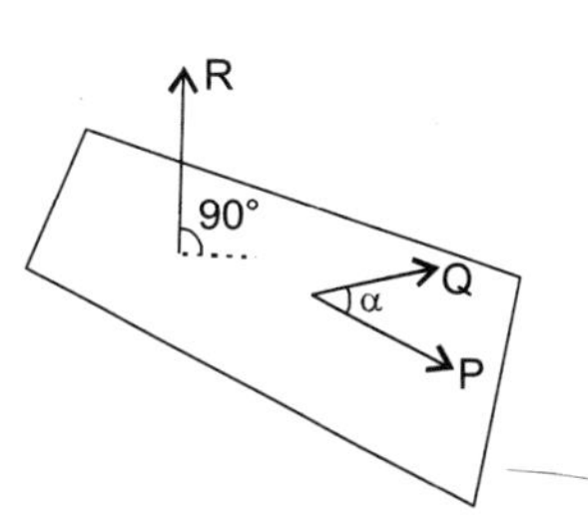

Fig. 1.14

$$\vec{P} \times \vec{Q} = \vec{R} \text{ and } |\vec{R}| = |\vec{P}||\vec{Q}| \sin\alpha \quad \text{... (1.24)}$$

Commutative law breaks down for this product, as

$$\vec{P} \times \vec{Q} = -\vec{Q} \times \vec{P} \quad \text{... (1.25)}$$

If there are three vectors $\vec{P}$, $\vec{Q}$, $\vec{R}$ then,

$$\vec{R} \times \left(\vec{P} + \vec{Q}\right) = \left(\vec{R} \times \vec{P}\right) + \left(\vec{R} \times \vec{Q}\right) \quad \text{... (1.26)}$$

Hence for this case *distributive property* holds good. Now comes the cross product of rectangular unit vectors. If *z*-axis is assumed in such a sense that it is consistent with right hand screw rule while rotating from *x* to *y* axis, the reference is called a *right-hand triad* and we can write,

$$\vec{i} \times \vec{j} = \vec{k} \quad \text{... (1.27a)}$$

In left-hand triad,

$$\vec{i} \times \vec{j} = -\vec{k} \quad \text{... (1.27b)}$$

Right-hand triad and Left-hand triad are illustrated in Figs. 1.15 (*a*) and (*b*).

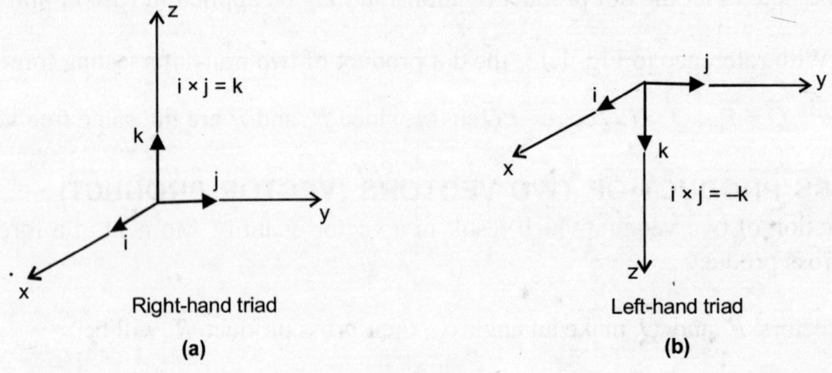

Fig. 1.15

In order to find out the cross product of two vectors in terms of their rectangular co-ordinates,

$$\vec{P}\times\vec{Q} = \left(P_x\vec{i}+P_y\vec{j}+P_z\vec{k}\right)\times\left(Q_x\vec{i}+Q_y\vec{j}+Q_z\vec{k}\right)$$

$$= \left(P_yQ_z - P_zQ_y\right)\vec{i}+\left(P_zQ_x - P_xQ_z\right)\vec{j}+\left(P_xQ_y - P_yQ_x\right)\vec{k} \qquad \text{... (1.28)}$$

This also can be found by evaluation of following determinant

$$\vec{P}\times\vec{Q} = \begin{vmatrix} i & j & k \\ P_x & P_y & P_z \\ Q_x & Q_y & Q_z \end{vmatrix} = \vec{i}\begin{vmatrix} P_y & P_z \\ Q_y & Q_z \end{vmatrix} - \vec{j}\begin{vmatrix} P_x & P_z \\ Q_x & Q_z \end{vmatrix} + \vec{k}\begin{vmatrix} P_x & P_y \\ Q_x & Q_y \end{vmatrix}$$

$$= \vec{i}\left(P_yQ_z - P_zQ_y\right) - \vec{j}\left(P_xQ_z - P_zQ_x\right) + \vec{k}\left(P_xQ_y - P_yQ_x\right)$$

$$= \left(P_yQ_z - P_zQ_y\right)\vec{i}+\left(P_zQ_x - P_xQ_z\right)\vec{j}+\left(P_xQ_y - P_yQ_x\right)\vec{k}$$

1.14 MOMENT OF A FORCE ABOUT A POINT

The moment of any force $\vec{F}$ about the point P can be calculated as shown in Fig. 1.16, using cross product, $\vec{M}_P = \vec{r}\times\vec{F}$, where position vector of any point Q on the force $\vec{F}$ with respect to point P is given by $\vec{r}$. So, $\vec{r} = \overrightarrow{PQ}$. Here P is the given point and Q any point on the force $\vec{F}$. Moment of a force is a vector quantity and it will act along the normal to the plane contained by the point P and the line action of $\vec{F}$.

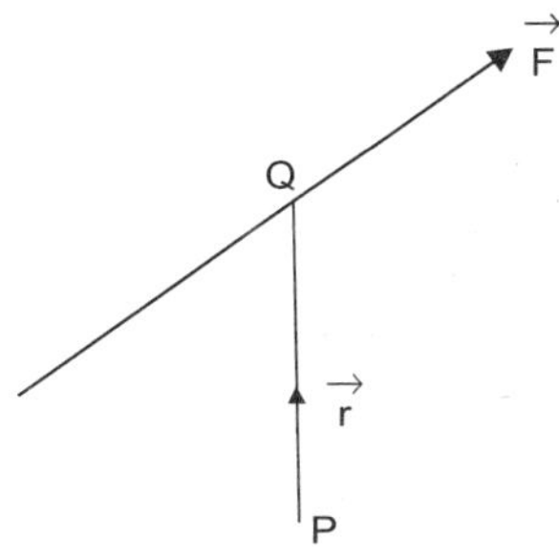

Fig. 1.16

1.15 MOMENT OF FORCE ABOUT A LINE

Moment of any force represented by $\vec{F}$ about the line AB can be determined as follows and shown in Fig. 1.17. On the line AB, any point P is taken and Q be any point on the force $\vec{F}$. Moment of the force $\vec{F}$ about the point P is $\vec{M}_P$ and $\vec{M}_P = \overrightarrow{PQ} \times \vec{F} = \vec{r} \times \vec{F}$. Magnitude of the moment of $\vec{F}$ about AB is equal to the projection of the moment $\vec{M}_P$ along the line AB. This projection is given by the dot product, $M_{AB} = \vec{M}_P \cdot \vec{r}_{AB}$ which is the magnitude of the moment and direction is along AB, where $\vec{r}_{AB}$ is unit vector along AB.

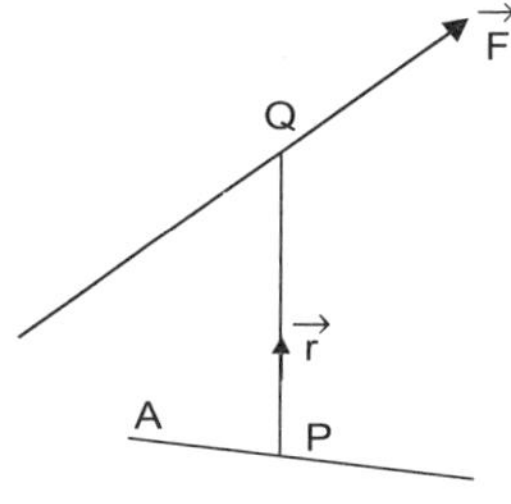

Fig. 1.17

Vector moment is $\vec{M}_{AB} = \left(\vec{M}_P \cdot \vec{r}_{AB}\right)\vec{r}_{AB}$. Thus, the moment of a force about a line is nothing but the projection of that moment along the same line. Again moment $\vec{M}_P$ means moment of the given force about any given point, say P on the same line AB.

Multiple Choice Questions

Select the best alternative (s):

1. A number of forces acting at a point will be in equilibrium, if
   (*a*) their total sum is zero
   (*b*) two resolved parts in two directions at right angles are equal
   (*c*) sum of resolved parts in any two perpendicular directions are both zero
   (*d*) none of the above
2. Two non-collinear parallel equal forces acting in opposite direction.
   (*a*) balance each other
   (*b*) constitute a moment
   (*c*) constitute a couple
   (*d*) constitute a moment of couple
   (*e*) constitute a result of a couple
3. If a rigid body is in equilibrium under the action of three forces, then,
   (*a*) these forces are equal
   (*b*) the lines of action of these forces meet at a point
   (*c*) the lines of action of these forces are parallel
   (*d*) (*b*) and (*c*) of the above
   (*e*) none of the above
4. According to principle of transmissibility of forces, the effect of a force upon a body is
   (*a*) maximum when it acts at the centre of gravity of a body
   (*b*) different at different points in its line of action
   (*c*) the same at every point in its line of action
   (*d*) minimum when it acts at the C.G of the body
   (*e*) none of the above
5. The magnitude of two forces, which when acting at right angle produce resultant force of $\sqrt{10}$ kg and when acting at 60° produce resultant of $\sqrt{13}$ kg. These forces are,
   (*a*) 2 and 6 kg
   (*b*) 3 and 1 kg
   (*c*) 5 and 5 kg
   (*d*) 2 and 5 kg
   (*e*) none of the above
6. According to law of triangle of forces
   (*a*) three forces acting at a point will be in equilibrium
   (*b*) three forces acting at a point can be represented by a triangle each side being proportional to the force
   (*c*) if three forces acting upon a particle are represented in magnitude and direction by the sides of a triangle taken in order, they will be in equilibrium
   (*d*) if three forces acting at a point are in equilibrium, each force is proportional to the sine of the angle between the other two
   (*e*) none of the above
7. If rain is falling in the opposite direction of the movement of a pedestrian. He has to hold his umbrella
   (*a*) more inclined when moving
   (*b*) less inclined when moving
   (*c*) more inclined when standing
   (*d*) less inclined when standing
   (*e*) none of the above
8. A single force and a couple acting in the same plane upon a rigid body,
   (*a*) balance each other
   (*b*) cannot balance each other
   (*c*) produce moment of a couple
   (*d*) are equivalent
   (*e*) none of the above

9. The weight of a body is due to
   (*a*) centripetal force of the earth
   (*b*) gravitational pull exerted by the earth
   (*c*) forces experienced by body in atmosphere
   (*d*) forces of attraction experienced by particles
   (*e*) gravitational force of attraction towards the centre of the earth

10. When trying to turn a key into a lock, the following is applied:
   (*a*) coplanar force (*b*) non-coplanar force
   (*c*) lever (*d*) moment
   (*e*) couple

11. Which of the following do not have identical dimension?
   (*a*) moment and impulse (*b*) torque and energy
   (*c*) torque and work (*d*) kinetic energy and potential energy

12. Two coplanar couples having equal and opposite moments
   (*a*) balance each other (*b*) produce a couple and unbalanced force
   (*c*) are equivalent (*d*) produce a moment of couple
   (*e*) cannot balance each other

13. A barge is pulled by two tugboats as in figure. What will be value of θ so that tension in the rope 2 is minimum?
   (*a*) 30° (*b*) 45° (*c*) 60° (*d*) 0°

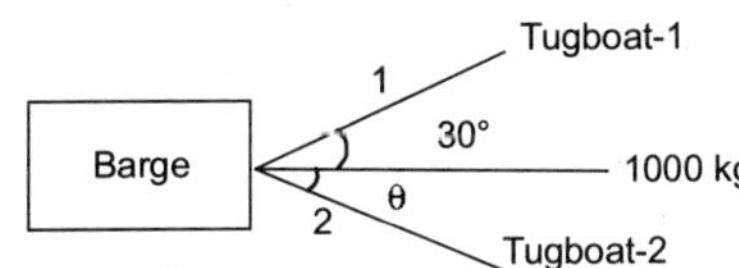

14. A force system is shown in figure. The axial force induced in *CD* will be
   (*a*) F (*b*) $F\sqrt{2}$ (*c*) $\frac{F}{\sqrt{2}}$
   (*d*) 0 (*e*) none of the above

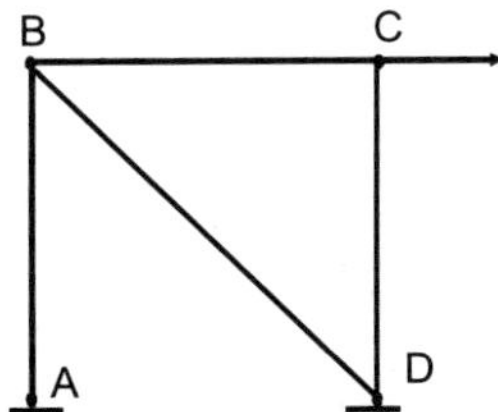

Answers

1. (*c*)	2. (*d*)	3. (*d*)	4. (*c*)	5. (*b*)	6. (*c*)	7. (*a*)
8. (*b*)	9. (*e*)	10. (*e*)	11. (*a*)	12. (*a*)	13. (*c*)	14. (*d*)

Numerical Examples

Example 1

Force 400 N acts along AB where A (3, 2, –4) and B (8, –5, 6). Write the force vector.

Solution Unit vector along AB

$$\overrightarrow{r_{AB}} = \frac{(8-3)i+(-5-2)j+(6+4)k}{\sqrt{(8-3)^2+(-5-2)^2+(6+4)^2}}$$

$$= \frac{5i-7j+10k}{\sqrt{174}}$$

∴ Force vector is

$$\vec{P} = P\times\overrightarrow{r_{AB}} = 400\times\frac{5i-7j+10k}{\sqrt{174}}$$

$$= 30.32(5i-7j+10k) \text{ Newton.}$$

Example 2

A Force vector is given by $\vec{P} = (3i+4j-5k)$ Newton. Find (*i*) magnitude of the force, (*ii*) component along *Z*-axis and along *Y*-axis, (*iii*) angle that the force makes with *Y*-axis.

Solution (*i*) Magnitude of force $\vec{P} = \sqrt{(3)^2+(4)^2+(-5)^2} = 5\sqrt{2}$ Newton

(*ii*) Component along Z-axis is –5 N and along Y-axis is 4N.

(*iii*) Angle makes with the *Y*-axis, $m = \cos\beta = \frac{4}{5\sqrt{2}} = 0.56568$

So, $\beta = 55.55°$.

Example 3

Forces $F_1 = 16$ N, $F_2 = 45$ N and $F_3 = 22$ N act along lines HP, EC and CB respectively. Write the three force vectors. Find also the sum of all three vectors.

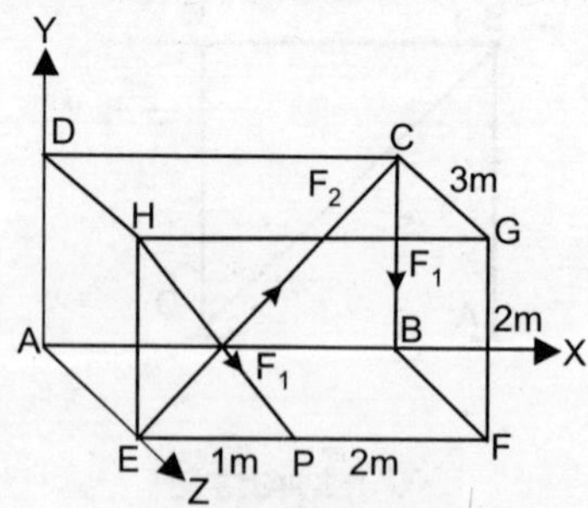

Solution From the figure,

For $\vec{F_1}$,

$$\vec{r_1} = \frac{(1-0)i+(0-2)j+(3-3)k}{\sqrt{(1-0)^2+(0-2)^2+(3-3)^2}} = \frac{i-2j}{\sqrt{5}}$$

For $\vec{F_2}$,
$$\vec{r_2} = \frac{(3-0)i+(2-0)j+(0-3)k}{\sqrt{(3-0)^2+(2-0)^2+(0-3)^2}} = \frac{3i+2j-3k}{\sqrt{22}}$$

For $\vec{F_3}$,
$$\vec{r_3} = \frac{(3-3)i+(0-2)j+(0-0)k}{\sqrt{(3-3)^2+(0-2)^2+(0-0)^2}} = -j$$

$\therefore$ Force vectors are
$$\vec{F_1} = 16\times\vec{r_1} = 16\times\frac{(i-2j)}{\sqrt{5}}$$
$$= 7.155(i-2j)\text{ Newton}$$
$$\vec{F_2} = 45\times\vec{r_2} = 45\times\frac{(3i+2j-3k)}{\sqrt{22}}$$
$$= 9.594(3i+2j-3k)\text{ Newton}$$
$$\vec{F_3} = 22\times\vec{r_3} = 22\times(-j) = (-22j)\text{ Newton}$$

$\therefore$
$$\vec{F_1}+\vec{F_2}+\vec{F_3} = 7.155(i-2j)+9.594(3i+2j-3k)+(-22j)$$
$$= (35.937i-17.122j-28.782k)\text{ Newton}$$

Example 4

The cartesian components of a force are $F_x = 16$ kN, $F_y = -12$ kN and $F_z = 20$ kN. Express the force in a vector form. Find the unit vector in the direction of the force.

Solution From the components of the force we can write the force vector.

As
$$\vec{F} = (16i - 12j + 20k)\text{kN}$$

$\therefore$ Magnitude of the force
$$= \sqrt{16^2+(-12)^2+20^2} = 28.284\text{ kN}$$

$\therefore$ Unit vector
$$= \frac{16i-12j+20k}{28.284} = 0.141(4i-3j+5k)\text{ kN}$$

Example 5

Force $\vec{F} = (8i+5j-6k)$N passes through the point Q (6, 2, 5), calculate the moment about the point P (3,1,1). The co-ordinate distances are in metres.

Solution As P(3, 1, 1) and Q (6, 2, 5), the position vector of Q with respect to P

$$\vec{r} = (6-3)i+(2-1)j+(5-1)k = 3i+j+4k \text{ and force } \vec{F} = (8i+5j-6k)$$

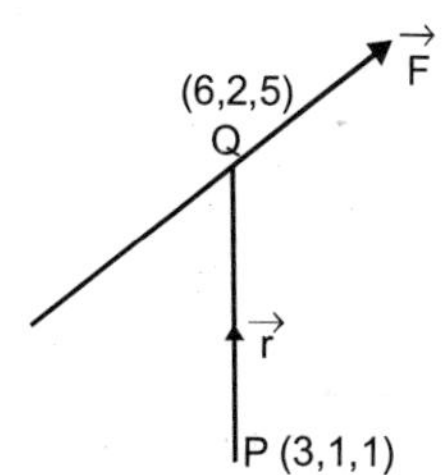

$\therefore$ Moment about P $\quad \overrightarrow{M_P} = \vec{r} \times \vec{F} = \begin{vmatrix} i & j & k \\ 3 & 1 & 4 \\ 8 & 5 & -6 \end{vmatrix} = (-26i + 50j + 7k)\ \text{N} \cdot \text{m}$

$\therefore$ Magnitude of the moment $\quad = \sqrt{(-26)^2 + 50^2 + 7^2} = 56.789\ \text{N} \cdot \text{m}$

EXAMPLE 6

Force $\vec{F} = (28i - 17j + 52k)$ Newton acts at a point P (–2, 1, 0). Find moment of this force about Y-axis.

SOLUTION Considering origin O (0, 0, 0) and point P (–2, 1, 0), the position vector of P,

$\vec{r} = (-2-0)i + (1-0)j + (0-0)k = (-2i + j)$ and given force $\vec{F} = (28i - 17j + 52k)\text{N}$

$\therefore$ Moment about O $\quad = \begin{vmatrix} i & j & k \\ -2 & 1 & 0 \\ 28 & -17 & 52 \end{vmatrix} = (52i + 104j + 6k)\text{N} \cdot \text{m}$

$\therefore$ Moment about Y-axis is 104 N · m.

EXAMPLE 7

A force of 6 kN acts along AG. Find the moment about the line FC.

SOLUTION Taking A (0, 0, 0) and G (4, 3, 3) as reference points

$$\overrightarrow{AG} = (4-0)i + (3-0)j + (3-0)k = 4i + 3j + 3k$$

$\therefore$ Unit vector along AG $\quad = \dfrac{4i + 3j + 3k}{\sqrt{4^2 + 3^2 + 3^2}} = 0.1715(4i + 3j + 3k)$

$\therefore$ Force vector along AG $\quad = 6 \times 0.1715(4i + 3j + 3k) = 1.029(4i + 3j + 3k)$

Again F (4, 0, 3) and C (4, 3, 0) are reference points.

$\therefore$ $\quad \overrightarrow{FC} = (4-4)i + (3-0)j + (0-3)k = 3j - 3k$

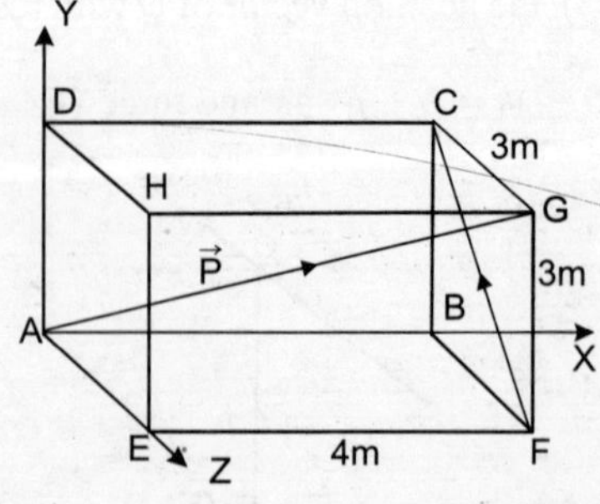

∴ Unit vector along *FC*

$$= \frac{3j-3k}{\sqrt{3^2+3^2}} = 0.707(j-k) \text{ also } \overrightarrow{FG} = 3j$$

∴ Moment about *FC*

$$= \begin{vmatrix} 0 & 3 & 0 \\ 4 & 3 & 3 \\ 0 & 1 & -1 \end{vmatrix} \times 1.029 \times 0.707 = 8.73 \text{ kN·m}$$

EXAMPLE 8

Moment of a certain force about the point *A* (3,7,–2)m is $(10i - 8j + 40k)$ kN·m. Find the moment of the same force about the line *AB*. Co-ordinate of *B* (5, 8, 1)m.

SOLUTION Moment of the same force about *AB* is the projection of the given moment along *AB*.

∴ Unit vector along *AB*

$$= \frac{(5-3)i+(8-7)j+(1+2)k}{\sqrt{2^2+1^2+3^2}} = 0.267(2i+j+3k)$$

∴ Moment of the force

$$= (10i-8j+40k)\cdot[0.267(2i+j+3k)]$$

$$= 35.284\text{N}\cdot\text{m}$$

∴ Vector moment

$$= 35.284 \times 0.267(2i+j+3k)$$

$$= 9.42(2i+j+3k)\text{N}\cdot\text{m}$$

EXAMPLE 9

Two vectors $\vec{A}$ and $\vec{B}$ are given as $\vec{A} = 2i - 6j - 3k$ and $\vec{B} = 4i + 3j - k$. Determine (*i*) their product (*ii*) the angle between them.

SOLUTION (*i*) $\vec{A}\cdot\vec{B} = A_xB_x + A_yB_y + A_zB_z = 2\times 4 + [(-6)\times 3] + [(-3)\times(-1)] = -7$

(*ii*) $|\vec{A}| = \sqrt{2^2+(-6)^2+(-3)^2} = 7$ and $|\vec{B}| = \sqrt{4^2+3^2+(-1)^2} = \sqrt{26}$

Angle between $\vec{A}$ and $\vec{B}$ is given by $\cos\theta = \dfrac{\vec{A}\cdot\vec{B}}{|\vec{A}||\vec{B}|} = \dfrac{-7}{7\times\sqrt{26}} = -0.196$

$\Rightarrow$ $\theta = 101.3°$.

EXAMPLE 10

The point of application of a force $\vec{F} = 5i + 10j - 15k$ is displaced from the point $P(1, 0, 3)$ to the point $Q(3, -1, -6)$. Find the work done by the force.

SOLUTION Here, $dx = 3 - 1 = 2,\ dy = (-1 - 0) = -1,\ dz = (-6 - 3) = -9$

$$\therefore \quad \vec{d} = 2i - j - 9k$$

We know work done $= \vec{F} \cdot \vec{d} = (5i + 10j - 15k) \cdot (2i - j - 9k) = 135$.

EXAMPLE 11

Find the projection of vector $\vec{A} = 4i - 3j + k$ on the line joining the points P (2, 3,–1) and Q(–2,–4, 3).

SOLUTION Here, $dx = (-2 - 2) = -4,\ dy = (-4 - 3) = -7,\ dz = 3 - (-1) = 4$

$$\therefore \quad \overrightarrow{PQ} = -4i - 7j + 4k = \vec{B}$$

and

$$\left|\vec{B}\right| = \sqrt{(-4)^2 + (-7)^2 + 4^2} = 9$$

We know $\vec{A} \cdot \vec{B} = AB\cos\theta$, the projection of vector $\vec{A}$ on vector $\vec{B}$ is

$$A\cos\theta = \frac{\vec{A} \cdot \vec{B}}{B} = \frac{[4 \times (-4)] + [(-3) \times (-7)] + [1 \times 4]}{9} = 1.$$

EXAMPLE 12

Determine the cross product of vector $\vec{A} = 4i - 3j - 2k$ and $\vec{B} = 2i + 2j - 3k$ and angle between them.

SOLUTION

$$\vec{A} \times \vec{B} = \begin{vmatrix} i & j & k \\ 4 & -3 & -2 \\ 2 & 2 & 3 \end{vmatrix} = -5i - 16j + 14k$$

$$\text{Unit vector } \vec{r} = \frac{\vec{A} \times \vec{B}}{\left|\vec{A} \times \vec{B}\right|} = \frac{-5i - 16j + 14k}{\sqrt{(-5)^2 + (-16)^2 + 14^2}}$$

$$= -0.229i - 0.733j + 0.641k$$

Angle between vector $\vec{A}$ and $\vec{B}$

$$\sin\theta = \frac{|\vec{A}\times\vec{B}|}{|\vec{A}||\vec{B}|} = \frac{\sqrt{(-5)^2+(-16)^2+14^2}}{\sqrt{4^2+(-3)^2+(-2)^2}.\sqrt{2^2+2^2+(-3)^2}} = 0.9835$$

$\therefore$ $$\theta = 79.57°$$

Example 13

Two forces are given $\vec{F_1} = (2i-3j+4k)\,\text{N}$ and $\vec{F_2} = (6i-9j+12k)\,\text{N}$. Prove that two forces are parallel. Also find the angle between them.

Solution (*i*)

$$\vec{F_1}\times\vec{F_2} = \begin{vmatrix} i & j & k \\ 2 & -3 & 4 \\ 6 & -9 & 12 \end{vmatrix} = 0$$

Hence $\vec{F_1}$ and $\vec{F_2}$ are parallel. Again we know

$$\cos\theta = \frac{\vec{F_1}\cdot\vec{F_2}}{F_1F_2} = \frac{2\times6+(-3)(-9)+4\times12}{\left(\sqrt{2^2+(-3)^2+4^2}\right)\left(\sqrt{6^2+(-9)^2+12^2}\right)} = 1$$

$\therefore$ $$\theta = 0°$$

$\therefore$ Angle between the given vectors = 0°

Example 14

Find the perpendicular distance from the point *A* (1, 2, 3) to the line joining the origin *O* and the point *B* (2, 10, 5).

Solution Position vector of point *A* (1, 2, 3) is given by $\overrightarrow{OA} = i+2j+3k = \vec{A}$ and the point *B* (2, 10, 5) is given by $\overrightarrow{OB} = 2i+10j+5k = \vec{B}$.

Perpendicular distance

$$AD = A\sin(\vec{A}\cdot\vec{B}) = \frac{A\cdot B\sin(\vec{A}\cdot\vec{B})}{B} = \frac{|\vec{A}\times\vec{B}|}{|B|}$$

Now,

$$\vec{A}\times\vec{B} = \begin{vmatrix} i & j & k \\ 1 & 2 & 3 \\ 2 & 10 & 5 \end{vmatrix} = -20i+j+6k$$

$\therefore$ $$|\vec{A}\times\vec{B}| = \sqrt{(-20)^2+1^2+6^2} = 20.9$$

$\therefore$ $$AD = \frac{20.9}{\sqrt{2^2+10^2+5^2}} = 1.84 \text{ Unit}$$

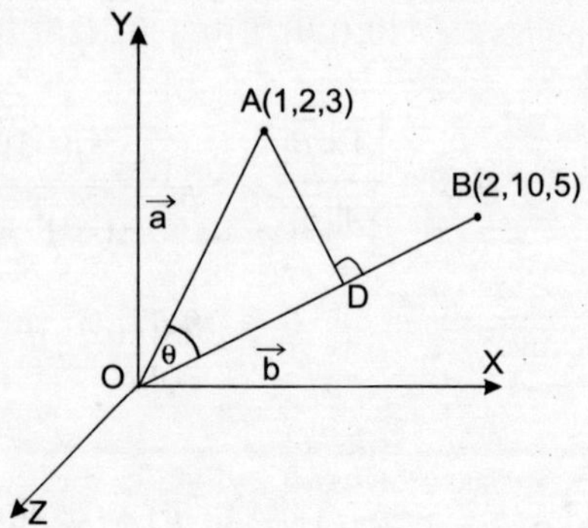

EXAMPLE 15

A force given by $\vec{F} = 3i + 2j - 4k$ is applied at the point $P(1, -1, 2)$. Find the moment of the force $\vec{F}$ about the point $O(2, -1, 3)$.

SOLUTION Position vector $\vec{r}$ of point P with respect to O

$$\vec{r} = [(1-2)]i + [(-1)-(-1)]j + [2-3]k = -i - k$$

We know moment

$$\vec{M} = \vec{r} \times \vec{F} = \begin{vmatrix} i & j & k \\ -1 & 0 & -1 \\ 3 & 2 & -4 \end{vmatrix} = 2i - 7j - 2k.$$

EXAMPLE 16

A rectangular block is acted upon by force as shown in figure. Replace this set of forces by an equivalent force-couple system acting at A.

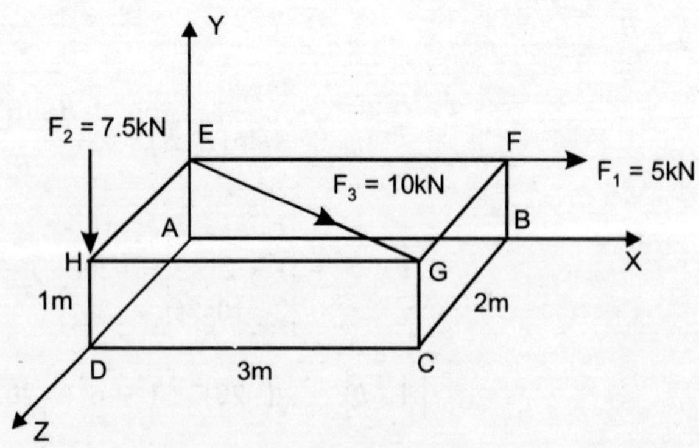

SOLUTION Co-ordinate of various points are $A\,(0,0,0)$, $E\,(0,1,0)$, $D\,(0,0,2)$, $F\,(3,1,0)$, $G\,(3,1,2)$, $H(0,1,2)$. Different force vectors

$$\vec{F_1} = \frac{5\left[(3-0)i+(1-1)j+(0-0)k\right]}{\sqrt{3^2}} = 5i$$

$$\vec{F_2} = \frac{7.5\left[(0-0)i+(0-1)j+(2-2)k\right]}{\sqrt{(-1)^2}} = -7.5j$$

$$\vec{F_3} = \frac{10\left[(3-0)i+(1-1)j+(2-0)k\right]}{\sqrt{3^2+2^2}}$$

$$= 8.32i + 5.547k$$

∴ Resultant force $\vec{R} = \vec{F_1}+\vec{F_2}+\vec{F_3} = 5i + (-7.5j) + (8.32i + 5.547k)$

$$= 13.32i - 7.5j + 5.547k$$

The resultant couple is found by summing the moment of each force $\vec{F_1}$, $\vec{F_2}$ and $\vec{F_3}$ about A. Position vector from A to convenient points on forces $\vec{F_1}$, $\vec{F_2}$, and $\vec{F_3}$ respectively being

$$\overrightarrow{r_{AE}} = (0-0)i + (1-0)j + (0-0)k = j$$

$$\overrightarrow{r_{AD}} = (0-0)i + (0-0)j + (2-0)k = 2k$$

∴ Moments of all the forces about A $\left(\overrightarrow{M_R}\right)_A = \overrightarrow{r_{AE}} \times \vec{F_1} + \overrightarrow{r_{AD}} \times \vec{F_2} + \overrightarrow{r_{AE}} \times \vec{F_3}$

$$= \begin{vmatrix} i & j & k \\ 0 & 1 & 0 \\ 5 & 0 & 0 \end{vmatrix} + \begin{vmatrix} i & j & k \\ 0 & 0 & 2 \\ 0 & -7.5 & 0 \end{vmatrix} + \begin{vmatrix} i & j & k \\ 0 & 1 & 0 \\ 8.32 & 0 & 5.547 \end{vmatrix}$$

$$= 20.547i - 13.32k$$

EXAMPLE 17

Determine the moment of the 100 N force as shown in figure, about point A and B respectively.

SOLUTION The force is collinear with vector $\overline{DE}$ where $\overrightarrow{DE} = 8i + 4j - 4k$

So, unit vector $\overrightarrow{r_{DE}} = \frac{\overrightarrow{DE}}{\left|\overrightarrow{DE}\right|} = \frac{8i+4j-4k}{\sqrt{8^2+4^2+(-4)^2}}$

$$= 0.816i + 0.408j - 0.408k$$

∴ $\overrightarrow{F_{DE}} = F \cdot \overrightarrow{r_{DE}} = 100(0.816i + 0.408j - 0.408k)$

$$= 81.6i + 40.8j - 40.8k$$

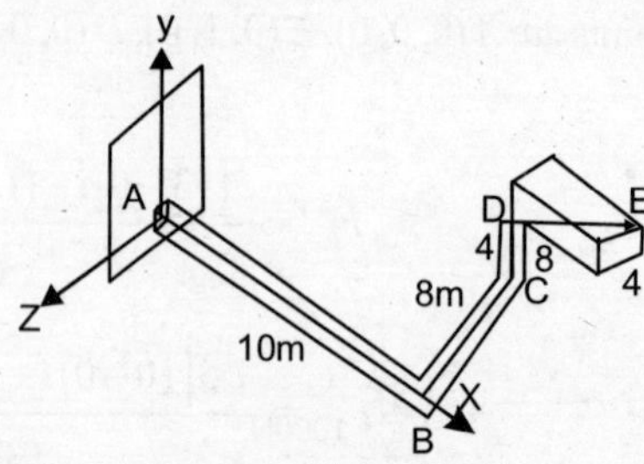

To get the moment M_A, we choose a position vector from point A to D, which is the line of action of force F.

So
$$\overrightarrow{AD} = 10i + 4j - 8k$$

$$\therefore \quad \overrightarrow{M_A} = \overrightarrow{AD} \times \vec{F} = \begin{vmatrix} i & j & k \\ 10 & 4 & -8 \\ 81.6 & 40.8 & -40.8 \end{vmatrix}$$
$$= 163.2i - 244.8j + 81.6k$$

$$\therefore \quad \left|\overrightarrow{M_A}\right| = \sqrt{163.2^2 + (-244.8)^2 + 81.6^2} = 305.32\,\text{N}\cdot\text{m}$$

Again position vector $\overrightarrow{BD} = 4j - 8k$

$$\therefore \quad \overrightarrow{M_B} = \overrightarrow{BD} \times \vec{F} = \begin{vmatrix} i & j & k \\ 0 & 4 & -8 \\ 81.6 & 40.8 & -40.8 \end{vmatrix}$$

$$= 163.2i - 652.8j - 326.4k$$

$$\therefore \quad \left|\overrightarrow{M_B}\right| = \sqrt{163.2^2 + (-652.8)^2 + (-326.4)^2}.$$
$$= 747.87\,\text{N}\cdot\text{m}$$

EXAMPLE 18

A load W of magnitude 1000N is supported by three cables AB, AC, and AD. Determine the tension in the cables.

SOLUTION This problem relates to concurrent force system in space. T_{AB}, T_{AC}, and T_{AD}, be the tension in the cable where directions are known. Co-ordinates of various points are A (0, –4.5, 0), B (2.8, 0, 0), C (0, 0, –2.4), D(–2.6, 0, 1.8).

Tension in the cable AB
$$\overrightarrow{T_{AB}} = \frac{T_{AB}\left[(2.8-0)\right]i + \left[0-(-4.5)\right]j + \left[0-0\right]k}{\sqrt{2.8^2 + 4.5^2}}$$
$$= T_{AB}(0.528i + 0.849j)$$

Tension in the cable AC
$$\overrightarrow{T_{AC}} = \frac{T_{AC}\left[(0-0)\right]i + \left[0-(-4.5)\right]j + \left[-2.4-0\right]k}{\sqrt{4.5^2 + 2.4^2}}$$
$$= T_{AC}(0.88j - 0.47k)$$

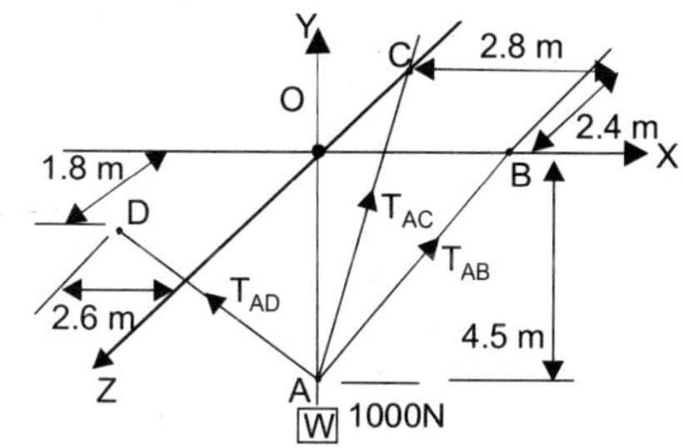

Tension in the cable *AD*

$$\overrightarrow{T_{AD}} = \frac{T_{AD}\left[(-2.6-0)\right]i+\left[0-(-4.5)\right]j+\left[1.8-0\right]k}{\sqrt{(-2.6^2)+4.5^2+1.8^2}}$$

$$= T_{AD}\left(-0.473i+0.818j+0.327k\right)$$

Weight *W* is acting vertically downwards

$\therefore$ $$\vec{W} = 1000\left(0i-4.5j+0k\right) = -4500j$$

Applying the equations of equilibrium,

$$\left.\begin{aligned} 0.528T_{AB}+0.0T_{AC} \quad -0.473T_{AD} \;-0 \\ 0.849T_{AB}+0.88T_{AC} \;+\; 0.818T_{AD} = 4500 \\ 0.0T_{AB} \quad -0.47T_{AC} \;+\; 0.327T_{AD} = 0 \end{aligned}\right\}$$

Solving above equations, $T_{AD} = 2058.47\text{N}, T_{AB} = 1832.04\text{N}, T_{AC} = 1432.69\text{N}$.

[Note: Some deviation in the magnitude of results appeared due to truncation effect.].

EXAMPLE 19

In figure a vertical boom *AE* is supported by guy wire from *A* to *B*, *C* and *D*. If the tensile load in *AD* = 252N, find the forces in *AC* and *AB* so that the resultant force on *A* will be vertical.

SOLUTION Co-ordinates of the points, *A* (0, 12, 0), *B* (0, 0, –9), *C* (–4, 0, 3), *D* (6, 0, 4).

$\therefore$ $$\overrightarrow{AB} = -12j-9k,\ \overrightarrow{AC} = -4i-12j+3k,$$

and $$\overrightarrow{AD} = 6i-12j+4k$$

$\therefore$ $$\overrightarrow{r_{AB}} = \frac{-12j-9k}{\sqrt{(-12)^2+(-9)^2}} = -0.8j-0.6k$$

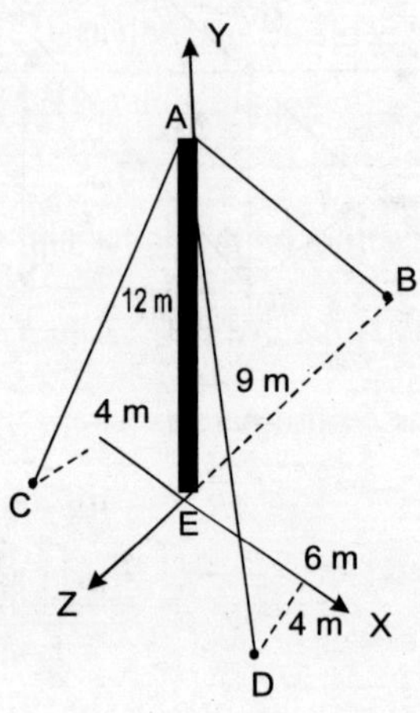

$$\overrightarrow{r_{AC}} = \frac{-4i-12j+3k}{\sqrt{(-4)^2+(-12)^2+3^2}}$$

$$= -0.308i - 0.923j + 0.231k$$

$$\overrightarrow{r_{AD}} = \frac{6i-12j+4k}{\sqrt{6^2+(-12)^2+4^2}}$$

$$= 0.429i - 0.857j + 0.286k$$

$$\therefore \quad \overrightarrow{F_{AB}} = F_{AB}(-0.8j-0.6k) = -0.8jF_{AB} - 0.6kF_{AB}$$

$$\therefore \quad \overrightarrow{F_{AC}} = F_{AC}(-0.308i - 0.923j + 0.231k)$$

$$= -0.308iF_{AC} - 0.923jF_{AC} + 0.231kF_{AC}$$

$$\therefore \quad \overrightarrow{F_{AD}} = 252(0.429i - 0.857j + 0.286k)$$

$$= 108.1i - 215.96j + 72.1k$$

So, resultant

$$\vec{R} = (108.1 - 0.308F_{AC})i + (-215.96 - 0.923F_{AC} - 0.8F_{AB})j + (72.1 - 0.6F_{AB} + 0.231F_{AC})k$$

As resultant is vertical, so

$$\left.\begin{aligned} \sum F_x &= 108.1 - 0.308F_{AC} = 0 \\ \sum F_z &= 72.1 - 0.6F_{AB} + 0.231F_{AC} = 0 \end{aligned}\right\}$$

Solving these two equations we have, $F_{AC} = 350.97\text{N}$, $F_{AB} = 255.29\text{N}$.

EXAMPLE 20

A tower guy wire is *anchored* by means of a bolt at A. The tension in the wire is 250 N. Determine (a) the components F_x, F_y, F_z of the force acting on the bolt (b) the angles $\theta_x, \theta_y, \theta_z$ defining the direction of the force.

SOLUTION The line of action of the force acting on the bolt A and B and force is directed from A to B. So

distance from A to B $= \sqrt{(0-40)^2 + (80-0)^2 + (0-(-30))^2} = 94.3$

Denoting by i, j, k, the unit vector along the co-ordinate axes,

$$\overrightarrow{AB} = -40i + 80j + 30k$$

$\therefore$ Unit vector

$$\overrightarrow{r_{AB}} = \frac{-40i + 80j + 30k}{94.3}$$

$\therefore$

$$\vec{F} = F \times \overrightarrow{r_{AB}} = \frac{2500}{94.3}(-40i + 80j + 30k) = -1060i + 2120j + 795k$$

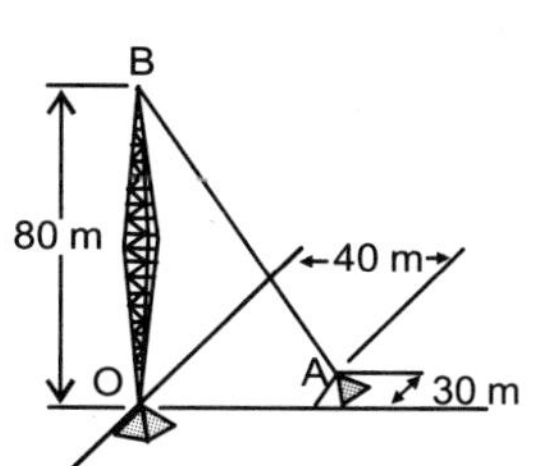

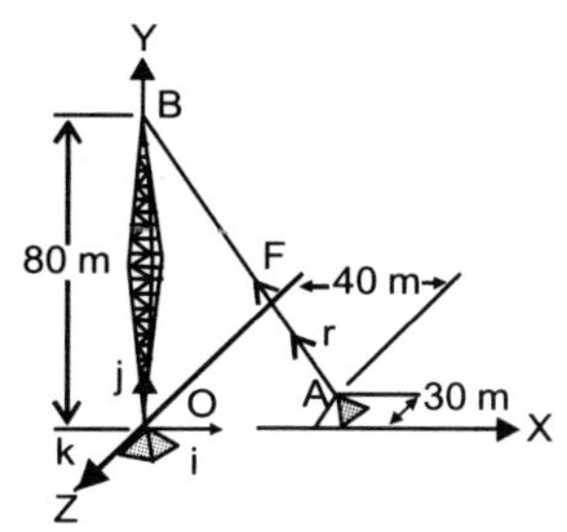

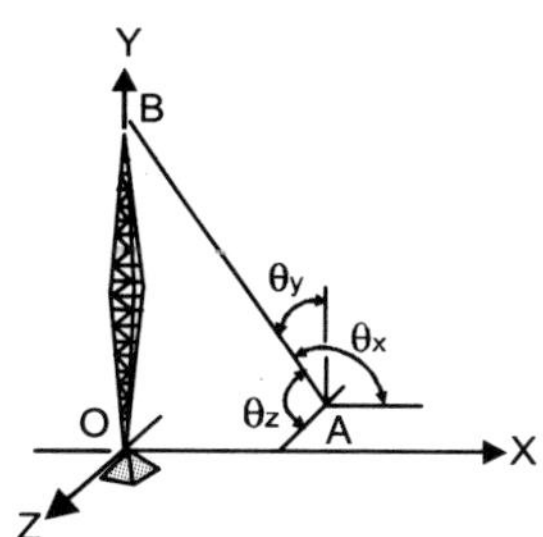

The components of $\vec{F}$ are $F_x = -1060\text{N}$, $F_y = 2120\text{N}$, $F_z = 795\text{N}$

Directions of the forces

$$\cos\theta_x = \frac{F_x}{F} = \frac{-1060}{2500}$$

$\Rightarrow$

$$\theta_x = 115.1^\circ$$

$$\cos\theta_y = \frac{F_y}{F} = \frac{2120}{2500}$$

$\Rightarrow$

$$\theta_y = 32.0^\circ$$

$$\cos\theta_z = \frac{F_z}{F} = \frac{795}{2500}$$

$\Rightarrow$

$$\theta_z = 71.5^\circ$$

Example 21

A wall of precast concrete is temporarily held by the cables AB and AC shown in figure. Tension in cable AB is 3.8kN and in cable AC is 5.5kN. Determine the magnitude and direction of the resultant of the forces exerted by the cable AB and AC on stake A.

Solution

$$AB = \sqrt{(0-5)^2+(2.5-0)^2+(0-(-3.5))^2} = 6.6\text{m}$$

$$AC = \sqrt{(0-5)^2+(2.5-0)^2+(-8.5-(-3.5))^2} = 7.5\text{m}$$

$\therefore$ $$\overrightarrow{AB} = -5i+2.5j+3.5k \text{ and } \overrightarrow{AC} = -5i+2.5j-5k$$

$\therefore$ $$\overrightarrow{F_{AB}} = \frac{3.8}{6.6}(-5i+2.5j+3.5k) = (-2.9i+1.4j+2k)\,\text{kN}$$

$\therefore$ $$\overrightarrow{F_{AC}} = \frac{5.5}{7.5}(-5i+2.5j-5k) = (-3.6i+1.8j-3.6k)\,\text{kN}$$

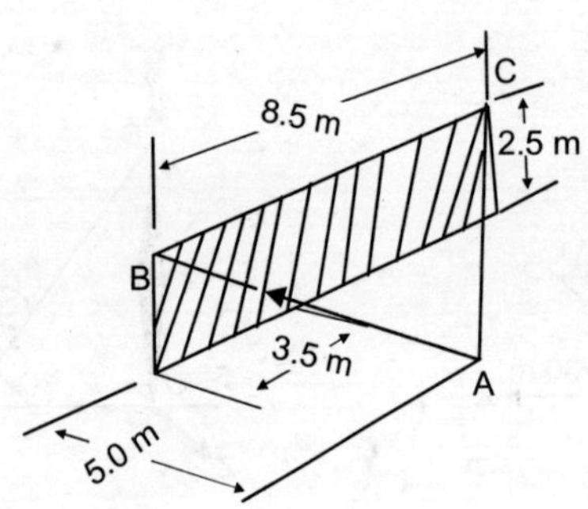

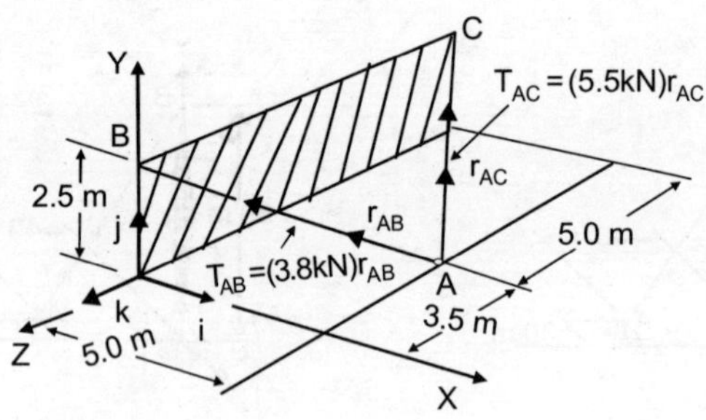

Resultant of the forces exerted by the two cables

$$\vec{R} = \overrightarrow{F_{AB}} + \overrightarrow{F_{AC}} = (-6.5i+3.2j-1.6k)\,\text{kN}$$

So, $$|\vec{R}| = \sqrt{(-6.5)^2+3.2^2+(-1.6)^2} = 7.4\text{kN}$$

Directions of resultant force

$$\cos\theta_x = \frac{-6.5}{7.4}$$

$\Rightarrow$ $$\theta_x = 151.4^\circ$$

$$\cos\theta_y = \frac{3.2}{7.4}$$

$\Rightarrow$ $$\theta_y = 64.3^\circ$$

$$\cos\theta_z = \frac{-1.6}{7.4}$$

$$\Rightarrow \qquad \theta_z = 102.48^\circ$$

EXAMPLE 22

A crane is supporting a 2000 N box through three cables *AB*, *CB*, and *DB*. The point *D* is at the centre of the outer edge of the box; *C* is 1.6m away from the corner of this edge, and *B* is directly above the center of the box. What are the force F_1, F_2, and F_3 transmitted by the cables.

SOLUTION Considering *B* as origin (0, 0, 0), the distance from *A* to *B*

$$= \sqrt{(0-(-1.5))^2 + (0-(-3))^2 + (0-1)^2} = 3.5\,\text{m}$$

$$\therefore \qquad \overrightarrow{AB} = 1.5i + 3j - 1k$$

So, unit vector

$$\overrightarrow{r_{AB}} = \frac{1.5i + 3j - k}{3.5}$$

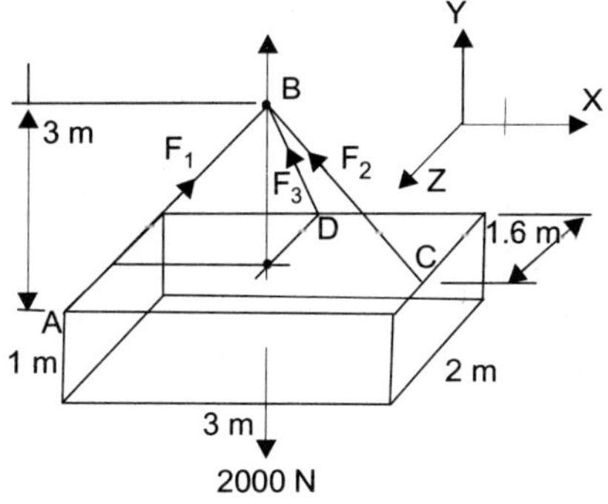

Hence, reactive forces

$$\overrightarrow{F_{AB}} = F_1 \times \frac{1.5i + 3j - k}{3.5} = F_1(0.4286i + 0.857j - 0.286k)$$

$$\overrightarrow{F_{DB}} = F_2 \times \frac{3j + k}{\sqrt{3^2 + 1^2}} = F_2(0.9487j + 0.3162k)$$

$$\overrightarrow{F_{CB}} = F_3 \times \frac{-1.5i + 3j - 0.6k}{\sqrt{(-1.5)^2 + 3^2 + (-0.6)^2}}$$

$$= F_3(-0.4402i + 0.8805j - 0.1761k)$$

Now, resultant of all these three forces is equal to 2000N and as 2000N acts vertically, downward from its C.G. We can write,

$$\vec{R} = (0.4286F_1 - 0.4402F_3)i + (0.857F_1 + 0.9487F_2 + 0.8805F_3)j + (-0.286F_1 + 0.3162F_2 - 0.1761F_3)k$$

and, $\vec{W} = 2000(0i - 3.5j + 0k)$

So, the equations of equilibrium will be

$$\left.\begin{aligned} 0.4286F_1 - 0.4402F_3 &= 0 \\ 0.857F_1 + 0.9487F_2 + 0.8805F_3 &= 7000 \\ -0.286F_1 + 0.3162F_2 - 0.1761F_3 &= 0 \end{aligned}\right\}$$

Solving above equations, we obtain, $F_1 = 2267.73$ N, $F_2 = 3280.59$ N, $F_3 = 2208.11$ N

Exercise

1. A force 1500 N acts along the line PQ, where $P(7,-2,4)$m and $Q(11,0,-3)$m. Calculate moment of this force about $A(6,-4,-3)$m.

   Ans. [$180.58(-28i+35j-6k)$ N·m]

2. A force $(800i+460j-380k)$N acts at $P(7,2,4)$. Find the moment about the line AB, $A(9,4,3)$m and $B(2,8,7)$m.

   Ans. [$9.63(-7i+4j+4k)$ N·m]

3. A force of 1200 N acts along PQ, where $P(4,5,-2)$ and $Q(-3,1,6)$. Calculate its moment about a line joining $A(3,2,0)$ to $B(2,2,1)$ m.

   Ans. [$52.83(-i+k)$ N·m]

4. A force $\vec{F_1} = (10i+6j+3k)$ N acts at a position (3, 0, 2). At point (0, 2, –3) an equal but opposite force $-\vec{F_1}$ acts. What is the couple moment? What are the direction cosine normal to the plane of the couple?

5. The force F of 20 kN is such that if shifted to any point A, the force will lie in planes perpendicular to x and z-axes thereby creating moments about these axes and in a sense opposite to the two moments already acting on the plate. Find x and z.

   Ans. [$x = 3.8$m, $z = 4.1$m]

6. A thin plate $ABCDEF$ of negligible weight is supported in x-z plane at points A, B, and C such that supports can provide vertical reactions only. It is acted upon by forces P=5kN and Q= 3kN and a couple of 5kN·m as shown. Find the reaction at supports A, B, and C.

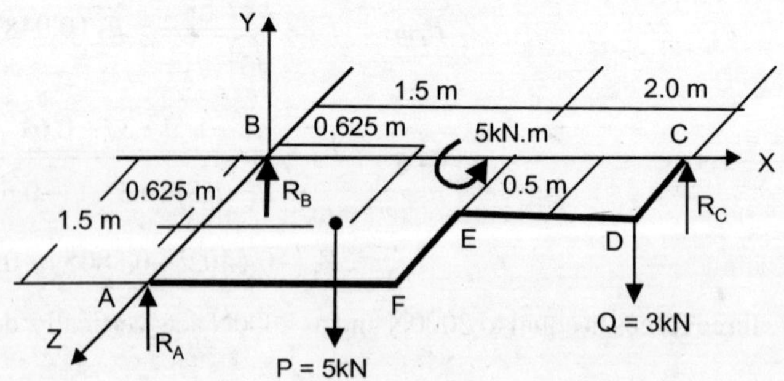

P-6

Ans. [$R_A = 6.417$ kN, $R_B = 2.31$ kN, $R_C = 3.893$ kN]

7. Determine the forces in the bars AB, AC, and AD when loaded at the joint point A by a force. $\vec{F} = (-30i - 20j + 40k)$ kN.

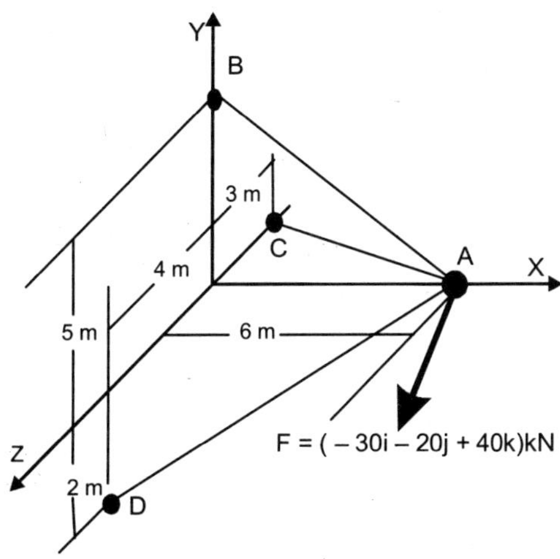

P-7

8. Find the resultant of the force system in figure in which $P = 280$ N, $T = 260$ N and $F = 210$ N. (All dimensions are in m).

Ans. [R = 674.46 N]

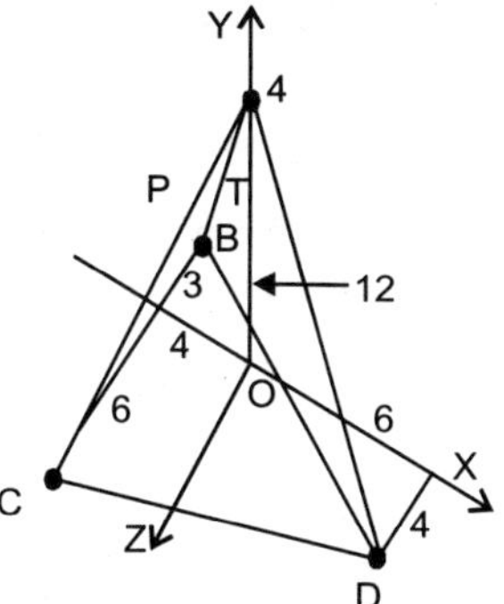

P-8

9. In figure a boom AC is supported by a ball and socket joint at C and by the cables BE and AD. If the force multiplier of a force F acting from B to E is $F_m = 10\text{N/m}$ and that of a force P acting from A to D is, find the component of each force along AC. Using the above data find the component of F and P along the direction DE.

Ans. [$P_{AC} = 99.75\text{N}, \;\; F_{DE} = 82.2\text{N}, \;\; P_{DE} = -56.916\text{N}$]

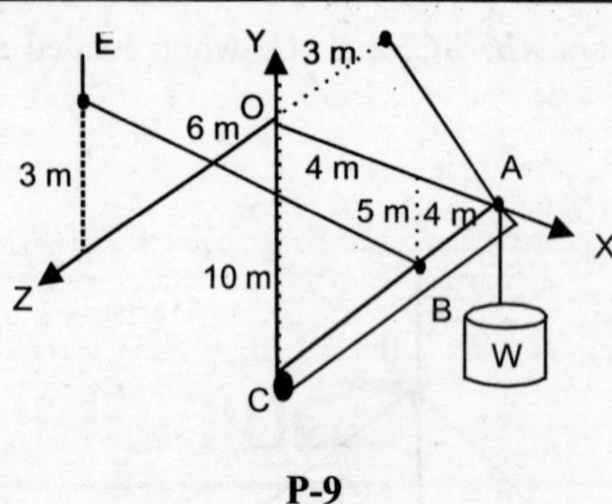

P-9

10. A force $\vec{F}$ acts from B (8, –3, 0) to D (0, 0, 6) produces a moment of 1000 N·m about an axis directed from A (12, 0, 0) to E (0, 4, –6), compute the magnitude of $\vec{F}$.

Ans. [$\vec{F} = 276.85\text{N}$]

11. In the adjacent figure, F = 1000 N while O (0, 0, 0), A (0, 10, 0) and B (5, 0, 4). Calculate the moment of force about O.

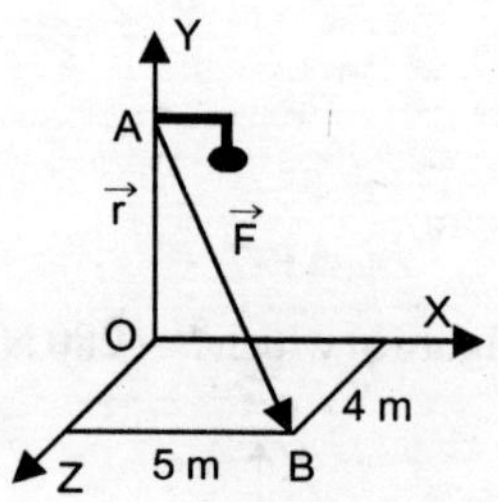

P-11

12. A vertical pole is guyed by a wire PQ which is anchored by means of a bolt Q as shown in figure. The tension in the wire is 2500N. Determine (*i*) the components of the force acting on the bolt (*ii*) the angles α_x, α_y, and α_z defining the direction of the force and (*iii*) specify the force vector.

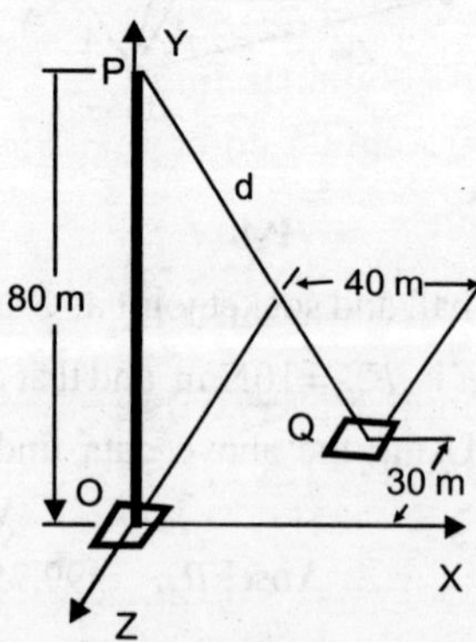

P-12

CHAPTER 2

EQUILIBRIUM OF FORCES

2.1 TWO AND THREE DIMENSIONAL FORCE SYSTEM

Any *coplanar force* can be represented in a two-dimensional plane. But if the nature of the force is *non-coplanar* it is not at all possible to represent the force in two-dimensional system. Detailed exposure of these we already have in the previous chapter. In this section a brief overview is presented.

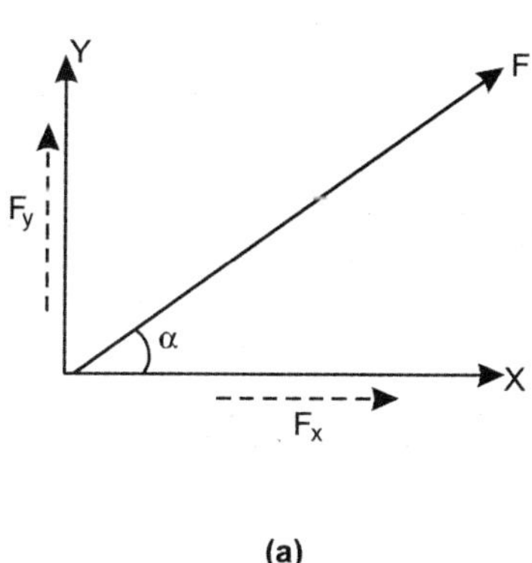

(a)

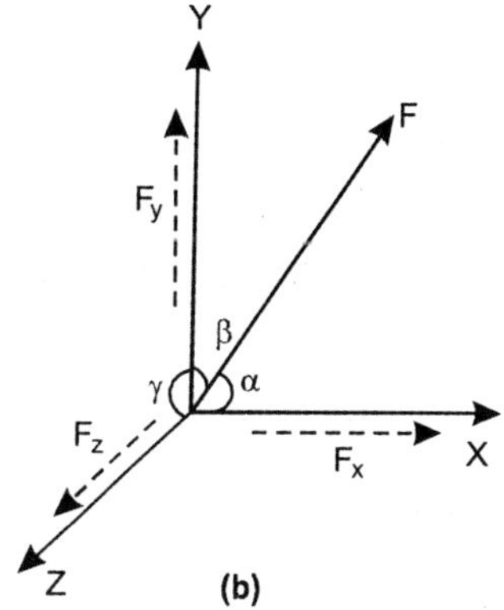

(b)

Fig. 2.1

In Figure 2.1 (*a*), the force F is a coplanar force and so is represented in 2-*D* coordinate system. Hence resolution of force along X and Y co-ordinate axes are as following:

$$\left.\begin{aligned} F_x &= F\cos\alpha \\ F_y &= F\sin\alpha \end{aligned}\right\} \qquad \text{... (2.1)}$$

In Figure 2.1 (*b*), force F makes angle α, β, γ with respect to X, Y, Z axis respectively. Along the co-ordinate axes, the force can be resolved as

$$\left.\begin{aligned} F_x &= F\cos\alpha \\ F_y &= F\cos\beta \\ F_z &= F\cos\gamma \end{aligned}\right\} \qquad \text{... (2.2)}$$

2.2 MOMENT OF A FORCE

Moment of a force about an axis through a point is equal to the product of the force (F) and perpendicular distance of the point from the line of action of the force (d). So, $M = Fd$. In other word, *moment is a measure of rotational tendency of a force on a body*. In Figure 2.2 (a), the force F is at right side of O and acting upwards thereby creating a tendency of rotation *counter clockwise.* And in Figure 2.2 (b), F is acting downward, by virtue of which rotational tendency is *clockwise*.

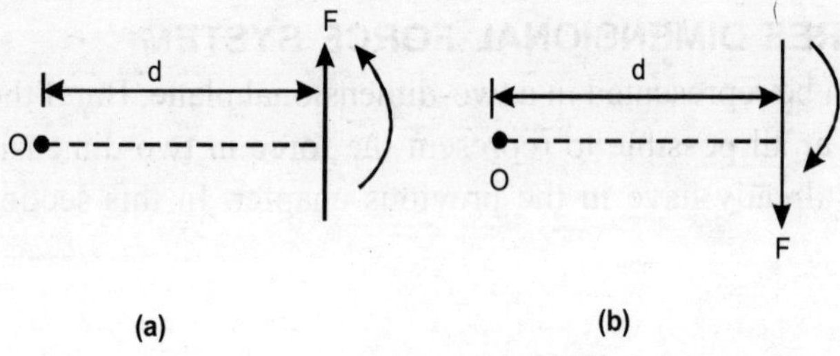

Fig. 2.2

Here the point O is defined as *moment centre* through which the axis of rotation passes and the distance d is called *lever arm distance* [Refer Figure 2.3(a)].

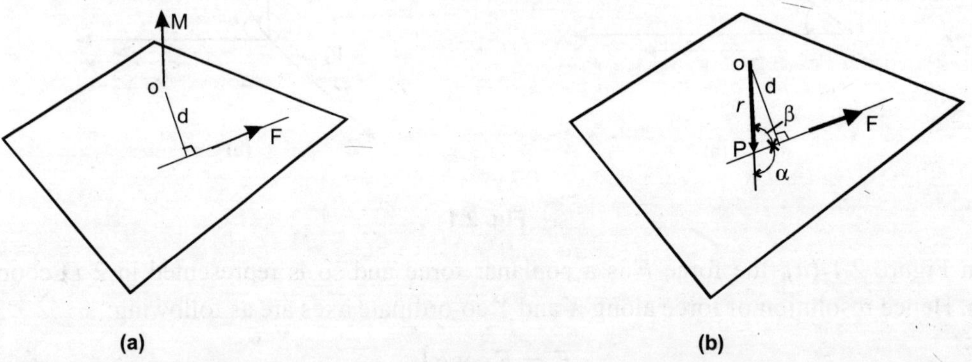

Fig. 2.3

Vectorially we can represent moment of a force as cross product vector which is shown in Figure 2.3(b) as

$$M = r \times F$$

So magnitude of moment,

$$|M| = |r \times F| = |r||F|\sin\alpha = |F||r|\sin\beta = Fd$$

2.3 THEOREM OF VARIGNON

This theorem relates the moment of a force and resolution of moments. This theorem states that,

The moment of force about an axis is equal to the algebraic summation of moments of its components about the same axis.

Let us consider, referring to Figure 2.4, a force F has two resolution of components F_1 and F_2 along any two directions, O is an arbitrarily chosen point which serves as monent centre on the Y-axis of mututally perpendicular X–Y co-ordinate system. d, d_1 and d_2 are the perpendicular distances from O on the line of action of F, F_1, and F_2 respectively. Lever arms d, d_1 and d_2 subtend an angle θ, θ_1, and θ_2 respectively with respect to Y-axis.

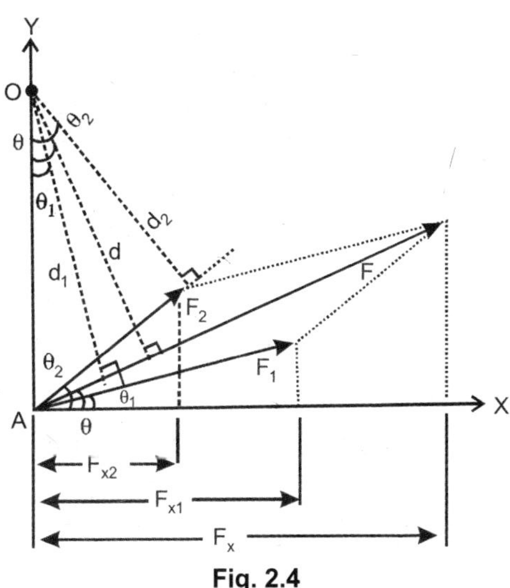

Fig. 2.4

Now, moment of the force F about O

$$F \cdot d = F(OA \cos \theta) = OA\,(F \cos \theta) = OA \cdot F_x \qquad \text{... (2.3)}$$

Moments of the force F_1 about O

$$F_1 \cdot d_1 = F_1\,(OA \cos \theta_1) = OA\,(F_1 \cos \theta_1) = OA \cdot F_{x1} \qquad \text{... (2.4)}$$

and moment of force F_2 about O

$$F_2 \cdot d_2 = F_2\,(OA \cos \theta_2) = OA\,(F_2 \cos \theta_2) = OA \cdot F_{x2} \qquad \text{... (2.5)}$$

Summation of (2.4) and (2.5) yields,

$$F_1 \cdot d_1 + F_2 \cdot d_2 = OA \cdot F_{x1} + OA \cdot F_{x2} = OA\,(F_{x1} + F_{x2}) = OA \cdot F_x = F \cdot d \qquad \text{... (2.6)}$$

A successive application of above approach leads to the extension of *Theorem of Varignon* to a system of forces.

2.4 MOMENT OF COUPLE

A system of two equal parallel forces acting in opposite direction cannot be reduced to one resultant force. Hence two equal parallel forces form a *couple*. On a rigid body a couple has only one effect, namely a *turning action*. Quantitatively this *turning action* is represented as the moment of forces about a point of axis. And this moment is termed as *moment of a couple* or simply *couple moment*. In vectors to point 1 and 2 respectively, along the respective line of action of each force. Adding moment of each force about O we can have couple moment M as,

$$M = r_1 \times F + r_2 \times (-F) = (r_1 - r_2) \times F = d \times F \quad \text{... (2.7)}$$

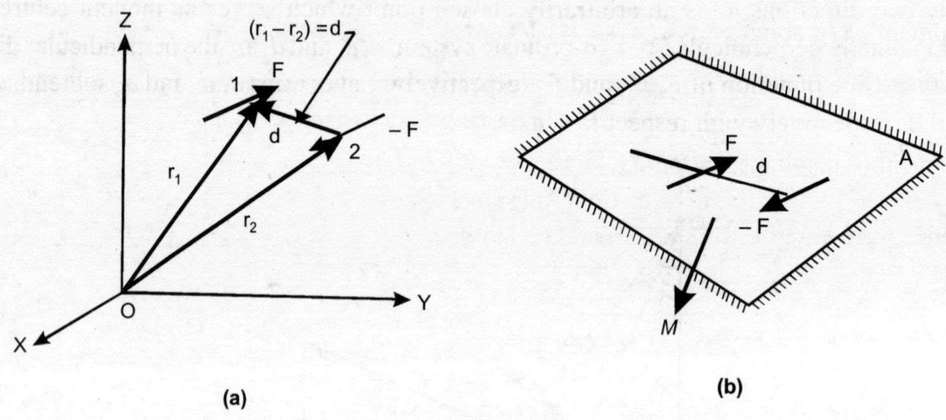

Fig. 2.5

Here d lies in the plane of the couple and so M has an orientation normal to the plane of the couple. And the sense of this moment is thus represented by *right hand screw rule*. It is shown in Figure 2.5(*b*).

To compute the magnitude of couple moment vector, it may be the easiest way to select position 1 and 2 so that d is perpendicular to the lines of action of forces.

Therefore, $$|M| = |d||F|\sin 90° = |d||F| = |F|d \quad \text{... (2.8)}$$

2.5 RESULTANT OF TWO LIKE PARALLEL FORCES

Let us consider, two forces F_1 and F_2 of unequal or equal magnitude are acting on a body in the same direction but at two different points A and B, shown in Figure 2.6.

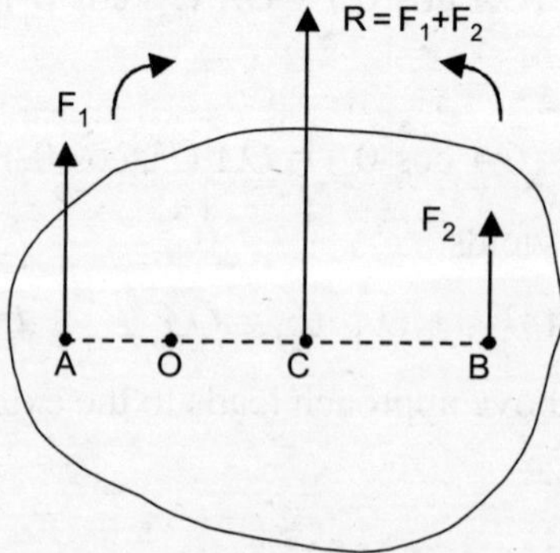

Fig. 2.6

Now the magnitude of the resultant of the force, $R = F_1 + F_2$.

Once the magnitude of the resultant force in computed, it is necessary to obtain its point of action. Let us assume point C be the point of action of resultant R. O here is an arbitrarily chosen reference point on ACB. Now,

Moment of F_1 about O $= F_1 \times AO$ (Clockwise) ... (2.9 *a*)

Moment of F_2 about O $= -F_2 \times BO$ (Anti clockwise) ... (2.9 *b*)

Adding (2.9 *a*) and (2.9*b*), we obtain, algebraic summation of moment of F_1 and F_2 about O

$$= F_1 \times AO - F_2 \times BO \text{ (Clockwise)} \quad \text{... (2.10)}$$

Moment of resultant R about O

$$= -R \times CO \text{ (Anti clockwise)} \quad \text{... (2.11)}$$

Hence we can write from (2.10) and (2.11) that,

$$F_1 \times AO - F_2 \times BO = -R \times CO$$

$$= -(F_1 + F_2) \times CO$$

$$\Rightarrow \quad F_1 \times (AO + CO) = F_2\,(BO - CO)$$

$$\Rightarrow \quad F_1 \times AC = F_2 \times BC$$

$$\Rightarrow \quad \frac{F_1}{F_2} = \frac{BC}{AC} \quad \text{... (2.12)}$$

2.6 RESULTANT OF TWO UNLIKE PARALLEL FORCES

In this case, the material body is subjected to two unequal forces F_1 and F_2 acting in opposite direction having point of action at A and B. As done earlier, a point O is chosen as reference point, shown in Figure 2.7. If $F_1 > F_2$, the F_2 magnitude of resultant force will be $R = F_1 - F_2$ which is assumed to act through the point C. Now,

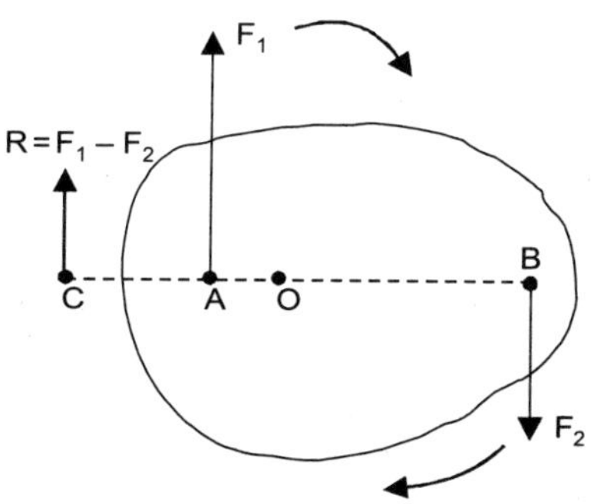

Fig. 2.7

Moment of F_1 about $O = F_1 \times AO$ (Clockwise) ... (2.13 *a*)

Moment of F_2 about $O = F_2 \times BO$ (Clockwise) ...(2.13 *b*)

Adding (2.13 *a*) and (2.13 *b*), we find the algebraic summation of moments of F_1 and F_2 about O.

$$= F_1 \times AO + F_2 \times BO \quad \text{... (2.14)}$$

Now, moment of resultant force R about O $= R \times CO = (F_1 - F_2) \times CO$ ... (2.15)

From (2.14) and (2.15), we can write,

$$\Rightarrow \quad F_1 \times AO + F_2 \times BO = (F_1 - F_2) \times CO$$

$$\Rightarrow \quad F_2 \times (BO + CO) = F_1 \,(CO - AO)$$

$$\Rightarrow \quad F_2 \times BC = F_1 \times AC$$

$$\Rightarrow \quad \frac{F_1}{F_2} = \frac{BC}{AC} \quad \text{... (2.16)}$$

2.7 FREE BODY DIAGRAM

When we consider the static equilibrium of a body, subjected to a number of forces, we assume that the movement of the body is *restricted in every sense*, either by *translation* or by *rotation* and then it is obvious that the body exerts some pressure against its supports. This in turn, causes an equal and opposite pressure on the body, which we call *support reaction.* Therefore, the force exerted by a body on its support and respective support reaction are two collinear forces, which remain always in equilibrium. A free body diagram shows all the forces in action on the body, including the support reactions, after the body is made free of all supports (i.e., supports are replaced by respective reactive forces).

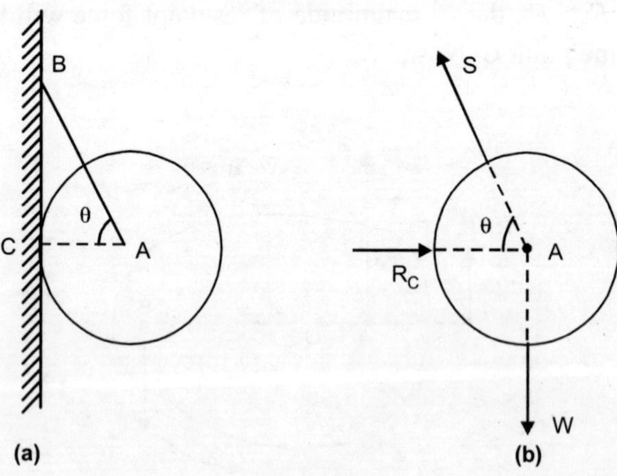

Fig. 2.8

Figure 2.8 (*a*) shows that a horizontal right circular cylindrical roller with weight W is supported against the vertical wall with the help of a tie rod AB and the roller touches the wall at C, which is considerd as a support.

On careful observation, we may conclude that

(*i*) The weight *W* of the roller acts vertically downward direction.

(*ii*) The tie bar *AB* will be subjected to a tensile force; let it be *S*.

(*iii*) The reactive force exerted by wall on the roller at *C* acts in normal direction to the plane of wall and let it be R_c.

Figure 2.8 (*b*) shows all the above mentioned forces on the roller after it has been made free of all supports and this represents the free body diagram of the roller.

2.8 RESOLUTION OF A FORCE BY EQUIVALENT FORCE AND COUPLE

Let us consider a force *F* acting on a body at *A*. It is to find out a combination of force and a moment at *B* which is *statically equivalent* to *F* at *A*.

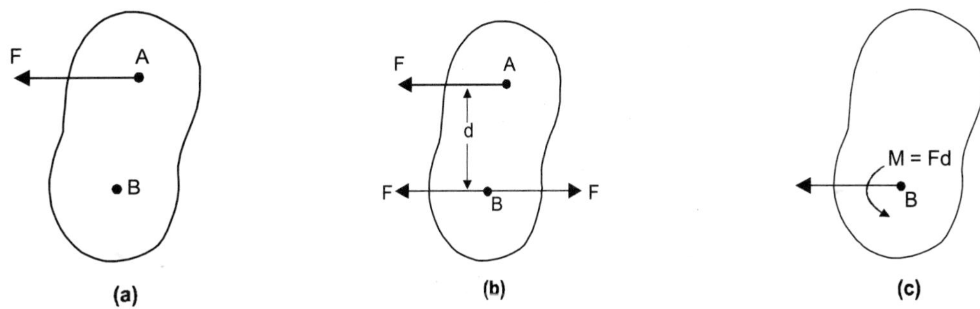

Fig. 2.9

In order to do so, two equal and opposite forces at *B* are introduced, of magnitude *F* and parallel to line of action of force at *A*. The perpendicular distance between two parallel lines of action is *d* as shown in Figure 2.9 (*b*). If we consider two forces *F*, acting at *A* towards left and *F* acting *B* towards right, it can be said that a moment of couple is formed, the magnitude of which,

$$M = F \times d \qquad \text{... (2.17)}$$

Now the only remaining forces is *F* at *B* leftwards.

Hence force *F* at *B* and a moment of couple *M* at *B*, which is shown in Figure 2.9(*c*).

Following the same principle, any force system can be replaced at any point by equivalent single force and single couple moment. In Figure 2.10, some arbitrary system of forces and couples are shown in solid lines. So the resultant force is

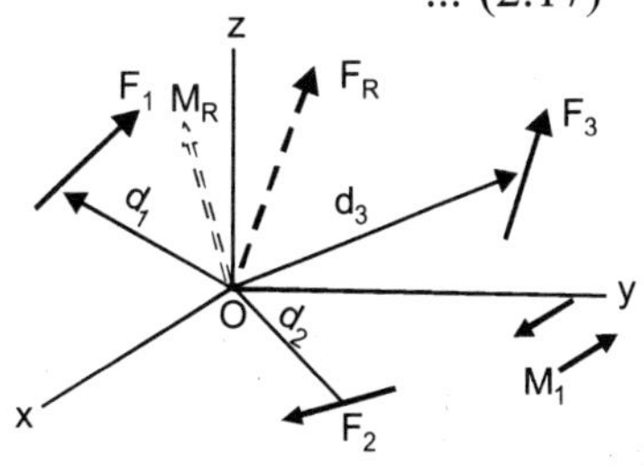

Fig. 2.10

$$F_R = \left[\sum_{i=1}^{n}(F_i)_x\right]i + \left[\sum_{i=1}^{n}(F_i)_y\right]j + \left[\sum_{i=1}^{n}(F_i)_z\right]k \qquad \text{...(2.18)}$$

The couple moment accompanying F_R for a chosen point *O* is

$$M_R = (d_1 \times F_1 + d_2 \times F_2 + \cdots\cdots) + (M_1 + M_2 + \cdots\cdots) = \sum_{i=1}^{n} d_i \times F_i + \sum_{i=1}^{n} M_i \qquad \text{...(2.19)}$$

2.9 EQUATION OF EQUILIBRIUM

Following are the necessary conditions for a rigid body to remain in equilibrium:

(*a*) *Resultant force F_R must be a zero vector, i.e., $F_R = 0$*

(*b*) *Resultant couple moment M_R for any point a be zero vector i.e., $M_R = 0$*

Referring to Figure 2.11 above conditions can be written as

$$\left.\begin{aligned}\sum_{i=1}^{n} F_i = 0 \\ \sum_{i=1}^{n} d_i \times F_i + \sum_{i=1}^{n} M_i = 0\end{aligned}\right\} \quad \text{...(2.20)}$$

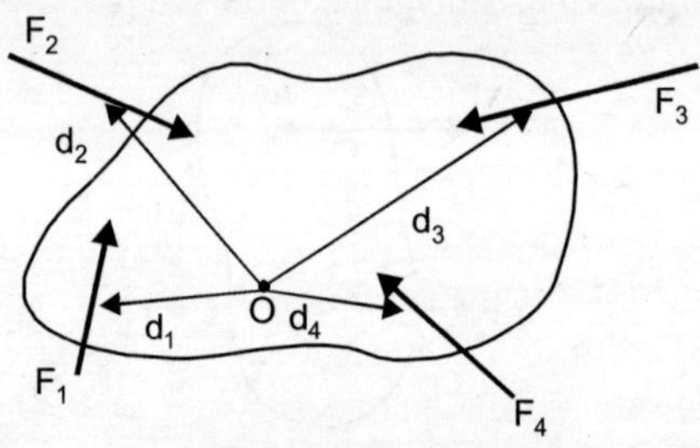

Fig. 2.11

Where d_i is the displacement vector from the common point O to any point on the line of action of the respective forces. So, it can be said that, for static equilibrium,

(*i*) *The vector summation of the forces must be zero.*

(*ii*) *The moment of the system of forces and couples about any point in space must be zero.*

In three-dimension orthogonal co-ordinate system, general equations of equilibrium are as following:

$$\left.\begin{aligned}\sum_{i=1}^{n} (F_x)_i = 0 \\ \sum_{i=1}^{n} (F_y)_i = 0 \\ \sum_{i=1}^{n} (F_z)_i = 0\end{aligned}\right\} \quad \text{... (2.21)} \qquad \left.\begin{aligned}\sum_{i=1}^{n} (M_x)_i = 0 \\ \sum_{i=1}^{n} (M_y)_i = 0 \\ \sum_{i=1}^{n} (M_z)_i = 0\end{aligned}\right\} \quad \text{... (2.22)}$$

Multiple Choice Questions

Select the best alternative (s):

1. Two non-collinear parallel equal forces acting in opposite direction:
   (*a*) balance each other (*b*) constitute a moment
   (*c*) constitute a couple (*d*) constitute a moment of couple
   (*e*) constitute a resultant couple
2. If a rigid body is in equlibrium under the action of three forces, then
   (*a*) these forces are equal
   (*b*) the lines of action of these forces meet in a point
   (*c*) the lines of action of these forces are parallel
   (*d*) (*b*) and (*c*) of the above
   (*e*) none of the above
3. A single force and a couple acting in the same plane upon a rigid body.
   (*a*) balance each other (*b*) cannot balance each other
   (*c*) produce moment of a couple (*d*) are equivalent
   (*e*) none of the above
4. A weight of 500 kg is held on a smooth plane, inclined at 30° to the horizontal by a force *P* acting 30° above the planer. The reaction of the plane on the weight will be.
   (*a*) 500 kg (*b*) 250 kg (*c*) 476.51 kg
   (*d*) 288.68 kg (*e*) none of the above
5. Reactive components in the free body diagram of a roller joint are:
   (*a*) only vertical force (*b*) only horizontal force (*c*) a vertical force and a moment
   (*d*) all of the above
6. Reactive components in the free body diagram of a clamped edge are:
   (*a*) vertical force (*b*) horizontal force (*c*) moment (*d*) all of the above
7. A loaded beam will be unstable if at its ends are:
   (*a*) one roller and one hinge supported (*b*) both hinge supported
   (*c*) both roller supported (*d*) both clamped edge
8. In an inclined supported roller joint, reactive force acts:
   (*a*) parallel to the support (*b*) perpendicular to the support
   (*c*) 60° horizontal to support (*d*) none of the above
9. If three forces acting in one plane upon a rigid body, kept it in equilibrium, then they must either
   (*a*) meet in a point (*b*) be all parallel
   (*c*) at least two of them must meet (*d*) all the above are correct
   (*e*) none of the above

Answers

1. (*d*)	2. (*d*)	3. (*b*)	4. (*d*)	5. (*a*)	6. (*d*)
7. (*c*)	8. (*b*)	9. (*d*)			

Numerical Examples

EXAMPLE 1

Referring to figure, $r = 12$cm, $Q = 500$ N and $h = 6$cm. Find magnitude of P required to start the roller over the curb.

SOLUTION As soon as roller moves over curb, it loses contact with floor.

Here
$$\sin\theta = \frac{12-6}{12} = 0.5$$

$\Rightarrow$
$$\theta = 30^o$$

Taking moment equilibrium at A, we obtain,

$$Q \times r\cos\theta = P \times (12+6)$$

$\Rightarrow$
$$18P = 500 \times 12\cos 30^o$$

$\Rightarrow$
$$P = 288.68\text{N}$$

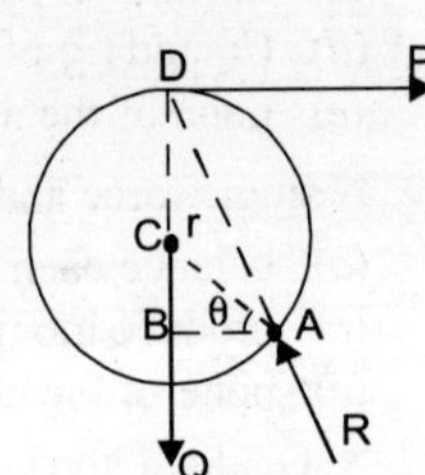

FBD of roller

EXAMPLE 2

Referring to figure, $Q = 100$N, $r = 6$cm, $AC = 12$cm and wall is smooth, determine tension in tie rod and reaction at wall.

SOLUTION This type of problem can be solved in two ways,

(1) *Geometrical Approach:*

From geometry,
$$\sin\theta = \frac{6}{12} = 0.5$$

$\Rightarrow$
$$\theta = 30^o$$

Applying Lami's theorem,
$$\frac{R}{\sin 150^o} = \frac{S}{\sin 90^o} = \frac{100}{\sin 120^o}$$

$\therefore$
$$\left.\begin{aligned} R &= 100 \times \frac{\sin 150^o}{\sin 120^o} = 57.73\text{N} \\ S &= 100 \times \frac{\sin 90^o}{\sin 120^o} = 115.47\text{N} \end{aligned}\right\}$$

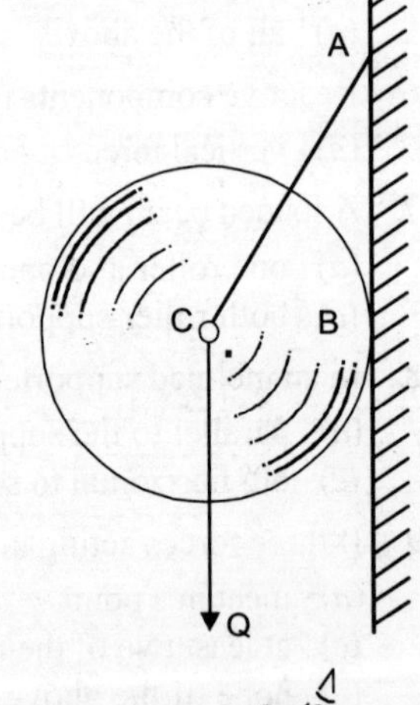

(2) *Force Approach:* Resolving all forces along X and Y axes, respectively,

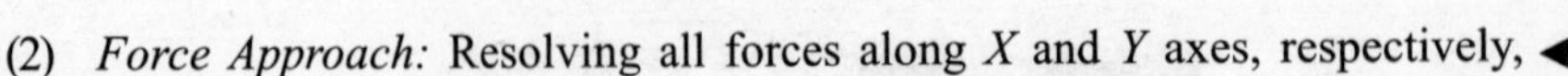

$$\left.\begin{aligned} S\cos 60^o - R &= 0 \\ S\sin 60^o - 100 &= 0 \end{aligned}\right\}$$

Solving above, we obtain, $R = 57.73$ N, $S = 115.47$ N.

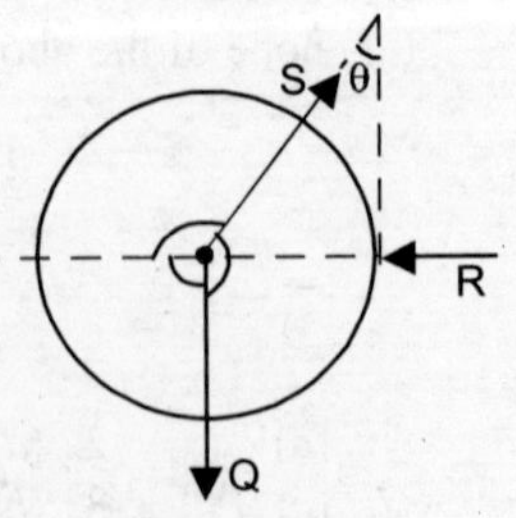

FBD of roller

EXAMPLE 3

Determine notch reactions at D and E as shown in figure, where weight of roller Q = 12N and all surfaces are smooth.

SOLUTION Applying Lami's theorem :

$$\frac{R_d}{\sin 120^\circ} = \frac{R_e}{\sin 150^\circ} = \frac{Q}{\sin 90^\circ}$$

$\Rightarrow$ $R_d = 10.4\text{N}$ and $R_e = 6\text{N}$

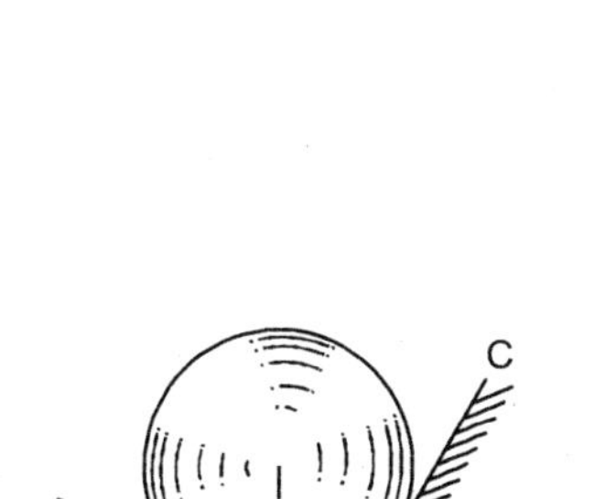

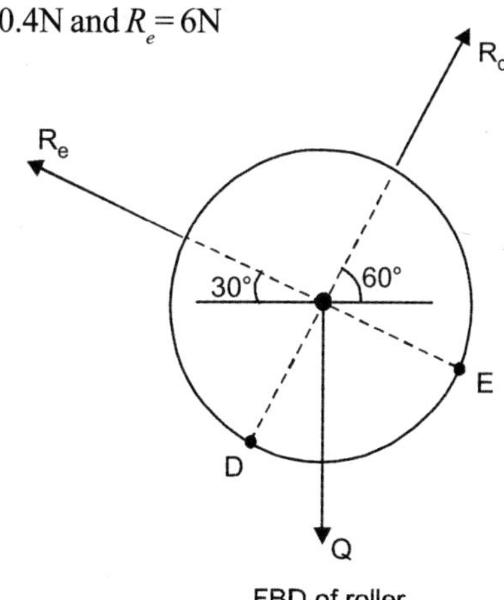

FBD of roller

EXAMPLE 4

Referring to the figure weight of each sphere Q = 100N and all contact surface are smooth, determine reactions at A, B, C.

SOLUTION Consider FBD of top sphere and apply Lami's theorem:

$$\frac{Q}{\sin 90^\circ} = \frac{R}{\sin 150^\circ} = \frac{R_a}{\sin 120^\circ}$$

$\Rightarrow$ $R = 50\text{N}$ and $R_a = 86.6\text{N}$

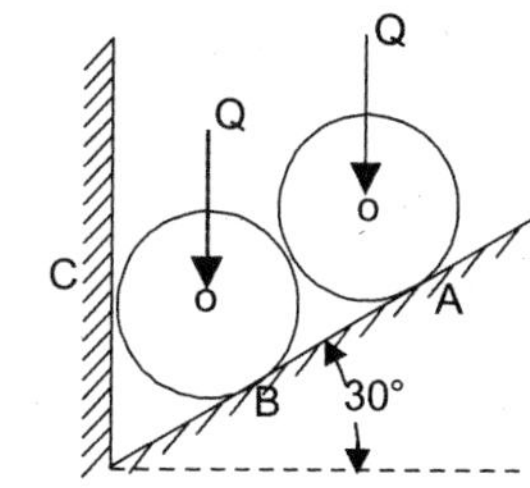

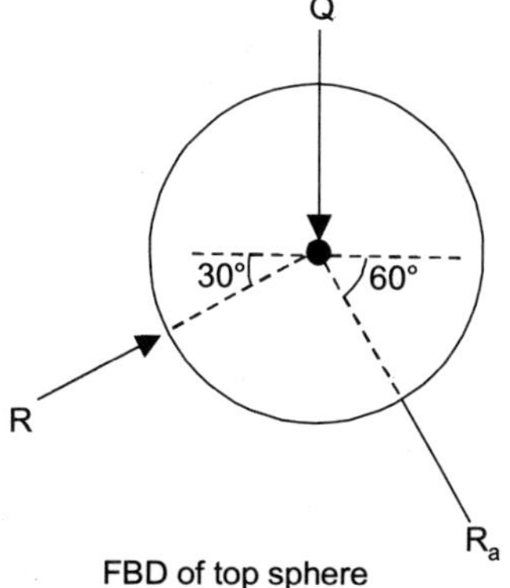

FBD of top sphere

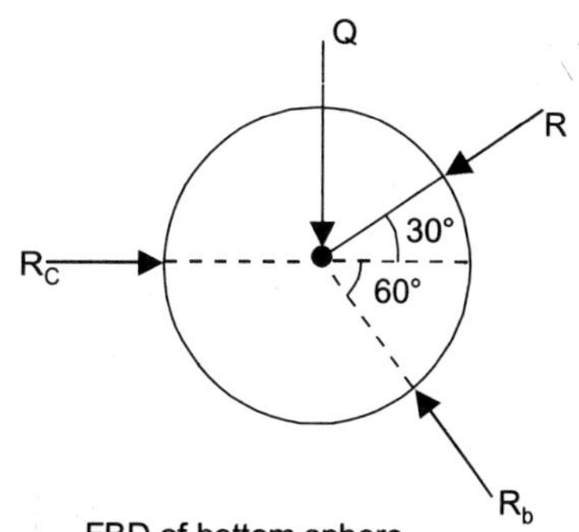

FBD of bottom sphere

Consider FBD of ottom sphere. Resolving all forces along X and Y directions, we have

$$\sum F_y = 0$$

$\Rightarrow$ $$R_b \sin 60^\circ - Q - R\sin 30^\circ = 0$$

$\Rightarrow$ $$R_b = 144.34\text{N}$$

$$\sum F_x = 0$$

$\Rightarrow$ $$R_c - R\cos 30^\circ - R_b \cos 60^\circ = 0$$

$\Rightarrow$ $$R_c = 115.47\text{N}$$

EXAMPLE 5

Assuming the beam weightless in figure, find tension in tie rod CD and reaction at A.

SOLUTION Considering FBD of the beam and taking moment equilibrium at A,

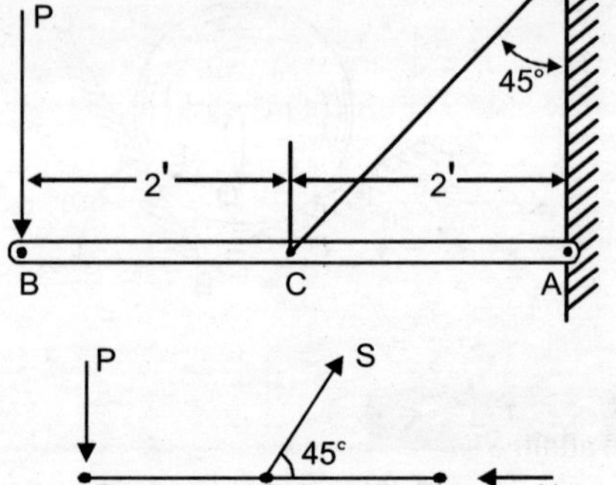

FBD of beam

$$S\sin 45^\circ \times 2 - P \times 4 = 0$$

$\Rightarrow$ $$S = 2.83P$$

Again considering FBD of the beam,

$$\sum F_x = 0$$

$\Rightarrow$ $$H_A = S\cos 45^\circ = 2P$$

$$\sum F_y = 0$$

$\Rightarrow$ $$-V_A + S\sin 45^\circ - P = 0$$

$\Rightarrow$ $$V_A = P$$

$\therefore$ $$R_A = \sqrt{H_A^2 + V_A^2} = 2.24P$$

EXAMPLE 6

Referring to the figure, neglecting friction at B, determine reactions at B and C, while applied force is 50 N.

SOLUTION Taking moment equilibrium about C, we have,

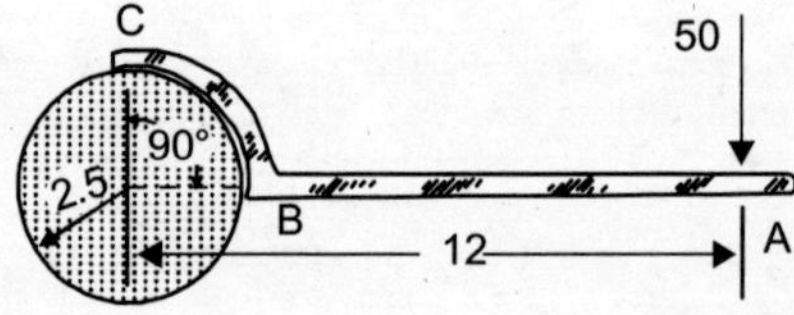

$$R_B \times 2.5 = 50 \times 12$$

$\Rightarrow$ $$R_B = 240\text{N}$$

Again considering equations of equilibrium along X and Y axis,

$$H_C = R_B = 240\text{ N and } V_C = P = 50\text{N}$$

$\therefore$ $$R_C = \sqrt{H_c^2 + V_c^2} = 245.15\text{N}$$

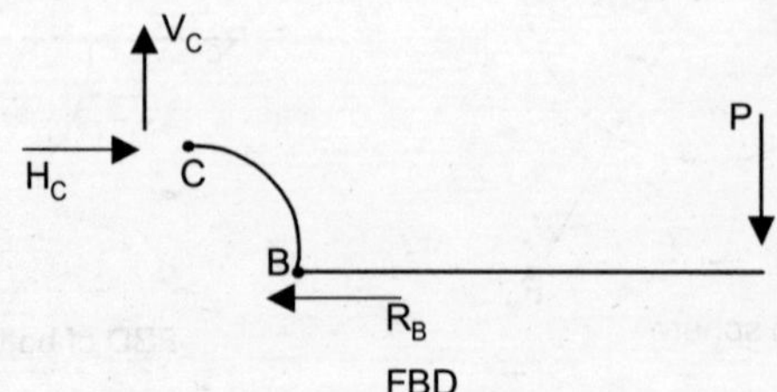

FBD

EXAMPLE 7

In adjacent figure bar and cables are assumed weightless. Determine the position of Q so that tension S in cable BC becomes maximum and also evaluate it.

SOLUTION Considering free body diagram of the bar, let the moment equilibrium is taken at A, while Q is assumed to act at a distance x along bar. So,

$\Rightarrow$ $Q \cdot x\cos 60° - S\sin 75° \times l\cos 60° + S\cos 75° \times l\sin 60° = 0$

$$\Rightarrow \quad S = 1.93\frac{Qx}{l}$$

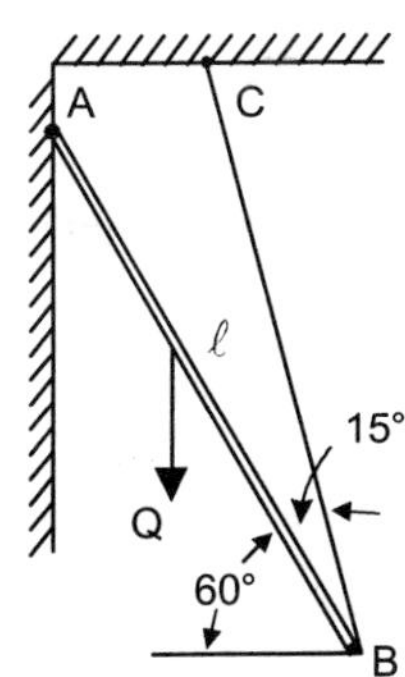

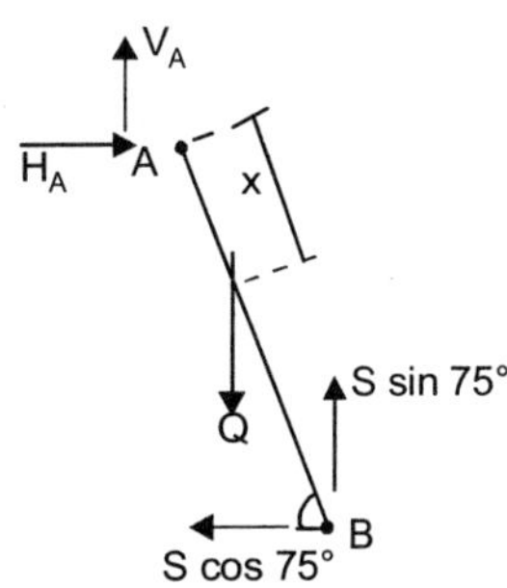

FBD of bar

To attain S a maximum value, $\frac{x}{l}$ will be maximum, which is possible only when $x = l$, i.e., Q acts at B. In that case

$$S = 1.93Q$$

EXAMPLE 8

Referring to figure determine Q, when the system is in equilibrium.

SOLUTION Referring to the FBD of the bracket we have $V_B = Q$.

Again from moment equilibrium at F, using FBD of beam

$$W \times l_1 = V_B \times (l - l_1) = Q \times (l - l_1)$$

$$\Rightarrow \quad Q = W \cdot \frac{l_1}{l - l_1}$$

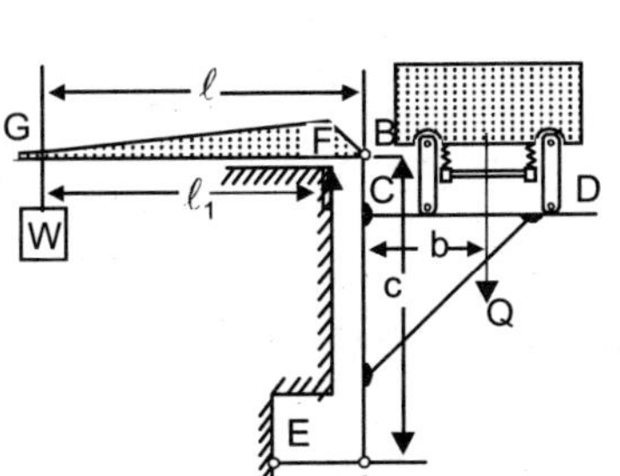

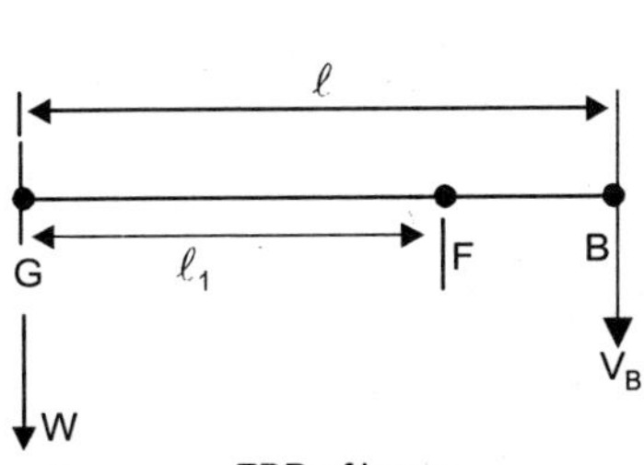

FBD of beam

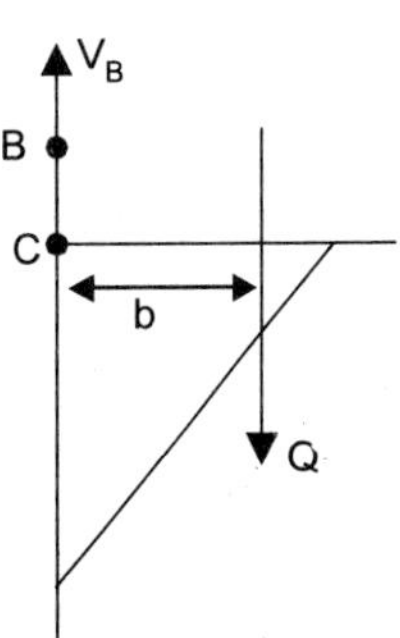

FBD of bracket

EXAMPLE 9

Referring to the attached figure determine reactions at A, D, C.

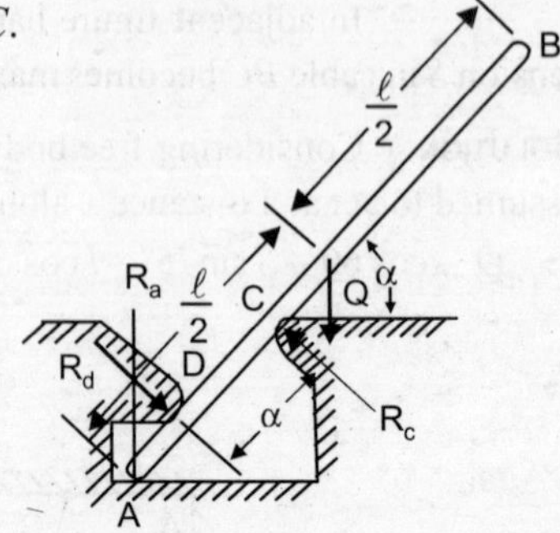

SOLUTION From vertical equilibrium we obtain, $R_a = Q$

Considering equilibrium along inclined line $R_c = R_d$

and from moment equilibrium, $R_d \times a = R_a \times \dfrac{\ell}{2} \cos \alpha = Q \dfrac{\ell}{2} \cos\alpha$

$\Rightarrow$ $R_d = \dfrac{Q\ell}{2a} \cos\alpha$

EXAMPLE 10

Referring to the figure, length of string $AC = 2r$, and bar $AB = 3r$ and weight of the bar $= Q$. Find the tension in the string.

SOLUTION We have $\sin \theta = \dfrac{r}{2r} = 0.5$

$\Rightarrow$ $\theta = 30^{\circ}$

From FBD of the bar $AE = EB = 1.5r,$

So, $AD = \sqrt{(2r)^2 - r^2}$

$= r\sqrt{3}$

Taking moment equilibrium at A

$R \times r\sqrt{3} = Q \times 1.5r \cos 60^{\circ}$

$\Rightarrow$ $R = 0.433Q$

Again, from FBD of sphere, $R\cos\theta = S \cos\theta$

$\Rightarrow$ $S = R = 0.433Q$

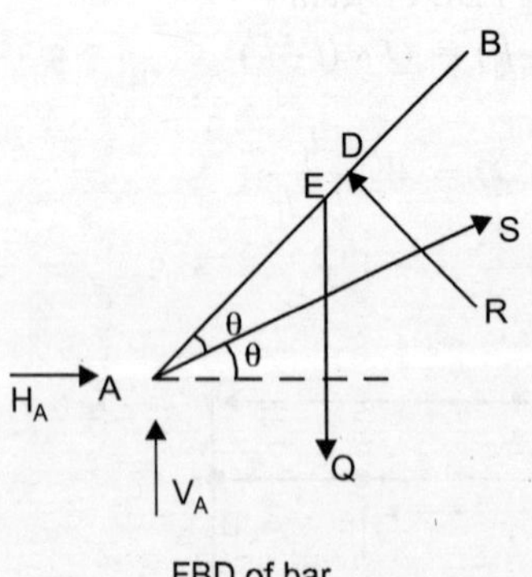

FBD of bar

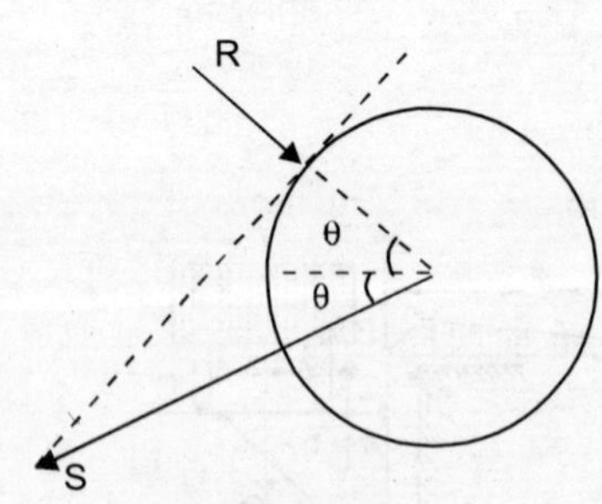

FBD of sphere

EXAMPLE 11

In the adjoining figure, the bolt B fits loosely in a vertical slot in the plate. Find forces transmitted to the bolts A and B.

SOLUTION As the bracket plate can do up-down movement through the vertical slot at B, $V_B = 0$ and only H_B exists.

Taking moment equilibrium at A.

$$H_B \times 0.8 = P \times 0.6$$

$\Rightarrow$ $$H_B = 0.75P$$

Again, from equations of force equilibrium,

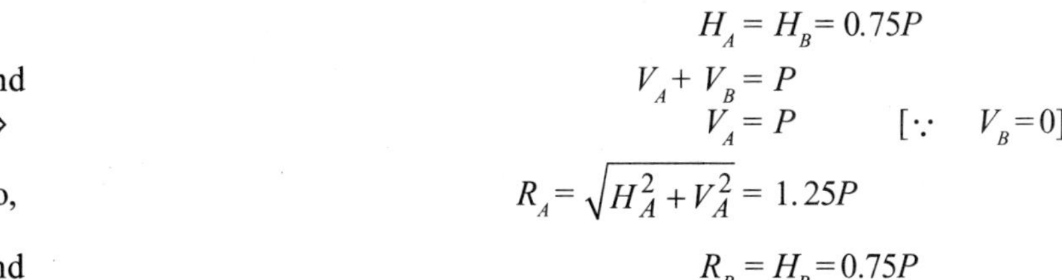

$$H_A = H_B = 0.75P$$

and $$V_A + V_B = P$$

$\Rightarrow$ $$V_A = P \qquad [\because \; V_B = 0]$$

So, $$R_A = \sqrt{H_A^2 + V_A^2} = 1.25P$$

and $$R_B = H_B = 0.75P$$

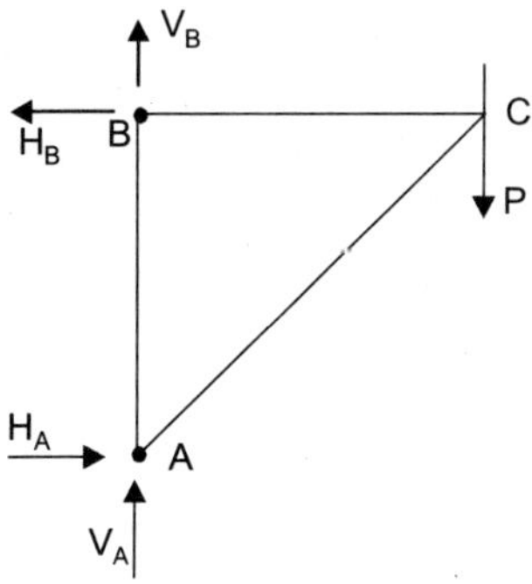

EXAMPLE 12

Two smooth spheres, each of radius r and weight Q rests in a horizontal channel having vertical walls, the distance between which is b. Find the pressure on the walls and floor at points of contact A, B, D. Use $r = 25$cm, $b = 90$cm, $Q = 450$N.

SOLUTION From equations of force equilibrium of top sphere,

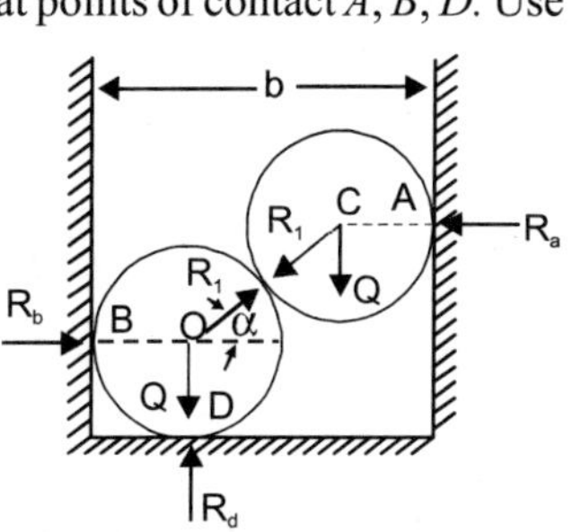

$$\left.\begin{aligned} R_1 \sin\alpha - Q &= 0 \\ R_1 \cos\alpha - R_a &= 0 \end{aligned}\right\}$$

or $$\left.\begin{aligned} R_1 \sin\alpha - 450 &= 0 \\ R_1 \cos\alpha - R_a &= 0 \end{aligned}\right\}$$

Solving these $$R_1 = \frac{450}{\sin\alpha}$$

and $$R_a = 450 \cot \alpha$$

From equations of force equilibrium of bottom sphere,

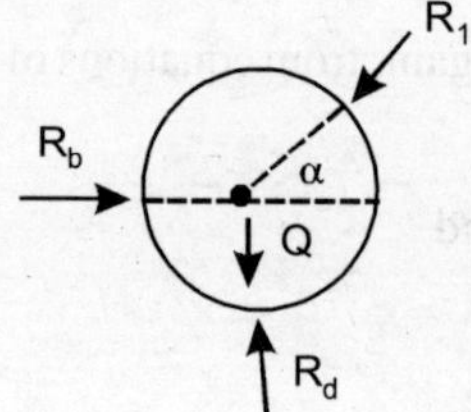

FBD of top sphere

$$\left.\begin{aligned} R_1 \cos\alpha - R_b &= 0 \\ -R_1 \sin\alpha + R_d &= Q \end{aligned}\right\}$$

or $$\left.\begin{aligned} \frac{450}{\sin\alpha}\cos\alpha - R_b &= 0 \\ -450 + R_b &= 450 \end{aligned}\right\}$$

Solving these, $R_b = 450 \cot \alpha$ and $R_b = 900\text{N}$

From geometry, $b = 2r + 2r \cos\alpha$

$\Rightarrow$ $90 = 50 + 50 \cos\alpha$

$\Rightarrow$ $\alpha = \cos^{-1}\left(\frac{4}{5}\right) = 36.869^{\circ}$

FBD of bottom sphere

Substituting value of α, $R_a = R_b = 600\text{N}$.

EXAMPLE 13

A right circular roller of weight W rests on a smooth horizontal plane and held in position by an inclined bar AC. Find tension S in the bar and vertical reaction R_b at B.

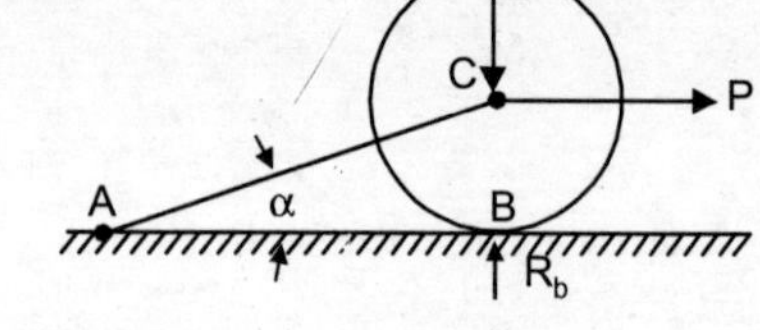

SOLUTION From equations of force equilibrium of roller,

$$\left.\begin{aligned} P &= S\cos\alpha \\ R_b &= W + S\sin\alpha \end{aligned}\right\}$$

Solving these, $S = P \sec \alpha$ and $R_b = W + P \tan\alpha$.

EXAMPLE 14

Two smooth circular cylinders, each of weight $W = 100\text{N}$ and radius $r = 6\text{cm}$ are connected at their centres by string AB of length $l = 16\text{cm}$ and rest upon a horizontal plane, supporting above them a third cylinder of weight $Q = 200\text{N}$ and radius $r = 6\text{cm}$. Find the force S in the string AB and the pressure produced by the floor at points of contact D and E.

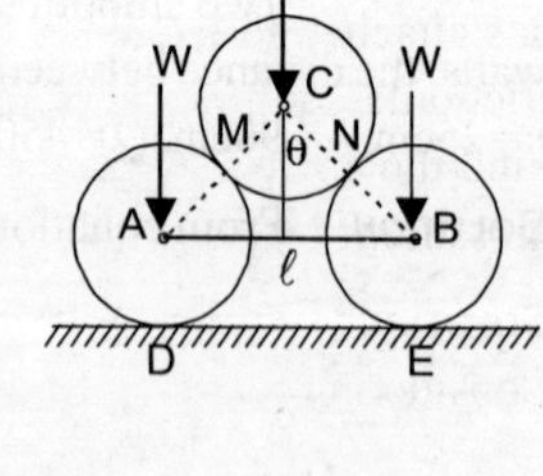

SOLUTION From geometry, $$\sin\theta = \frac{l/2}{2r} = \frac{8}{2\times 6} = \frac{2}{3}$$

$\therefore$ $$\theta = \sin^{-1}\left(\frac{2}{3}\right) = 41.81^{\circ}$$

From symmetry it can be concluded that reactive force at points of contact at M and N are equal say

$$R_M = R_N = R \text{ and } R_D = R_E$$

From equilibrium of vertical forces in sphere C, we obtain

$$2R\cos\theta = Q$$

$$\therefore \qquad R = \frac{200}{2\cos 41.81^\circ} = 134.163\text{N}$$

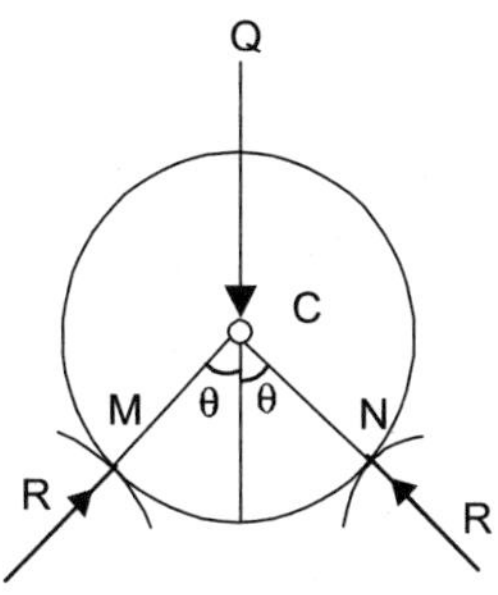

FBD of top sphere

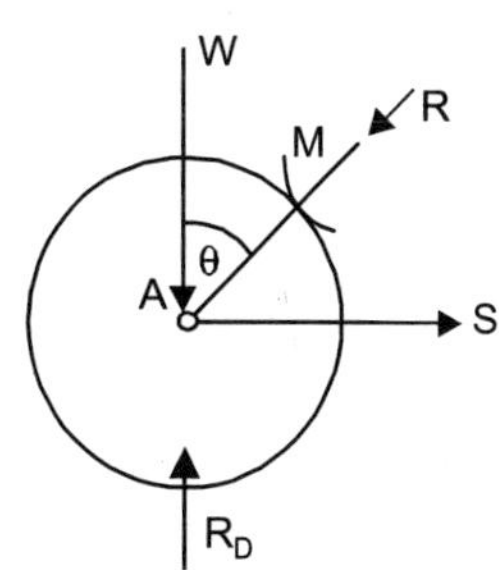

FBD of bottom left sphere

From equations of force equilibrium of sphere A

$$\left.\begin{aligned} S &= R\sin\theta \\ W + R\cos\theta &= R_D \end{aligned}\right\}$$

or

$$\left.\begin{aligned} S &= 134.163\sin 41.81^\circ \\ 100 + 134.163\cos 41.81^\circ &= R_D \end{aligned}\right\}$$

Solving these, $S = 89.44$N, $R_D = 199.99\text{N} = R_E$.

EXAMPLE 15

A ball of weight W rests upon a smooth horizontal plane and has attached to its center by two strings AB and AC which pass over frictionless pulleys at B and C and carry loads P and Q, respectively. Find the angle α when the ball in a position of equilibrium. Also find pressure R between ball and plane.

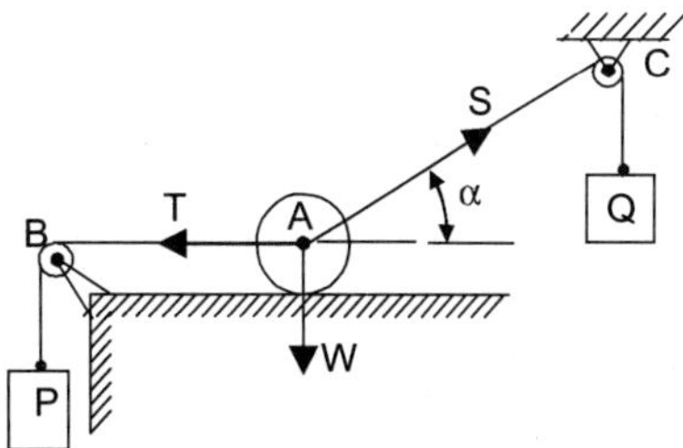

SOLUTION From equilibrium of load $T = P$ and that of load Q, $S = Q$. Considering equations of force equilibrium of ball,

$$\left.\begin{aligned} S\cos\alpha &= T \\ S\sin\alpha + R &= W \end{aligned}\right\}$$

or

$$\left.\begin{aligned} Q\cos\alpha &= P \\ Q\sin\alpha - R &= W \end{aligned}\right\}$$

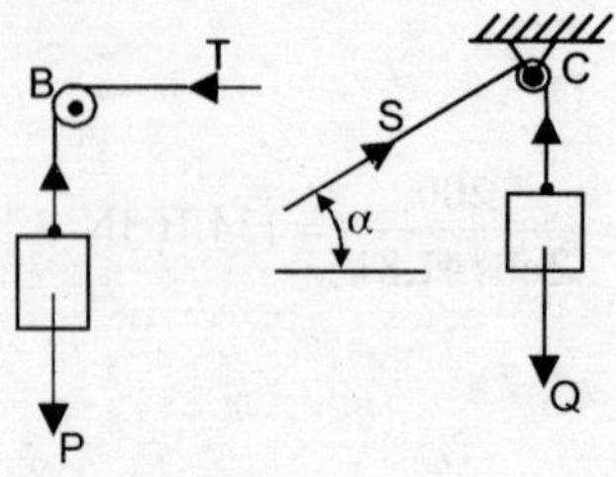

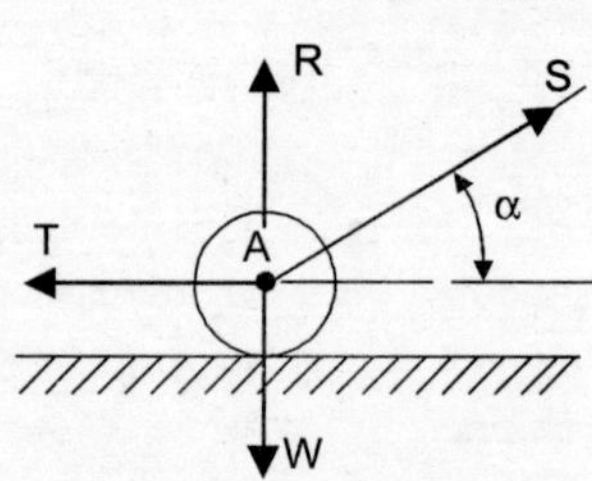

From first equation of above, $\cos\alpha = \dfrac{P}{Q}$

From second equation of above, $R = W - Q\sin\alpha = W - Q\sqrt{1-\dfrac{P^2}{Q^2}} = W - \sqrt{Q^2 - P^2}$

EXAMPLE 16

A 75kg man stands on the middle rung of a 25 kg ladder. Assuming a smooth wall at *B* and a stop at *A* to prevent slipping, find reactions at *A* and *B*.

SOLUTION Considering moment equilibrium of entire system at *A*,

$$R_B \times 12 = (75+25)\times 3$$

$$\Rightarrow \quad R_B = 25\text{kg}$$

Considering vertical and horizontal force equilibrium of the system,

$$\left.\begin{aligned} R_{AH} &= R_B = 25\text{kg} \\ R_{AV} &= 75+25 = 100\text{kg} \end{aligned}\right\}$$

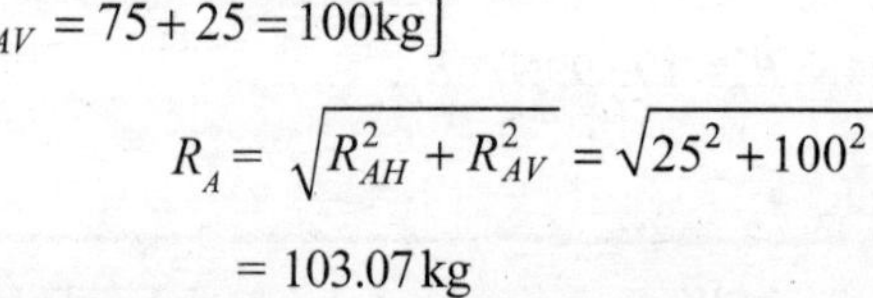

So, $$R_A = \sqrt{R_{AH}^2 + R_{AV}^2} = \sqrt{25^2 + 100^2}$$

$$= 103.07\,\text{kg}$$

EXAMPLE 17

A horizontal prismatic bar *AB* of negligible weight is hinged at *A* and supported at *B* by tie rod *BC*. *A* weight *P* can have any position along the bar as defined by *x* from the wall. Determine tensile force *S* in *BC*.

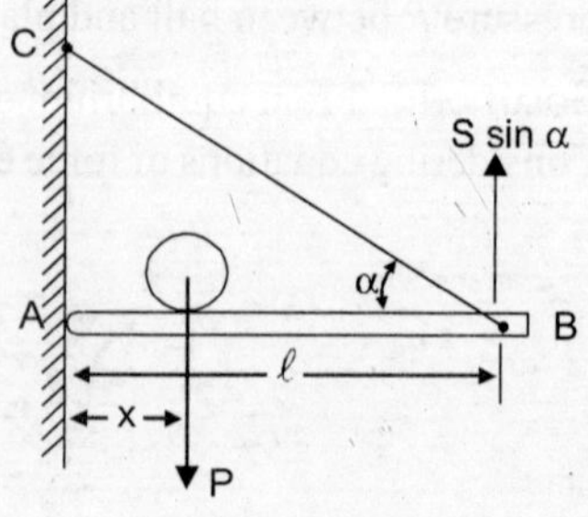

SOLUTION Considering moment equilibrium at *A* of entire system we obtain,

$$S\sin\alpha \times \ell = P \times x$$

$$\therefore \quad S = \frac{Px}{l\sin\alpha}$$

EXAMPLE 18

Two rollers $P = 60N$, $Q = 100N$ are connected by a flexible inextensible string DE and rest on two mutually perpendicular planes AB and BC. Find tension in the string AB and BC. Find tension in the string S and BC. Find tension in the string S and angle ϕ. Assume the string passes freely through slots in smooth inclined planes AB and BC. $\alpha = 30°$

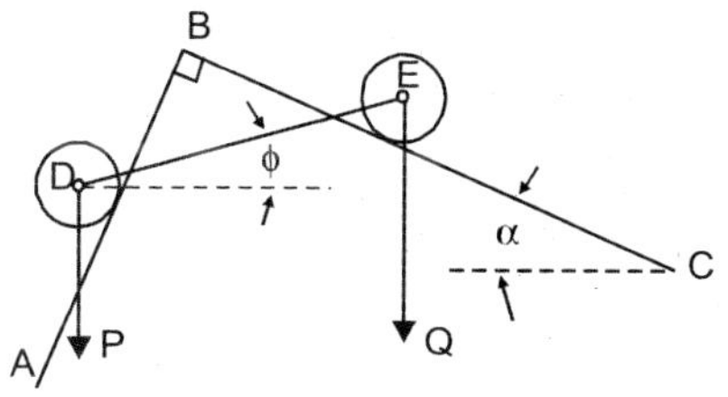

SOLUTION *Approach I*: Considering equilibrium of roller P,

$$\left.\begin{aligned} S\cos\phi &= R_P\cos 30° \\ S\sin\phi + R_P\sin 30° &= 60 \end{aligned}\right\}$$

or

$$S\cos\phi = R_P\frac{\sqrt{3}}{2} \qquad \text{...(1)}$$

and

$$S\sin\phi + \frac{1}{2}R_P = 60 \qquad \text{....(2)}$$

Considering equilibrium of roller Q,

$$\left.\begin{aligned} R_Q\cos 60° &= S\cos\phi \\ R_Q\sin 60° &= 100 + S\sin\phi \end{aligned}\right\}$$

or

$$\frac{1}{2}R_Q = S\cos\phi \qquad \text{...(3)}$$

$$R_Q\frac{\sqrt{3}}{2} = 100 + S\sin\phi \qquad \text{...(4)}$$

From (1) and (3)

$$\frac{1}{2}R_Q = R_P\frac{\sqrt{3}}{2}$$

$\Rightarrow$

$$R_Q = R_P\sqrt{3}$$

Making (2) – (4)

$$S\sin\phi + \frac{1}{2}R_P = 60$$

$$\frac{S\sin\phi + 100 = R_P\sqrt{3}\frac{\sqrt{3}}{2}}{\frac{1}{2}R_P - 100 = 60 - \frac{3}{2}R_P}$$

$\Rightarrow$

$$R_P = 80\,\text{N}$$

$\Rightarrow$

$$R_Q = 138.56\text{N}$$

Again from (1),

$$S\cos\phi = 80 \times \frac{\sqrt{3}}{2} = 69.282 \qquad \text{... (5)}$$

and from (2),

$$S\sin\phi = 60 - \frac{1}{2} \times 80 = 20 \qquad \text{... (6)}$$

So, from (5) and (6),

$$\tan\phi = \frac{20}{69.282} = 0.28867$$

$\therefore$

$$\phi = 16.102^\circ$$

and

$$S = \frac{20}{\sin\phi} = 72.11\text{N}$$

Approach II: Applying Lami's theorem in roller *P*.

$$\frac{S}{\sin(90^\circ + 30^\circ)} = \frac{R_P}{\sin(90^\circ + \phi)} = \frac{60}{\sin(60^\circ + 90^\circ - \phi)}$$

$\Rightarrow$

$$1.1547\,S = \frac{R_P}{\cos\phi} = \frac{60^\circ}{\cos(60^\circ - \phi)}$$

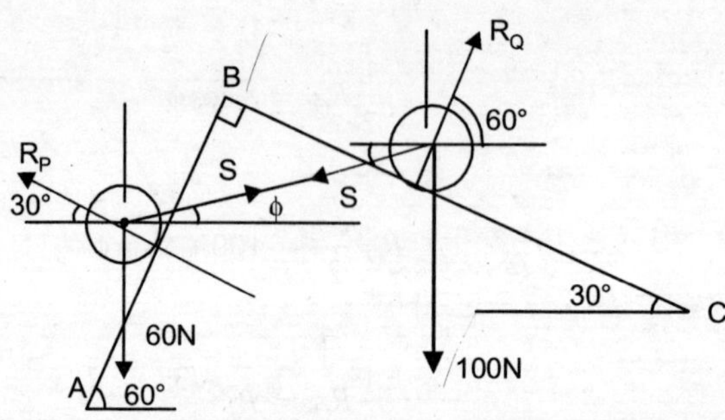

$\Rightarrow$

$$S = \frac{51.962}{\cos(60^\circ - \phi)} \qquad \text{... (7)}$$

Applying Lami's theorem in roller *Q*,

$$\frac{S}{\sin(90^\circ + 60^\circ)} = \frac{R_Q}{\sin(90^\circ - \phi)} = \frac{100}{\sin(90^\circ + 30^\circ + \phi)}$$

$\Rightarrow$

$$2S = \frac{R_Q}{\cos\phi} = \frac{100}{\cos(30^\circ + \phi)}$$

$\Rightarrow$

$$S = \frac{50}{\cos(30^\circ + \phi)} \qquad \text{... (8)}$$

Equating two values of S, in (7) and (8), we obtain,

$$\frac{51.962}{\cos(60^\circ - \phi)} = \frac{50}{\cos(30^\circ + \phi)}$$

Solving above, we obtain, ϕ=16.102° and tension in the string

$$S = \frac{50}{\cos(30^\circ + 16.102^\circ)} = 72.11\text{N}$$

EXAMPLE 19

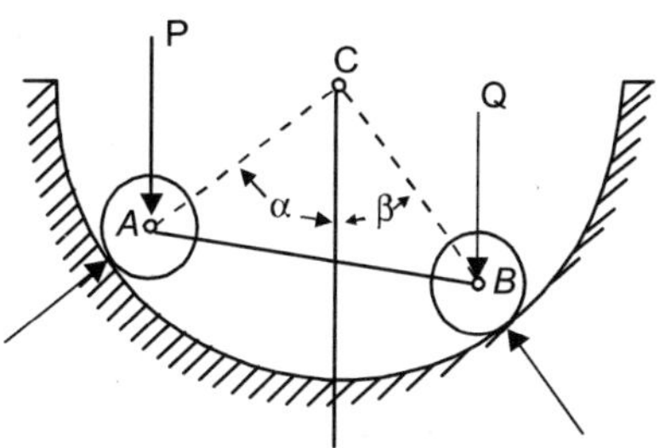

A rigid bar AB with rollers of weights P = 50N and Q= 100N at ends is supported inside a circular ring as in figure. The radius of the ring and AB are such that radii AC and BC form 90° at C i.e., $\alpha + \beta = 90^\circ$. Neglecting friction and weight of AB, find configuration of equilibrium as defined by angle $(\alpha - \beta)/2$ that AB makes with horizontal. Find also R_A, R_B and compressive force S in AB.

SOLUTION From $\Delta ACB, AC = BC$

and $\angle ACB = 90^\circ$

So, $\angle CAB = \angle CBA = 45^\circ$

Again from ΔDBE, considering given condition,

$$\theta = \frac{\alpha - \beta}{2}$$

$$\Rightarrow \quad \theta + \beta = \alpha - \theta$$

Considering equation of force equilibrium of roller A

$$\left.\begin{aligned} S\cos\theta &= R_A \sin\alpha \\ S\sin\theta + R_A\cos\alpha &= P \end{aligned}\right\}$$

or

$$\left.\begin{aligned} R_A &= \frac{S\cos\theta}{\sin\alpha} \\ S\sin\theta + R_A\cos\alpha &= 50 \end{aligned}\right\}$$

From above two equations,

$$50 = S\sin\theta + \frac{S\cos\theta}{\sin\alpha}\cos\alpha = S\frac{\cos(\alpha - \theta)}{\sin\alpha} \quad \text{...(1)}$$

Considering equations of force equilibrium of roller B

$$\left.\begin{aligned} S\cos\theta &= R_B \sin\beta \\ R_B\cos\beta - S\sin\theta &= Q \end{aligned}\right\}$$

or

$$\left.\begin{aligned} R_B &= S\frac{\cos\theta}{\sin\beta} \\ R_B\cos\beta - S\sin\theta &= 100 \end{aligned}\right\}$$

From above two equations,

$$100 = \frac{S\cos\theta}{\sin\beta}\cos\beta - S\sin\theta = S\frac{\cos(\theta+\beta)}{\sin\beta} \quad \text{... (2)}$$

From (1) and (2)

$$\frac{100}{50} = \frac{\cos(\theta+\beta)}{\sin\beta}\times\frac{\sin\alpha}{\cos(\alpha-\theta)} = \frac{\cos(\theta+\beta)}{\sin\beta}\times\frac{\sin\alpha}{\cos(\theta+\beta)}$$

$$= \frac{\sin\alpha}{\sin\beta} = \frac{\sin\alpha}{\sin(90^\circ-\alpha)} = \tan\alpha$$

$\Rightarrow$ $$\tan\alpha = \frac{100}{50} = 2$$

$\Rightarrow$ $$\alpha = \tan^{-1}2 = 63.435^\circ$$

$\Rightarrow$ $$\beta = 90^\circ - 63435^\circ = 26.565^\circ$$

$\Rightarrow$ $$\theta = \frac{1}{2}(\alpha-\beta) = 18.435^\circ$$

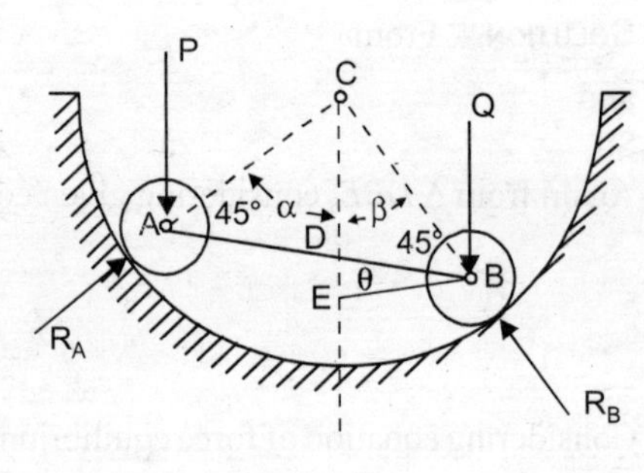

From (1), $$S = \frac{50\sin\alpha}{\cos(\alpha-\theta)} = 63.24\text{ N}$$

Again $$R_A = \frac{S\cos\theta}{\sin\alpha} = 67.076\text{N}$$

$$R_B = \frac{S\cos\theta}{\sin\beta} = 134.152\text{N}$$

EXAMPLE 20

Two equal loads of 2500N are supported by a flexible string *ABCD* at points *B* and *D*. Find the tensions in the portions *AB*, *BC*, *CD* of the string.

SOLUTION Let the tension in *AB*, *BC* and *CD* be T_1, T_2, and T_3 respectively. Considering equilibrium of point *B*,

$$\left.\begin{aligned} -T_1\sin 30^\circ + T_2\cos 30^\circ = 0 \\ T_1\cos 30^\circ - 2500 - T_2\sin 30^\circ = 0 \end{aligned}\right\}$$

or

$$\left.\begin{aligned} -\frac{T_1}{2} + T_2\frac{\sqrt{3}}{2} = 0 \\ T_1\frac{\sqrt{3}}{2} - \frac{T_2}{2} = 2500 \end{aligned}\right\}$$

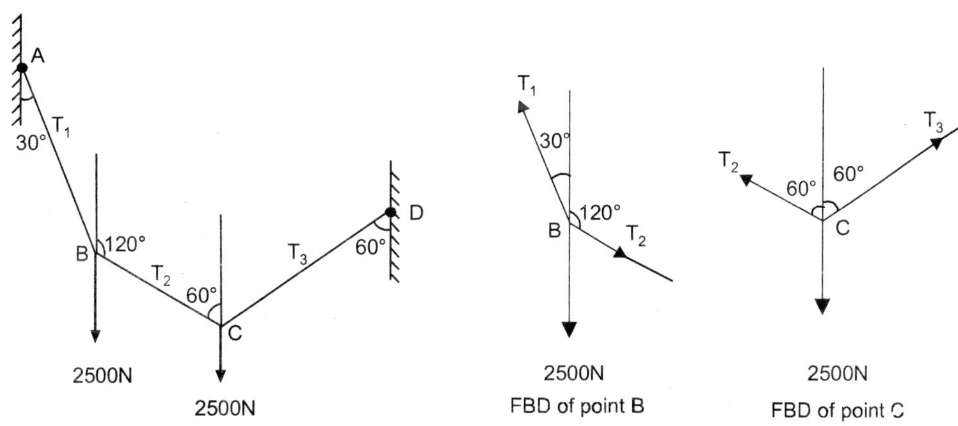

FBD of point B FBD of point C

Solving these $\qquad T_1 = 4330\text{N}, T_2 = 2500\text{N}$

Again considering equilibrium of point C,

$$-T_2 \cos 30^\circ + T_3 \cos 30^\circ = 0$$

$\Rightarrow \qquad T_3 = T_2 = 2500\text{N}$

EXAMPLE 21

A uniform rod AB of negligible weight is hinged at the end A and supported at B by a string. Find value of θ corresponding to position of equilibrium if $Q = P/2$.

SOLUTION Considering moment equilibrium of AB at A.

$$Pl\cos\theta - Ql\sin\left(90^\circ - \frac{\theta}{2}\right) = 0$$

$\Rightarrow \qquad Pl\cos\theta - Q\cos(\theta/2) = 0$

$\Rightarrow \qquad P(2\cos^2\theta/2 - 1) - Q\cos(\theta/2) = 0$

$\Rightarrow \qquad 2P\cos^2(\theta/2) - P - Q\cos(\theta/2) = 0$

$\Rightarrow \qquad \cos(\theta/2) = \dfrac{Q \pm \sqrt{Q^2 + 8P^2}}{4P} = \dfrac{1}{4}\left(\dfrac{Q}{P} \pm \sqrt{\dfrac{Q^2}{P^2} + 8}\right)$

Substituting $\dfrac{Q}{P} = \dfrac{1}{2}$, $\cos(\theta/2) = 0.843, -0.593$. Considering (+) ve value, $\theta = 65.08^\circ$.

EXAMPLE 22

A rigid bar is subjected to a system of parallel forces as shown in figure. Reduce the system to (*i*) a single force (*ii*) a single force-moment system at A, (*iii*) a single force-moment system at B.

SOLUTION (*i*) If the resultant single force is R.

$$R = 30 - 120 + 20 - 50 = -120\text{N}$$

If R acts at a distance x m from A, consider moment equilibrium at B.

$$-120 \times x = -120 \times 0.8 + 20 \times 1.4 - 50 \times 2.4$$

$$\Rightarrow \quad x = 1.567\text{m}$$

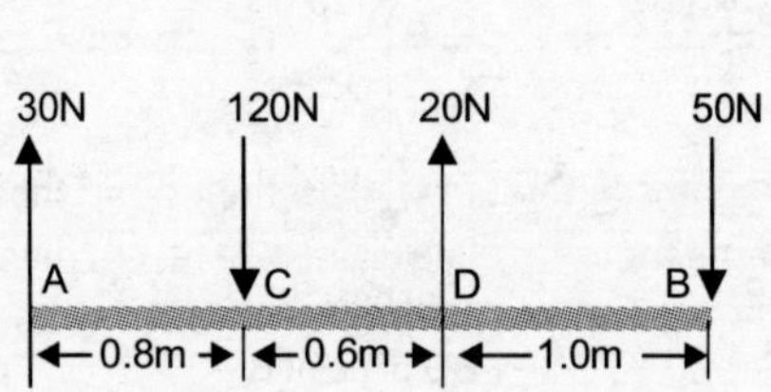

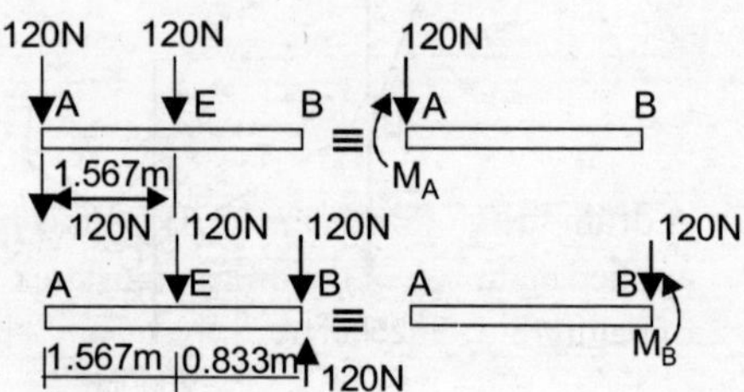

(*ii*) At point A, 120N force is provided both from upward and downward direction.

$$\therefore \quad M_A = -120 \times 1.567 = -188.04 \text{ N·m}$$

(*iii*) At point B, 120N force is provided both from upward and downward direction. So, point of application of resultant (–120N) force is (2.4 – 1.567) = 0.833m away from B.

$$\therefore \quad M_B = 120 \times 0.833 = 99.96 \text{ N·m}$$

Exercise

1. A uniform rod rests on a fixed smooth sphere with its lower end pressing against a smooth vertical wall which touches the sphere. If θ is the inclination of the rod with the vertical in the position of equilibrium show that $r = 2l \sin\frac{\theta}{2} \cos^3\frac{\theta}{2}$ where l is the length of the base and r the radius of the sphere.

2. A cone has a base radius r and altitude h. A string is fastened to the vertex and to a point on the circumference of its circular base and is then put over a smooth peg. If the cone remains in equilibrium with its axis horizontal show that the length of the string must be $\sqrt{h^2 + 4r^2}$.

3. A solid hemisphere having a radius r and weight W is placed with its curved surface on a smooth horizontal floor and a string of length l (less than the radius) is attached to a point on the circumference of the base and to a point on the table. Determine the position of equilibrium and show that in the position of equilibrium the tension in the string is given by $\frac{3}{8} \frac{r - l}{\sqrt{2rl - l^2}} W$.

4. Two equal smooth spheres each of weight W and radius r are placed within a hollow vertical cylinder of a radius a ($a < 2r$) open at both ends and resting on a horizontal plane. Show that least weight the cylinder can have so as not to be upset is given by $\frac{2(a - r)}{a} W$.

5. A roller of weight $W = 1000$N rests on a smooth 45° incline. It is held on incline by a string AC at 15° with inclined plane. Find tension in the string and reaction at point of contact of D.

Ans. [T=733 N, R_D = 897N]

6. A sphere of mass M rests in a V-groove whose sides are inclined at angles α and β to horizontal. Another identical sphere of mass M rests on the first sphere and in contact with the side inclined at angle α. Find reaction of lower sphere at point of contact with β incline.

Ans. $\left[\dfrac{2Mg\sin\alpha}{\sin(\alpha+\beta)}\right]$

7. A drum full of oil of density 1kg/lit. is resting against a foot step 10cm high. Find the force required to be applied at the top of the drum of just turn it over the foot stop. The drum is of 50 cm diameter, 1.5m long and of negligible weight.

Ans. [1444.6N]

CHAPTER
3

FRICTION

3.1 INTRODUCTION

Friction is a phenomenon occurs at the interface of two surfaces due to *shearing effect*. Due to friction, the force generated is called *frictional force* or *frictional resistance* and it acts as tangent to the contact surfaces, in *opposition* to either *motion* or *trend of motion*. Sometimes this frictional force is found to be very much effective in order to maintain equilibrium. As an example, walking on ice is not easy due to the absence of sufficient amount of frictional resistance.

One thing is very much important to remember that problems related to two surfaces separated by a film of greese or oil are not at all problems of friction. So in this context friction means *dry friction* or *Coulomb friction*.

Dry friction is a very complicated phenomenon. The probable cause of dry friction is believed to be the *interlocking* of protrusion of associated surfaces, at *microscopic level*. In Fig. 3.1, (*a*) the roughness of surfaces are shown by protrusion while Fig. 3.1(*b*) shows their interlocking stage. When sliding is present between the surfaces, these *protruberances* either *sheared off* or *melted* due to high local temperature.

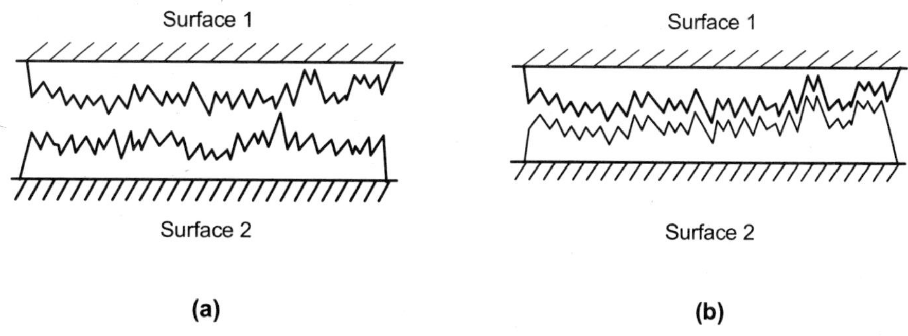

Fig. 3.1

Generally dry friction is classified as following:

(*a*) *Static Friction*

(*b*) *Dynamic Friction or Kinetic Friction*

(*c*) *Rolling Friction*

Figure 3.2 (*a*) assumes a block of weight *W* placed on a rough surface. A horizontal force *P* is applied to it to start sliding. The variation of force as a function of time is plotted in Fig. 3.2 (*b*). It is observed that up to a certain time *t*, force *P* is gradually increasing in straight line and at point *M* the plot falls down by a small amount and then tries to continue almost as a straight line. The point *M* is the instant where force *P* just overcomes all frictional resistance to start motion, and after point *M* the body continues to move. This point is defined as *point of impending motion*, or *point of limiting friction*, before which the friction is *static friction* and after which the friction is *dynamic friction*. Up to point *M*, the frictional force is a *self-adjusting* force as it is proportional to the external force *P*. Sometimes frictional force is termed as *pseudo force*.

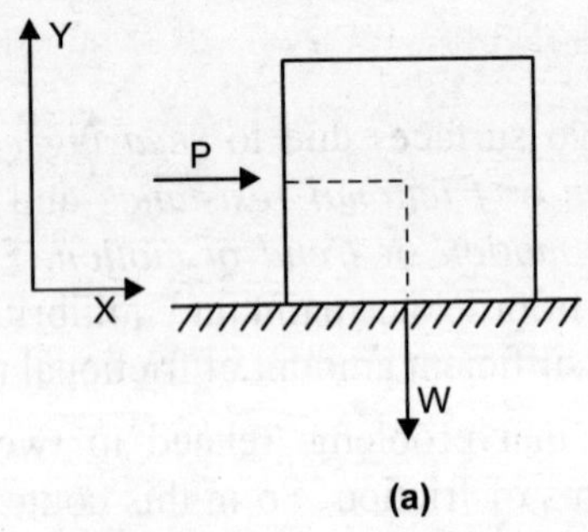

(a)

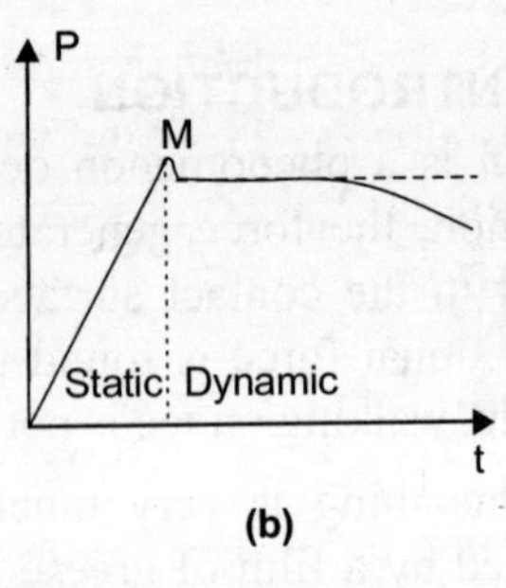

(b)

Fig. 3.2

3.2 LAWS OF FRICTION

In 1781 C.A. Coulomb presented summarised outcome of a large number of experiments related to dry friction that become famous as *Coulombs Laws of Friction* or simply *Laws of Friction* which are as following:

(*i*) *The total force of friction that can be developed is independent of magnitude of area of contact.*

(*ii*) *The total force of friction that can be developed is proportional to the normal reaction transmitted across the surface of contact.*

(*iii*) *For low relative velocities of sliding, the total force of friction is practically independent of velocity.*

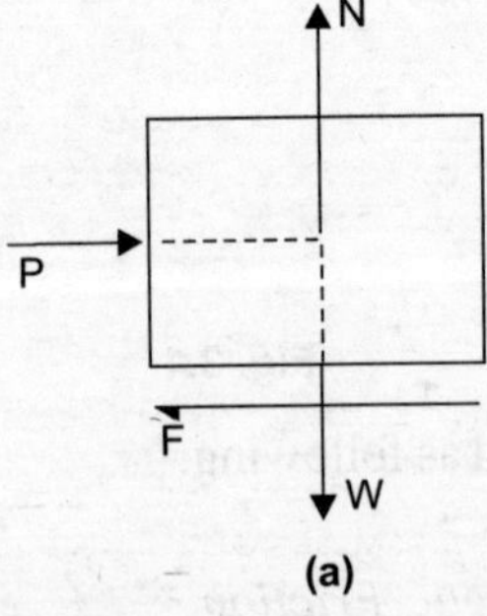

(a)

Fig. 3.2

Conclusion (*ii*) as above can be mathematically represented as

$$F \propto N$$

$$\Rightarrow \qquad F = \mu N \qquad ...(3.1)$$

where F = frictional resistance,

N = normal reaction,

P = applied force, and

W = self weight.

Here, the constant of proportionality μ, is defined as coefficient of friction. In static condition, it is expressed as μ_s and in dynamic condition μ_k. For same set of contact surfaces, it has been observed that μ_k is almost 75% of μ_s. (*Friction force better to be shown with half arrow, as it is a sort of shear force*)

3.3 ANGLE OF FRICTION

At the time of *impending motion*, the force of friction F is defined as *force of limiting friction*. In this condition Fig. 3.3 (*b*) will illustrate as how the resultant of *normal reaction N* and *force of limiting friction F* can be obtained. If the resultant force R makes an angle (independent of magnitude of area of contact), ϕ (phi) with normal reaction N

$$\tan\phi = \frac{\mu N}{N} = \mu \qquad ...(3.2)$$

Here this angle ϕ is defind as *limiting angle of friction* or simply *the angle of friction.*

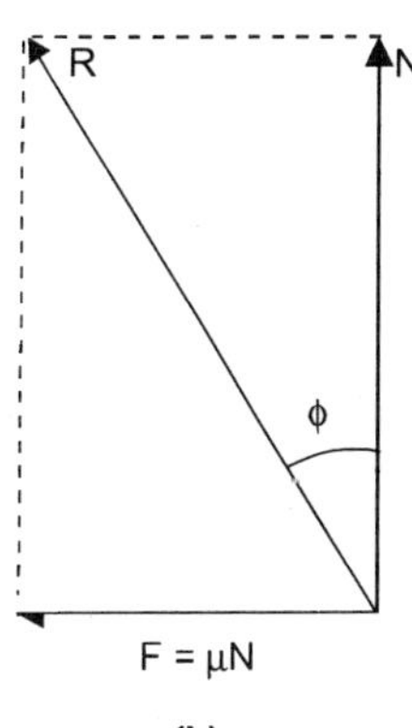

(b)

Fig. 3.3

Coefficient of Friction for Different Materials

Materials	Coefficient of Static Friction(μ_s)	Coefficient of Kinetic Friction(μ_k)
Steel on Steel (Dry)	0.74	0.57
Aluminium on Steel	0.61	0.47
Copper on Steel	0.53	0.36
Rubber on Concrete	1.00	0.80
Wood on Wood	0.25–0.50	0.20
Waxed wood on Wet snow	0.14	0.10
Glass on Glass	0.94	0.40
Waxed wood on Dry snow	—	0.04
Metal on Metal (lubricated)	0.15	0.06
Ice on Ice	0.10	0.03

Contd...

Teflon on Teflon	0.04	0.04
Synovial joints in human body	0.01	0.003
Steel on Steel (greasy)	0.10	0.05
Teflon on Steel	0.041	0.04
Rubber tyre on Dry pavement	0.90	0.80
Metal on Ice	0.022	0.02

3.4 ANGLE OF REPOSE

The physical interpretation of *angle of repose* can well be understood from Fig. 3.4 (*a*). Here the block on the inclined surface won't move until the inclination reaches at ϕ, up to which frictional force becomes higher than inclined component of self weight of block. At inclination ϕ, frictional force just balances, inclined component of self weight, during when the angle exceeding ϕ overrules frictional force and cause motion. This angle is termed as *limiting angle of friction in inclined plane* or *angle of repose*. Taking freebody diagram of block as shown in Fig. 3.4 (*b*) and resolving forces along the incline and its normal direction, we have

$$\left.\begin{aligned} F = \mu R = W \sin\phi \\ R = W \cos\phi \end{aligned}\right\} \qquad \text{...(3.3)}$$

or

$$\mu = \tan\phi \qquad \text{...(3.4)}$$

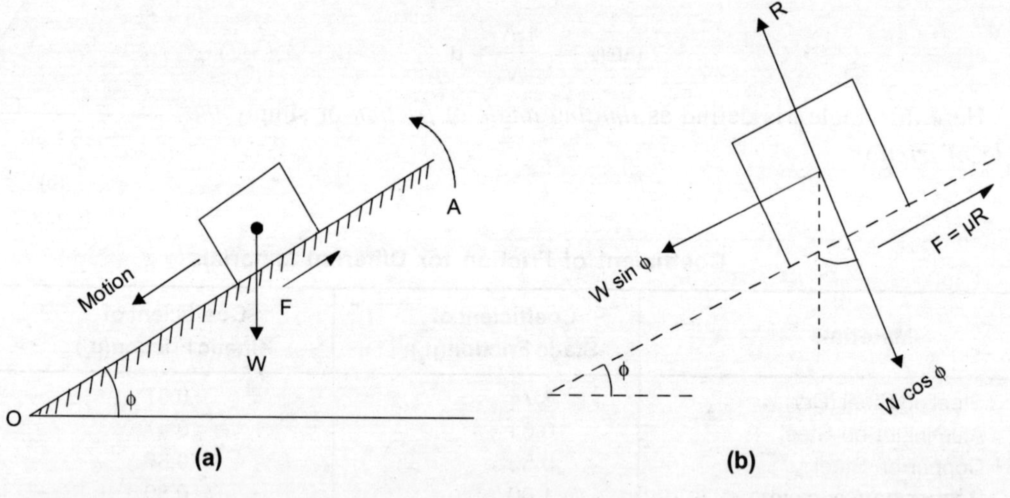

Fig. 3.4

3.5 ROLLING FRICTION

This situation is encountered when a car wheel *moves without slip* along horizontal surface. Here the resistance to the rolling motion exists due to the presence of extenally applied force *P*. As the surface gets deformed due to self weight of body, a heaping of material occurs in front of wheel. By virtue of this phenomenon, contact between wheel and surface ceases to remain as a point but extends over an area, which is called *area of influence* or *influence area*. Ground reaction acts through each point of influence area in the form of distributed loading, the resultant of which is N acting at an angle ϕ with the vertical direction.

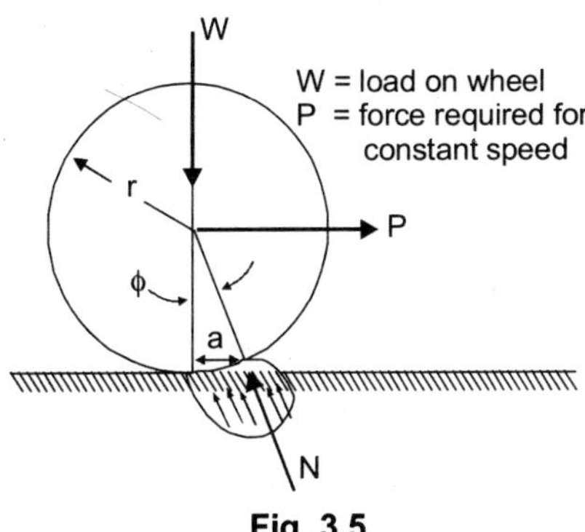

Fig. 3.5

Now from equations of equlibrium along vertical and horizontal direction, we have

$$\left.\begin{aligned} W &= N\cos\phi \\ P &= N\sin\phi \end{aligned}\right\} \qquad \ldots(3.5)$$

Therefore, $$\frac{P}{W} = \tan\phi \qquad \ldots(3.6)$$

As the influence area is substantially small, ϕ becomes a small angle and so, $\tan\phi \approx \sin\phi$. Again, if r be radius of roller, it can be said,

$$\sin\phi = \frac{a}{r} \qquad \ldots(3.7)$$

Clubbing (3.6), (3.7) we obtain,

$$\frac{P}{W} = \frac{a}{r}$$

$\Rightarrow$ $$a = \frac{P}{W}r \qquad \ldots(3.8)$$

The distance a, above is defined as *coefficeint of rolling resistance* and the ratio $\left(\frac{a}{r}\right)$ is defined as *coefficient of rolling friction.*

Multiple Choice Questions

Select the best alternative (s):

1. The coefficient of friction depends on
   (*a*) area of contact (*b*) shape of surfaces (*c*) strength of surfaces
   (*d*) nature of surfaces (*e*) all of the above
2. The ratio of limiting friction and normal reaction is known as
   (*a*) coefficient of friction (*b*) angle of friction (*c*) angle of repose
   (*d*) sliding friction (*e*) frictional resistance
3. Frictional force encountered after commencement of motion is called,
   (*a*) post friction (*b*) limiting friction (*c*) kinematic friction
   (*d*) frictional resistance (*e*) dynamic friction
4. Coefficient of friction is the
   (*a*) angle between normal reaction and the resultant of normal reaction and the limiting friction
   (*b*) ratio of limiting friction and normal reaction
   (*c*) the friction force acting when the body is just about to move
   (*d*) the frictional force acting when the body is in motion
   (*e*) tangent of angle of repose
5. A body of weight *W* on inclined plane of α being pulled up by a horizontal force *P* will be on the point of motion up the plane when *P* is equal to
   (*a*) W (*b*) $W \sin(\alpha + \phi)$ (*c*) $W \tan(\alpha + \phi)$
   (*d*) $W \tan(\alpha - \phi)$ (*e*) $W \tan\alpha$
6. Kinetic friction is the
   (*a*) tangent of angle between normal reaction and resultant of normal reaction and the limiting friction
   (*b*) ratio of limiting friction and normal reaction
   (*c*) the friction force acting when the body is in motion
   (*d*) the friction force when the body is just about to move
   (*e*) dynamic friction
7. The maximum frictional force which comes into play when a body just begins to slide over another surface is called
   (*a*) limiting friction (*b*) sliding friction (*c*) rolling friction
   (*d*) kinematic friction (*e*) dynamic friction
8. Two bodies of 100 kg and 400 kg are resting on two inclined planes α and β towards each other and the bodies are joined together by a string passing over a pulley connected at the top of inclined planes. Coefficient of friction for two bodies with their inclined planes are μ_1 and μ_2. Tension in string will be
   (*a*) 100 kg (*b*) 300 kg (*c*) 400 kg
   (*d*) 500 kg (*e*) 600 kg
9. The effort required to be applied parallel to plane, to move a body of weight *W* upon rough inclined plane $(\mu = \text{coeff. of friction} = \tan\phi)$ with inclination to horizontal is
   (*a*) $W \tan\alpha$ (*b*) $W \tan(\alpha + \phi)$ (*c*) $W \tan(\alpha - \phi)$
   (*d*) $W(\sin\alpha + \mu\cos\alpha)$ (*e*) $W(\cos\alpha + \mu\sin\alpha)$
10. A particle inside a hollow sphere of radius, *r* having coefficient of friction $\frac{1}{\sqrt{3}}$ can rest up to a height of
    (*a*) *r*/2 (*b*) *r*/4 (*c*) *r*/8 (*d*) 0.134*r* (*e*) 3*r*/8

11. The angle which an inclined plane makes with the horizontal when a body placed on it is about to move down is known as angle of
(*a*) friction (*b*) limiting friction (*c*) repose
(*d*) kinematic friction (*e*) static friction

12. The minimum coefficient of friction between a sphere and inclined plane of θ, so that the sphere may roll without slipping is
(*a*) $\tan\theta$ (*b*) $1/2 \tan\theta$ (*c*) $2/7 \tan\theta$
(*d*) $1/7 \tan\theta$ (*e*) $3/7 \tan\theta$

13. A ladder resting against a wall will never slip irrespective of where man stands on it, if the ladder makes an angle
(*a*) not greater than friction angle with vertical (*b*) equal to friction angle with vertical
(*c*) greater than friction angle with vertical (*d*) any angle irrespective of friction angle
(*e*) none of the above

14. A locomotive of weight W is pulled by a force P just equal to the total friction at the points of contact A and B. The vertical reaction R_A and R_B respectively will be
(*a*) $[W/2 + Pb/(2a)]$, and $[W/2 - Pb(2a)]$ (*b*) $W/2$ each
(*c*) $[W/2 - Pb(2a)]$, and $[W/2 + Pb/(2a)]$ (*d*) $[W/2 + Pb\,(2a)]$ each
(*e*) $[W/2 - Pb/(2a)]$ each

15. A rectangular block of width W and height h is resting on a horizontal floor. It is to be avoided from overturning when a horizontal pull acts at any height on the block. This will be possible when
(*a*) $W/h > \mu$ (*b*) $W/h < \mu$ (*c*) $W/(2h) > \mu$ (*d*) $W/(2h) < \mu$

16. Two rectangular blocks of weight W each are connected by a flexible cord and rest upon a horizontal and an inclined plane with the cord passing over a pulley as shown in figure. If μ is the coefficient of friction for all contiguous surfaces, angle θ for motion of system to impend will be
(*a*) $\tan\theta = \mu$ (*b*) $\tan(\theta/2) = \mu$ (*c*) $\tan 2\theta = \mu$
(*d*) $\tan\theta = 2\mu$ (*e*) $\tan\theta = \mu/2$

17. A circular disc of weight W rolls down an inclined plane of inclination θ. If force of friction be F, then the total net force on the disc parallel to plane is equal to
(*a*) $W - F\sin\theta$ (*b*) $W\sin\theta - F$ (*c*) $W\cos\theta - F$
(*d*) $F\cos\theta - W$ (*e*) $W\tan\theta - F$

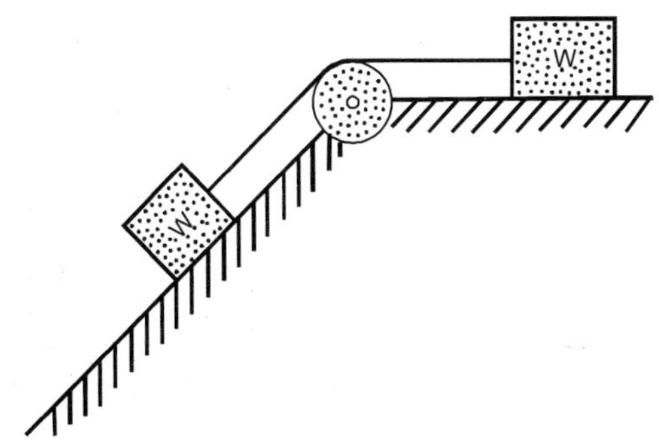

18. A body is resting on a plane inclined at angle of 30° to horizontal. What force would be required to slide it down, if the coefficient of friction between body and plane is 0.3.

(*a*) zero
(*b*) 1 kg
(*c*) 5 kg
(*d*) would depend on weight of body
(*e*) none of the above

19. Least force that starts a body along a plane acts at an angle with the plane

(*a*) equal to the angle of friction
(*b*) little more than angle of friction
(*c*) little less than angle of friction
(*d*) of zero degree
(*e*) none of the above

Answers

1. (*d*)	2. (*a*)	3. (*e*)	4. (*b*)
5. (*c*)	6. (*c*)	7. (*a*)	8. (*b*)
9. (*d*)	10. (*d*)	11. (*c*)	12. (*a*)
13. (*e*)	14. (*b*)	15. (*b*)	16. (*d*)
17. (*b*)	18. (*a*)	19. (*c*)	

Numerical Examples

Two block of weight W_1 and W_2, connected with a string are at rest as in figure. If the angle of friction of each block be ϕ, find magnitude and direction of least force P necessary for upper block to induce sliding.

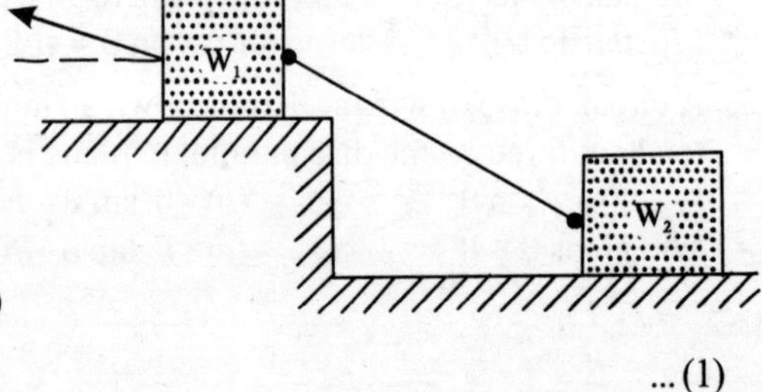

Solution From FBD of W_1, we obtain

$$\Sigma F_x = 0$$

$$\Rightarrow \quad S\cos\theta + \mu R_1 - P\cos\alpha = 0 \quad \text{...(1)}$$

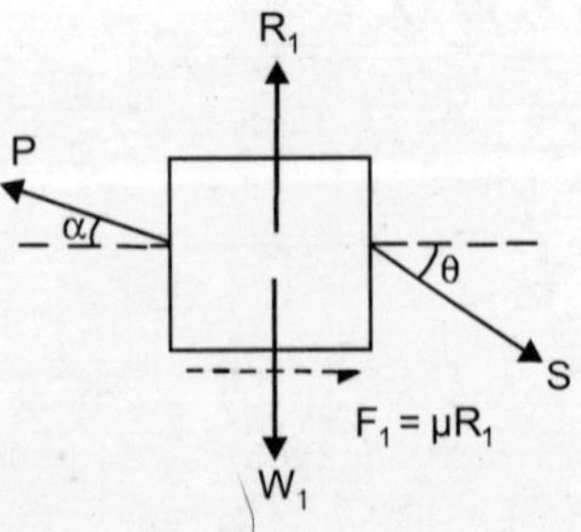

FBD of W

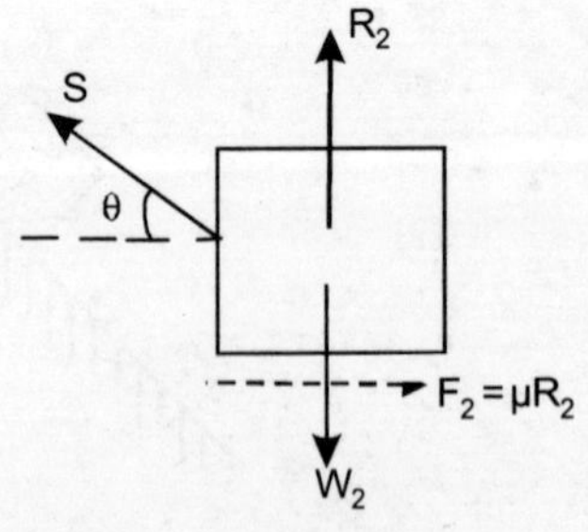

FBD of W_2

and $$\sum F_y = 0$$

$$\Rightarrow \quad R_1 - W_1 + P\sin\alpha - S\sin\theta = 0 \quad \text{...(2)}$$

From FBD of W_2, we obtain $$\sum F_x = 0$$

$$\Rightarrow \quad \mu R_2 - S\cos\theta = 0 \quad \text{...(3)}$$

and $$\sum F_y = 0$$

$$\Rightarrow \quad S\sin\theta + R_2 - W_2 = 0 \quad \text{...(4)}$$

Adding equations (1) and (3) $$\mu(R_1 + R_2) = P\cos\alpha$$

$$\Rightarrow \quad R_1 + R_2 = \frac{P\cos\alpha}{\mu}$$

Adding equations (2) and (4)

$$(R_1 + R_2) - (W_1 + W_2) + P\sin\alpha = 0$$

$$\Rightarrow \quad \frac{P\cos\alpha}{\mu} - P\sin\alpha = W_1 + W_2$$

$$\Rightarrow \quad P = \frac{\mu(W_1 + W_2)}{(\cos\alpha + \mu\sin\alpha)}$$

If P is to be minimum $$\frac{dP}{d\alpha} = 0$$

Hence $(\cos\alpha + \mu\sin\alpha)$ will be maximum.

$$\Rightarrow \quad \frac{d}{d\alpha}(\cos\alpha + \mu\sin\alpha) = 0$$

$$\Rightarrow \quad \mu = \tan\alpha$$

As per given condition $$\mu = \tan\phi$$

$$\Rightarrow \quad \alpha = \phi$$

$$\therefore \quad P_{\min} = \frac{\tan\phi(W_1 + W_2)}{(\cos\phi + \tan\phi\sin\phi)} = (W_1 + W_2)\sin\phi$$

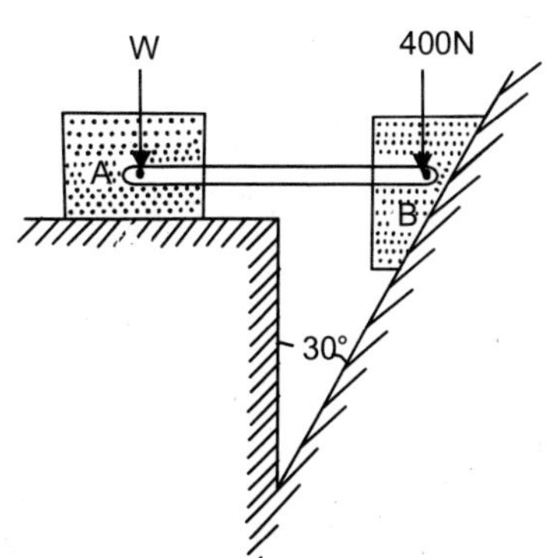

In the adjacent figure $\mu_A = 0.4$, $\phi_B = 15°$. Find smallest W for the equilibrium of the system.

Solution Considering equilibrium of B

$$\sum F_y = 0$$

$$\Rightarrow \quad F_B\cos 30° + R_B\sin 30° - 400 = 0$$

Here, $$F_B = \mu_B \cdot R_B = \tan\phi_B \cdot R_B$$

So, $$(\tan\phi_B \cdot R_B)\cos 30° + R_B \sin 30° - 400 = 0$$

$$\Rightarrow \quad R_B = 546\text{N}$$

Again, $$\sum F_x = 0$$

$$\Rightarrow \quad T - R_B \cos 30° + F_B \sin 30° = 0$$

$$\Rightarrow \quad T = 400\text{N}$$

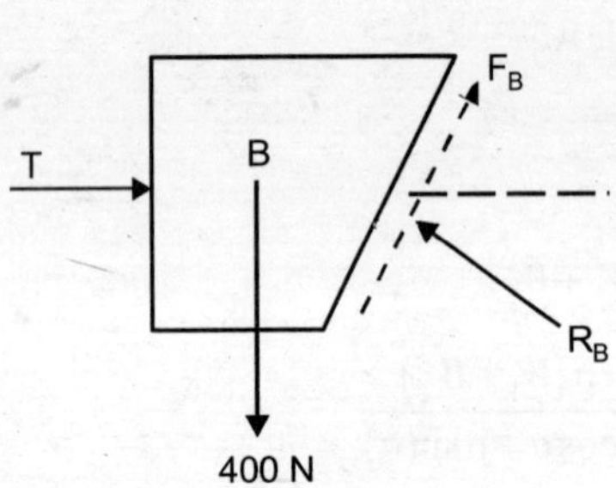

FBD of inclined block

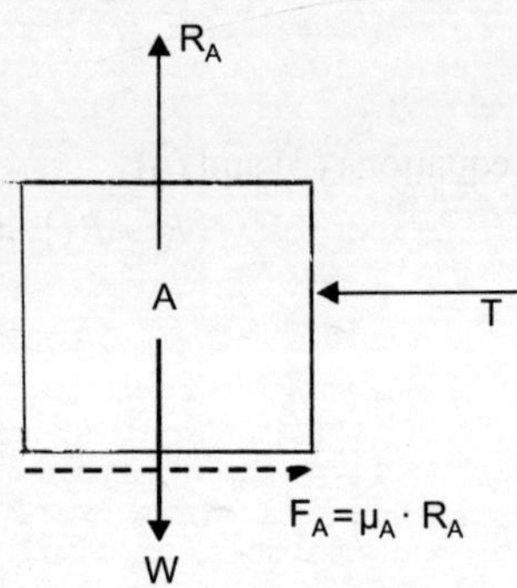

FBD of left block

Considering equilibrium of A

$$\sum F_x = 0$$

$$T = \mu_A R_A = \mu_A \cdot W$$

$$\Rightarrow \quad W = \frac{T}{\mu_A} = \frac{400}{0.4} = 1000\text{N}$$

In the adjoining Fig., $W_1 = W_2 = 50\text{N}$, $\mu_1 = 0.2$, $\mu_2 = 0.3$, find α for which sliding will impend.

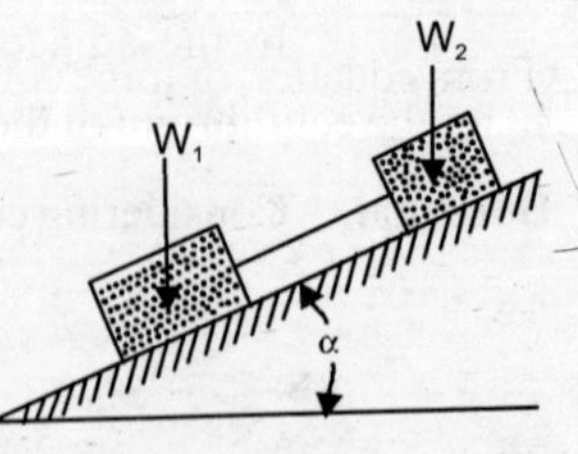

SOLUTION For block W_2, force equilibrium perpendicular to the plane,

$$R_2 = W_2 \cos\alpha$$

and along the plane, $$\mu_2 R_2 - S - W_2 \sin\alpha = 0$$

$$\Rightarrow \quad S = \mu_2 W_2 \cos\alpha - W_2 \sin\alpha \qquad ...(1)$$

For block W_1, force equilibrium perpendicular to plane,

$$R_1 = W_1 \cos\alpha$$

and along the plane, $S + \mu_1 R_1 - W_1 \sin\alpha = 0$

$$\Rightarrow \quad S = W_1 \sin\alpha - \mu_1 W_1 \cos\alpha \qquad ...(2)$$

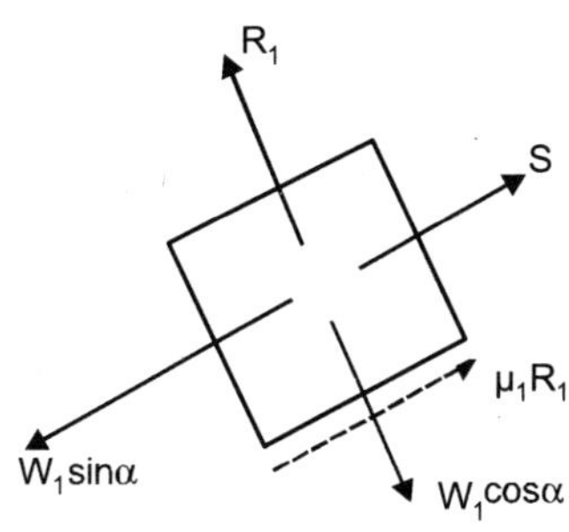

FBD of W_1

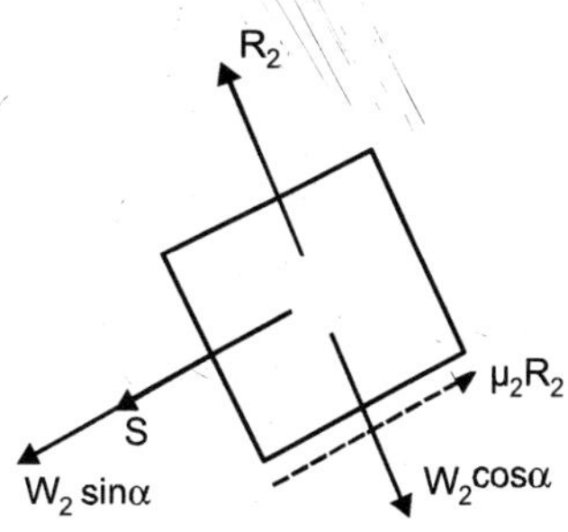

FBD of W_2

Equating above two equations (1) and (2) for S,

$$\Rightarrow \quad \mu_2 W_2 \cos\alpha - W_2 \sin\alpha = W_1 \sin\alpha - \mu_1 W_1 \cos\alpha$$

$$\sin\alpha(W_1 + W_2) = \cos\alpha(\mu_1 W_1 + \mu_2 W_2)$$

$$\Rightarrow \quad \tan\alpha = \frac{W_1 + W_2}{\mu_1 W_1 + \mu_2 W_2} = 0.25$$

For this problem in this figure $W_1 = 200\text{N}$, $W_2 = 50\text{N}$, $\mu = 0.3$. Find necessary P to impend slipping.

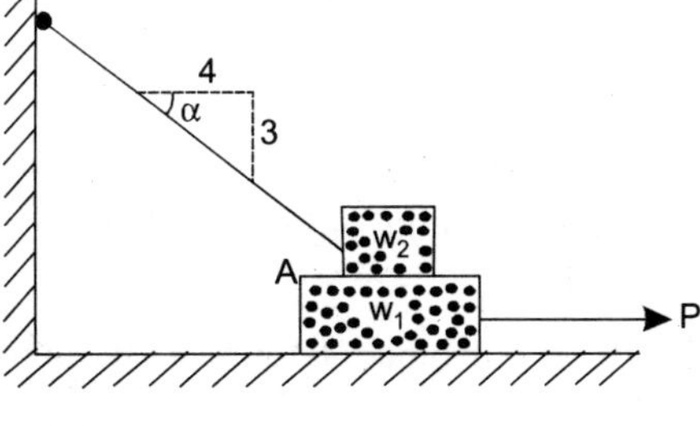

SOLUTION Here, $\sin\alpha = \frac{3}{5}$, $\cos\alpha = \frac{4}{5}$

From equation of force equilibrium of W_2

$$\sum F_x = 0$$

$$\Rightarrow \quad \mu R_2 = S\cos\alpha$$

$$\Rightarrow \quad S = \frac{\mu R_2}{\cos\alpha}$$

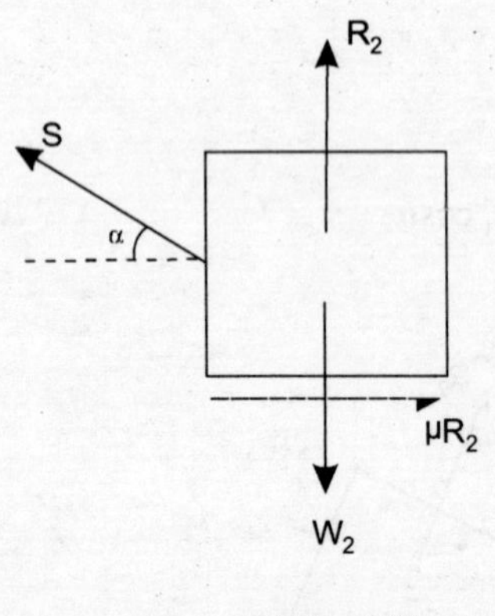

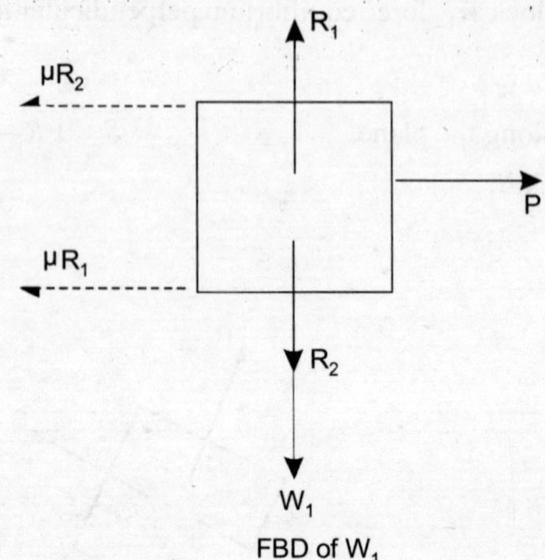

$$\sum F_y = 0$$

$$\Rightarrow \quad R_2 + S\sin\alpha - W_2 = 0$$

$$\Rightarrow \quad R_2 + \mu R_2 \tan\alpha = W_2$$

$$\Rightarrow \quad R_2 = \frac{W_2}{1+\mu\tan\alpha} = 40.8\text{N}$$

From equation of force equilibrium of W_1

$$\sum F_y = 0$$

$$\Rightarrow \quad R_1 = R_2 + W_1 = 240.8\text{N}$$

and

$$\sum F_x = 0$$

$$\Rightarrow \quad P = \mu R_1 + \mu R_2 = 84.48\text{N}$$

EXAMPLE 5

For this problem as shown in figure, find μ, where the two blocks are identical.

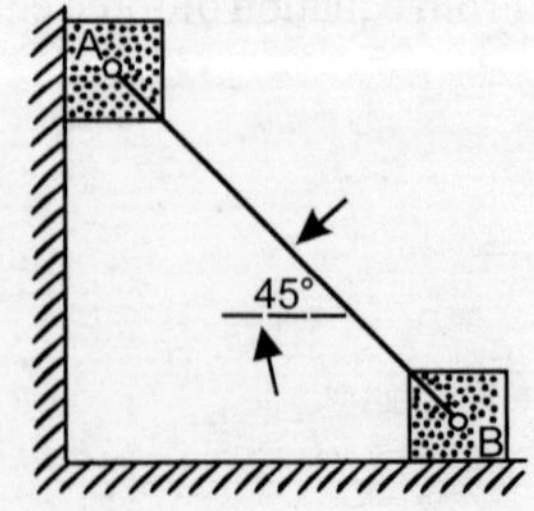

SOLUTION In block A, force equilibrium equations are

$$\sum F_x = 0$$

$$\Rightarrow \quad R_A = S\cos 45^\circ$$

and

$$\sum F_y = 0$$

$$\Rightarrow \quad S\sin 45° + \mu R_A - W = 0$$

$$\Rightarrow \quad S = \frac{\sqrt{2}}{1+\mu}W \quad ...(1)$$

In block B, force equilibrium equations are

$$\sum F_x = 0$$

$$\Rightarrow \quad S\cos 45° - \mu R_B = 0$$

$$\Rightarrow \quad R_B = \frac{S}{\mu\sqrt{2}}$$

and

$$\sum F_y = 0$$

$$\Rightarrow \quad R_B - S\sin 45° - W = 0$$

$$\Rightarrow \quad \frac{S}{\mu\sqrt{2}} - \frac{S}{\sqrt{2}} - W = 0$$

$$\Rightarrow \quad S = \frac{W \cdot \mu\sqrt{2}}{1-\mu} \quad ...(2)$$

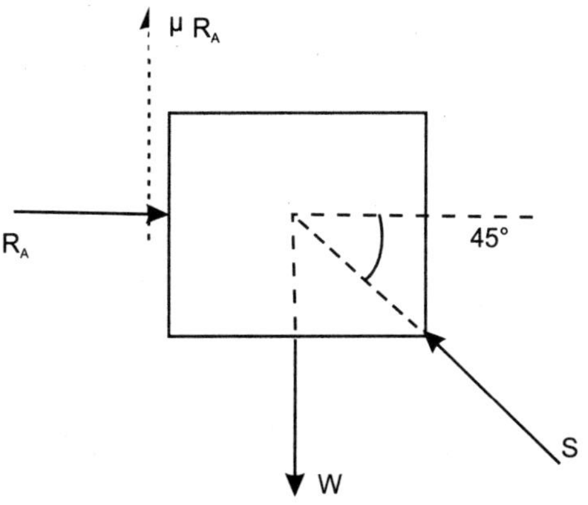

FBD of block A

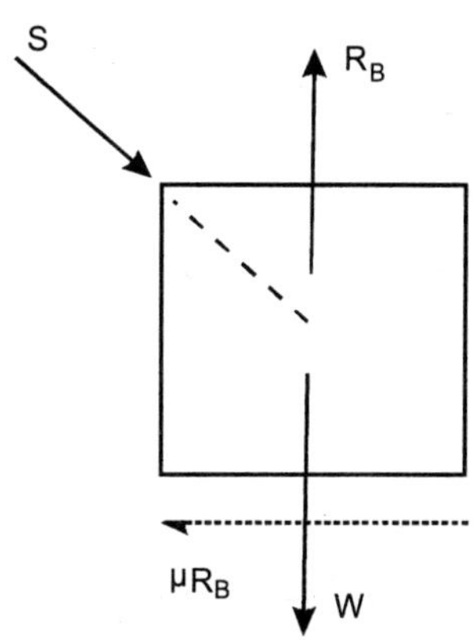

FBD of block B

Equating (1) and (2)

$$\frac{1}{1+\mu} = \frac{\mu}{1-\mu}$$

$$\Rightarrow \quad \mu^2 + 2\mu - 1 = 0$$

$$\Rightarrow \quad \mu = 0.414$$

EXAMPLE 6

For this problem in this figure $W_1 = W_2$ and coefficient of static friction for all contiguous surfaces is μ. Find α for impending motion. Consider pulley frictionless.

SOLUTION From block W_2 force equiiibrium perpendicular to the incline,

$$R_2 = W_2 \cos\alpha$$

And along the inclined plane, $S + \mu R_2 - W_2 \sin\alpha = 0$

$$\Rightarrow \quad S = W_2 \sin\alpha - \mu W_2 \cos\alpha \quad \text{... (1)}$$

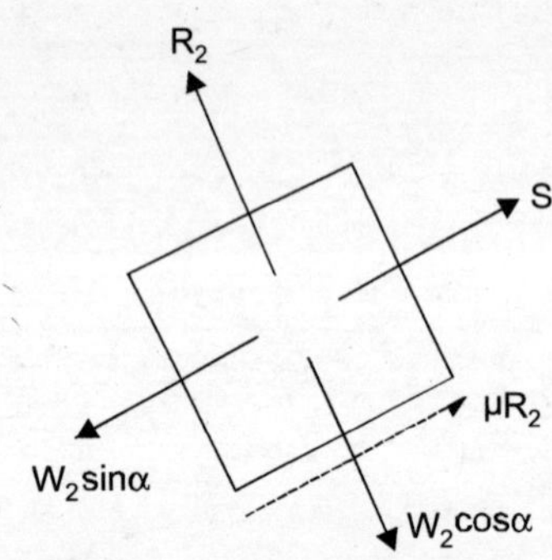

FBD of W_2

FBD of W_1

From block W_1,

$$S = \mu R_1 = \mu W_1 \quad \text{... (2)}$$

Equating (1) and (2) while

$$W_1 = W_2$$

$$\mu = \sin\alpha - \mu\cos\alpha$$

$$\Rightarrow \quad \mu = \tan(\alpha/2)$$

EXAMPLE 7

For the problem in this figure, find magnitude of P causing motion of cylinder to impend.

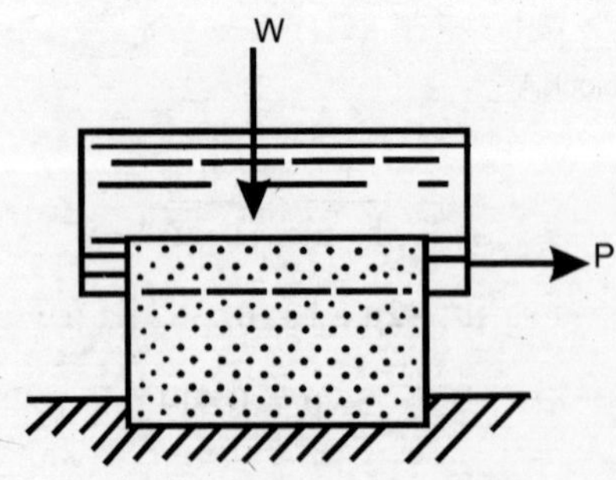

SOLUTION From symmetry we can say that the reaction coming from notch on the cylinder on either side will be equal. Considering equilibrium of cylinder, resolving forces along W, we obtain,

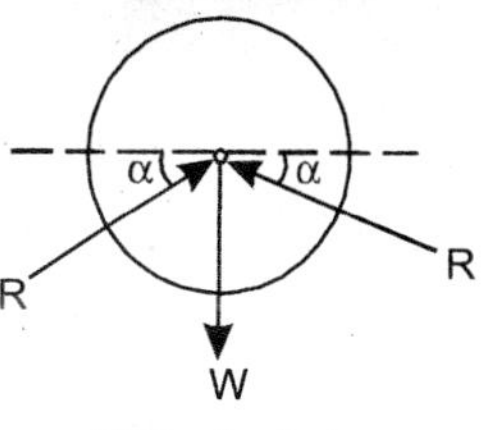

FBD of cylinder

$$2R\sin\alpha = W$$

$$\Rightarrow \qquad R = \frac{W}{2\sin\alpha}$$

As the horizontal friction force induced on inside surface of notch be $F = \mu R$ on either side,

$$P = 2F = 2\mu R = \frac{\mu W}{\sin\alpha}$$

EXAMPLE 8

In the adjacent figure weight of the block 1600N and $\mu = 0.2$. Find the value of P for impending motion.

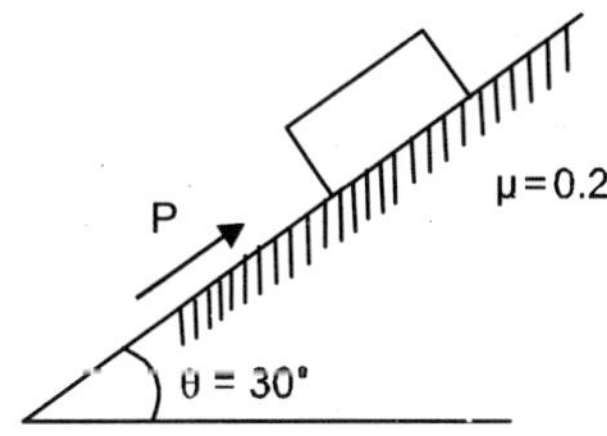

SOLUTION Resolution of forces in direction perpendicular to the inclined plane

$$R = 1600\cos 30°$$

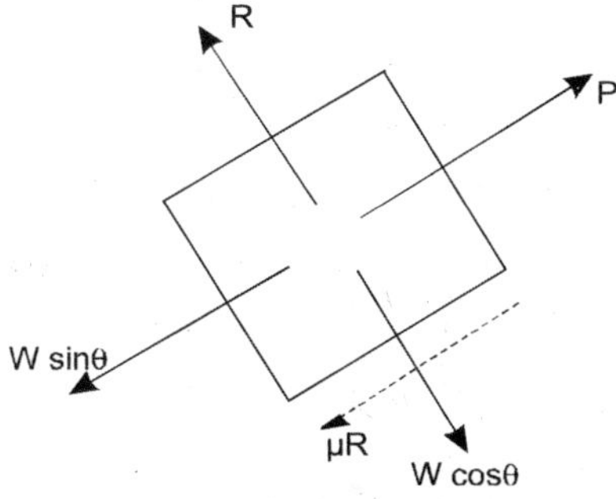

FBD of block A

And resolution of forces along the plane

$$\begin{aligned} P &= \mu R + 1600\sin 30° \\ &= 0.2\,(1600\cos 30° + 1600\sin 30°) \\ &= 1077.13\text{N} \end{aligned}$$

EXAMPLE 9

For the adjacent figure find θ just before sliding begins.

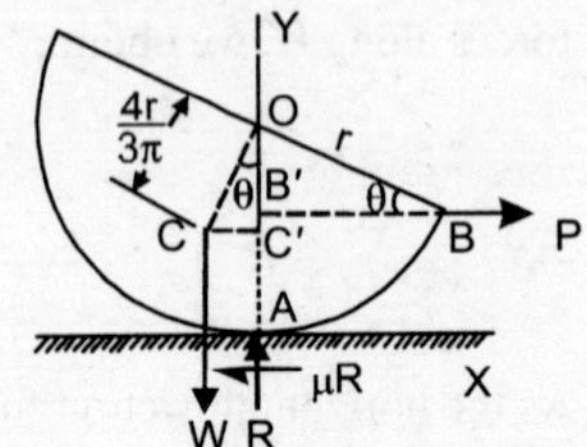

SOLUTION From the equations of equilibrium we can write $R = W$ and $P = \mu R = \mu W$. Taking moment equilibrium about A

$$\sum M_A = 0$$

$$\Rightarrow \quad W\frac{4r}{3\pi}\sin\theta = P(r - r\sin\theta) = \mu Wr(1-\sin\theta)$$

$$\Rightarrow \quad \frac{4}{3\pi}\sin\theta + \mu\sin\theta = \mu$$

$$\Rightarrow \quad \sin\theta = \frac{3\pi\mu}{(4+3\pi\mu)}$$

EXAMPLE 10

A heavy prismatic bar AB of mass M is supported by a circular ring as shown in figure. Length of AB is such that it subtends 90° in the ring. If the angle of friction be ϕ, find greatest θ that the bar can make during equilibrium.

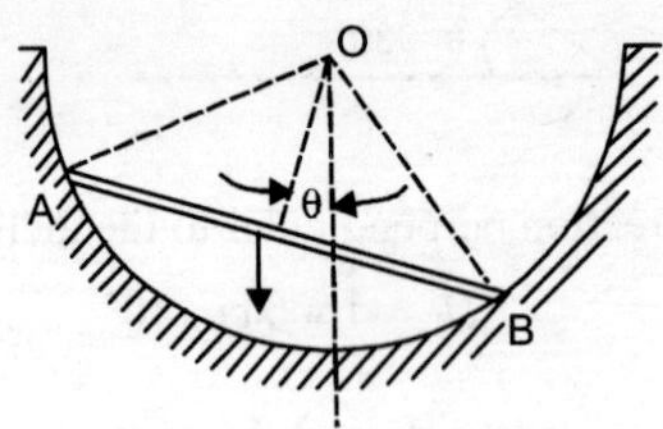

SOLUTION (*i*) *Force Approach*: Resolution of forces along specified x and y axis results,

$$Mg\cos(90° - \theta) - N_B\cos 45° + N_A\cos 45° - \mu N_B\cos 45° - \mu N_A\cos 45° = 0$$

$$\Rightarrow \quad \sqrt{2}Mg\sin\theta = (N_B - N_A) + \mu(N_B + N_A) \qquad \text{...(1)}$$

and,

$$-Mg\cos\theta + N_B\sin 45° + N_A\sin 45° - \mu N_B\sin 45° + \mu N_A\sin 45° = 0$$

$$\Rightarrow \quad \sqrt{2}Mg\cos\theta = (N_B + N_A) + \mu(N_A - N_B) \qquad \text{...(2)}$$

Taking moment equilibrium about C,

$$N_B \cdot CB\sin 45° - \mu N_B \cdot CB\sin 45° - N_A \cdot AC\sin 45° - \mu N_A \cdot AC\sin 45° = 0$$

$$\Rightarrow \quad (N_A - N_B) + \mu(N_A + N_B) = 0 \qquad \text{...(3)}$$

Dividing (1) by (2)

$$\tan\theta = \frac{(N_B - N_A) + \mu(N_A + N_B)}{(N_A + N_B) + \mu(N_A - N_B)} \qquad \text{...(4)}$$

Substituting (3) into (4), we obtain,

$$\tan\theta = \frac{2\mu}{1-\mu^2} = \frac{2\tan\phi}{1-\tan^2\phi} = \tan 2\phi$$

$$\Rightarrow \qquad \theta = 2\phi$$

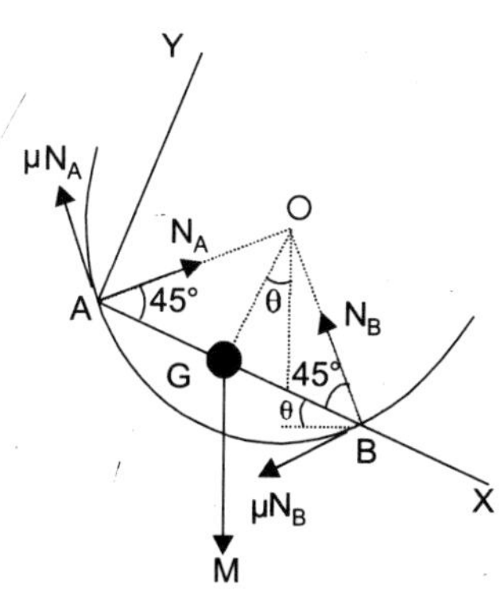

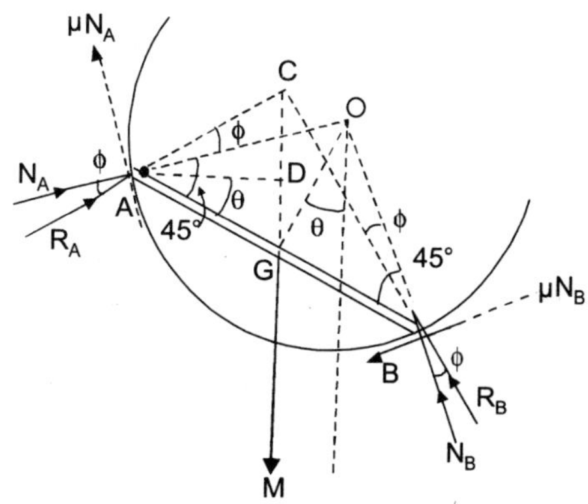

(*ii*) *Geometrical Approach* : For equilibrium, R_A, R_B and M should be concurrent at C. Here,

$$\angle CAB = (45° + \phi),\ \angle CBA = (45° - \phi),$$

$$\angle CAD = (45° + \phi - \theta),\ \angle C = 90°,$$

$$AB = L,\ AG = L/2$$

From ΔDAG, $$AD = \frac{L}{2}\cos\theta \qquad \text{...(1)}$$

From Δ CAB, $$AC = L\cos(45° + \phi) \qquad \text{...(2)}$$

From Δ CAD, $$AD = AC\cos(45° + \phi - \theta)$$

$$= L\cos(45° + \phi)\cos(45° + \phi - \theta) \qquad \text{...(3)}$$

From equations (1) and (3)

$$\frac{L}{2}\cos\theta = L\cos(45° + \phi)\cos(45° + \phi - \theta)$$

$$\Rightarrow \qquad \cos\theta = 2\cos\alpha\cdot\cos(\alpha - \theta) = \cos\theta + \cos(2\alpha - \theta)$$

$$\Rightarrow \qquad \cos(2\alpha - \theta) = 0 = \cos 90°$$

$$\Rightarrow \qquad 90° + 2\phi - \theta = 90°$$

$$\Rightarrow \qquad \theta = 2\phi$$

EXAMPLE 11

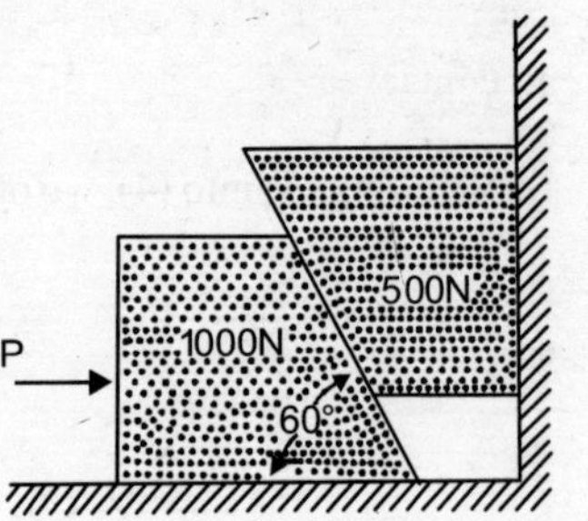

Referring to figure, μ at floor $(\mu_f) = 0.25$, at wall $(\mu_w) = 0.30$, between blocks $(\mu_b) = 0.20$. Find minimum P to hold the system in equilibrium.

SOLUTION From block W_1 we find

$$\sum F_x = 0$$

$$\Rightarrow \quad R_b \cos 30^\circ - \mu_b R_b \cos 60^\circ - R_1 = 0$$

$$\Rightarrow \quad R_1 = 0.766 R_b$$

$$\sum F_y = 0$$

$$\Rightarrow \quad \mu_b R_b \sin 60^\circ + \mu_w R_1 + R_b \sin 30^\circ - W_1 = 0$$

$$\Rightarrow \quad 0.673 R_b + 0.3 R_1 = 500$$

$$\Rightarrow \quad R_b = 553.8\text{N}$$

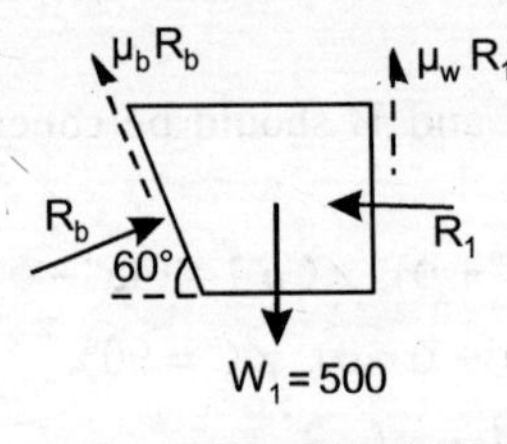

FBD of block W_1

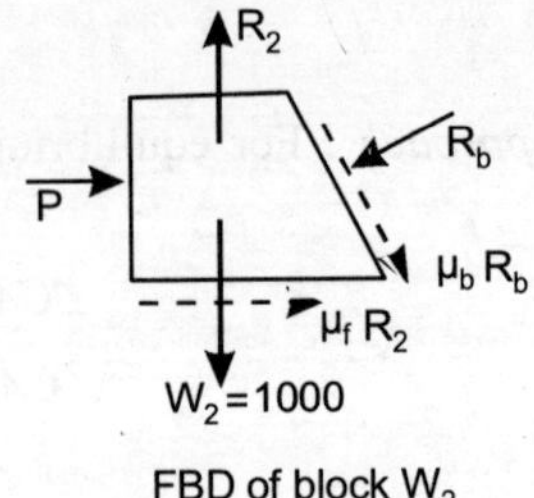

FBD of block W_2

From block W_2 we obtain,

$$\sum F_y = 0$$

$$\Rightarrow \quad R_2 - R_b \sin 30^\circ - \mu_b R_b \sin 60^\circ - W_2 = 0$$

$$\Rightarrow \quad R_2 = 1372.7\text{N}$$

$$\sum F_x = 0$$

$$\Rightarrow \quad P + \mu_f R_2 + \mu_b R_b \cos 60^\circ - R_b \cos 30^\circ = 0$$

$$\Rightarrow \quad P = 80.8\text{N}$$

EXAMPLE 12

Referring to the figure, force P is applied to string wound around large roller. Assume μ same for all contact surfaces, determine necessary condition for larger roller can be pulled over the smaller one.

SOLUTION From FBD of larger roller,

$$\sum F_x = 0$$

$$\Rightarrow \quad P + F\cos\theta - R\cos(90° - \theta) = 0$$

$$\Rightarrow \quad R = \frac{P}{\sin\theta - \mu\cos\theta} \qquad ...(1)$$

Consider moment equilibrium $\sum M_{C_1} = 0$

$$\Rightarrow \quad F \times \frac{D}{2} - P \times \frac{D}{2} = 0$$

$$\Rightarrow \quad P = F = \mu R$$

$$\Rightarrow \quad R = \frac{P}{\mu} \qquad ...(2)$$

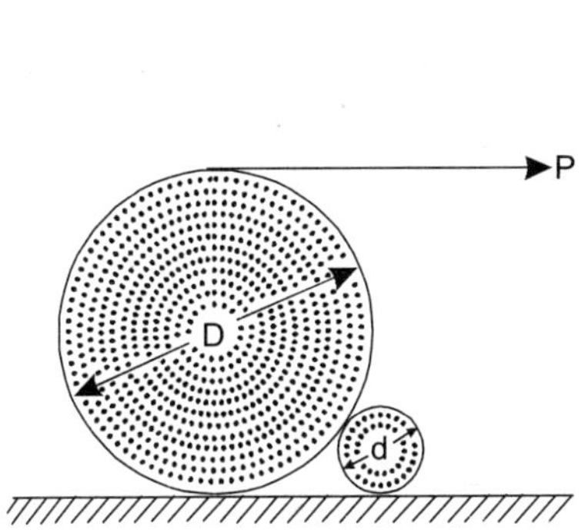

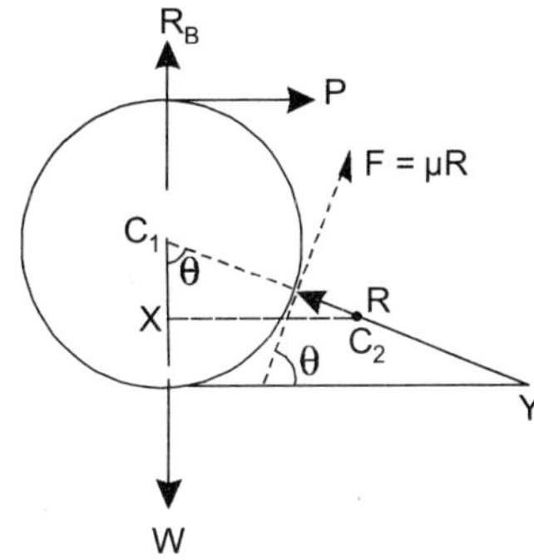

FBD of large roller

Equating the values of R, in equations (1) and (2) we obtain,

$$\mu = \sin\theta - \mu\cos\theta$$

$$\Rightarrow \quad \mu = \frac{\sin\theta}{1+\cos\theta} = \frac{2\sin\theta/2\cos\theta/2}{\cos^2\theta/2} = \tan\theta/2 \qquad ...(3)$$

From, ΔC_1XC_2, $\cos\theta = \dfrac{D/2 - d/2}{D/2 + d/2} = \dfrac{D-d}{D+d}$

So, $\sin\theta = \sqrt{1 - \left(\dfrac{D-d}{D+d}\right)^2} = 2\dfrac{\sqrt{Dd}}{D+d}$

Substituting these values in equation (3) we obtain, $\mu = \sqrt{\dfrac{d}{D}}$

EXAMPLE 13

Block $A = 100$N rests on a rough plane, inclined at 45°. A is connected to block $B = 300$N resting on rough horizontal plane, by a weightless rigid bar inclined 30° to horizontal. Find the horizontal force required to be applied to B to just move A in upward direction. Assume $\mu = 0.2679$ for all rough surfaces.

SOLUTION Considering equilibrium of A and resolving forces along the inclined plane and normal to the inclined plane,

$$\left.\begin{aligned}100\sin 45° + \mu R_A &= T\cos 15° \\ 100\cos 45° + T\sin 15° &= R_A\end{aligned}\right\}$$

or

$$\left.\begin{aligned}100\frac{1}{\sqrt{2}} + 0.2679R_A &= 0.9659T \\ 100\frac{1}{\sqrt{2}} + 0.2588T &= R_A\end{aligned}\right\}$$

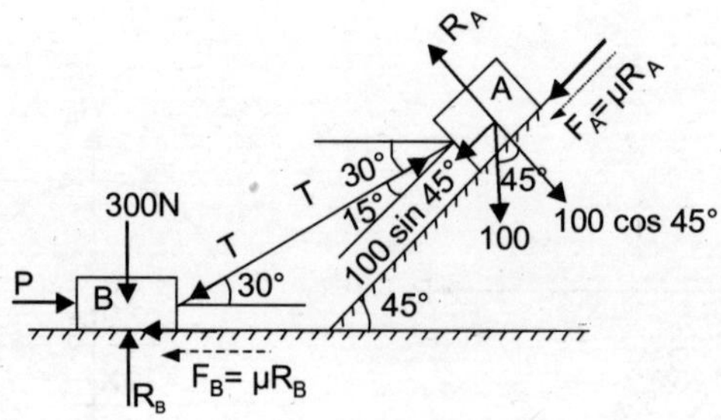

Solving these, $T = 99.97$N. Considering equilibrium of B and resolving force along horizontal and vertical directions.

$$\left.\begin{aligned}T\cos 30° + \mu R_B &= P \\ T\sin 30° + 300 &= R_B\end{aligned}\right\}$$

or

$$\left.\begin{aligned}T\frac{\sqrt{3}}{2} + 0.2679R_B &= P \\ T\frac{1}{2} + 300 &= R_B\end{aligned}\right\}$$

Solving these, $P = 180.334$N

EXAMPLE 14

Referring to the figure, determine least value of P to cause motion to impend rightwards. Assume the pulley be frictionless and coefficient of friction of blocks and contact surfaces be 0.2.

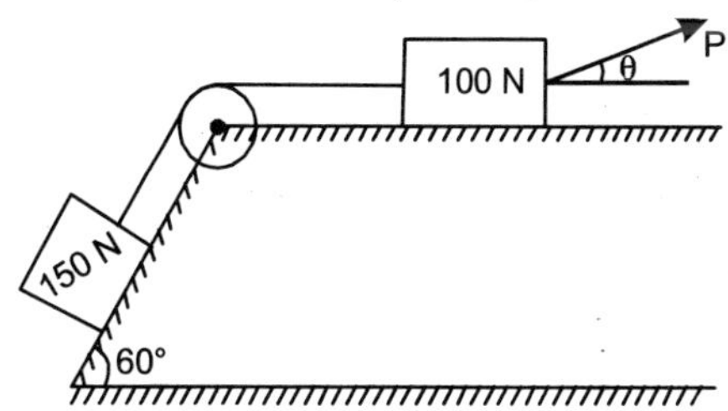

SOLUTION Considering equilibrium of block of weight 150 N and resolving forces along inclined plane and normal to inclined plane,

$$\left.\begin{aligned} 150\sin 60^\circ + \mu R &= T \\ 150\cos 60^\circ &= R \end{aligned}\right\}$$

or

$$\left.\begin{aligned} 150\frac{\sqrt{3}}{2} + 0.2R &= T \\ 150\frac{1}{2} &= R \end{aligned}\right\}$$

Solving these, $R = 75$ N, and $T = 144.9$N

Considering equilibrium of block of weight 100N and resolving forces along horizontal and vertical directions,

$$\left.\begin{aligned} T + \mu R_1 &= P\cos\theta \\ R_1 + P\sin\theta &= 100 \end{aligned}\right\}$$

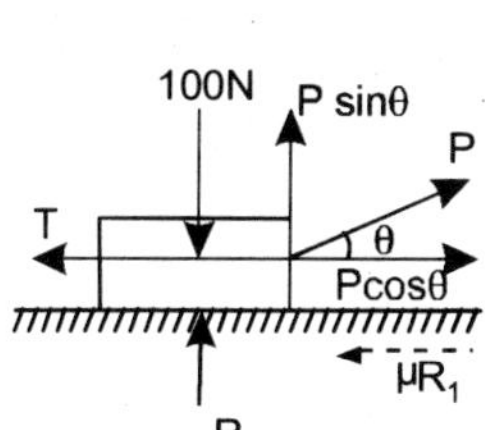

or

$\therefore$

$$\left.\begin{aligned} 144.9 + 0.2R_1 &= P\cos\theta \\ R_1 + P\sin\theta &= 100 \end{aligned}\right\}$$

Solving these, we obtain $P = \dfrac{164.9}{(\cos\theta + 0.2\sin\theta)}$

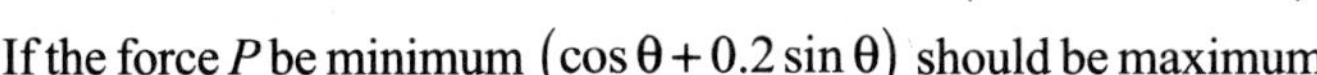

If the force P be minimum $(\cos\theta + 0.2\sin\theta)$ should be maximum

$\therefore$ $$\frac{d}{d\theta}(\cos\theta + 0.2\sin\theta) = 0$$

$\Rightarrow$ $$-\sin\theta + 0.2\cos\theta = 0$$

$\Rightarrow$ $$\tan\theta = 0.2$$

$\Rightarrow$ $$\theta = \tan^{-1}(0.2) = 11.309^\circ$$

So, $$P_{MIN} = \frac{164.9}{(\cos 11.309^\circ + 0.2\sin 11.309^\circ)} = 161.88\text{N}$$

EXAMPLE 15

A uniform ladder of length 13 m weighing 25N is placed against a smooth vertical wall with its lower end 5m from the wall. The coefficient of friction between ladder and floor is 0.3. Show that the ladder will remain in equilibrium in this position. What is the frictional force acting on the ladder at the point of contact between ladder and floor?

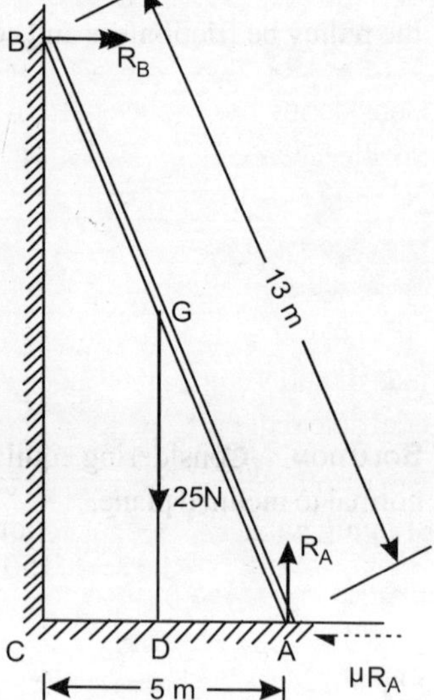

SOLUTION The weight of ladder 25N acts at middle point of AB, in vertically downward direction. If the ladder just deviates from equilibrium, A will have a tendency to move towards right and frictional resistance will act leftward. Now, equating vertical forces,

$$R_A = 25\text{N}$$

So, frictional force at A, $F_A = \mu R_A = 0.3 \times 25 = 7.5\text{N}$

When the ladder is in equilibrium, let the frictional resistance F'_A, reaction at B is R'_B and that at A is R'_A

From geometry, $AD = 5/2 = 2.5\text{m}$,

$$BC = \sqrt{AB^2 - AC^2} = \sqrt{13^2 - 5^2} = 12 \text{ m}$$

The ladder will remain in equilibrium, if summation of moment of all the forces about point A is zero

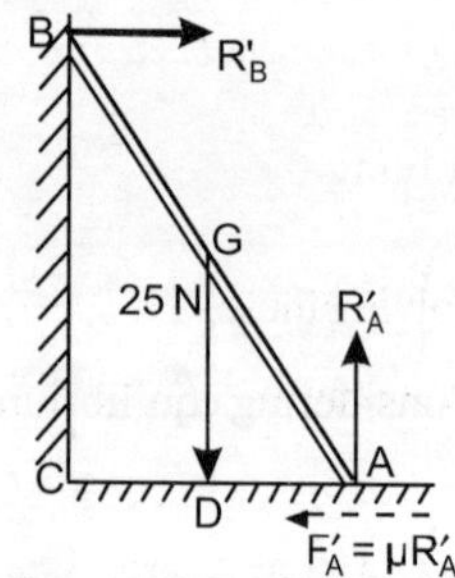

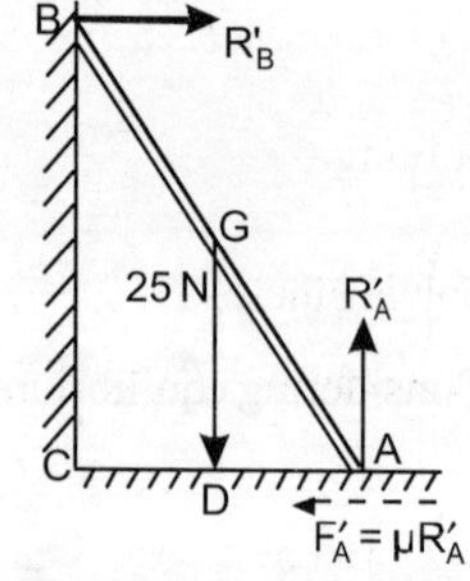

$\therefore$ $25 \times AD = R'_B \times BC$

$\Rightarrow$ $25 \times 2.5 = R'_B \times 12$

$\Rightarrow$ $R'_B = 5.21\text{N}$

Equating horizontal forces, $(F'_A = R'_B = 5.21\text{N}) < (F_A = 7.5\text{N})$

So, at A frictional resistance in equilibrium condition is less that the frictional resistance required just to impend motion.

EXAMPLE 16

The weight of 14 m long bar is 600 N and may be concentrated at point 6 m from bottom. Supporting wall is smooth. Supporting floor is rough with coefficient of static friction (1/3). Establish by calculations, if the bar would stand in the 60° position, as shown.

SOLUTION Here we have to compute actual frictional resistance at B and frictional resistance required for equilibrium at B. To find actual frictional resistance, let us resolve forces horizontally and vertically,

$$\therefore \quad \left. \begin{aligned} R_A &= F_B = \mu R_B = \frac{1}{3} R_B \\ R_B &= 600\text{N} \end{aligned} \right\}$$

Solving above, $F_B = 200\text{N}$

To find frictional resistance in equilibrium, summation of moments of all forces about B should be zero. Now, from geometry,

$$BD = 6 \text{ m}, AD = 14 - 6 = 8 \text{ m}, BE = BD \cos 60° = 3\text{m}, AC = AB \sin 60° = 12.124\text{m}.$$

$$\therefore \quad 600 \times BE = R_A \times AC$$

$$\Rightarrow \quad 600 \times 3 = R_A \times 12.124$$

$$\Rightarrow \quad R_A = 148.46 \text{ N}$$

Considering horizontal equilibrium, $F_B = R_A = 148.46 \text{ N} < 200 \text{ N}$

So, the ladder will stand stable at 60° inclination, as force required to impend motion (200N) is higher than 148.46 N.

EXAMPLE 17

A ladder 5m long 250N weight is resting as shown in figure. A man of 800N climbs the ladder. At what position will be induce slipping? Take coefficient of friction for all contact surfaces 0.2.

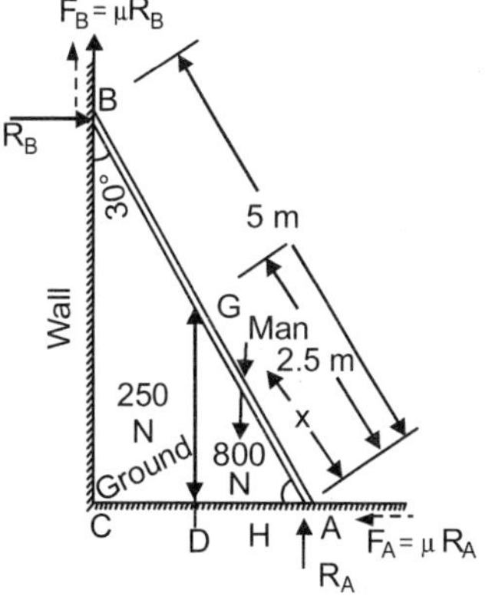

SOLUTION Let x is the distance along ladder where the man climbs and slipping impends. Resolving forces horizontally and vertically,

$$\left.\begin{aligned} R_B &= \mu R_A \\ R_A + \mu R_B &= 250 + 800 \end{aligned}\right\}$$

or

$$\left.\begin{aligned} R_B &= 0.2 R_A \\ R_A + 0.2 R_B &= 1050 \end{aligned}\right\}$$

Solving these, $R_A = 1009.6 \text{ N}$

and $R_B = 201.92 \text{ N}$

From geometry, $AD = AG \cos 60° = 1.25 \text{ m}$

$$BC = AB \cos 30° = 4.33 \text{ m}$$

$$AH = x \cos 60° = 0.5x,$$

$$AC = AB \cos 60° = 2.5 \text{ m}$$

In order to find the position x, summation of moments of all forces about A should be equated to zero.

$$\therefore \quad 800 \times AH + 250 \times AD = R_B \times BC + \mu R_B \times AC$$

$$\Rightarrow \quad 800 \times 0.5x + 250 \times 1.25 = 201.92 \times 4.33 + 0.2 \times 201.92 \times 2.5$$

$$\Rightarrow \quad x = 1.6569 \text{ m}$$

EXAMPLE 18

A block overlying a 10° wedge on a horizontal floor and leaning against a vertical wall and weighing 1500N is to be raised by applying a horizontal force to the wedge. Assuming coefficient of friction for all contact surfaces be 0.3, determine the minimum horizontal force to be applied to raise the block.

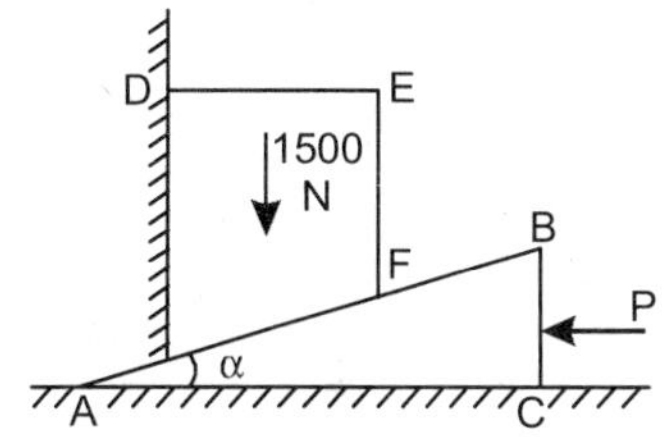

SOLUTION Here as $\mu = \tan \phi = 0.3,$

$$\phi = \tan^{-1}(0.3) = 16.699°$$

Method I : Resolution of forces acting on the block horizontally, gives

$$R_3 \cos \phi = R_2 \sin(\alpha + \phi)$$

$$\Rightarrow \qquad R_3 \cos 16.699^\circ = R_2 \sin(10^\circ + 16.699^\circ)$$

$$\Rightarrow \qquad R_2 = 2.1318 R_3$$

Again, resolution of forces in vertical direction gives

$$R_3 \sin\phi + 1500 = R_2 \cos(\alpha + \phi)$$

$$\Rightarrow \qquad R_3 \cos 16.699^\circ + 1500 = 2.1318 R_3 \cos(10^\circ + 16.699^\circ)$$

$$\Rightarrow \qquad R_3 = 927.55\text{N}$$

Considering equilibrium of the wedge and resolving forces vertically,

$$R_2 \cos(\alpha + \phi) = R_1 \cos\phi$$

$$\Rightarrow \qquad 2.1318 \times 927.55 \cos(10^\circ + 16.699^\circ) = R_1 \cos 16.699^\circ$$

$$\Rightarrow \qquad R_1 = 1844.3 \text{ N}$$

Again resolving forces horizontally for wedge,

$$P = R_1 \sin\phi + R_2 \sin(\alpha + \phi)$$

$$= 1844.3 \sin 16.699^\circ + 2.1318 \times 927.55 \sin(10^\circ + 16.699^\circ)$$

$$= 1418.378 \text{ N}$$

Method II : The block will remain in equilibrium if 1500 N, R_2 and R_3 be concurrent at point O. Applying Lami's theorem at point O,

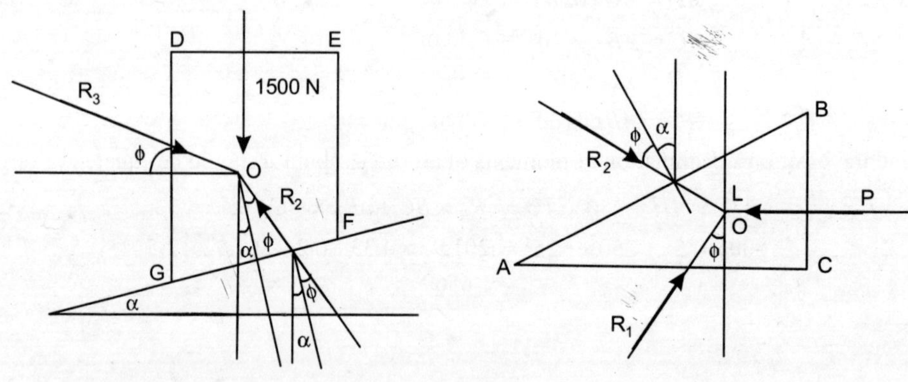

$$\frac{1500}{\sin(\phi + 90^\circ + \alpha + \phi)} = \frac{R_2}{\sin(90^\circ - \phi)} = \frac{R_3}{\sin[180^\circ - (\alpha + \phi)]}$$

$$\Rightarrow \qquad \frac{1500}{\sin 133.398^\circ} = \frac{R_2}{\sin 73.301^\circ} = \frac{R_3}{\sin 153.301^\circ}$$

$$\Rightarrow \qquad R_2 = 1500 \times \frac{\sin 73.301^\circ}{\sin 133.398^\circ} = 1977.352\text{N}$$

$$\Rightarrow \qquad R_3 = 1500 \times \frac{\sin 153.301^\circ}{\sin 133.398^\circ} = 927.548\text{N}$$

For equilibrium of wedge P, R_1, R_2 should be concurrent at L. Applying Lami's theorem at L,

$$\frac{P}{\sin\left[180^\circ - \phi - (\alpha + \phi)\right]} = \frac{R_1}{\sin(90^\circ + \alpha + \phi)} = \frac{R_2}{\sin(90^\circ + \phi)}$$

$$\Rightarrow \qquad \frac{P}{\sin 136.602^\circ} = \frac{R_1}{\sin 116.699^\circ} = \frac{1977.352}{\sin 106.699^\circ}$$

$$\Rightarrow \qquad P = 1977.352 \times \frac{\sin 136.602^\circ}{\sin 106.699^\circ} = 1418.38\text{N}$$

EXAMPLE 19

A horizontal bar of 1.2m length and of negligible weight rests on rough inclined plane. If angle of friction 15°, determine minimum value of x at which $Q = 250$N may be applied before slipping impends.

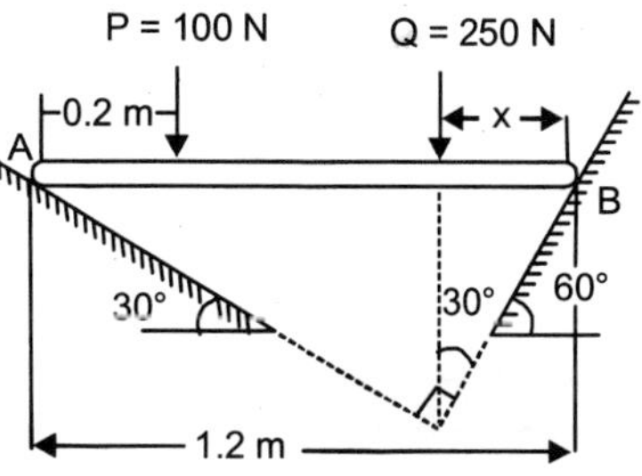

SOLUTION Here at point A, normal reaction N_a and frictional resistance (μN_a) are replaced by R_a, similarly at B, N_b and (μN_b) are replaced by R_b. Now, considering equilibrium in horizontal and vertical directions, we have,

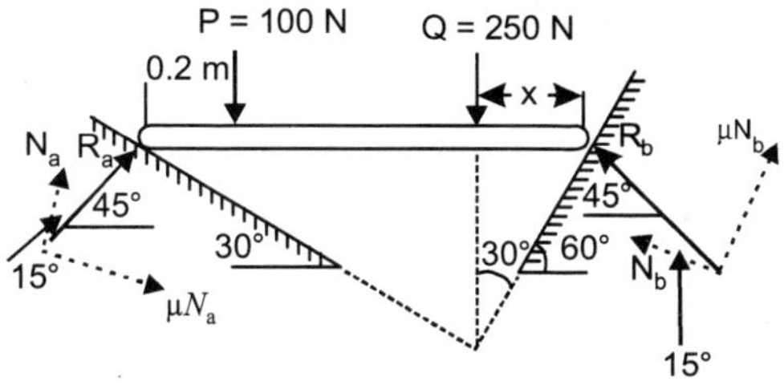

$$\left.\begin{aligned} R_a \sin 45^\circ + R_b \sin 45^\circ &= 100 + 250 \\ R_a \cos 45^\circ &= R_b \cos 45^\circ \end{aligned}\right\}$$

Solving these, $\qquad R_a = 247.487\text{N} = R_b$

Again, considering moment equilibrium at B,

$$250x + 100(1.2 - 0.2) - R_a \sin 45^\circ \times 1.2 = 0$$

$$\Rightarrow \qquad 250x + 100 - 247.487 \times \frac{1}{\sqrt{2}} \times 1.2 = 0$$

$$\Rightarrow \qquad x = 0.4399 \text{ m}$$

EXAMPLE 20

The tongs are used to handle hot steel tubes that are being heat treated in an oil bath. For a 20º jaw opening, what is the minimum coefficient of static friction between the jaws and the tube that will enable the tongs to grip the tube without slipping?

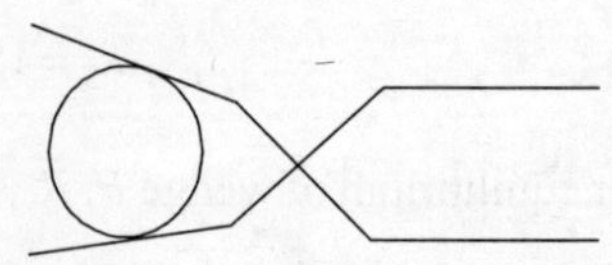

SOLUTION Considering the equilibrium of the tube and resolving forces about horizontal axis, we obtain,

$$N_A \sin 10^\circ + N_B \sin 10^\circ = \mu N_A \cos 10^\circ + \mu N_B \cos 10^\circ$$

$$\Rightarrow \quad \mu \cos 10^\circ (N_A + N_B) = \sin 10^\circ (N_A + N_B)$$

$$\Rightarrow \quad \mu = \tan 10^\circ = 0.1763$$

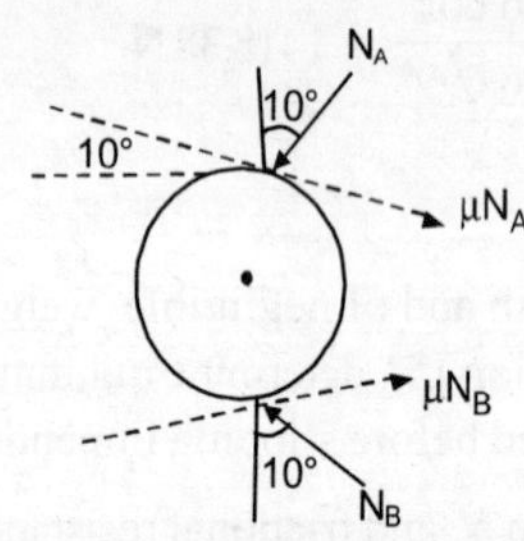

FBD of tube

EXAMPLE 21

A smooth circular cylinder of weight Q and radius r is supported by two semicircular cylinders each of the same radius r and weight $Q/2$ as shown. Find maximum value of distance b for which motion will impend. Take $\mu = \frac{1}{2}$ for horizontal plane.

SOLUTION Considering equilibrium of cylinder A and resolving forces along horizontal and vertical directions,

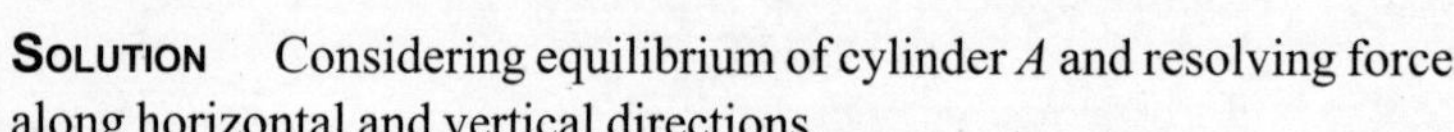

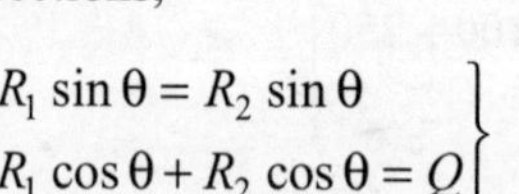

$$\left. \begin{aligned} R_1 \sin\theta &= R_2 \sin\theta \\ R_1 \cos\theta + R_2 \cos\theta &= Q \end{aligned} \right\}$$

Solving these, $$R_1 = R_2 = \frac{Q}{2\cos\theta}$$

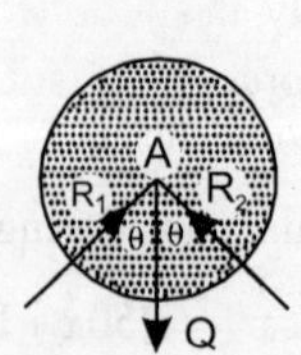

FBD of top cylinder

Considering equilibrium of half cylinder B or C and resolving vertically and horizontally,

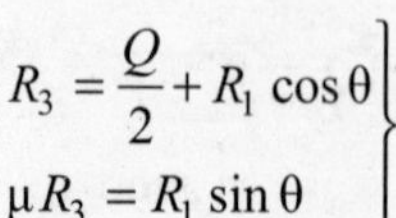

$$\left. \begin{aligned} R_3 &= \frac{Q}{2} + R_1 \cos\theta \\ \mu R_3 &= R_1 \sin\theta \end{aligned} \right\}$$

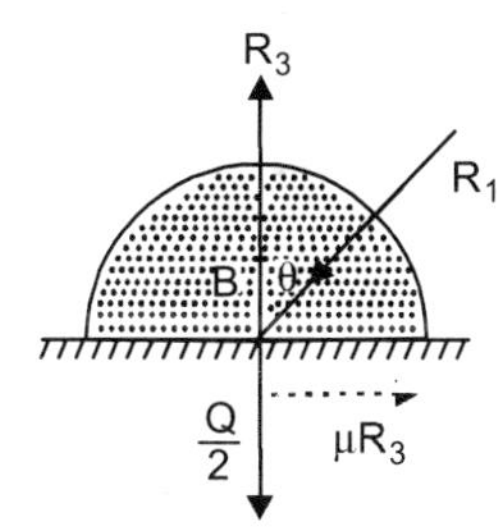

FBD of cylinder

$$\Rightarrow \quad R_3 = \frac{Q}{2} + \frac{Q}{2\cos\theta}\cos\theta$$

$$\Rightarrow \quad \frac{1}{2}R_3 = \frac{Q}{2\cos\theta}\sin\theta$$

Solving these, $\quad R_3 = Q$ and $\tan\theta = 1$

From geometry, $\quad \tan\theta = \dfrac{b/2}{x}$ and

$$x = \sqrt{(2r)^2 - (b/2)^2} = \frac{1}{2}\sqrt{16r^2 - b^2}$$

Comparing these two expressions of $\tan\theta$

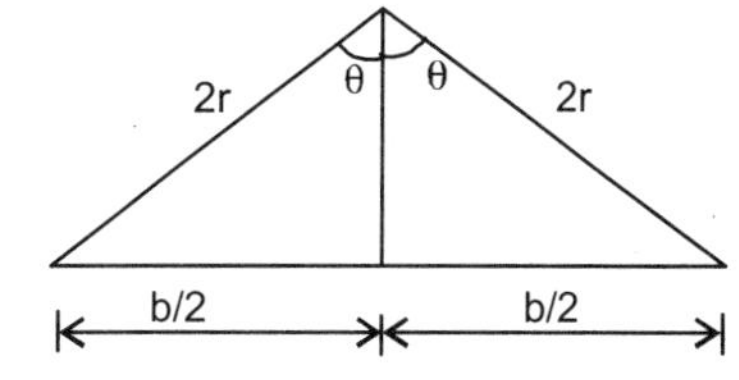

$$\frac{b}{2} = x = \frac{1}{2}\sqrt{16r^2 - b^2}$$

$$\Rightarrow \quad b = \sqrt{16r^2 - b^2}$$

$$\Rightarrow \quad 2b^2 = 16r^2$$

$$\Rightarrow \quad b = 2.828r$$

EXAMPLE 22

A solid right circular cone of altitude h =12 cm and radius of base r = 3cm has its center of gravity C on its geometric axis at the distance $\dfrac{h}{4} = 3$ cm above the base. This cone rests on an inclined plane AB, which makes an angle 30° with the horizontal and for which the coefficient of friction is $\mu = 0.5$. A horizontal force P is applied to the vertex O of the cone and acts in the vertical plane of the figure as shown. Find the maximum and minimum values of P consistent with equilibrium of the cone, if the weight W = 10N.

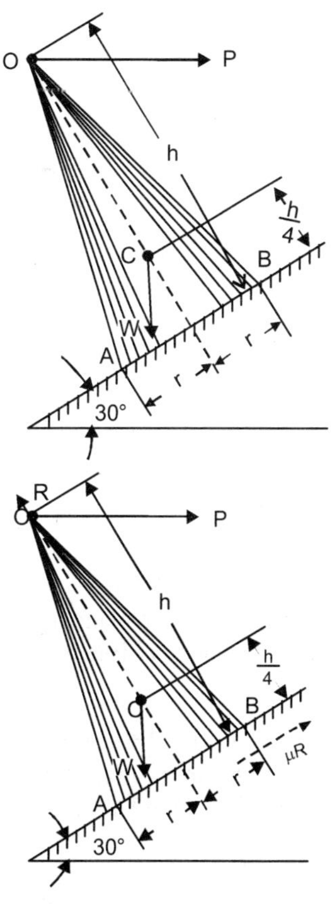

SOLUTION Movement of cone on inclined plane are of two types. Firstly, it may *slide down* along the incline, which is resisted by a horizontal force P, say P_1. Secondly, the cone may *overturn*, which is resisted by horizontal P, say P_2. Careful observation suggests that, force required to stop sliding (P_1) will be pretty lower than force required to stop overturning (P_2). So, P_1 = minimum permissible value of P and P_2 = maximum permissible value of P. Firstly, in sliding down case, considering the equilibrium of cone and resolving forces along incline and its normal direction,

$$\mu R + P_1 \sin 60° = W \sin 30°$$

$$P_1 \cos 60° + W \cos 30° = R$$

or

$$\Rightarrow \quad 0.5R + P_1\frac{\sqrt{3}}{2} = 10\times\frac{1}{2}$$

$$\Rightarrow \quad P_1 \times \frac{1}{2} + 10\times\frac{\sqrt{3}}{2} = R$$

Solving above, $P_1 = 0.6\text{N} = P_{MIN}$.

Secondly, in overturning case, summation of moments of forces about point B is zero.

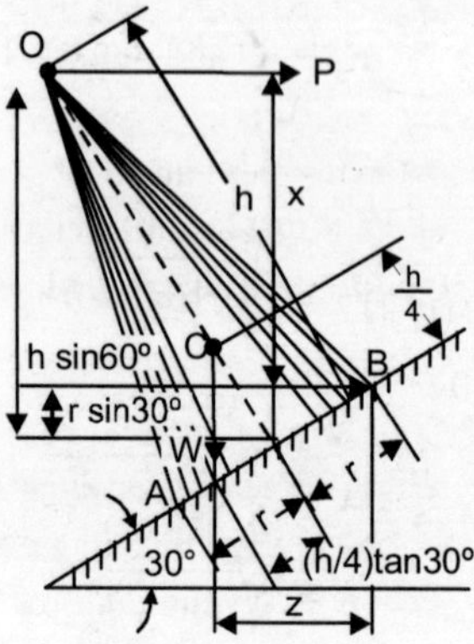

From geometry,

$$x = h\sin 60° - r\sin 30° = \left(6\sqrt{3} - 1.5\right)\text{cm}$$

$$z = r\cos 30° + \frac{h}{4}\cos 60° = \left(1.5\sqrt{3} + 1.5\right)\text{cm}$$

$$\therefore \quad P_2 \times x = W \times z$$

$$\Rightarrow \quad P_2\left(6\sqrt{3} - 1.5\right) = 10\left(1.5\sqrt{3} + 1.5\right)$$

$$\Rightarrow \quad P_2 = 4.608\text{N} = P_{\max}$$

Exercise

1. A pull of 60N inclined at 25° to the horizontal plane, is required just to move body placed on a rough horizontal plane. But the push required to move the body is 75N. If the push is inclined at 25° to the horizontal, find the weight of the body and coefficient of friction.

   Ans. [253. 83N, 0.238]

2. A body of weight 400N is pulled up along an inclined plane having inclination 30° to the horizontal at a steady speed. If the coefficient of friction between the body and the plane is 0.3 and force is applied parallel to the inclined plane, find the force required. Find also the work done on the body if the distance travelled by the body is 10m along the plane.

   Ans. [303.92N, 3039.2N·m]

3. An effort of 180N is required just to move a certain body up an inclined plane of angle 15° the force being parallel to the plane. If the angle of the plane is made 20°, the effort required, again applied parallel to the plane, is found to be 210N. Find the weight of the body and coefficient of friction.

   Ans. [386. 68N, 0.214]

4. A cord connects two bodies of weights 500N and 1000N. The two bodies are placed on an inclined plane and cord is parallel to inclined plane. The coefficients of friction for the weight of 500N is 0.20 and that of 1000N is 0.4. Determine the inclination of the plane to the horizontal and tension in the cord when the motion is about to take place, down the inclined plane. The body weight 500N is below the body weighing 1000N.

Ans.[18.417°, 63.086 N]

5. A uniform ladder of length 13m and weighing 30N is placed against a smooth vertical wall with its lower end 10m from the wall. In this position the ladder is just to slip. Determine: (*i*) the coefficient of friction between the ladder and the floor, and (*ii*) frictional force acting on the ladder at the point of contact between ladder and floor.

[**Ans.** (*i*) 0.602, (*ii*) 18.06N]

6. A uniform ladder of length 10m and weighing 20N is placed against a smooth vertical wall with its lower end 6m from the wall. The coefficient of friction between the ladder and floor is 0.45. Show that the ladder will remain in equilibrium in this position. What is the frictional force acting on the ladder at the point of contact between the ladder and floor.

Ans. [7.5N]

7. A uniform ladder of weight 800N and of length 7m rests on a horizontal ground and leans against a smooth vertical wall. The angle made by the ladder with the horizontal is 60°. When a man of weight 600N stands on the ladder at a distance 4m from the top of the ladder, the ladder is at the point of sliding. Determine the coefficient of friction between the ladder and the floor.

Ans. [0.237]

8. A uniform ladder of weight 250N and of length 5m rests on a horizontal ground and leans against a rough vertical wall. The coefficient of friction between the ladder and floor is 0.3 and between the ladder and vertical wall is 0.2. When a weight of 900N is placed on the ladder at a distance of 2m from the top of the ladder; the ladder is at the point of sliding. Find (*i*) The angle made by the ladder with horizontal (*ii*) Reaction at the foot of the ladder, and (*iii*) Reaction at the top of the ladder.

Ans. [(*i*) 61.52°, (*ii*) 1084.9N, (*iii*) 325.47N]

9. A uniform ladder of length 15m rests against a vertical wall making an angle of 60° with the horizontal. Coefficient of friction between wall and ladder, and ground and ladder are 0.3 and 0.25. A man weighing 500N ascends the ladder. How high will he be able to go before the ladder slips? Find the weight necessary to be put at the bottom of the ladder so as to be just sufficient to permit the man to go to top. Assume weight of ladder as 850N.

Ans. [6.39m, 607.6N]

10. A uniform ladder 3m long weighs 18N. It is placed against a wall making an angle 60° with the floor. The coefficient of friction between the wall and ladder is 0.25 and that between the floor and ladder is 0.35. The ladder, in addition to its own weight, has to support a man weighing 90N at its top at *A*. Calculate: (*i*) The horizontal force *F* to be applied to the ladder at the floor level to prevent slipping. (*ii*) If the force *F* is not applied what would be the minimum inclination of the ladder with the horizontal, so that there is no slipping of it with the man at the top?

11. A uniform rod rests in limiting equilibrium within a rough vertical circle. If the rod subtends an angle 2a at the centre of the circle and if ϕ is the angle of limiting friction show that the inclination of the rod to the horizontal is $\tan^{-1}\left[\dfrac{\sin 2\phi}{\cos 2\phi + \cos 2\alpha}\right]$.

12. A uniform rod rests in a vertical plane within a fixed hemispherical bowl whose radius is equal to the length of the rod. Show that in limiting equilibrium the inclination of the rod to the horizontal is $\tan^{-1}\left(\dfrac{4\pi}{3-\pi^2}\right)$ where μ is the coefficient of friction.

13. If the ratio of the greatest to the least force which acting parallel to a rough inclined plane can support a weight on it be equal to the ratio of the weight to the normal reaction show that the coefficient of friction is equal to $\left(\tan\alpha \tan^2\dfrac{\alpha}{2}\right)$ where α is the angle of the plane with the horizontal.

14. A uniform ladder of length *l* rests on a rough horizontal ground with its upper end projecting slightly over a smooth rail at a height *a*. If the ladder is about to slip down and μ be the coefficient of friction, show that

$$\mu = \frac{a\sqrt{l^2 - a^2}}{l^2 + a^2}.$$

15. A semi-circular disc rests in a vertical plane with its curved surface on a rough horizontal plane and against an equally rough vertical plane, the coefficient of friction being μ. Show that the greatest angle that the bounding diameter can make with the horizontal is $\sin^{-1}\left(\dfrac{3\pi}{4}\dfrac{\mu+\mu^2}{1+\mu^2}\right)$.

16. A solid hemisphere rests in equilibrium on a rough ground and against an equally rough vertical wall. Show that if the equilibrium is limiting, the inclination q of the base to the horizontal is $\sin^{-1}\left(\dfrac{8}{3}\dfrac{\phi(1+\phi)}{1+\mu^2}\right)$.

17. A uniform thin hemispherical bowl rests with its curved surface on a rough horizontal plane whose coefficient of friction is μ. Prove that when the bowl is at the point of slipping the inclination of the axis of the bowl with the vertical is $\sin^{-1}(2\mu)$.

18. A hemispherical shell rests on a rough inclined plane whose angle of friction is ϕ. Show that the inclination of the plane base of the rim with the horizontal cannot be greater than $\sin^{-1}(\sin\phi)$.

19. A uniform hemisphere of radius *a* and weight W_1 rests with its spherical surface on a horizontal plane and a rough particle of weight W_2 rests on the plane face. Show that the distance of the particle from the centre of the plane face is not greater than $\dfrac{3\mu a}{8}\left(\dfrac{W_1}{W_2}\right)$ where μ is the coefficient of friction.

PROPERTIES OF SURFACE AND PHYSICAL BODIES

4.1 INTRODUCTION

In this chapter, we are going to have some ideas regarding certain aspects of plane figure as well as of solid bodies. There are some mathematical expressions that convey the meaning about shape and disposition of a surface or body relative to some references. These are explained in the following articles.

4.2 CENTRE OF GRAVITY, CENTROID AND CENTRE OF MASS

Any material body can be of three types, namely (*i*) one-dimensional, measurable only along length. (*ii*) two-dimensional having length and width (*iii*) three-dimensional having length, width and thickness. If all these above bodies are considered *prismatic* in geometry and having material *homogeneous*, the magnitude of the gravity force for each prismatic element will be *proportional to length* for one-dimensional body while *cross-section is uniform and proportional to cross-sectional area* while *thickness is uniform*.

The centre of gravity of a material body is that point through which the resultant of the distributed gravity force passes, irrespective of the orientation of the body in space. So, *centre of gravity is a term which is applicable only for bodies having the property of weight.*

Instead of material bodies, if we consider geometric figures, like a plane were or a plane figure, the above concept can be applied likewise and term aptly used is centriod instead of centre of gravity. Hence, *centroid is the term applicable only for any plane figure or curve in plane which does not possess the property of weight.*

Centre of mass is a point where the entire mass of any body may be assumed to remain concentrated. *Centre of mass* and *centre of gravity* are identical points in place where the intensity of gravitational field remains uniform. For a very large object and objects where gravitational acceleration accepts the effect of height or the other cases where the field of gravitation is not uniform, centre of mass vary with centre of gravity.

4.3 ANALYTICAL EXPRESSION OF CENTROID

To determine analytically, the position of centroid of any geometrical figure or a curve in plane, let us assume reference co-ordinate axes X and Y. If a very small length 'ΔL_i' along the length of the curve is assumed, as shown in Fig. 4.1 (*a*), we determine analytically the centroid as,

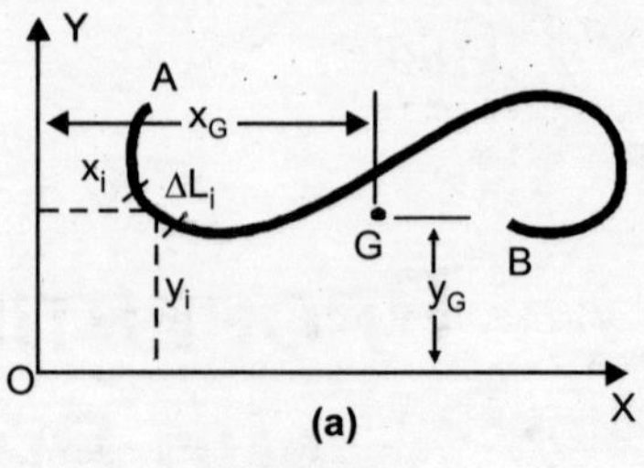

(a)

Fig. 4.1

$$x_G = \frac{\sum_{i=1}^{n} \Delta L_i x_i}{\sum_{i=1}^{n} \Delta L_i} \text{ and } y_G = \frac{\sum_{i=1}^{n} \Delta L_i y_i}{\sum_{i=1}^{n} \Delta L_i} \quad \text{....(4.1)}$$

Similarly a very small area ΔA_i is assumed within the plane figure, as shown Fig. 4.1 (*b*). So, the centroid of the plane area will be

$$x_G = \frac{\sum_{i=1}^{n} \Delta A_i x_i}{\sum_{i=1}^{n} \Delta A_i} \text{ and } y_G = \frac{\sum_{i=1}^{n} \Delta A_i y_i}{\sum_{i=1}^{n} \Delta A_i} \quad \text{....(4.2)}$$

Within limits, ΔL_i and ΔA_i reduce to infinitesimal in dimension and then the exact expressions for centroid will be

$$x_G = \frac{\int_L x dL}{\int_L dL} \text{ and } y_G = \frac{\int_L y dL}{\int_L dL} \quad \text{... (4.3)}$$

$$x_G = \frac{\int_A x dA}{\int_A dA} \text{ and } y_G = \frac{\int_A y dA}{\int_A dA} \quad \text{... (4.4)}$$

If a rectangular co-ordinate system is assumed through centroid for each of the figure, the axes are designated as *centroidal axes* if centroid is defined as centriodal co-ordinates.

In equation (4.4) the term are often called as *first moment of area* about *X*-axis or *Y*-axis respectively.

When an area of a definite shape has *substantial thickness,* it is termed as a *solid.* To determine centre of gravity of any solid, followings relationships are used.

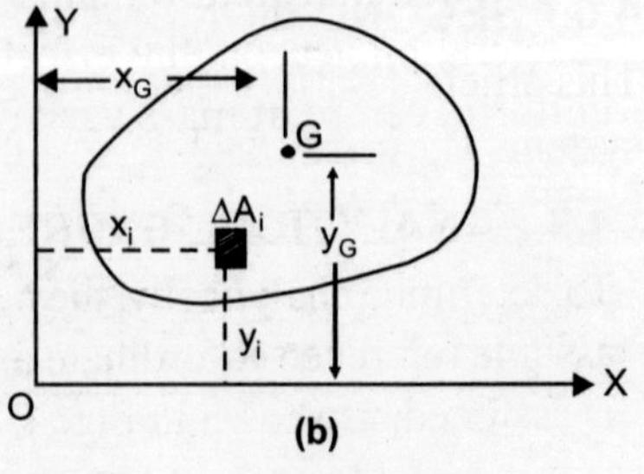

(b)

Fig. 4.1

$$x_G = \frac{\int\limits_W x dW}{\int\limits_W dW} \text{ and } y_G = \frac{\int\limits_W y dW}{\int\limits_W dW} \qquad ...(4.5)$$

Here dW stands for weight of differential solid element.

4.4 PAPPUS-GULDINUS THEOREMS

These theorems, first proposed by Pappus about 300 A.D. and restated by Guldinus about 1640 A.D. relate to the revolution of a curve or area is plane, about a non-intersecting axis. The theorems are:

(*i*) *The area of the surface generated by the revolution of any plane curve about a non-intersecting axis in its plane is equal to the product of the length of the curve and the distance travelled by its centroid.*

(*ii*) *The volume of the solid generated by the revolution of any plane surface about a non-intersecting axis in its plane is equal to the product of the area of the surface and the distance travelled by its centroid.*

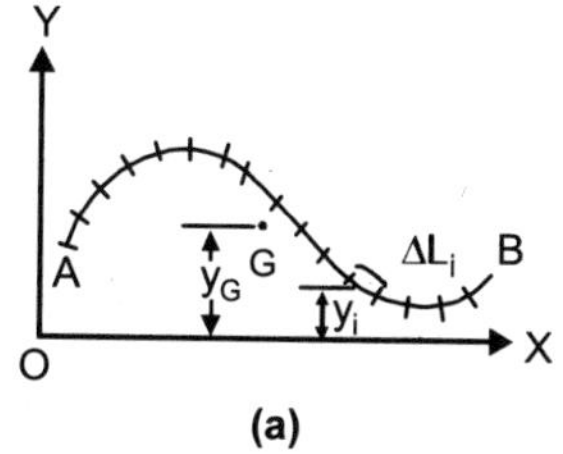

(a)

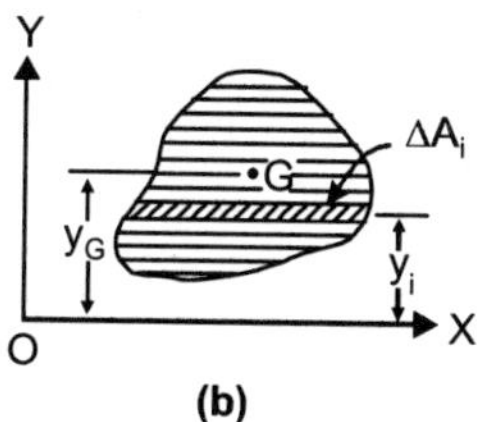

(b)

Fig. 4.2

In Fig. 4.2 (*a*), the plane curve of length L can revolve about X-axis which does not intersect the length of the curve. As, the first theorem states, the area generated by ΔL_i will be,

$$\Delta A_i = \Delta L_i \times 2\pi y_i$$

and within limit, $dA = 2\pi y dL$

So, $$A = \int dA = 2\pi \int y dL = (2\pi y_G)L \qquad ...(4.6)$$

Similarly, for second theorem also, the relation obtained as,

$$\forall = \int d\forall = 2\pi \int y dA = (2\pi y_G)A \qquad ...(4.7)$$

4.5 SECOND MOMENT OF PLANE AREA

This article describes other properties of a plane figure with reference to some reference co-ordinate system. The *second moment* of the area A in Fig. 4.3 about reference X and Y axes respectively are

$$I_x = \int\limits_A y^2 dA \text{ and } I_y = \int\limits_A x^2 dA \qquad ... (4.8)$$

A term *moment of inertia* is found to be used synonymously with *second moment of area.* The term *inertia* has some relation with gravitational acceleration and thus applicable in material bodies, not in geometrical figures. Hence, the use of the term moment of inertia in case of a plane area is not appropriate in the true sense of the term, though used commonly and more frequently.

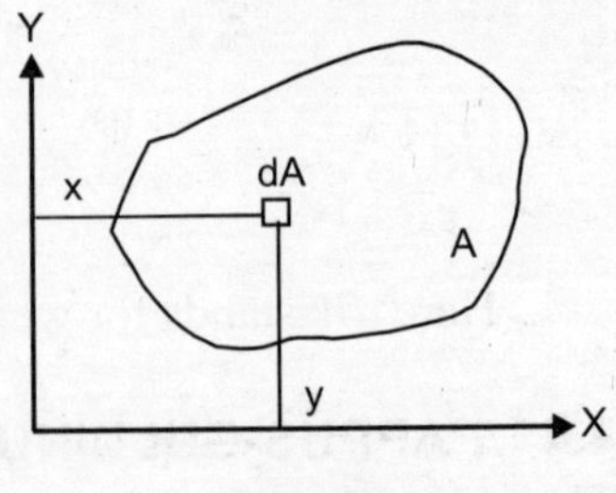

Fig. 4.3

Another mathematical parameter in this context is *radius of gyration,* which possesses values both with reference to X and Y axis, those are

$$r_x = \sqrt{\frac{I_x}{A}} \text{ and } r_y = \sqrt{\frac{I_y}{A}} \qquad \text{... (4.9)}$$

Radius of gyration depends not only on shape of the area but also on the position of reference.

4.6 POLAR MOMENT OF AREA OR PERPENDICULAR AXIS THEOREM

In Fig. 4.4, let us imagine through point O an axis perpendicular to reference X–Y axis is chosen. Now as the distance of dA from orthogonal axis is r, the second moment of area about orthogonal axis,

$$J = \int_A r^2 dA = \int_A (x^2 + y^2) dA = \int_A x^2 dA + \int_A y^2 dA = I_y + I_x \qquad \text{... (4.10)}$$

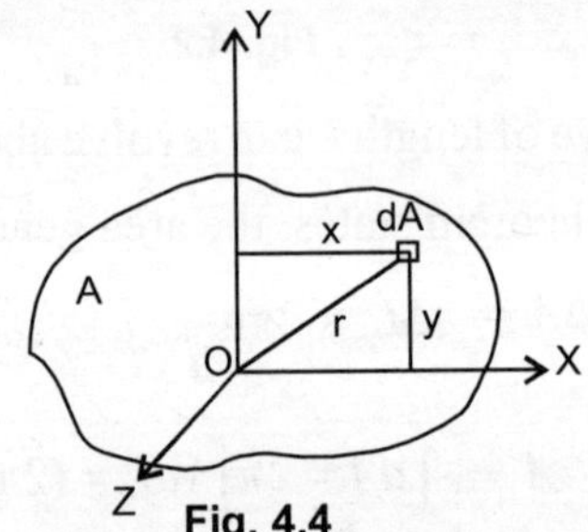

Fig. 4.4

This parameter J is defined as polar moment of area. Sometimes, J_p is synonymously used with J. Hence, we can state that *sum of the pairs of second moment of area is a constant for all such references at a point.* This statement is often described as the statement of perpendicular axis theorem of second moment of area.

4.7 PARALLEL AXIS THEOREM

In Fig. 4.5, X–Y be the rectangular co-ordinate axes through any point O in the plane of the area and x–y the centroidal axis of the area while X is parallel to x and Y is parallel to y. Now by definition, we can write,

$$I_x = \int_A (y+b)^2 dA = \int_A y^2 dA + 2b \int_A y\, dA + b^2 \int_A dA = I_{Gx} + Ab^2 \qquad \text{...(4.11}a\text{)}$$

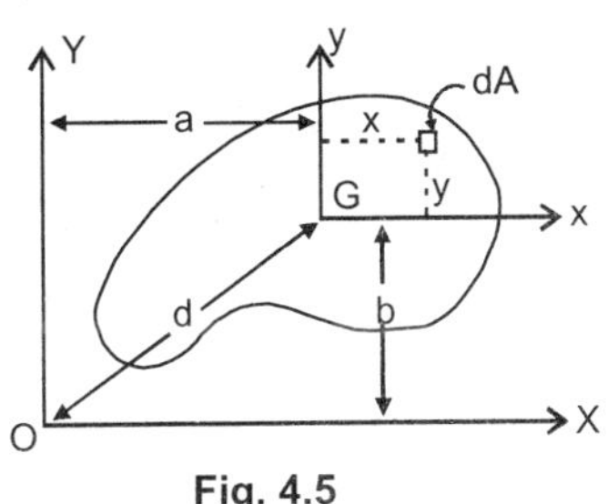

Fig. 4.5

Here x being centroidal axis, $\int_A y dA = 0$, since the x axis passes through the centroid G of the area. I_{Gx} denotes the second moment of area about centroidal x axis. Similarly it can be written that,

$$I_y = I_{Gy} + Aa^2 \quad ...(4.11b)$$

These two equations are considered as the expression for parallel axis theorem. The theorem states that, *the second moment of area of any geometric figure with respect to any axis in its plane is equal to the second moment of area with respect to a parallel centroidal axis plus the product of the total area and the square of the distance between the two parallel axes, one of which must be a centroidal axis.*

4.8 MASS MOMENT OF INERTIA

Let us assume a small mass Δm which can rotate freely about any axis AA'. Now if r be perpendicular distance of Δm from AA', first moment of Δm will be $r\,\Delta m$. Subsequently second moment of Δm will be $r(r\,\Delta m) = r^2\,\Delta m$. Eventually second moment of mass is defined as mass moment of inertia.

So, $$\Delta I = r^2 \Delta m \quad ...(4.12)$$

Now for a rigid body, if the total mass (m) is assumed to be a combination of such infinitesimally small masses, then the mass moment of inertia of a rigid body,

$$I = \int dI = \int r^2 dm \quad ... (4.13)$$

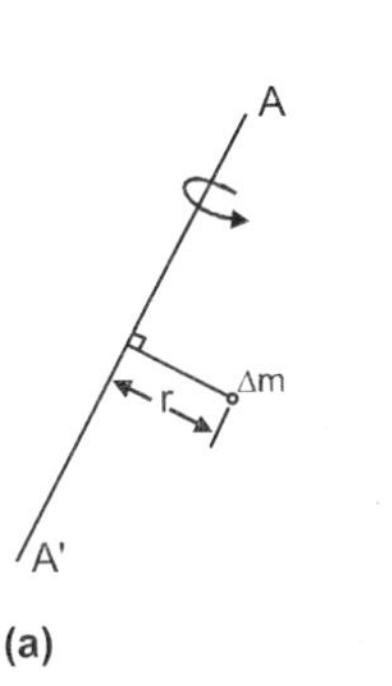

(a)

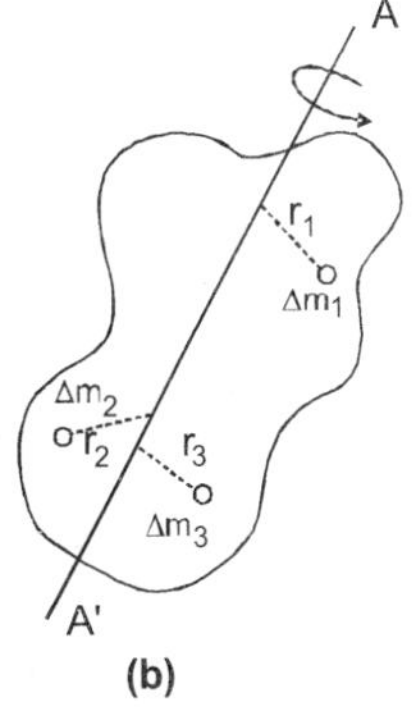

(b)

Fig. 4.6

The radius of gyration (k) of the rigid body,

$$k = \sqrt{\frac{I}{m}} \qquad \text{... (4.14)}$$

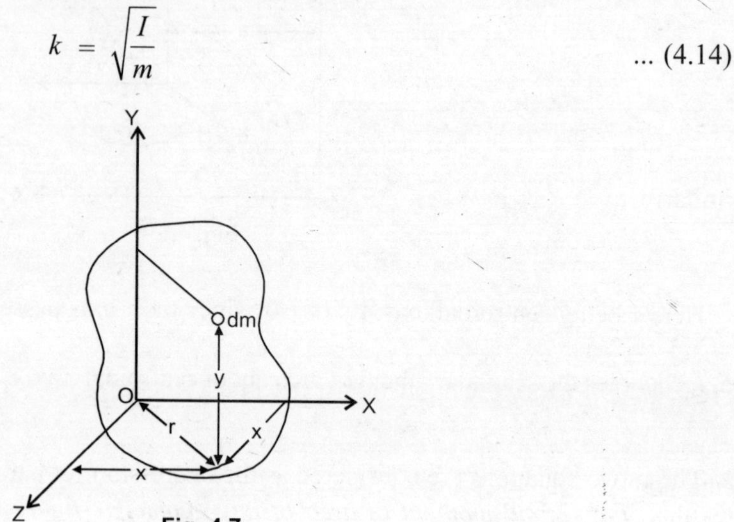

Fig. 4.7

To be more precise, in a 3D-rectangular axes system, as shown in Figure 4.7, the mass moment of inertia will be

$$\left.\begin{aligned} I_x &= \int (y^2 + z^2)\,dm \\ I_y &= \int (x^2 + z^2)\,dm \\ I_z &= \int (x^2 + y^2)\,dm \end{aligned}\right\} \qquad \text{... (4.15)}$$

Subsequently, we also find radii of gyration about three different axes, as

$$r_x = \sqrt{\frac{I_x}{m}},\; r_y = \sqrt{\frac{I_y}{m}},\; r_z = \sqrt{\frac{I_z}{m}} \qquad \text{... (4.16)}$$

4.9 PARALLEL AXIS THEOREM FOR MASS MOMENT OF INERTIA

Referring to Fig. 4.8, the body has centroid at G with centroidal axes system $Gx'y'z'$. It has an arbitrary axes system $Oxyz$ having origin at O. Now applying parallel axis theorem for x axis,

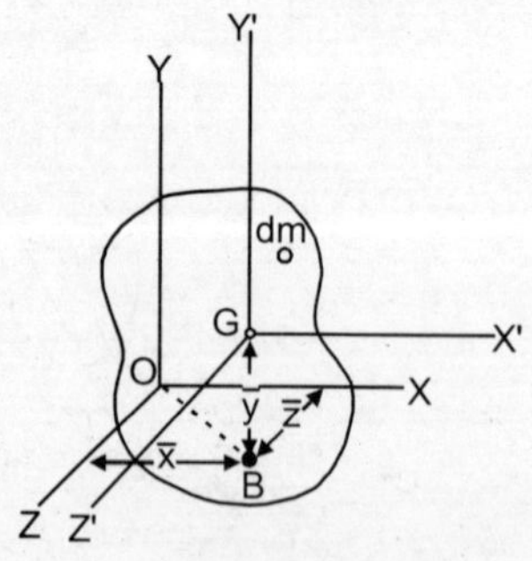

Fig. 4.8

$$I_x = \int (y^2 + z^2) dm = \int \left[(y' + \bar{y})^2 + (z' + \bar{z})^2 \right] dm$$

$$= \int \left[(y')^2 + (z')^2 \right] dm + 2\bar{y} \int y' dm + 2\bar{z} \int z' dm + \int \left[(\bar{y})^2 + (\bar{z})^2 \right] dm$$

$$= I_{x'} + 0 + 0 + m\left((\bar{y})^2 + (\bar{z})^2 \right) = I_{x'} + m\left((\bar{y})^2 + (\bar{z})^2 \right) \quad \text{...(4.17}a\text{)}$$

Similarly,

$$I_y = I_{y'} + m\left[(\bar{x})^2 + (\bar{z})^2 \right] \quad \text{(4.17}b\text{)}$$

$$I_z = I_{z'} + m\left[(\bar{x})^2 + (\bar{y})^2 \right] \quad \text{... (4.17}c\text{)}$$

4.10 MASS MOMENT OF INERTIA OF THIN PLATES

Let us consider a thin plate of thickness t, has centre of gravity at C. AA' and BB' are two perpendicular centroidal axes. Now infinitesimally small mass dm can be interpreted as, $dm = \rho \cdot t \cdot dA$, where mass density of the material is ρ.

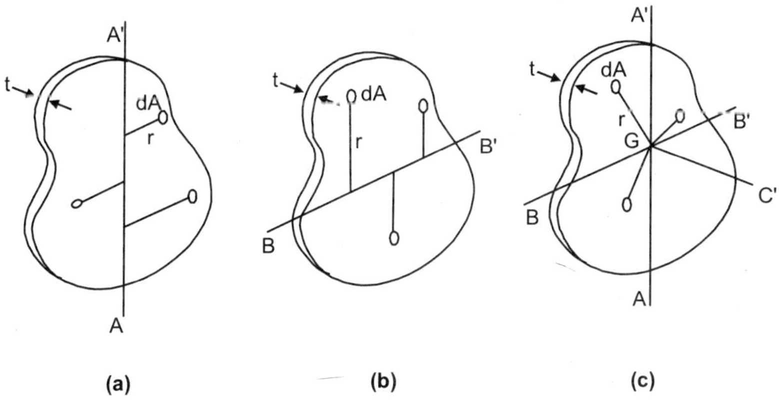

Fig. 4.9

Now, $$I_{AA', \text{mass}} = \int r^2 dm = \rho t \int r^2 dA = \rho t \cdot I_{AA', \text{area}} \quad \text{... (4.18}a\text{)}$$

Similarly, $$I_{BB', \text{mass}} = \rho t \cdot I_{BB', \text{area}} \quad \text{... (4.18}b\text{)}$$

$$I_{CC', \text{mass}} = \rho t \cdot I_{CC', \text{area}} = \rho t \left(I_{AA', \text{area}} + I_{BB', \text{area}} \right)$$

$\therefore$ $$I_{CC', \text{mass}} = I_{AA', \text{mass}} + I_{BB', \text{mass}} \quad \text{... (4.18}c\text{)}$$

Multiple Choice Questions

Select the best alternative (s):

1. The centroid of a uniform lamina lies at
   (*a*) the centre of heavy portion
   (*b*) the bottom surface
   (*c*) the mid point of its axis
   (*d*) all of the above
   (*e*) none of the above
2. Centre of gravity of a solid cone lies on the axis at the height
   (*a*) one-fourth of the total height above base
   (*b*) one-third of the total height above base
   (*c*) one-half of the total height above base
   (*d*) three-eighth of the total height above the base
   (*e*) none of the above
3. M.I of a thin circular ring of radius r and mass M about an axis perpendicular to plane of ring is
   (*a*) Mr^2
   (*b*) $\pi r^4/2$
   (*c*) $2/5\,Mr^2$
   (*d*) $2/3\,Mr^2$
   (*e*) $Mr^2/2$
4. The C.G of a solid hemisphere lies on the central radius
   (*a*) at distance $3r/2$ from the plane base
   (*b*) at distance $3r/4$ from the plane base
   (*c*) at distance $3r/5$ from the plane base
   (*d*) at distance $3r/8$ from the plane base
   (*e*) at distance $r/2$ from the plane base
5. The centriod of a trapezium of base b *height* h and upper side a lies at following distance from the base.
   (*a*) $(h/3)\{(2a+b)/(a+b)\}$
   (*b*) $(h/3)\{(a+b)/(2a+b)\}$
   (*c*) $(h/3)\{(a+2b)/(a+b)\}$
   (*d*) $(h/2)\{(2a+b)/(a+b)\}$
   (*e*) $(h/4)\{(2a+b)/(a+b)\}$
6. The centroid of an isosceles triangle with base a and other sides b lies at following distance from the base.
   (*a*) $\sqrt{(4a^2-b^2)}/6$
   (*b*) $\sqrt{(a^2-b^2)}/6$
   (*c*) $\sqrt{(2a^2-b^2)}/6$
   (*d*) $\sqrt{(a^2-2b^2)}/6$
   (*e*) $\sqrt{(4a^2-b^2)}/3$
7. According to theorem of perpendicular axes, if I_{xx} and I_{yy} be the M.I of a lamina about xx and yy axes, then M.I of the lamina about axis zz which is perpendicular to xx and yy equal to
   (*a*) $I_{xx}+I_{yy}$
   (*b*) $I_{xx}\times I_{yy}$
   (*c*) I_{xx}/I_{yy}
   (*d*) I_{yy}/I_{xx}
   (*e*) $\sqrt{I_{yy}^2+I_{xx}^2}$
8. The centroid of a plane lamina will not be at its geometrical centre in case of a
   (*a*) right angled triangle
   (*b*) equilateral triangle
   (*c*) square
   (*d*) circle
   (*e*) rectangle
9. M.I of a rectangular area of base b and height d about x axis at base is given by
   (*a*) $bd^3/3$
   (*b*) $bd^3/4$
   (*c*) $bd^3/6$
   (*d*) $bd^3/12$
   (*e*) $bd^3/8$

10. The C.G of a right circular solid cone of height h lies at the following distance from the base.

(*a*) $h/2$ (*b*) $h/3$ (*c*) $h/6$
(*d*) $h/4$ (*e*) $3h/5$

11. M.I of circular area whose diameter is d about an axis perpendicular to the area passing through its centre is given by

(*a*) $\pi d^4/64$ (*b*) $\pi d^4/32$ (*c*) $\pi d^4/12$
(*d*) $\pi d^4/16$ (*e*) $\pi d^4/24$

12. M.I of a hollow circular section about a central axis perpendicular to the section as compared to its M.I about horizontal axis is

(*a*) same (*b*) double (*c*) half
(*d*) four times (*e*) one-fourth

13. M.I of a triangle of base a and height h about the base is given by

(*a*) $ah^3/6$ (*b*) $ah^3/12$ (*c*) $ah^2/6$
(*d*) $ah^2/12$ (*e*) $ah^3/3$

14. If I_x and I_y be the moment of inertia about any two axes at right angles to each other in the plane of the area and intersecting at the pole, then the polar moment of inertia I_P will be

(*a*) $I_x + I_y$ (*b*) $(I_x + I_y)/2$ (*c*) $I_x \times I_y$
(*d*) $(I_x + I_y)^{1/2}$ (*e*) $\int I_x \times I_y$

15. M.I of a rectangular area of base b and d about centroidal x-axis is given by

(*a*) $bd^3/3$ (*b*) $bd^3/4$ (*c*) $bd^3/6$
(*d*) $bd^3/12$ (*e*) $bd^3/8$

Answers

1. (*c*)	2. (*a*)	3. (*a*)	4. (*d*)
5. (*a*)	6. (*a*)	7. (*a*)	8. (*a*)
9. (*a*)	10. (*d*)	11. (*b*)	12. (*b*)
13. (*b*)	14. (*a*)	15. (*d*)	

Numerical Examples

EXAMPLE 1

Find the center of gravity of a thin uniform rod.

SOLUTION The figure shows a thin uniform rod legth L whose weight per unit length w is the same throughout, because the rod is uniform. Centre of gravity of the rod G and $\bar{x}$ = distance of G from the end O, an arbitrarily chosen origin. Let us consider a small length dx at a distance x from O. From definition of C.G

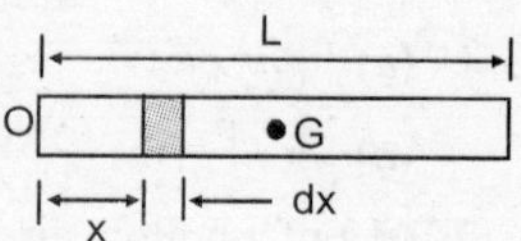

$$x_G = \frac{\int_0^L x\,dx}{\int_0^L dx} = \frac{\left[x^2/2\right]_0^L}{\left[x\right]_0^L} = \frac{L}{2}$$

Hence, C.G of a uniform rod lies at its middle point.

EXAMPLE 2

Determine the centroid of a quadrant AB of a circle of radius a as shown in figure.

SOLUTION (*a*) *Method -I* : The curve AB which is an arc of a circle of radius a and considering an element of length dL. The equation of curve AB is circle and is given by $x^2 + y^2 = a^2$. Differentiating this equation we have,

$$2xdx + 2ydy = 0,$$

$$\Rightarrow \qquad 2ydy = -2xdx$$

$$\Rightarrow \qquad dy = \frac{-xdx}{y}$$

Consider the centroid of length dL is at a distance x and y from Y axis and X axis respectively. Now using equation (4.3) for y_G, we get,

$$y_G = \frac{\int y \cdot dL}{\int dL}$$

Again

$$dL = \sqrt{dx^2 + dy^2} = \sqrt{dx^2 + \left(\frac{-xdx}{y}\right)^2} = dx\sqrt{\frac{y^2 + x^2}{y^2}} = \frac{a}{y} \cdot dx$$

Substituting the value of dL

$$\therefore \qquad y_G = \frac{\int_0^a y \cdot (a/y)dx}{\int dL} = \frac{a\left[x\right]_0^a}{(2\pi a/4)} = \frac{2a}{\pi}$$

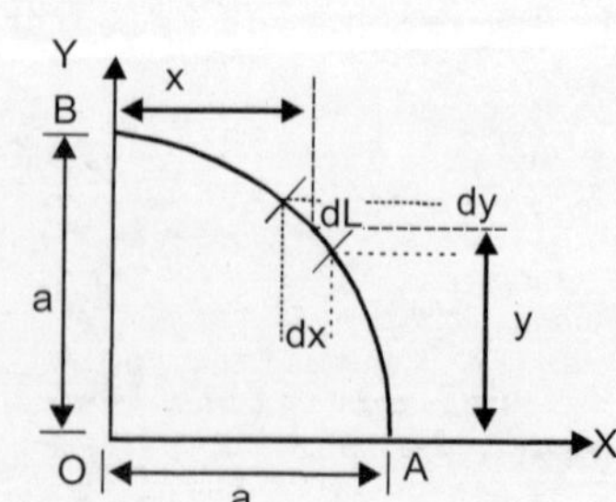

From symmetry of the figure about X-axis and Y-axis,

$$x_G = y_G = \frac{2a}{\pi}$$

(*b*) *Method - II*

Here $\qquad dL = a\,d\theta,\ y = a\sin\theta,\ x = a\cos\theta$

Now,

$$y_G = \frac{\int y\,dL}{\int dL} = \frac{\int_0^{\pi/2} a\sin\theta \cdot a d\theta}{\int_0^{\pi/2} a\,d\theta} = \frac{a^2[-\cos\theta]_0^{\pi/2}}{a[\theta]_0^{\pi/2}} = \frac{2a}{\pi}$$

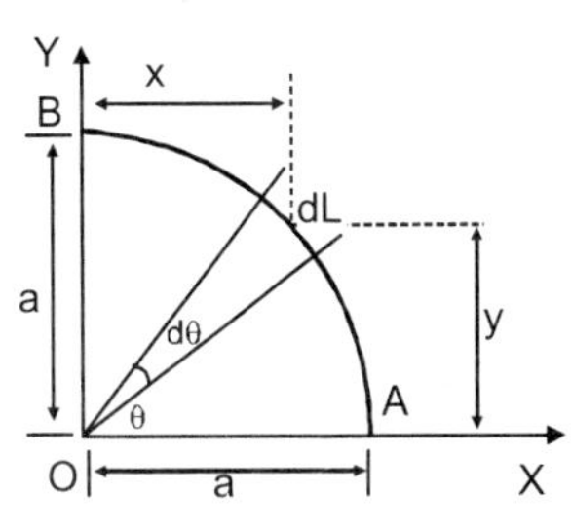

And

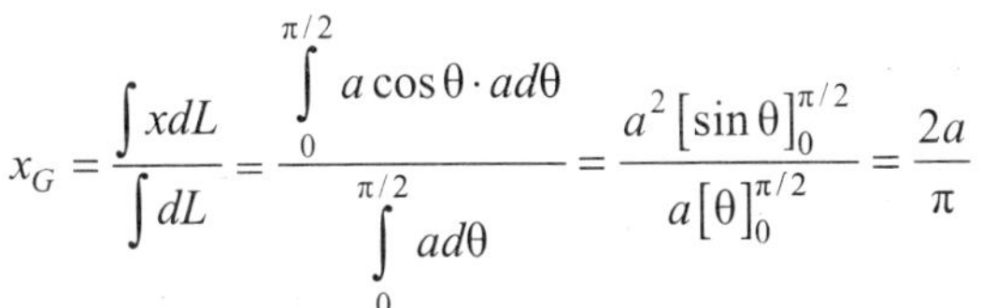

$$x_G = \frac{\int x\,dL}{\int dL} = \frac{\int_0^{\pi/2} a\cos\theta \cdot a d\theta}{\int_0^{\pi/2} a d\theta} = \frac{a^2[\sin\theta]_0^{\pi/2}}{a[\theta]_0^{\pi/2}} = \frac{2a}{\pi}$$

EXAMPLE 3

Determine the centroid of a uniform triangular lamina.

SOLUTION Consider a triangle of base b and height h. Base coincides with x-axis. Let us consider an elementary strip of thickness dy from a distance of y from the base. Let l = length of this strip and area of this strip $dA = ldy$.

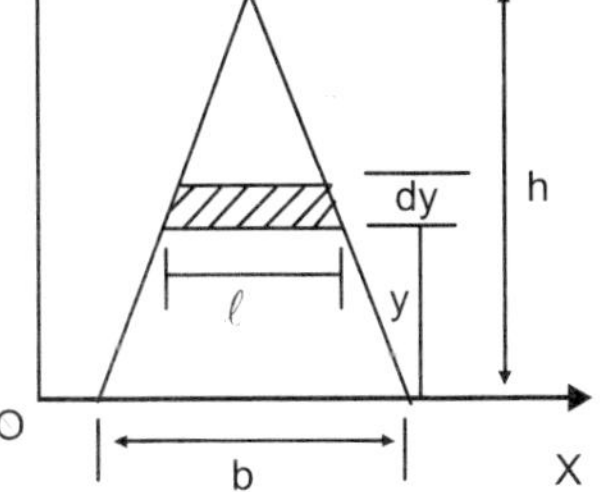

From similar triangle

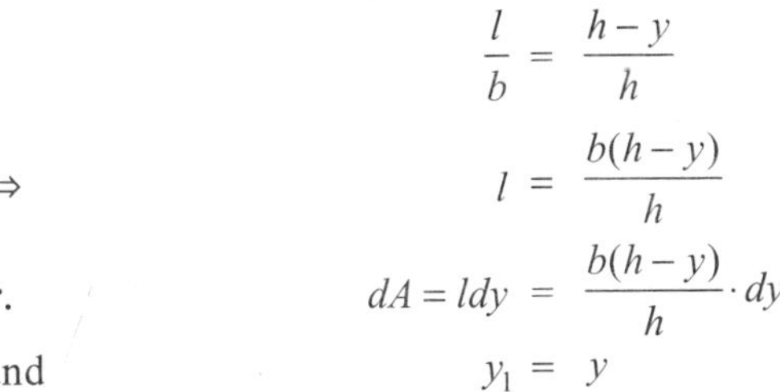

$$\frac{l}{b} = \frac{h-y}{h}$$

$$\Rightarrow \qquad l = \frac{b(h-y)}{h}$$

$$\therefore \qquad dA = ldy = \frac{b(h-y)}{h}\cdot dy$$

and $\qquad y_1 = y$

Now using the equation for y_G we get,

$$y_G = \frac{\int y_1\,dA}{\int dA} = \frac{\int_0^h y\cdot b\frac{(h-y)}{h}\cdot dy}{\int_0^h b\frac{(h-y)}{h}\cdot dy} = \frac{\left[\frac{hy^2}{2} - \frac{y^3}{3}\right]_0^h}{\left[hy - \frac{y^2}{2}\right]_0^h} = \frac{h}{3}$$

So, centroid of the triangle lies from base 1/3 of height and from vertex $2/3$ of height.

EXAMPLE 4

Determine the centroid of a uniform rectangular lamina.

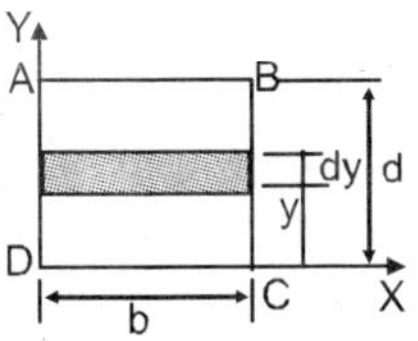

SOLUTION Consider a rectangular section $ABCD$ having width b and depth d. Let x–x is the horizontal axis passing through base DC of the rectangular section. Consider a small rectangular strip of thickness dy at a distance y from x–x axis. So, area of strip $dA = bdy$, and $y_1 = y$. From the equation of the C.G we get,

$$y_G = \frac{\int y_1 \, dA}{\int dA} = \frac{\int_0^d y \cdot b \cdot dy}{\int_0^d b dy} = \frac{\left[\frac{y^2}{2}\right]_0^d}{[y]_0^d} = \frac{d}{2} = \frac{\text{depth}}{2}$$

Similarly $$x_C = \frac{b}{2} = \frac{\text{width}}{2}.$$

EXAMPLE 5

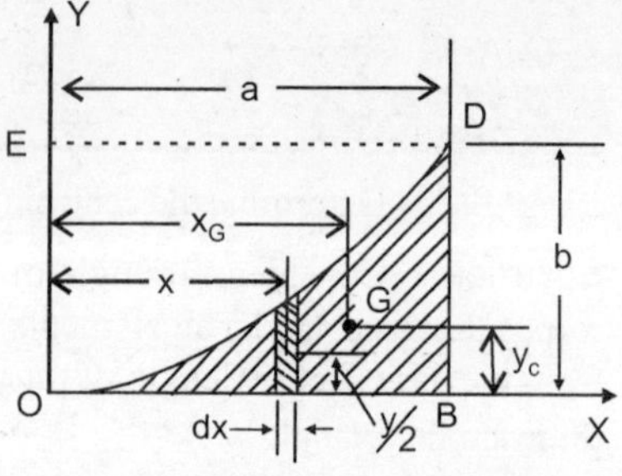

Determine the co-ordinate of the centroid of the area *OBD* in the figure, if the curve *OD* represents the equation of a parabola, given by $x^2 = 4ky$ in which $OB = x = a$ and $BD = y = b$.

SOLUTION Substituting $x = a, y = b$ in given equation, we have

$$a^2 = 4kb$$

or $$k = \frac{a^2}{4b}$$

So, the equation is $$x^2 = 4(a^2/4b)\, y = (a^2/b)y$$

$\Rightarrow$ $$y = (b/a^2)x^2$$

Consider an elementary strip of thickness dx at distance x from X-axis and height y. Area of strip $dA = y \cdot dx$ and centroid of the strip $x_1 = x$, $y_1 = y/2$

$\therefore$ $$y_G = \frac{\int y_1 dA}{\int dA} = \frac{\int (y/2) \cdot y \cdot dx}{\int y \cdot dx} = \frac{1/2 \int y^2 dx}{\int y \cdot dx} = \frac{1/2 \int_0^a (b^2/a^4) x^4 dx}{\int_0^a (b/a^2) x^2 dx} = \frac{3b}{10}$$

Similarly, $$x_G = \frac{\int x_1 dA}{\int dA} = \frac{\int x \cdot y \cdot dx}{\int y \cdot dx} = \frac{\int_0^a x(b/a^2)x^2 dx}{\int_0^a (b/a^2)x^2 dx} = \frac{3a}{4}.$$

EXAMPLE 6

Determine the co-ordinates x_G and y_G of centroid *G* of the area of the circular sector *OBD* of radius *r* and central angle is shown in figure.

SOLUTION *Method - I*: The *x*-axis is chosen as the axis of symmetry and therefore y_G is automatically zero. Consider an elementary strip of angle $d\theta$ at a distance θ from X-axis. Height of triangular element r and base $rd\theta$ and centroid of triangle from vertex is $2r/3$. So, centroid of elementary area from *O* along *X*-axis

$$x_1 = \frac{2}{3} r \cdot \cos\theta$$

Area of strip $$dA = \frac{1}{2} \cdot r \cdot d\theta \cdot r = \frac{1}{2} r^2 d\theta.$$

From equation of centroid we get

$$x_G = \frac{\int x_1 dA}{\int dA} = \frac{2\int_0^{\alpha/2} \frac{2}{3} r\cos\theta \cdot \frac{1}{2} r^2 d\theta}{2\int_0^{\alpha/2} \frac{1}{2} r^2 d\theta} = \frac{\frac{2}{3} r[\sin\theta]_0^{\alpha/2}}{[\theta]_0^{\alpha/2}} = \frac{4r\sin(\alpha/2)}{3\alpha}.$$

Method II : Consider an elementary strip in the form of a partial circular ring of thickness dr_0 at a distance r_0 from O. Area of strip $dA = r_0\, \alpha\, dr_0$ and centroid of strip

$$x_1 = [r_0 \sin\ (\alpha/2)/(\alpha/2)]$$

From the equation of centroid we get $$x_G = \frac{\int x_1 dA}{\int dA} = \frac{\int_0^r \frac{2r_0}{\alpha} \sin(\alpha/2) \cdot r_0 \cdot \alpha \cdot dr_0}{\int_0^r r_0 \cdot \alpha \cdot dr_0} = \frac{4r\sin(\alpha/2)}{3\alpha}.$$

For a semicircular area $\alpha = \pi$, which gives $$x_G = \frac{4r}{3\pi} \text{ and } y_G = 0$$

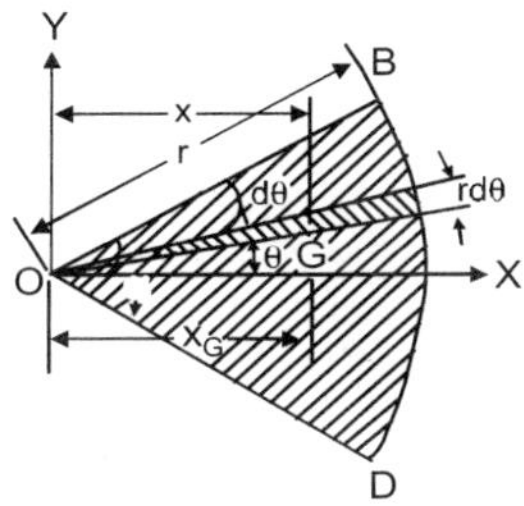

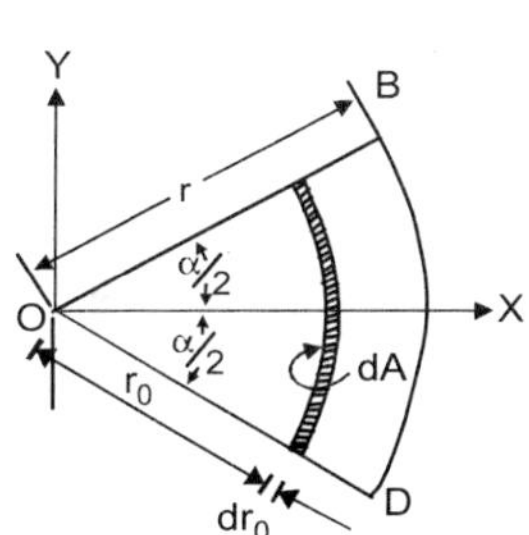

For the quarter circular area, by symmetry, we see immediately that this result also is applicable to quarter circle where the measurements are

$$x_G = \frac{4r}{3\pi} = y_G$$

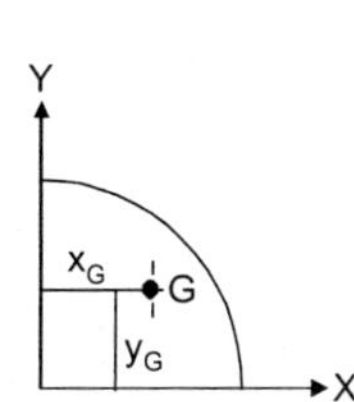

EXAMPLE 7

Determine the co-ordinates of the centroid of the shaded area between the parabola $x^2 = ay$ and the straight line $y = x$ as shown in figure.

SOLUTION The equation of parabola and straight line are

$$y = \frac{x^2}{a}$$

and $$y = x$$

$$\therefore \quad x = \frac{x^2}{a}$$

$$\Rightarrow \quad a = x, \text{ so } y = a$$

Hence, the co-ordinates of point A are (a, a). Consider an elementary strip of thickness dx and height y as shown in figure, y_1 = co-ordinate of point D which lies on the straight line, OA. y_2 = co-ordinate of point E which lies on the parabola OA. The horizontal co-ordinates of the points D and E are same which is x. Then area of the strip is given by $dA = y \cdot dx = (y_1 - y_2)dx$

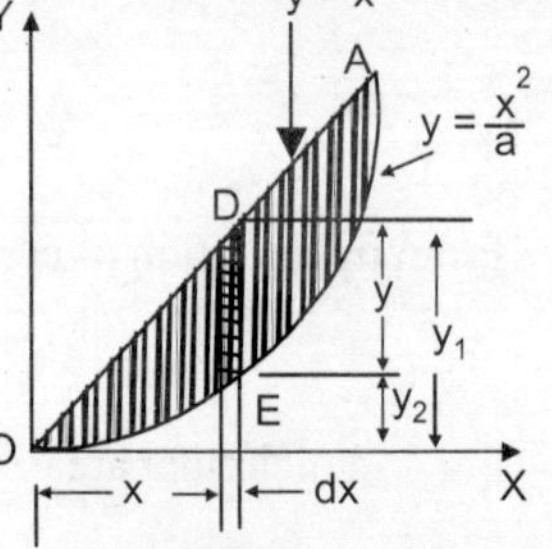

Again, $$y_1 = x$$

and $$y_2 = \frac{x^2}{a}$$

$$\therefore \quad dA = \left(x - \frac{x^2}{a}\right)dx$$

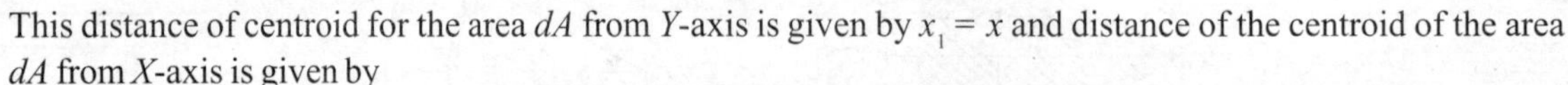
This distance of centroid for the area dA from Y-axis is given by $x_1 = x$ and distance of the centroid of the area dA from X-axis is given by

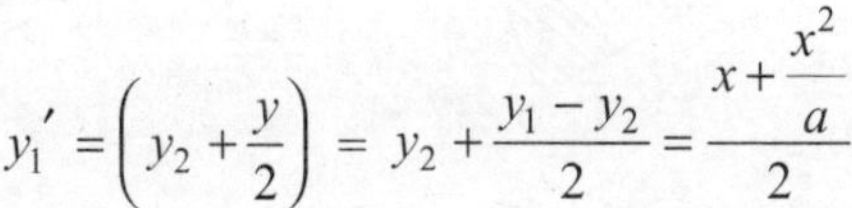
$$y_1' = \left(y_2 + \frac{y}{2}\right) = y_2 + \frac{y_1 - y_2}{2} = \frac{x + \frac{x^2}{a}}{2}$$

From the equation of centroid of area, we get,

$$x_G = \frac{\int x_1\, dA}{\int dA} = \frac{\int_0^a x\left(x - \frac{x^2}{a}\right)dx}{\int_0^a \left(x - \frac{x^2}{a}\right)dx} = \frac{a}{2}$$

and $$y_G = \frac{\int y_1'\, dA}{\int dA} = \frac{\int_0^a \frac{1}{2}\left(x + \frac{x^2}{a}\right)\left(x - \frac{x^2}{a}\right)dx}{\int_0^a \left(x - \frac{x^2}{a}\right)dx} = \frac{2a}{5}.$$

EXAMPLE 8

Determine by direct integration method the co-ordinates of the centroid of the area formed the quarter of an ellipse as shown in figure.

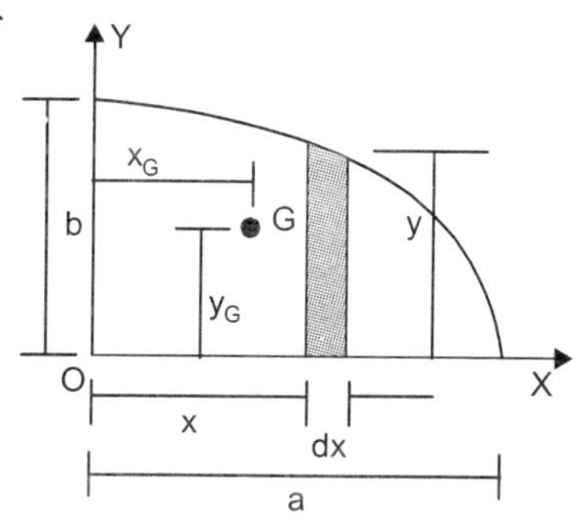

SOLUTION Consider an elementary strip of shaded area at a distance x from Y-axis and thickness dx, height y

Equation of ellipse,

$$\frac{x^2}{a^2} + \frac{y^2}{b^2} = 1$$

$\Rightarrow$

$$y = \frac{b}{a}\left[a^2 - x^2\right]^{1/2}$$

Area of the strip $dA = y \cdot dx$.

From the equation of centroid, we have

$$x_G = \frac{\int x_1 dA}{\int dA} = \frac{\int_0^a x \cdot \frac{b}{a}\left(a^2 - x^2\right)^{1/2} dx}{\int_0^a \frac{b}{a}\left(a^2 - x^2\right)^{1/2} dx} = \frac{-\frac{1}{3}\left[\sqrt{(a^2 - x^2)^3}\right]_0^a}{\frac{1}{2}\left[\left(x\sqrt{a^2 - x^2} + a^2 \sin^{-1}\frac{x}{a}\right)\right]_0^a} = \frac{4a}{3\pi}$$

$$y_G = \frac{\int_0^a \frac{1}{2} y \cdot y \cdot dx}{\int dA} = \frac{\frac{1}{2}\int_0^a y^2 dx}{\frac{\pi ab}{4}} = \frac{\frac{b^2}{2^2}\left[a^2 x - \frac{x^3}{3}\right]_0^a}{\frac{\pi ab}{4}} = \frac{4b}{3\pi}$$

EXAMPLE 9

Locate the centroid of the area under the curve $x = ky^3$ from $x = 0$ to $x = a$.

SOLUTION Consider an elementary strip of thickness of dx, height y at a distance x from y-axis.

Centroid of strip $x_1 = x$ and $y_1 = y/2$

Area of strip $dA = y\, dx$.

Substituting $x = a$ and $y = b$ in the equation, $a = kb^3$,

So $k = a/b^3$.

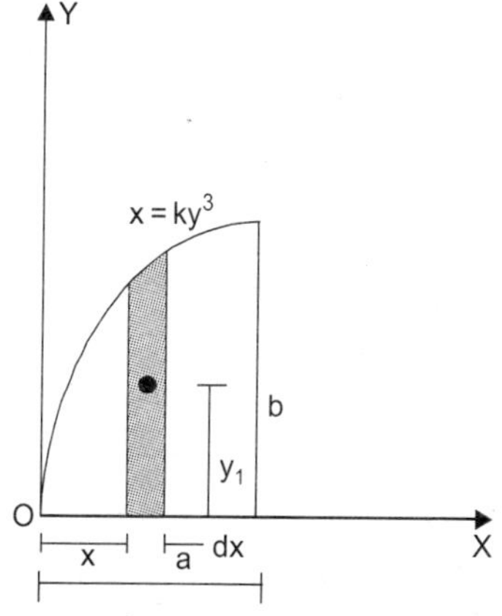

Finally $$y = \left(\frac{b^3 x}{a}\right)^{1/3}$$

From equation of centroid, we get

$$x_G = \frac{\int x_1\, dA}{\int dA} = \frac{\int_0^a x\left(\frac{b^3 x}{a}\right)^{1/3} \cdot dx}{\int_0^a \left(\frac{b^3 x}{a}\right)^{1/3} \cdot dx} = \frac{4a}{7}$$

$$y_G = \frac{\int y_1\, dA}{\int dA} = \frac{\frac{1}{2}\int_0^a \left(\frac{b^3 x}{a}\right)^{2/3} \cdot dx}{\int_0^a \left(\frac{b^3 x}{a}\right)^{1/3} \cdot dx} = \frac{2b}{5}$$

EXAMPLE 10

Determine the co-ordinate x_G and y_G of the centroid G of the area between the X-axis and the half sine wave ODB as shown in figure. The equation of the sine wave is $y = a \cdot \sin\left(\frac{\pi x}{l}\right)$ from $x = 0$ to $x = l$.

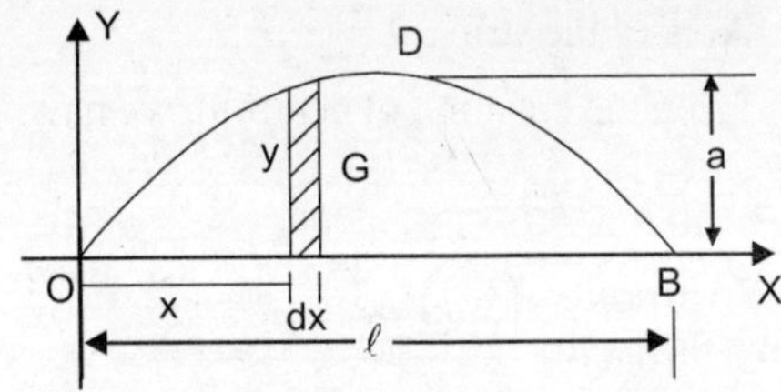

SOLUTION Consider an element dx at a distance x from Y-axis and height y.

Elemental area $dA = ydx$

Centriod from y-axis $x_1 = x$

and centriod from x-axis $y_1 = y/2$

From the equation of centroid, we get

$$x_G = \frac{\int x_1\, dA}{\int dA} = \frac{\int_0^l x \cdot a \cdot \sin\left(\frac{\pi x}{l}\right) dx}{\int_0^l a \cdot \sin\left(\frac{\pi x}{l}\right) dx} = \frac{l}{2}$$

$$y_G = \frac{\int y_1\, dA}{\int dA} = \frac{\frac{1}{2}\int_0^l a^2 \cdot \sin^2\left(\frac{\pi x}{l}\right) dx}{\int_0^l a \cdot \sin\left(\frac{\pi x}{l}\right) dx} = \frac{\pi a}{8}.$$

EXAMPLE 11

Determine co-ordinate of the centroid of the area as shown in figure.

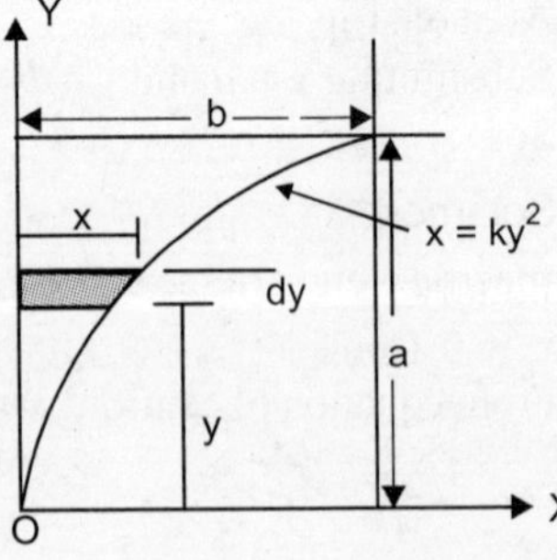

SOLUTION Equation of the curve $x = ky^2$

where $x = b$, $y = a$

Now, $x = ky^2$

therefore $b = ka^2$

$\Rightarrow$ $k = b/a^2$

Finally $y^2 = \frac{a^2 x}{b}$

Consider an elementary strip dy at a distance y from X-axis and length x.

Centroid of the strip $x_1 = x/2,\ y_1 = y.$

Area of strip $dA = xdy.$

From the equation of centroid, we get,

$$x_G = \frac{\int x_1 dA}{\int dA} = \frac{\int_0^a \frac{x}{2} x dy}{\int_0^a x\, dy} = \frac{\frac{b^2}{2a^4}\int_0^a y^4 dy}{\frac{b}{a^2}\int_0^a y^2 dy} = \frac{3b}{10}$$

$$y_G = \frac{\int y_1 dA}{\int dA} = \frac{\int_0^a y \cdot x dy}{\int_0^a x\, dy} = \frac{\int_0^a y^3 dy}{\int_0^a y^2 dy} = \frac{3a}{4}.$$

EXAMPLE 12

Determine the co-ordinates of the centriod of the area under the curve $y = b\sin\left(\frac{\pi x}{2a}\right)$, from $x = 0$ to $x = a$, when $x = a, y = b$.

SOLUTION Consider an elementary strip dx at a distance x from Y-axis, height y.

Centroid of strip $\quad x_1 = x, y_1 = y/2$

From the equation of centriod

$$x_G = \frac{\int x_1\, dA}{\int dA} = \frac{\int_0^a x \cdot y \cdot dx}{\int_0^a y \cdot dx} = \frac{\int_0^a x \cdot b\sin\left(\frac{\pi x}{2a}\right) dx}{\int_0^a b\sin\left(\frac{\pi x}{2a}\right) dx} = \frac{2a}{\pi}$$

$$y_G = \frac{\int y_1\, dA}{\int dA} = \frac{\int_0^a \frac{y}{2} y \cdot dx}{\int_0^a y \cdot dx} = \frac{\int_0^a \frac{1}{2} b\sin\left(\frac{\pi x}{2a}\right)^2 dx}{\int_0^a b\sin\left(\frac{\pi x}{2a}\right) dx} = \frac{\pi b}{8}$$

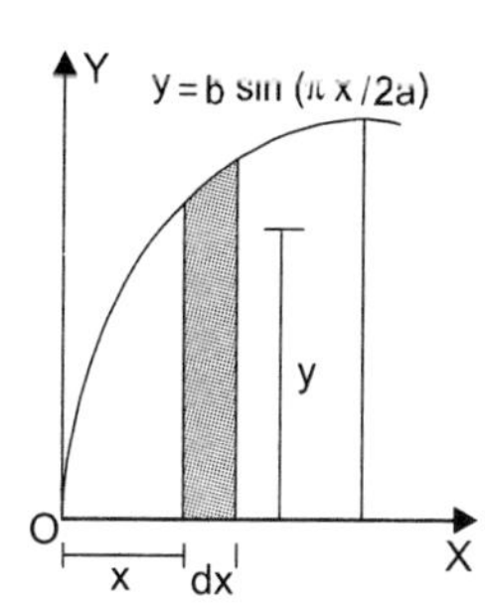

EXAMPLE 13

Determine the C.G of solid circular cylinder.

SOLUTION Let height of the cylinder = h, cross-sectional area of the cylinder = A, specific weight of the cylinder = γ. Since the cylinder is symmetrical about Y-axis, its $x_G = 0$. Consider an elementary strip of width dy at a distance y from X-axis. Therefore, volume of strip = $A\, dy$. Weight of this strip $dw = A\gamma dy$.

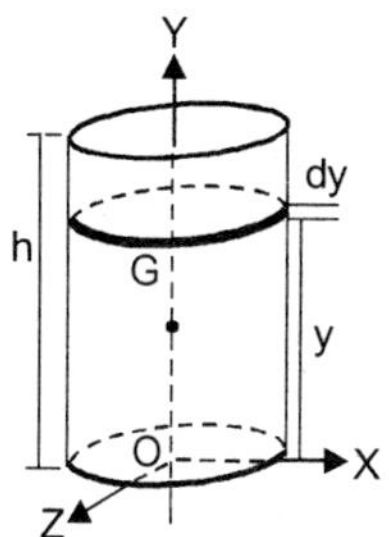

$$\therefore \quad y_G = \frac{\int y dw}{\int dw} = \frac{\int_0^h y \cdot A\gamma dy}{\int_0^h A\gamma dy} = \frac{h}{2}$$

EXAMPLE 14

Determine the C.G of a right circular cone of height h, radius R.

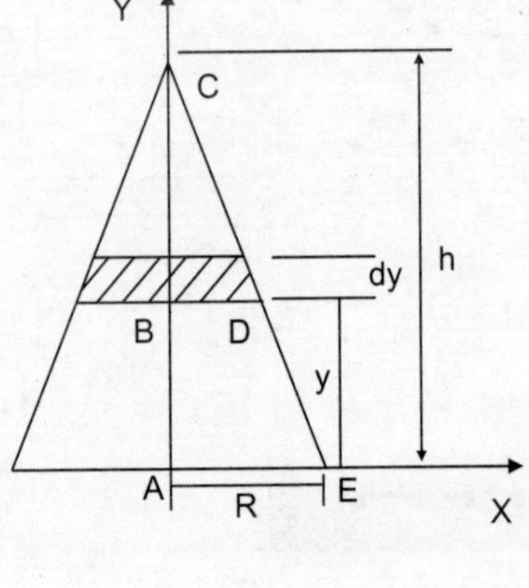

SOLUTION Consider an elementary strip of width dy at a distance y from X-axis. Radius of strip $BD = r$. Cross-sectional area $A = \pi r^2$

From triangle ACE and BCD, we get

$$\frac{BD}{BC} = \frac{AE}{AC}$$

or

$$BD = (h-y)\times\frac{R}{h} = \frac{R}{h}(h-y)$$

Volume of the strip = $\pi r^2\, dy$

Weight of strip

$$dw = \gamma\pi r^2 dy = \frac{\gamma\pi R^2}{h^2}(h-y)^2\, dy$$

As it is symmetrical about Y-axis, $x_G = 0$

$$y_G = \frac{\int_0^h y\cdot dw}{\int_0^h dw} = \frac{\int_0^h y\cdot\pi\gamma\cdot\frac{R^2}{h^2}(h-y)^2\, dy}{\int_0^h \pi\gamma\cdot\frac{R^2}{h^2}(h-y)^2\, dy} = \frac{h}{4}$$

EXAMPLE 15

Locate the C.G of a hemisphere of radius r with respect to the base.

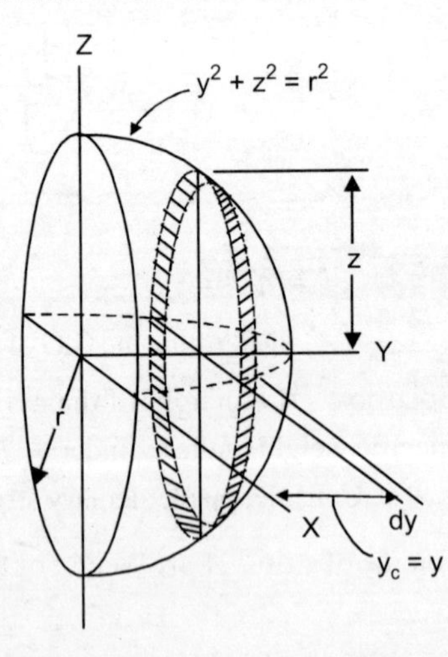

SOLUTION The axes chosen are shown in the figure.

Here, $x = 0$, $z = 0$, by symmetry about X-axis and Y-axis. The most convenient elementary strip is a circular slice of thickness dy parallel to the x–z plane. Since, the hemisphere intersects the y–z plane in the circle $y^2 + z^2 = r^2$,

The radius of the circular slice is $\quad z = \sqrt{r^2 - y^2}$.

The volume of the strip becomes $\quad dv = \pi\left(r^2 - y^2\right)dy$

Again we know that

$$y_G = \frac{\int y_1 dv}{v} = \frac{\int_0^r y\pi\left(r^2 - y^2\right)dy}{\frac{2}{3}\pi r^3} = \frac{3}{8}r$$

EXAMPLE 16

Determine the x and y co-ordinates of the centroid of the shaded area.

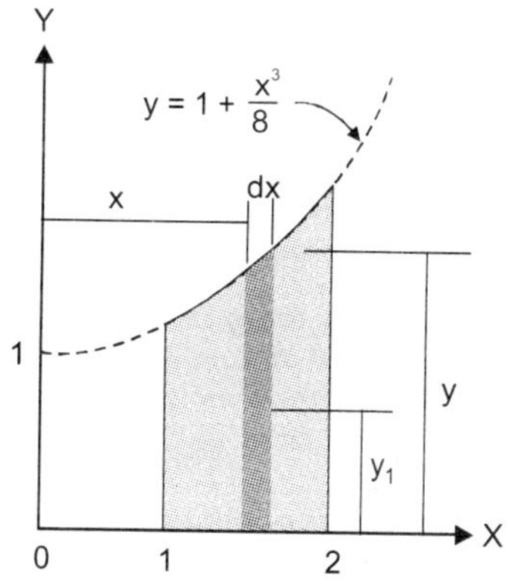

SOLUTION Consider an elementary strip dx at a distance x from Y-axis and length of strip = y. Area of strip $dA = ydx$, length of centroid of the strip, $x_1 = x$ *and* $y_1 = y/2$.

$$x_G = \frac{\int x_1\, dA}{\int dA} = \frac{\int_1^2 x\left(1+\frac{x^3}{8}\right)dx}{\int_1^2\left(1+\frac{x^3}{8}\right)dx} = \frac{\frac{(x_2-x_1)^2}{2}+\frac{(x_2-x_1)^5}{40}}{(x_2-x_1)+\frac{(x_2-x_1)^4}{32}}$$

$$y_G = \frac{\int y_1\, dA}{\int dA} = \frac{\frac{1}{2}\int_1^2\left(1+\frac{x^3}{8}\right)^2 dx}{\int_1^2\left(1+\frac{x^3}{8}\right)dx} = \frac{(x_2-x_1)+\frac{(x_2-x_1)^7}{448}+\frac{(x_2-x_1)^4}{16}}{(x_2-x_1)+\frac{(x_2-x_1)^4}{32}}.$$

EXAMPLE 17

Determine the location of the centroid of this circular arc.

SOLUTION Since the arc is symmetrical with respect to the X-axis, $y_G = 0$. Consider an elementary strip subtending an elemental angle $d\theta$ at an angle θ from x-axis.

$\therefore$
$$dL = r \cdot d\theta$$

and
$$x_1 = x = r\cos\theta$$

$\therefore$
$$x_G = \frac{\int x_1 dL}{\int dL} = \frac{\int_{-\alpha}^{\alpha}(r\cdot\cos\theta)\,r\cdot d\theta}{\int_{-\alpha}^{\alpha} r\cdot d\theta} = \frac{r\cdot\sin\alpha}{\alpha}$$

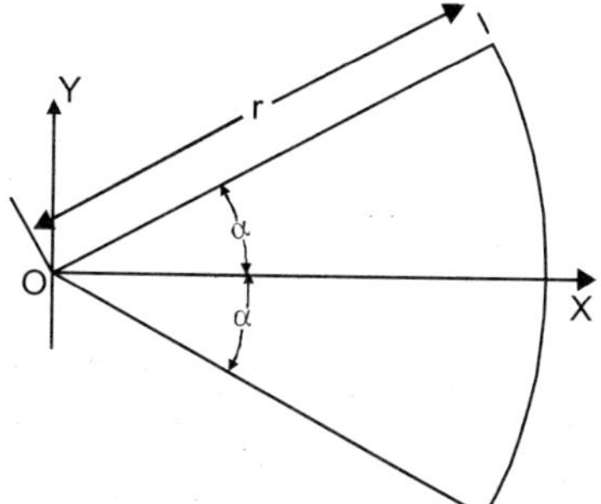

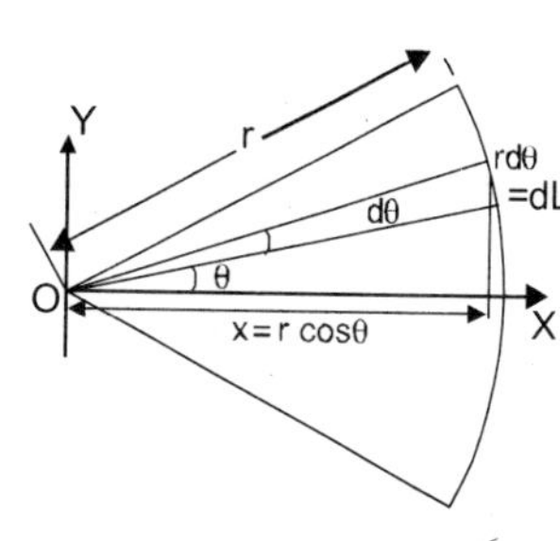

EXAMPLE 18

Determine the location of the centroid of the quarter of a circle of a radius r, when axes x and y are semi diameter of the circular area.

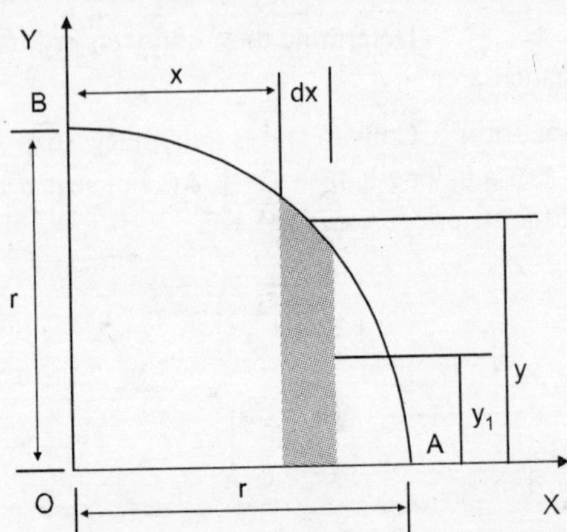

SOLUTION Equation of circle is $x^2 + y^2 = r^2$. A differential element is chosen as shown. Area of strip $dA = y \cdot dx$ and centroid of strip $x_1 = x$, and $y_1 = y/2$

where $$y = \sqrt{r^2 - x^2}$$

Area of quarter circle $$A = \frac{\pi r^2}{4}$$

$$x_G = \frac{\int x\, dA}{\int dA} = \frac{\int_0^r x \cdot y dx}{\frac{\pi r^2}{4}} = \frac{4}{\pi r^2}\left[-\frac{1}{3}\sqrt{\left(r^2 - x^2\right)^3}\right]_0^r = \frac{4r}{3\pi}$$

$$y_G = \frac{\int y_1 dA}{\int dA} = \frac{\int_0^r \frac{y}{2} y\, dx}{\frac{\pi r^2}{4}} = \frac{2}{\pi r^2}\left[r^2 x - \frac{x^3}{3}\right]_0^r = \frac{4r}{3\pi}.$$

EXAMPLE 19

Determine the position of the centroid of the shaded area in figure created by cutting a semicircle of diameter a and from a quarter circle of radius a.

SOLUTION x_G and y_G be the location of the centroid of the shaded area with respect to O as origin. Quarter circle is *element* 1 and unshaded semicircle is *element* 2.

$$\therefore \quad A_1 = \frac{\pi a^2}{4},\ x_1 = \frac{4a}{3\pi} = y_1$$

$$\therefore \quad A_1 x_1 = \frac{\pi a^2}{4} \times \frac{4a}{3\pi} = \frac{a^3}{3} = A_1 y_1$$

Again $$A_2 = \frac{\pi a^2}{8},\ x_2 = \frac{2a}{3\pi},\ y_2 = \frac{a}{2}$$

$$\therefore \quad A_2 x_2 = \frac{\pi a^2}{8} \times \frac{2a}{3\pi} = \frac{a^3}{12},$$

$$A_2 y_2 = \frac{\pi a^2}{8} \times \frac{a}{2} = \frac{\pi a^3}{16}$$

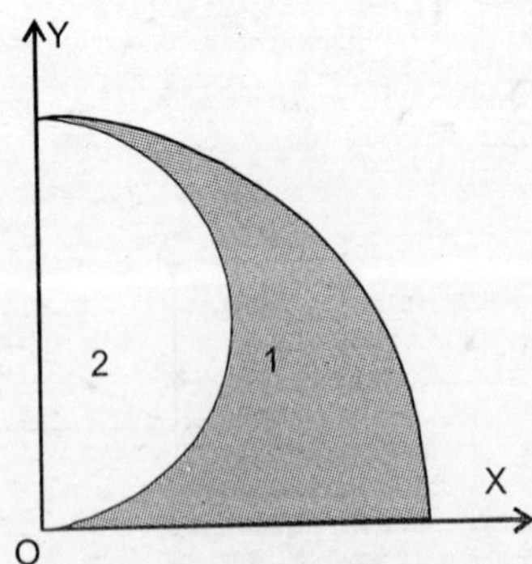

Element	Area (A_i)	x_i	y_i	A_ix_i	A_iy_i
1	$\frac{\pi a^2}{4}$	$\frac{4a}{3\pi}$	$\frac{4a}{3\pi}$	$\frac{a^3}{3}$	$\frac{a^3}{3}$
2	$\frac{-\pi a^2}{8}$	$\frac{2a}{3\pi}$	$\frac{a}{2}$	$-\frac{a^3}{12}$	$-\frac{\pi a^3}{16}$
Total	$\frac{\pi a^2}{8}$			$0.25a^3$	$0.13698a^3$

$$\therefore \quad x_G = \frac{\sum A_i x_i}{\sum A_i} = \frac{0.25a^3}{\left(\frac{\pi a^2}{8}\right)} = 0.636a,$$

$$y_G = \frac{\sum A_i y_i}{\sum A_i} = \frac{0.13698a^3}{\left(\frac{\pi a^2}{8}\right)} = 0.3488a$$

EXAMPLE 20

Locate the centroid of the shaded area of plate of uniform thickness.

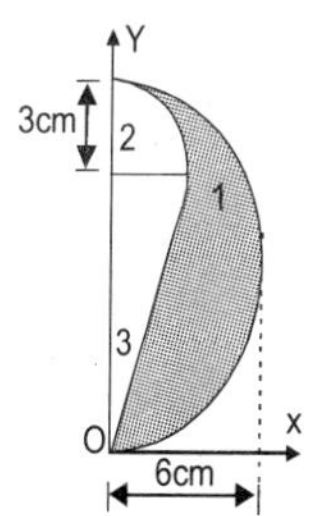

SOLUTION x_G and y_G be the location of the centroid of the shaded area with respect to B as origin. From semicircular figure,

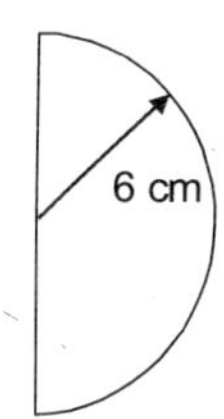

$$a_1 = \frac{\pi \times 6^2}{2} = 56.55\text{cm}^2, x_1 = \frac{4\times 6}{3\pi} = 2.546\text{cm}, y_1 = 6\text{cm}$$

$$a_1x_1 = (56.55\times 2.546)\,\text{cm}^3, a_1y_1 = (56.55\times 6)\,\text{cm}^3$$

From quarter circle figure,

$$a_2 = \frac{\pi\times 3^2}{4} = 7.068\text{cm}^2, x_2 = \frac{4\times 3}{3\pi} = 1.273\text{cm}$$

$$y_2 = 9 + \frac{4\times 3}{3\pi} = 10.273\text{cm}$$

$$a_2x_2 = (7.068\times 1.273)\,\text{cm}^3, a_2y_2 = (7.068\times 10.273)\,\text{cm}^3$$

From triangular figure,

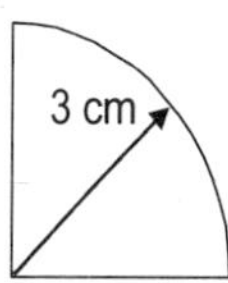

$$a_3 = \frac{1}{2}\times 3\times 9 = 13.5\text{cm}^2, x_3 = \frac{1}{3}\times 3 = 1\text{cm}, y_3 = \frac{2}{3}\times 9 = 6\text{cm}$$

$$a_3x_3 = (13.5\times 1)\,\text{cm}^3, a_3y_3 = (13.5\times 6)\,\text{cm}^3$$

$$x_G = \frac{\sum a_i x_i}{\sum a_i} = \frac{a_1x_1 - a_2x_2 - a_3x_3}{a_1 - a_2 - a_3}$$

$$= \frac{(56.55\times2.546)-(7.068\times1.273)-(13.5\times1)}{56.55-7.068-13.5} = 3.376\text{cm}$$

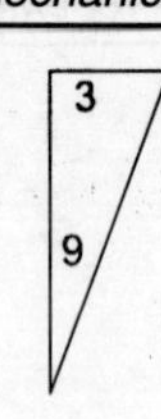

$$y_G = \frac{\sum a_i y_i}{\sum a_i} = \frac{a_1 y_1' - a_2 y_2 - a_3 y_3}{a_1 - a_2 - a_3}$$

$$= \frac{(56.55\times6)-(7.068\times10.273)-(13.5\times6)}{56.55-7.068-13.5} = 5.16\text{cm}$$

EXAMPLE 21

Referring to the figure, determine the x_G and y_G of the centroid of 4cm diameter circular hole cut in a thin plate, so that this point will be the centroid of the remaining area of *ABCDO*.

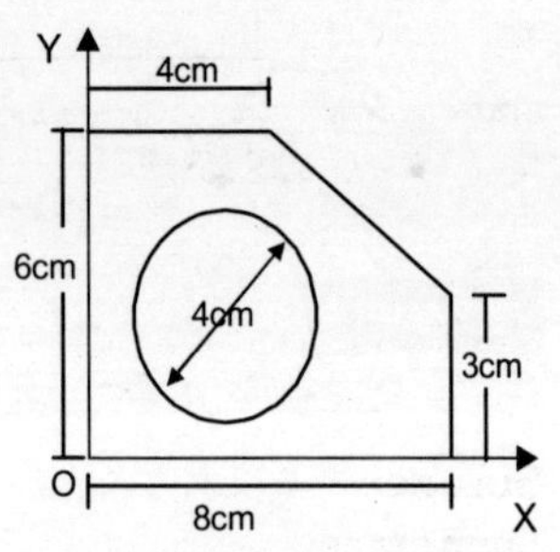

SOLUTION x_G and y_G be the co-ordinates with respect to origin *O*. From rectangle *ABEO*,

$$a_1 = 4\times6 = 24\text{cm}^2, x_1 = \frac{4}{2} = 2\text{cm}, y_1 = \frac{6}{2} = 3\text{cm}$$

$$a_1x_1 = (24\times2) = 48\text{cm}^3, a_1y_1 = (24\times3) = 72\text{cm}^3$$

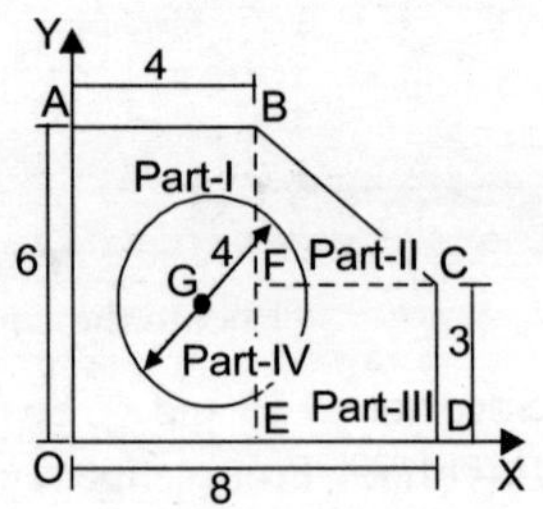

From triangle *BCF*,

$$a_2 = \frac{1}{2}\times4\times3 = 6\text{cm}^2, \; x_2 = 4+\frac{1}{3}\times4 = 5.33\text{cm},$$

$$y_2 = 3+\frac{1}{3}\times3 = 4\text{cm}, \; a_2x_2 = (6\times5.33) = 31.98\text{cm}^3,$$

$$a_2y_2 = (6\times4) = 24\text{cm}^3$$

From rectangle *FCDE*,

$$a_3 = 4\times3 = 12\text{cm}^2, x_3 = 4+\frac{4}{2} = 6\text{cm}, y_3 = \frac{3}{2} = 1.5\text{cm}$$

$$a_3x_3 = (12\times6) = 72\text{cm}^3, a_3y_3 = (12\times1.5) = 18\text{cm}^3$$

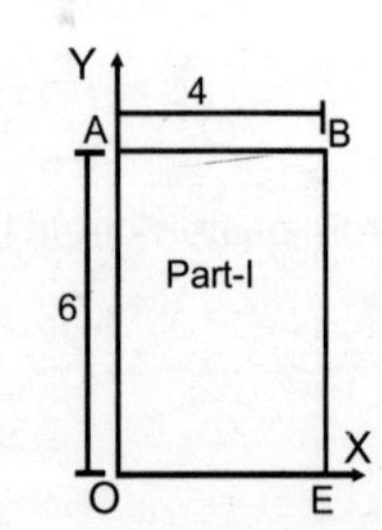

From circle of centre *G*,

$$a_4 = \frac{\pi}{4}\times4^2 = 12.566\text{cm}^2, x_4 = x_G, y_4 = y_G$$

$$a_4x_4 = 12.566x_C\ \text{cm}^3, a_4y_4 = 12.566y_G\ \text{cm}^3$$

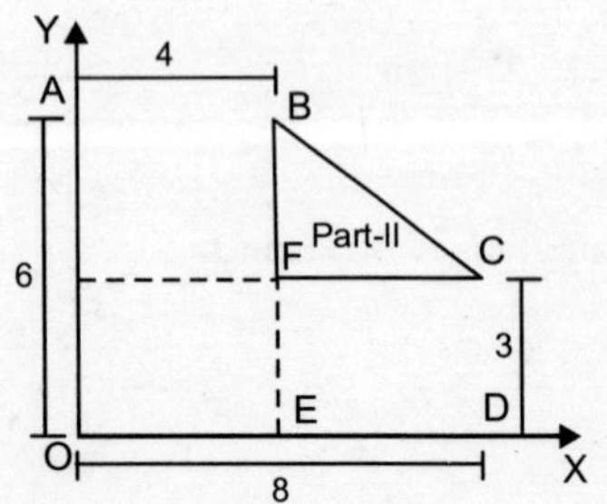

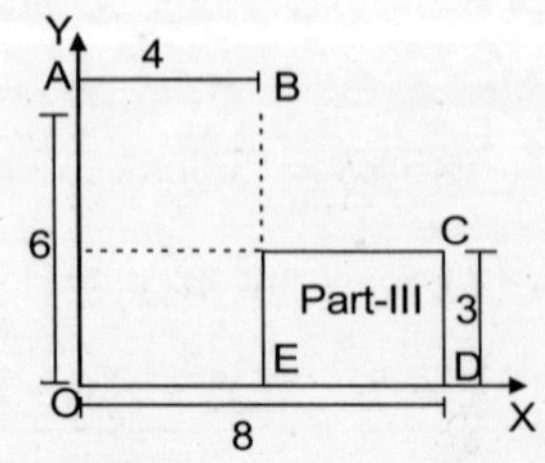

Element	Area (A_i)	x_i	y_i	A_ix_i	A_iy_i
1	24	2	3	48	72
2	6	5.33	4	31.98	24
3	12	6	1.5	72	18
4	–12.566	x_G	y_G	$-12.566x_G$	$-12.566y_G$
Total	29.434			$151.98-12.566x_G$	$114-12.566y_G$

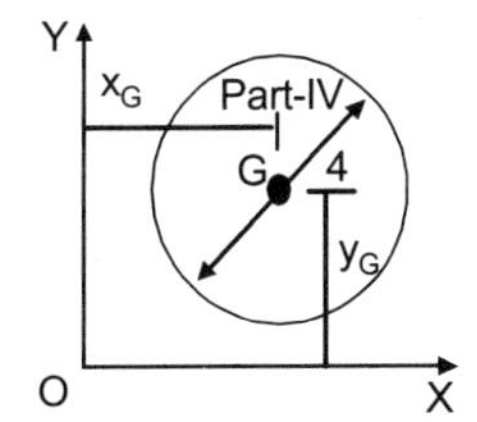

$$x_G = \frac{151.98-12.566x_G}{29.434}$$

and $$y_G = \frac{114-12.566x_G}{29.434}$$

∴ $x_G = 3.618$ cm and $y_G = 2.714$ cm

EXAMPLE 22

Locate the centroid of the *L*-Section shown in figure.

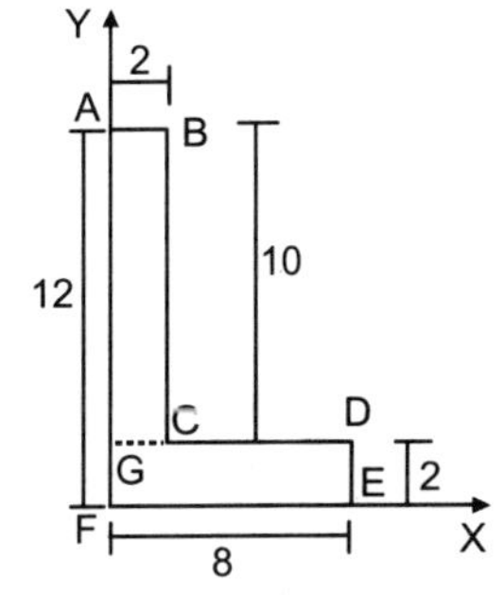

SOLUTION x_G and y_G be the co-ordinates with respect to origin *F*.
From rectangle *ABCG*,

$$a_1 = 2\times 10 = 20\text{cm}^2,\ x_1 = \frac{2}{2} = 1\text{cm},$$

$$y_1 = 2+\frac{10}{2} = 7\text{cm}$$

$$a_1x_1 = 20\times 1 = 20\text{cm}^3,\ a_1y_1 = 20\times 7 = 140\text{cm}^3$$

From rectangle *CDEF*,

$$a_2 = 8\times 2 = 16\text{cm}^2,\ x_2 = \frac{8}{2} = 4\text{cm},\ y_2 = \frac{2}{2} = 1\text{cm}$$

$$a_2x_2 = 16\times 4 = 64\text{cm}^3,\ a_2y_2 = 16\times 1 = 16\text{cm}^3$$

$$x_G = \frac{a_1x_1+a_2x_2}{a_1+a_2} = \frac{20+64}{20+16} = 2.33\text{cm}$$

$$y_G = \frac{a_1y_1+a_2y_2}{a_1+a_2} = \frac{140+16}{20+16} = 4.33\text{cm}$$

EXAMPLE 23

Locate the centroid of the Channel-Section shown in figure.

SOLUTION x_G and y_G be the co-ordinates with respect to origin *H*. From rectangle *ABCG*,

$$a_1 = 5\times 1.5 = 7.5\text{cm}^2,\ x_1 = \frac{5}{2} = 2.5\text{cm},\ y_1 = 10-\frac{1.5}{2} = 9.25\text{cm}$$

$$a_1x_1 = 7.5\times 2.5 = 18.75\text{cm}^3,\ a_1y_1 = 7.5\times 9.25 = 69.375\text{cm}^3$$

From rectangle *JDEI*,

$$a_2 = 7\times1.5 = 10.5\text{cm}^2,\ x_2 = \frac{1.5}{2} = 0.75\text{cm},\ y_2 = 1.5 + \frac{7}{2} = 5\text{cm}$$

$$a_2x_2 = 10.5\times0.75 = 7.875\text{cm}^3,\ a_2y_2 = 10.5\times5 = 52.5\text{cm}^3$$

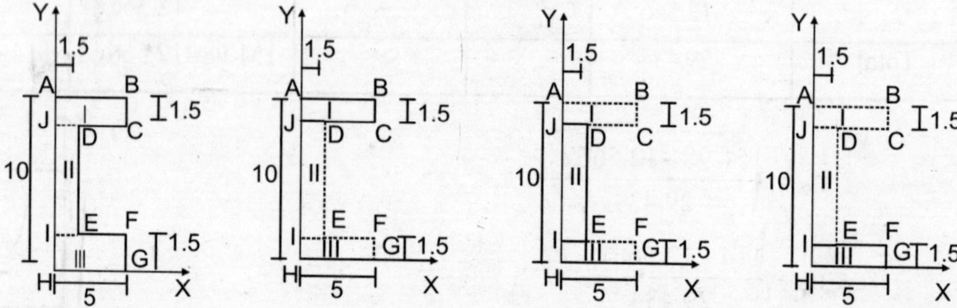

From rectangle *FGHI*,

$$a_3 = 5\times1.5 = 7.5\text{cm}^2,\ x_3 = \frac{5}{2} = 2.5\text{cm},\ y_3 = \frac{1.5}{2} = 0.75\text{cm}$$

$$a_3x_3 = 7.5\times2.5 = 18.75\text{cm}^3,\ a_3y_3 = 7.5\times0.75 = 5.625\text{cm}^3$$

$$x_G = \frac{a_1x_1 + a_2x_2 + a_3x_3}{a_1 + a_2 + a_3} = \frac{18.75 + 7.875 + 18.75}{7.5 + 10.5 + 7.5} = 1.779\text{cm}$$

$$y_G = \frac{a_1y_1 + a_2y_2 + a_3y_3}{a_1 + a_2 + a_3} = \frac{69.375 + 52.5 + 5.625}{7.5 + 10.5 + 7.5} = 5\text{cm}$$

EXAMPLE 24

Find the co-ordinates of the centroid of a semi circular area as shown in figure.

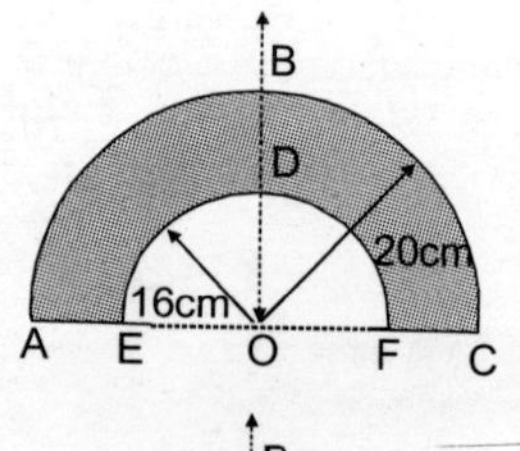

SOLUTION x_G and y_G be the co-ordinates of the centroid. It is symmetrical about *Y*-axis and hence $x_G = 0$. From semi-circle *OABCO*,

$$a_1 = \frac{\pi\times20^2}{2} = 628.32\text{cm}^2,\ y_1 = \frac{4\times20}{3\pi} = 8.488\text{cm}$$

$$a_1y_1 = 628.32\times8.488 = 5333.33\text{cm}^3$$

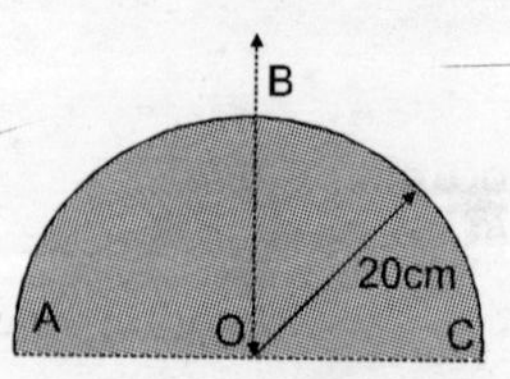

From semi-circle *OEDFO*,

$$a_2 = \frac{\pi\times16^2}{2} = 402.124\text{cm}^2,\ y_2 = \frac{4\times16}{3\pi} = 6.791\text{cm}$$

$$a_2y_2 = 402.124\times6.791 = 2730.67\text{cm}^3$$

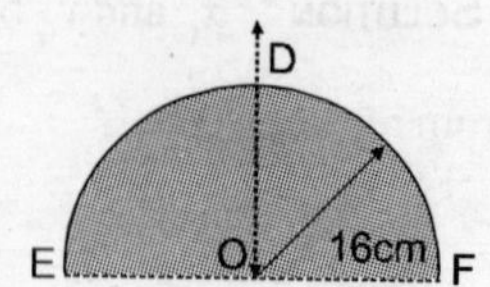

$$\therefore \quad y_G = \frac{a_1y_1 - a_2y_2}{a_1 - a_2} = \frac{5333.33 - 2730.67}{628.32 - 402.124} = 11.506\text{cm}$$

EXAMPLE 25

Determine the dimension W that will locate the centroidal axis at 4cm above the base of the section shown in figure.

SOLUTION As symmetrical about Y-axis, $x_G = 0$, where X-axis as bottom line of the figure. From rectangle $ABCH$,

$$a_1 = 8 \times 1 = 8\text{cm}^2,\ y_1 = 2 + \frac{8}{2} = 6\text{cm}$$

$$a_1 y_1 = 8 \times 6 = 48\text{cm}^3$$

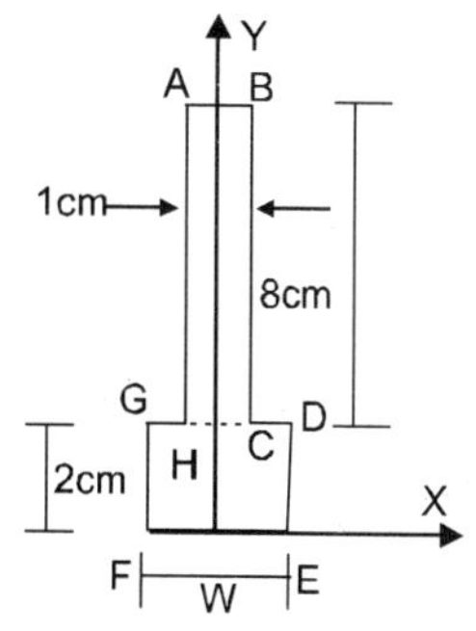

From rectangle $DEFG$,

$$a_2 = (W \times 2)\text{cm}^2,\ y_2 = \frac{2}{2} = 1\text{cm}$$

$$a_2 y_2 = (2W \times 1) = 2W\text{cm}^3$$

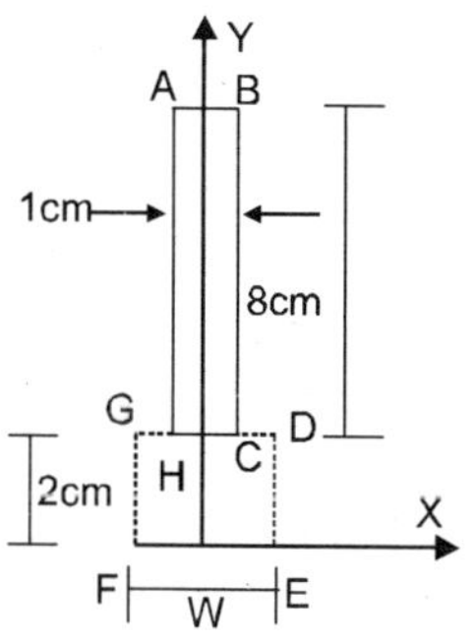

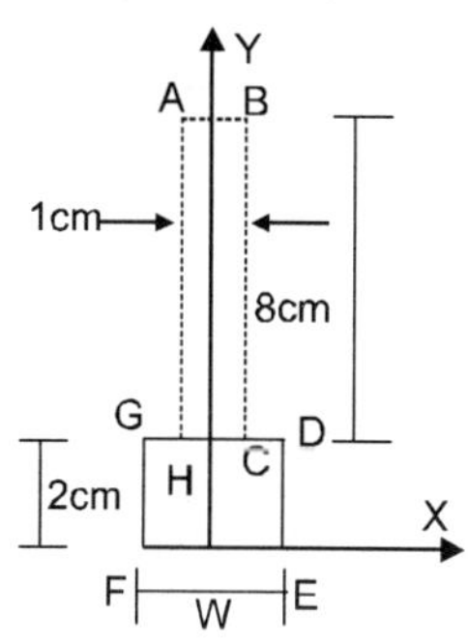

$$\therefore \quad y_G = \frac{a_1 y_1 + a_2 y_2}{a_1 + a_2} = \frac{48 + 2W}{8 + 2W}$$

$$\therefore \quad 4 = \frac{48 + 2W}{8 + 2W}$$

$$\therefore \quad W = 2.67\text{cm}$$

EXAMPLE 26

Find the centroid of the I- Section shown in figure.

SOLUTION As symmetrical about Y-axis, $x_G = 0$, where X-axis as bottom line of the figure. From rectangle $ABCD$,

$$a_1 = 10 \times 2 = 20\text{cm}^2,\ y_1 = 17 + \frac{2}{2} = 18\text{cm},$$

$$\therefore \quad a_1 y_1 = 20 \times 18 = 360\text{cm}^3$$

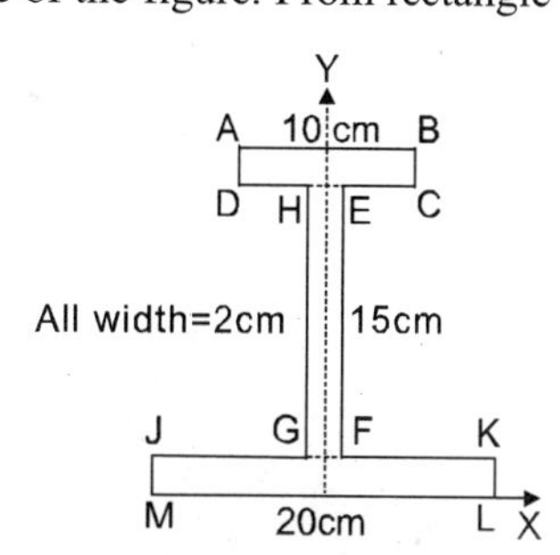

From rectangle $EFGH$,

$$a_2 = 2 \times 15 = 30\text{cm}^2,\ y_2 = 2 + \frac{15}{2} = 9.5\text{cm},$$

$$a_2 y_2 = 30 \times 9.5 = 285\text{cm}^3$$

From rectangle *JKLM*,

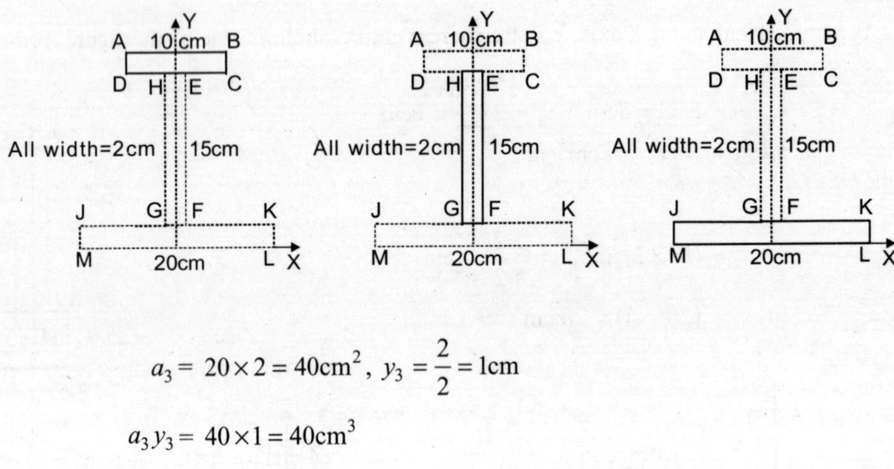

$$a_3 = 20\times 2 = 40\text{cm}^2,\ y_3 = \frac{2}{2} = 1\text{cm}$$

$$a_3 y_3 = 40\times 1 = 40\text{cm}^3$$

$$\therefore \quad y_G = \frac{a_1 y_1 + a_2 y_2 + a_3 y_3}{a_1 + a_2 + a_3} = \frac{360+285+40}{20+30+40} = 7.61\text{cm}$$

EXAMPLE 27

Determine the centroid of the plane uniform lamina as shown in the figure.

SOLUTION x_G and y_G be the co-ordinates of the centroid of the lamina with respect to origin as *D*. From rectangle *ABEF*,

$$a_1 = 10\times 5 = 50\text{cm}^2,\ x_1 = 2.5 + \frac{10}{2} = 7.5\text{cm},$$

$$\Rightarrow \quad a_1 x_1 = 50\times 7.5 = 375\text{cm}^3$$

$$y_1 = \frac{5}{2} = 2.5\text{cm}$$

$$\Rightarrow \quad a_1 y_1 = 50\times 2.5 = 125\text{cm}^3$$

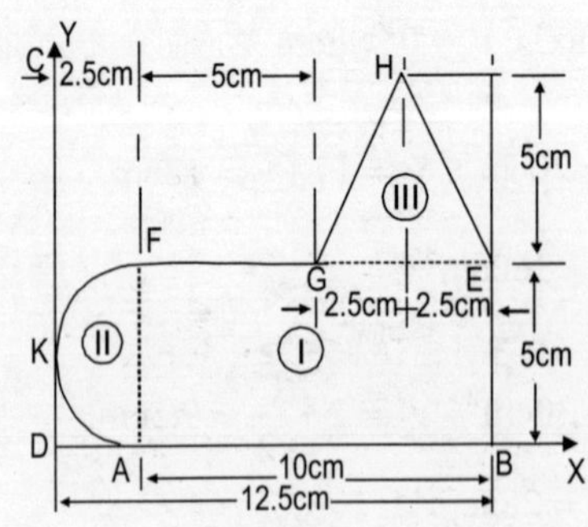

From semi-circle *AKF*,

$$a_2 = \frac{\pi \times 2.5^2}{2} = 9.8174\text{cm}^2,\ x_2 = 2.5 - \frac{4 \times 2.5}{3\pi} = 1.438\text{cm},\ y_2 = \frac{5}{2} = 2.5\text{cm}$$

$$a_2x_2 = 9.8174 \times 1.438 = 14.117\text{cm}^3,\ a_2y_2 = 9.8174 \times 2.5 = 24.543\text{cm}^3$$

From triangle *GHE*,

$$a_3 = \frac{1}{2} \times 5 \times 5 = 12.5\text{cm}^2,\ x_3 = 2.5 + 5 + 2.5 = 10\text{cm},\ y_3 = 5 + \frac{5}{3} = 6.67\text{cm}$$

$$a_3x_3 = 12.5 \times 10 = 125\text{cm}^3,\ a_3y_3 = 12.5 \times 6.67 = 83.375\text{cm}^3$$

$$x_G = \frac{a_1x_1 + a_2x_2 + a_3x_3}{a_1 + a_2 + a_3} = \frac{375 + 14.117 + 125}{50 + 9.8174 + 12.5} = 7.109\text{cm}$$

$$y_G = \frac{a_1y_1 + a_2y_2 + a_3y_3}{a_1 + a_2 + a_3} = \frac{125 + 24.543 + 83.375}{50 + 9.8174 + 12.5} = 3.2208\text{cm}$$

EXAMPLE 28

A right circular cylindrical tank containing water spins about its vertical geometric axis *O–O* at such speed that the free water surface is a paraboloid *ACB*. What will be the depth of water in the tank when it comes to rest?

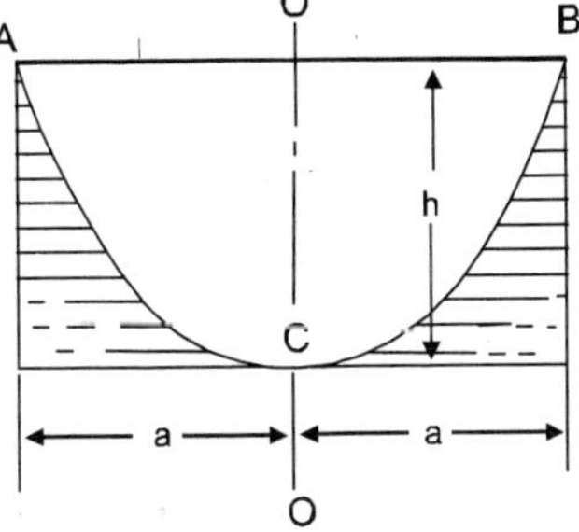

SOLUTION Let the equation of parabola is $x^2 = 4ky$.

For $\quad x = a$ and $y = h$,

$$k = \frac{a^2}{4h}$$

So, $\quad y = \frac{h}{a^2}x^2$

Elementary area chosen $dA = y\,dx = \left(\frac{h}{a^2}x^2\right)dx$

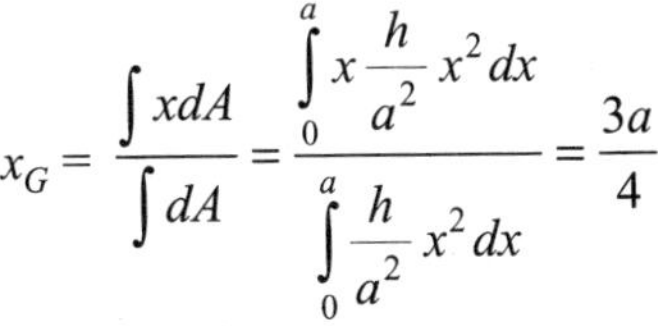

$$x_G = \frac{\int x\,dA}{\int dA} = \frac{\int_0^a x\frac{h}{a^2}x^2\,dx}{\int_0^a \frac{h}{a^2}x^2\,dx} = \frac{3a}{4}$$

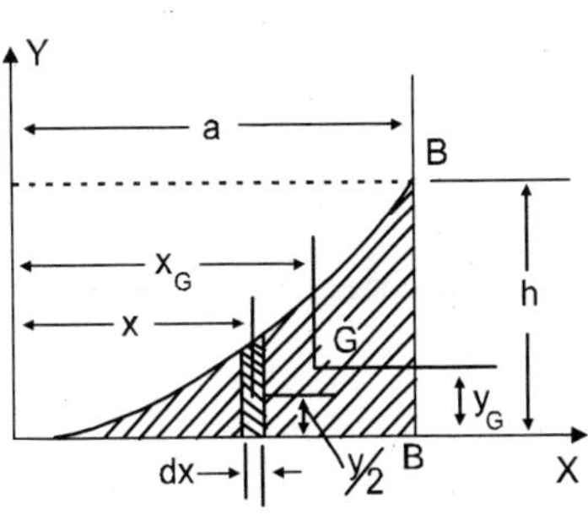

Let *H* be the height of water at rest. So, volume of water $\forall = \pi a^2 H$

Again $\quad \forall = 2\pi x_G A$

or, $\quad \pi a^2 H = 2\pi \cdot \frac{3a}{4} \cdot \frac{ah}{3}$

$\therefore \quad H = \frac{h}{2}.$

EXAMPLE 29

Determine by applying the Pappus theorem (*i*) the surface area, (*ii*) volume of a hemisphere of radius *a*.

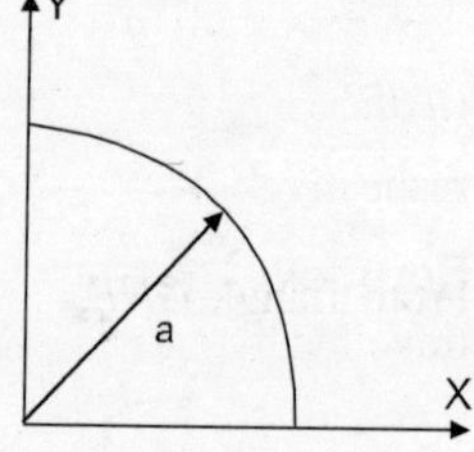

SOLUTION (*i*) The surface area of the hemisphere is obtained by rotating the quarter of circle of radius *a* about the *X*-axis as shown in figure, through 2π radian. The co-ordinate of the centroid from *X*-axis of the arc, $y_G = 2a/\pi$ and length of arc

$$L = 2\pi a / 4$$

∴ Surface area of hemisphere,

$$S = 2\pi L y_G = 2\pi\left(\frac{2\pi a}{4}\right)\times\frac{2a}{\pi} = 2\pi a^2$$

(*ii*) The volume of the hemisphere will be obtained by revolving the quadrant of a circle of radius *a* about the *X*-axis through 2π radian as shown in figure. The area of the quadrant $A = \frac{\pi a^2}{4}$, and co-ordinate y_G of centroid $= \frac{4a}{3\pi}$.

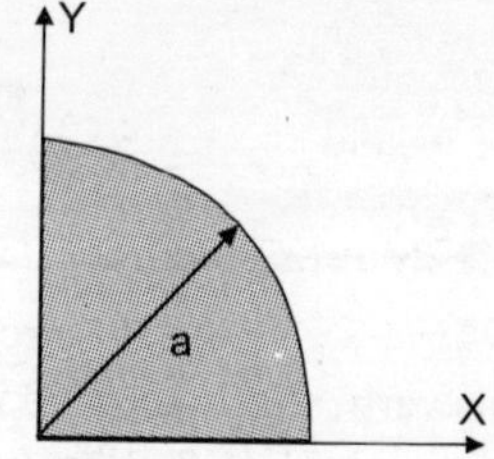

∴ Volume of hemisphere,

$$\forall = 2\pi A y_G = 2\pi\left(\frac{\pi a^2}{4}\right)\times\frac{4a}{3\pi} = \frac{2}{3}\pi a^3.$$

EXAMPLE 30

Determine the centroid of a quadrant of a circle, using the theorem of Pappus.

SOLUTION The volume of the hemisphere will be obtained by revolving the quadrant of a circle of radius *a* about the *X*-axis through 2π radian as shown in figure.

The area of the quadrant $A = \frac{\pi a^2}{4}$, and co-ordinate $= y_G$

∴ Volume of hemisphere, $\forall = 2\pi A y_G = 2\pi\left(\frac{\pi a^2}{4}\right)\times y_G$

Again using mensuration $\forall = \frac{2}{3}\pi a^3$

Equating these two,

$$2\pi y_G \frac{\pi a^2}{4} = \frac{2}{3}\pi a^3$$

∴ $$y_G = \frac{4a}{3\pi}.$$

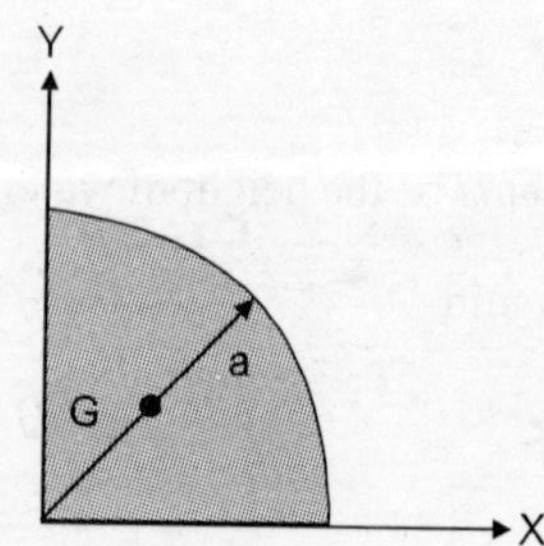

EXAMPLE 31

Using the theorems of Pappus-Guldinus, determine (*i*) the centroid of a semi-circular area (*ii*) the centroid of a semi-circular arc. We recall that the volume and the surface area of sphere are $\frac{4}{3}\pi r^3$ and $4\pi r^2$, respectively.

SOLUTION (*i*) The volume of a sphere is equal to the product of the area of a semi-circle and the distance travelled by the centroid of the semi-circle in one revolution about *X*-axis.

$$\therefore \quad \forall = 2\pi y_G A$$

$$\text{or} \quad \frac{4}{3}\pi r^3 = 2\pi y_G\left(\frac{\pi r^2}{2}\right)$$

$$\therefore \quad y_G = \frac{4r}{3\pi}$$

(*ii*) Likewise, the area, of a sphere is equal to the product of the length of the generating semi-circle and the distance travelled by its centroid in one revolution

$$\therefore \quad A = 2\pi y_G L$$

$$\text{or,} \quad 4\pi r^2 = 2\pi y_G(\pi r)$$

$$\therefore \quad y_G = \frac{2r}{\pi}$$

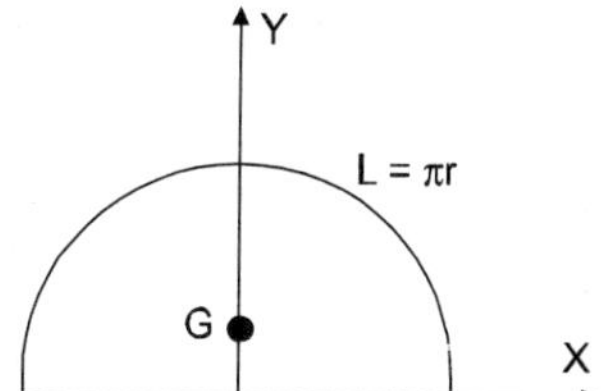

EXAMPLE 32

Determine the area of the surface of revolution, obtained by rotating a quarter circular arc about a vertical axis.

SOLUTION According to first theorem of Pappus-Guldinus, the area generated is equal to the product of the length of the arc and the distance travelled by its centroid. Referring to figure

$$x_G = 2r - \frac{2r}{\pi} = 2r\left(1 - \frac{1}{\pi}\right) \text{ and length } L = \frac{\pi r}{2}.$$

$$\therefore \text{ Area} \quad (A) = 2\pi x_G L$$

$$= 2\pi\left[2r\left(1 - \frac{1}{\pi}\right)\right]\left(\frac{\pi r}{2}\right) = 13.456r^2$$

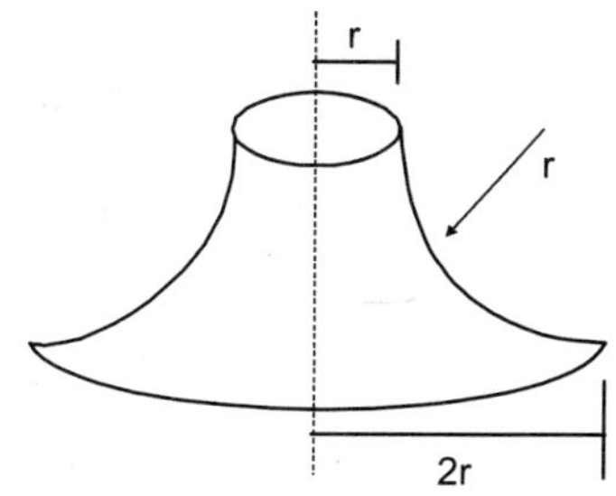

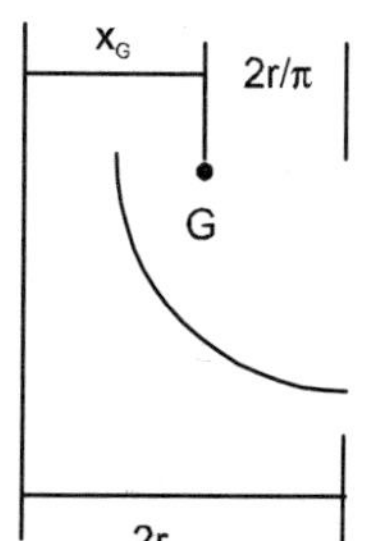

EXAMPLE 33

The circular arc is rotated through 360° about the Y-axis. Determine the outer surface area S of the resulting body, which is a portion of a sphere.

SOLUTION According to first theorem of Pappus-Guldinus, the area generated is equal to the product of the length of the arc and the distance travelled by its centroid. Referring to figure, $\alpha = 180° - 2\times 45° = 90° = \pi/2$ and length of the arc $L = a\cdot(\pi/2)$.

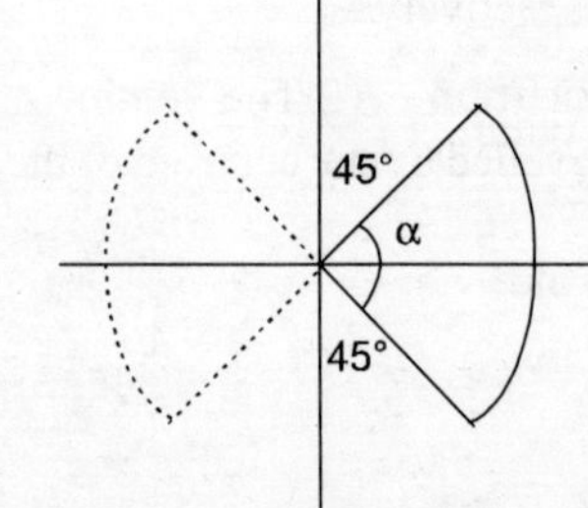

$$\therefore \qquad x_G = \frac{a\cdot\sin(\alpha/2)}{\alpha/2} = \frac{2a\cdot\sin(\pi/2)}{\pi/2} = \frac{2\sqrt{2}a}{\pi}$$

$\therefore$ Surface area
$$S = 2\pi x_G L = 2\pi\left(\frac{2\sqrt{2}a}{\pi}\right)\times\left(\frac{a\pi}{2}\right) = 2\sqrt{2}\pi a^2$$

EXAMPLE 34

Second moment of area of a rectangular section (a) about the centroidal axis, (b) about any edge of the section.

SOLUTION (a) Consider a small strip of thickness dy at a distance y from centroidal axis x_G and length b. The area of the small strip $dA = bdy$. Centroid of the strip $y_1 = y$. The second moment of whole area about centroidal x_G-axis,

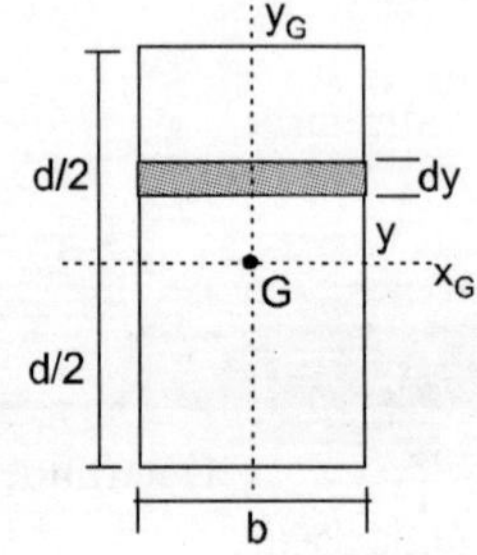

$$I_{xG} = \int_{-d/2}^{d/2} y_1^2 dA = \int_{-d/2}^{d/2} by^2 dy = \frac{1}{12}bd^3$$

Similarly, we have $\quad I_{yG} = \frac{1}{12}db^3$

Above relationship can also be found by considering small strip parallel to Y-axis at a distance x from Y-axis and integrating accordingly.

(b) Using the parallel axis theorem,

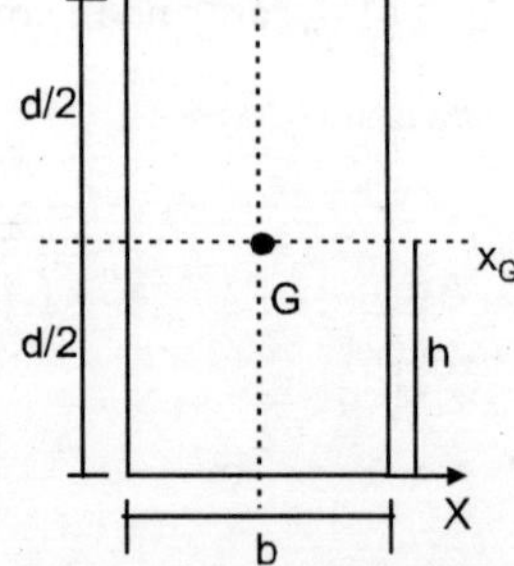

$$I_x = I_{x_G} + Ah^2 = \frac{1}{12}bd^3 + (b\cdot d)\cdot(d/2)^2 = \frac{1}{3}bd^3$$

Similarly,
$$I_y = \frac{1}{12}db^3 + (b\cdot d)\cdot(b/2)^2 = \frac{1}{3}db^3$$

EXAMPLE 35

Find the second moment of area of a hollow rectangular section of the figure (a) about centroidal axis (b) about edge of the section.

SOLUTION (a) Second moment of area about centroidal axis,

$$I_{xG} = I_{x1} + I_{x2} = \left(\frac{1}{12}bd^3 - \frac{1}{12}b_1d_1^3\right)$$

$$= \frac{1}{12}\left(bd^3 - b_1d_1^3\right)$$

Similarly, $$I_{yG} = \frac{1}{12}\left(db^3 - d_1 b_1^{\ 3}\right)$$

(*b*) Second moment of area about *X*-axis

$$I_x = I_{x1} - I_{x2}$$

Again, $$I_{x1} = I_{xG1} + A_1 h_1^{\ 2} = \frac{1}{12}bd^3 + (b \cdot d)\cdot(d/2)^2 = \frac{1}{3}bd^3$$

And, $$I_{x2} = I_{xG2} + A_2 h_2^{\ 2}$$

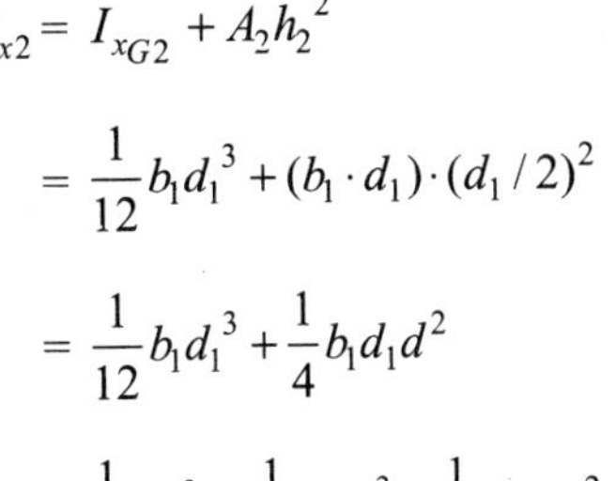

$$= \frac{1}{12}b_1 d_1^{\ 3} + (b_1 \cdot d_1)\cdot(d_1/2)^2$$

$$= \frac{1}{12}b_1 d_1^{\ 3} + \frac{1}{4}b_1 d_1 d^2$$

$$\therefore \quad I_x = \frac{1}{3}bd^3 - \frac{1}{12}b_1 d_1^{\ 3} - \frac{1}{4}b_1 d_1 d^2$$

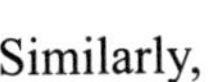

Similarly,

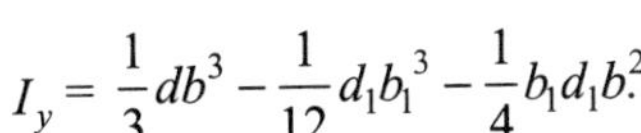

$$I_y = \frac{1}{3}db^3 - \frac{1}{12}d_1 b_1^{\ 3} - \frac{1}{4}b_1 d_1 b^2$$

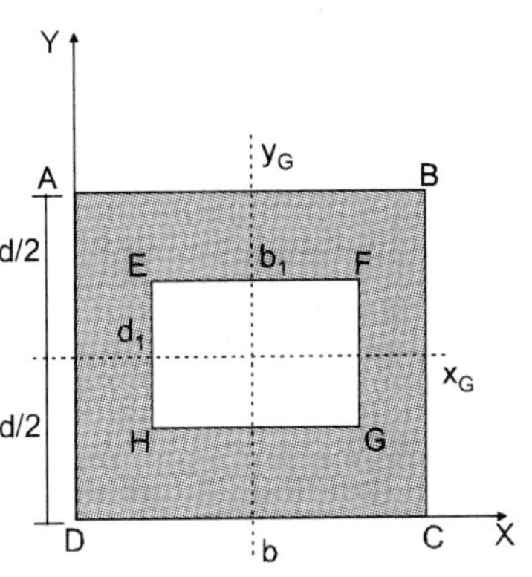

EXAMPLE 36

Find the second moment of area of an isosceles triangle of base *b* and height *h*, (*a*) about the axes *x* and *y*, (*b*) about an axis passing through the centroidal axis parallel to the base, (*c*) about an axis passing through vertex parallel to centroidal axis.

SOLUTION (*a*) Let us assume the axis of reference *x–y* shown in figure. Consider a small strip of thickness *dy* at a distance *y* from *X*-axis. Let length of strip be *x*. Now from similar triangles *ABC* and *ECF*,

$$\frac{x}{b} = \frac{h-y}{h} \qquad \Rightarrow \qquad x = \frac{b(h-y)}{h}$$

Area of elementary strip

$$dA = x \cdot dy = \frac{b(h-y)}{h}\cdot dy$$

$$\therefore \quad I_x = \int_0^h y^2 \frac{b}{h}(h-y)dy = \frac{1}{12}bh^3$$

This result is valid for any type of triangles. Similarly, the M.I of the triangles *OCB* about *Y*-axis

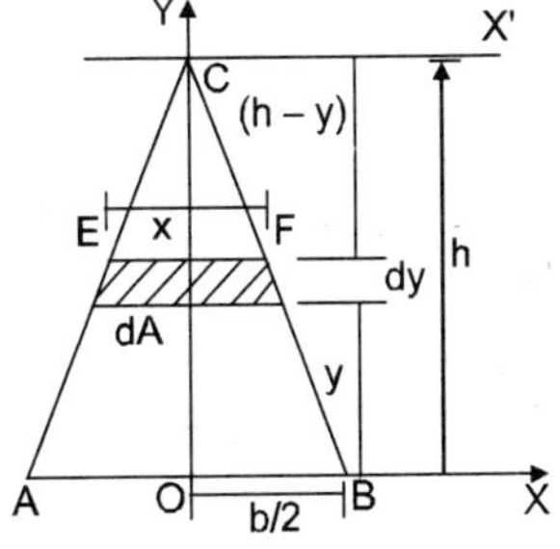

$$= \frac{1}{12}h(b/2)^3 = \frac{1}{96}hb^3$$

$$\therefore \quad I_y = 2\times\frac{1}{96}hb^3 = \frac{1}{48}hb^3$$

(*b*) From parallel axis theorem we have,

$$I_x = I_{xG} + Ah_1^2$$

$$\Rightarrow \qquad \frac{1}{12}bh^3 = I_{xG} + \frac{1}{2}bh(h/3)^2$$

$$I_{xG} = \frac{1}{12}bh^3 - \frac{1}{18}bh^3 = \frac{1}{36}bh^3$$

This result is valid for any type of triangles.This is the second moment of the triangle about the axis passing through centroidal axis parallel to base.

(*c*) Now applying parallel axis theorem, we have

$$I_{x'} = I_{xG} + Ah_2^2 = \frac{1}{36}bh^3 + \frac{1}{2}bh\left(\frac{2h}{3}\right)^2 = \frac{1}{4}bh^3$$

This result is also valid for any type of triangles.

EXAMPLE 37

Determine the second moment of a circular area of radius a (*i*) about an axis perpendicular to both x and y-axis i.e., z-axis (polar moment of inertia), (*ii*) about any diameter i.e., x or y-axis, (*iii*) about any tangential axis parallel to x or y-axis.

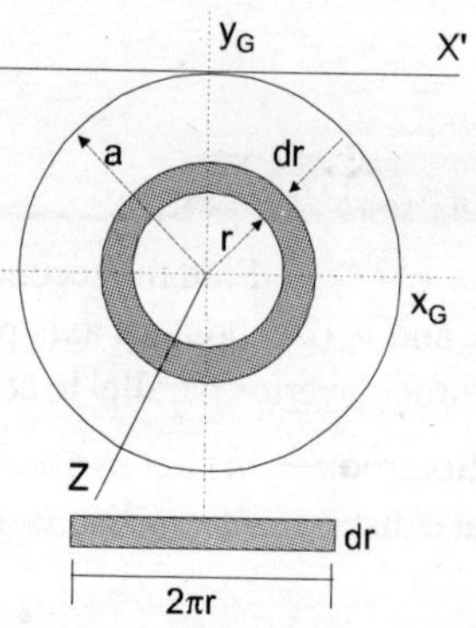

SOLUTION *Method I*: (*i*) The circular area together with the reference axis is shown in the figure. Let us consider a ring of width dr at a radius r from z axis, perpendicular to both x and y axis. The elementary area of the ring $dA = 2\pi r dr$.

$\therefore$ Polar moment of inertia about z-axis

$$I_z = J_O = \int r^2 dA = \int_0^a r^2 \cdot 2\pi r dr = \frac{\pi a^4}{2} = \frac{\pi d^4}{32}$$

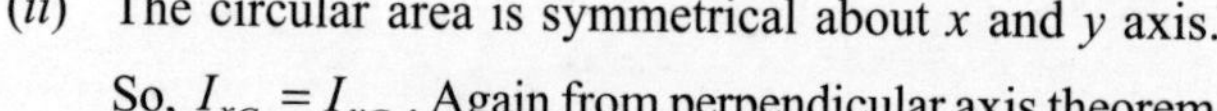

(*ii*) The circular area is symmetrical about x and y axis.

So, $I_{xG} = I_{yG}$. Again from perpendicular axis theorem,

$$I_{zG} = I_{xG} + I_{yG}$$

$$\therefore \qquad \frac{\pi a^4}{2} = 2I_{xG} = 2I_{yG}$$

$$I_{xG} = I_{yG} = \frac{\pi a^4}{4} = \frac{\pi d^4}{64}$$

(*iii*) Second moment of area about any tangential axis,

$$I_{x'} = I_{xG} + Ah^2 = \frac{\pi a^4}{4} + \pi a^2 \cdot a^2$$

$$= \frac{5\pi a^4}{4} = \frac{5\pi d^4}{64} = I_{y'}$$

Note: For hollow circular area having d_i and d_o as inside and outside diameter respectively,

$$I_z = \frac{\pi}{32}\left(d_o^4 - d_i^4\right) \text{ and } I_{xG} = I_{yG} = \frac{\pi}{64}\left(d_o^4 - d_i^4\right)$$

Method II : Let us consider a small element of area dA having width dr and angle subtended $d\theta$ at centre O, at a rotation θ.

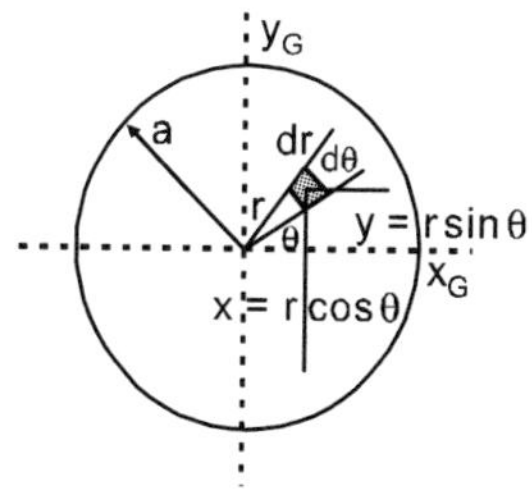

$$\therefore \quad dA = (rd\theta)\cdot dr \text{ and } x = r\cos\theta,\ y = r\sin\theta$$

$$\therefore \quad I_{yG} = \int x^2 dA = \int_0^{2\pi}\int_0^{a} (r\cos\theta)^2 \cdot (rd\theta)dr$$

$$= \int_0^{2\pi} \frac{a^4}{4}\cos^2\theta d\theta = \frac{\pi a^2}{4} = \frac{\pi d^4}{64} = I_{xG}$$

$$\therefore \quad J_O = I_{zG} = 2I_{xG} = 2I_{yG} = \frac{\pi a^4}{2} = \frac{\pi d^4}{32}.$$

EXAMPLE 38

Find second moment of the area of semi-circle of radius a (*i*) about Z-axis, Y-axis and X-axis, (*ii*) about centroidal axis.

SOLUTION (*i*) Area of element $dA = \pi r dr$

$$\therefore \quad I_z = \int r^2 dA = \int_0^a r^2 \pi r dr = \frac{\pi a^4}{4} = \frac{\pi d^4}{64}$$

But $$I_z = I_x + I_y = 2I_x = 2I_y$$

$$\therefore \quad I_x = \frac{\pi d^4}{128} = \frac{\pi a^4}{8} = I_y$$

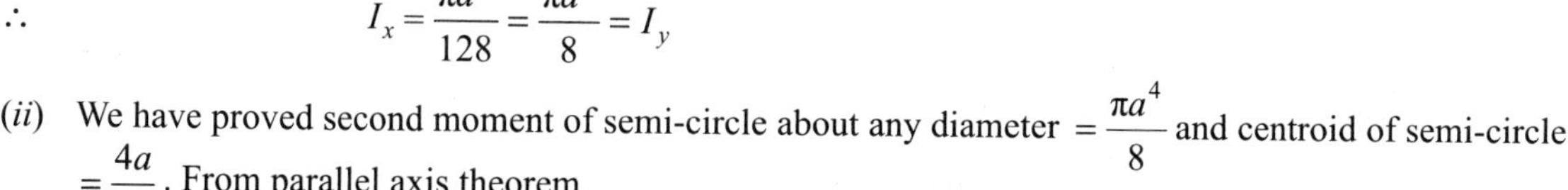

(*ii*) We have proved second moment of semi-circle about any diameter $= \frac{\pi a^4}{8}$ and centroid of semi-circle $= \frac{4a}{3\pi}$. From parallel axis theorem,

$$I_x = I_{xG} + Ah^2$$

$$\therefore \quad I_{xG} = I_x - Ah^2 = \frac{\pi a^4}{4} - \frac{\pi a^2}{2}\cdot\left(\frac{4a}{3\pi}\right)^2 = 0.5024a^4$$

$$I_y = I_{yG} + Ah_1^2$$

$$\therefore \quad I_{yG} = I_y - Ah_1^2 = \frac{\pi a^4}{4} - \frac{\pi a^2}{2}\cdot 0 = \frac{\pi a^4}{4}.$$

EXAMPLE 39

Find second moment of the area of quarter circle of radius a (*i*) about Y-axis and X-axis, (*ii*) about centroidal axis.

SOLUTION (*i*) Area of element $dA = \frac{\pi r}{2} dr$

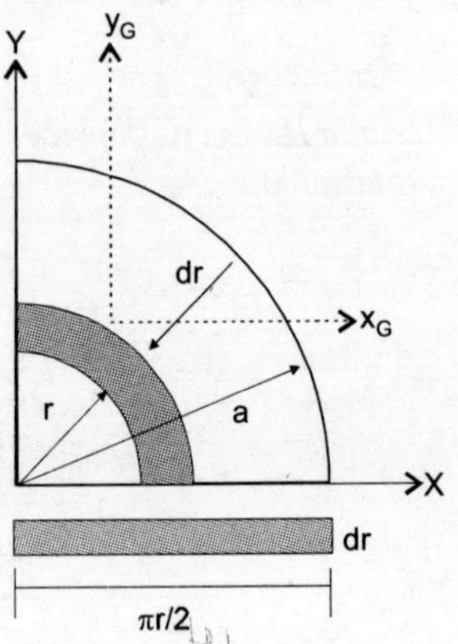

$$\therefore \quad I_z = \int_0^a r^2 dA = \int_0^a r^2 \frac{\pi r}{2} dr = \frac{\pi a^4}{8} = \frac{\pi d^4}{128}$$

Again, $I_x = I_y$ and $I_z = I_x + I_y = 2I_x = 2I_y$

$$\therefore \quad I_y = I_x = \frac{\pi a^4}{16} = \frac{\pi d^4}{256}$$

(*ii*) Centroid distance of the quarter circle $= \frac{4a}{3\pi}$

We know $I_x = I_{x_G} + Ah^2$

$$\therefore \quad I_{x_G} = I_x - Ah^2 = \frac{\pi a^4}{16} - \frac{\pi a^2}{4} \cdot \left(\frac{4a}{3\pi}\right)^2 = 0.0548a^4 = I_{y_G}$$

EXAMPLE 40

Determine second moment of area enclosed by the ellipse $\frac{x^2}{a^2} + \frac{y^2}{b^2} = 1$ with respect to X-axis. Also determine the radius of gyration.

SOLUTION Area of element $(dA) = 2ydx$

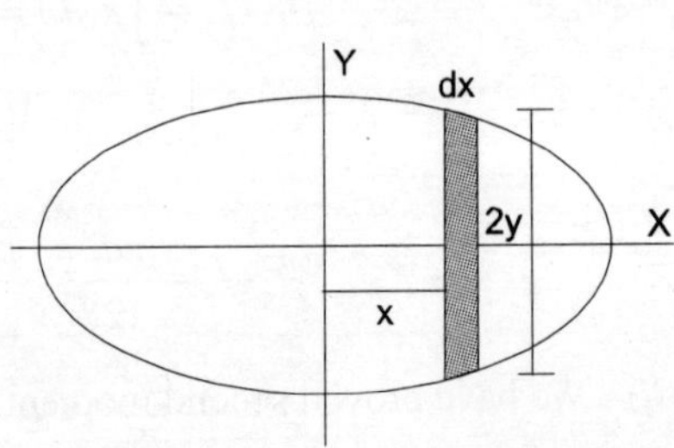

Again, $1 = \frac{x^2}{a^2} + \frac{y^2}{b^2}$

$$\therefore \quad y = \left[\left(1 - \frac{x^2}{a^2}\right) b^2\right]^{\frac{1}{2}} = \frac{b}{a}\left(a^2 - x^2\right)^{\frac{1}{2}}$$

$$\therefore \quad dA = \frac{2b}{a}\left(a^2 - x^2\right)^{\frac{1}{2}} dx$$

$$\therefore \quad I_x = \int y^2 dA = \int_{-a}^{a} \frac{b^2}{a^2}\left(a^2 - x^2\right) \frac{2b}{a}\left(a^2 - x^2\right)^{\frac{1}{2}} dx$$

$$= \frac{2b^3}{a^3} \int_{-a}^{a} \left(a^2 - x^2\right)^{\frac{3}{2}} dx$$

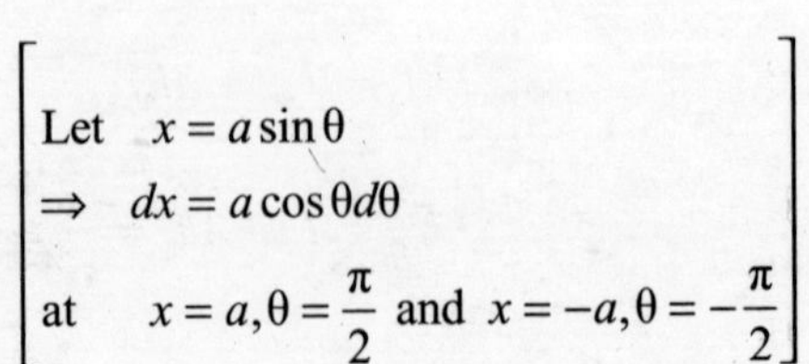

$$\left[\begin{array}{l} \text{Let } x = a\sin\theta \\ \Rightarrow dx = a\cos\theta d\theta \\ \text{at } x = a, \theta = \frac{\pi}{2} \text{ and } x = -a, \theta = -\frac{\pi}{2} \end{array}\right]$$

$$= \frac{2b^3}{a^3} \int_{-\pi/2}^{\pi/2} a^3 \cos^3\theta \cdot a\cos\theta d\theta$$

$$= 2ab^3 \int_{-\pi/2}^{\pi/2} \cos^4\theta d\theta$$

$$= 2ab^3 \int_{-\pi/2}^{\pi/2} \left(\frac{3}{8} + \frac{\cos 2\theta}{2} + \frac{\cos 4\theta}{8}\right) d\theta$$

$$= 2ab^3 \left[\frac{3\theta}{8} + \frac{\sin 2\theta}{4} + \frac{\sin 4\theta}{32}\right]_{\frac{-\pi}{2}}^{\frac{\pi}{2}} = \frac{\pi ab^3}{4}$$

We know $I_x = A \cdot k_x^{\ 2}$ where k_x = radius of gyration.

So, $$\frac{\pi ab^3}{4} = \pi ab \cdot k_x^{\ 2} \quad \Rightarrow \quad k_x = \frac{b}{2}$$

EXAMPLE 41

Find the second moment of the shaded area under the curve $y = kx^2$ with respect to the X and Y axis.

SOLUTION Substituting $x = a$, and $y = b$ in the equation, we find $b = ka^2$.

So, $$k = \frac{b}{a^2} \text{ and } y = \frac{b}{a^2} \cdot x^2.$$

Area of the element $dA = ydx = \frac{b}{a^2} x^2 dx$

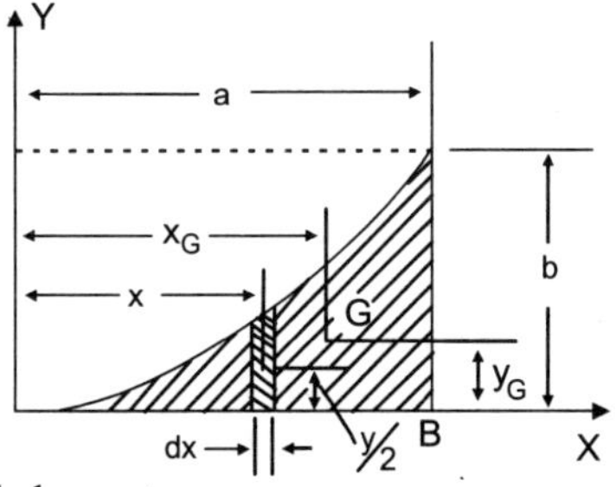

Second moment of area of differential element

$$dI_x = \frac{1}{3}(dx \cdot y)y^2 = \frac{1}{3}y^2 dA$$

$\therefore$ $$I_x = \int dI_x = \int_0^a \frac{1}{3} \cdot \frac{b^2}{a^4} \cdot x^4 \cdot \frac{b}{a^2} x^2 dx = \frac{b^3}{3a^6} \int_0^a x^6 dx = \frac{ab^3}{21}$$

Again, $$I_y = \int_0^a x^2 \cdot \frac{b}{a^2} \cdot x^2 dx = \frac{b}{a^2} \int_0^a x^4 dx = \frac{ba^3}{5}$$

Example 42

Determine the M.I of a uniform thin rod with respect to *Y*-axis as shown in figure.

Solution Let m = mass per unit length of rod. and

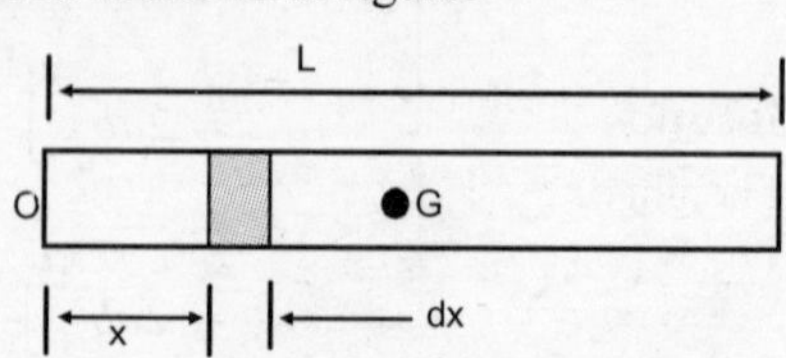

M = total mass of the rod = mL

Mass of the elemental strip = $m \cdot dx$

$$\therefore \qquad I_y = \int_0^L (mdx)x^2 = m\left[\frac{x^3}{3}\right]_0^L = \frac{mL^3}{3} = \frac{ML^2}{3}$$

Example 43

If the second moment of thin strip of area about x-axis is $2.56\times10^6\,\text{mm}^4$, determine the area A of the strip within a close approximation.

Solution Here, $I_{xG} = 0,\ I_x = 2.56\times10^6\,\text{mm}^4,\ h = 40\text{mm}$

From parallel axis theorem

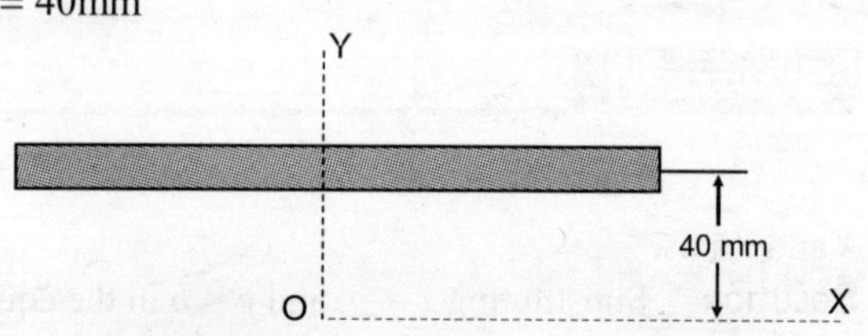

$$I_x = I_{xG} + Ah^2$$

$$\therefore \qquad 2.56\times10^6 = 0 + A\times40^2$$

$$\therefore \qquad A = 1600\text{mm}^2$$

Example 44

The shaded area shown in figure is equal to 125 cm². Its I_x=35000 cm⁴, $I_{x'}$=70000cm⁴. Determine the distance d_1 and the second moment about centroidal axis parallel to X-axis and about x'' axis parallel to centroidal axis.

Solution From parallel axis theorem,

$$I_x = I_{xG} + Ad_2^2 \text{ and } I_{x'} = I_{xG} + A(d_1 + d_2)^2$$

$$70000 = I_{xG} + A(d_1 + 7.5)^2$$

$$35000 = I_{xG} + A\times7.5^2$$

$$35000 = A\left[(d_1 + 7.5)^2 - 7.5^2\right]$$

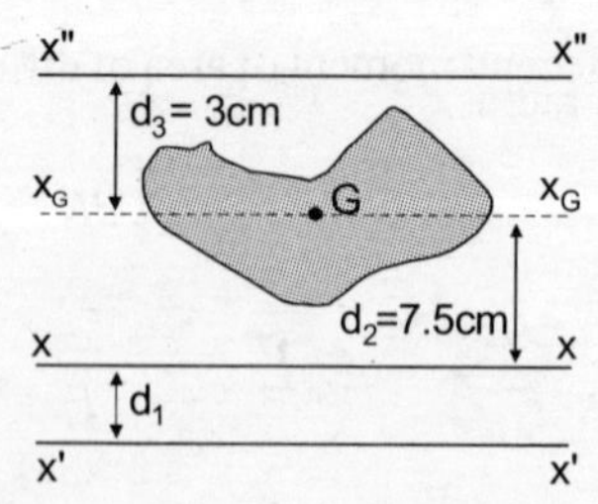

$$\Rightarrow \qquad (d_1 + 7.5)^2 = 336.25$$

$$\Rightarrow \qquad d_1 + 7.5 = 18.337$$

$$\therefore \qquad d_1 = 10.837\text{cm}$$

Again,

$$35000 = I_{xG} + 125\times7.5^2$$

$$\therefore \qquad I_{xG} = 35000 - 125\times7.5^2 = 27968.75\text{cm}^4$$

$$\therefore \qquad I_{x''} = I_{xG} + Ad_3^2 = 27968.75 + 125\times3^2 = 29093.75\text{cm}^4$$

EXAMPLE 45

Find the second moment of area of a rectangle having dimensions a and b with respect to a diagonal.

SOLUTION Here base $DB = \sqrt{a^2 + b^2}$

If height of triangle ABD be h,

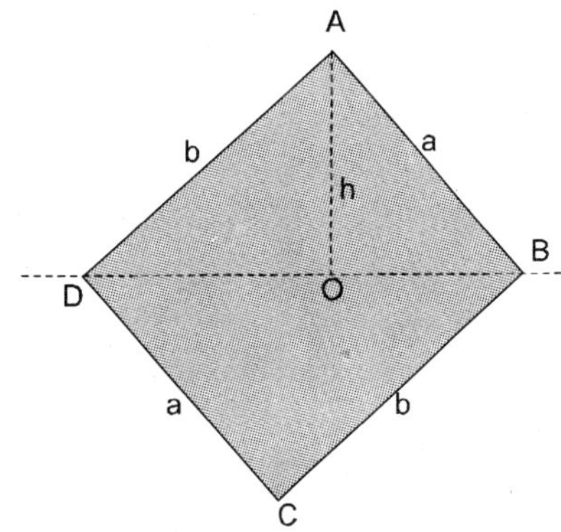

$$\frac{1}{2} DB \times h = \frac{1}{2} \times ab$$

$$\Rightarrow \quad \frac{1}{2}\sqrt{a^2 + b^2} \times h = \frac{1}{2} \times ab$$

$$\Rightarrow \quad h = \frac{ab}{\sqrt{a^2 + b^2}}$$

Second moment of triangle ABD about base DB,

$$I_{DB} = \frac{1}{12} \times DB \times h^3 = \frac{\sqrt{a^2+b^2}}{12} \times \frac{a^3b^3}{\left(\sqrt{a^2+b^2}\right)^3} = \frac{a^3b^3}{12\left(a^2+b^2\right)}$$

$\therefore$ M.I of rectangle about diagonal $I_d = 2I_{DB} = \dfrac{a^3b^3}{6\left(a^2+b^2\right)}$.

EXAMPLE 46

Determine radius of gyration about a polar axis (z-axis), through the midpoint of the hypotenuse of the right-triangular area shown in figure.

SOLUTION M.I of a rectangle $a \times b$ about polar axis, passing through P

$$I_z = \frac{ab}{12}\left(a^2 + b^2\right)$$

M.I of a triangle about polar axis passing through P

$$I_{z1} = \frac{1}{2} \times \frac{ab(a^2+b^2)}{12}$$

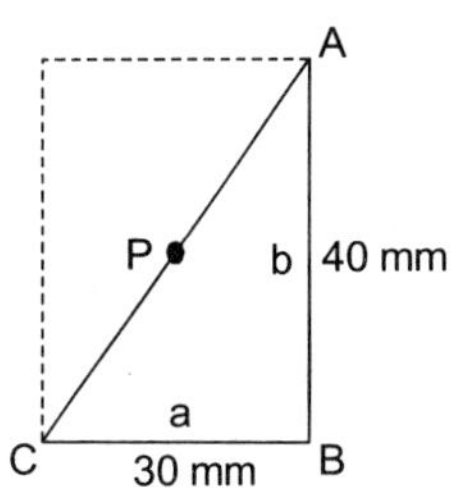

If radius of gyration about polar axis be k_z

$$Ak_z^2 = I_{z1}$$

$$\therefore \quad \frac{ab}{2} \times k_z^2 = \frac{ab\left(a^2+b^2\right)}{24}$$

$$\therefore \quad k_z = \sqrt{\frac{a^2+b^2}{12}} = \sqrt{\frac{30^2+40^2}{12}} = 14.43\text{mm}$$

EXAMPLE 47

Find the second moment of the shaded area about *AB*.

SOLUTION Second moment of rectangle *ABCD* about *AB*

$$I_{x1} = I_{xG1} + A_1 h_1^2 = \frac{1}{12} \times 20 \times 25^3 + 20 \times 25 \times \left(\frac{25}{2}\right)^2$$

$$= 104166.67 \text{cm}^4$$

For semi-circle

$$I_{x2} = I_{xG2} + Ah^2$$

$$\therefore \quad I_{xG2} = I_{x2} - Ah^2 = \frac{\pi a^4}{4} - \frac{\pi a^2}{2}\left(\frac{4a}{3\pi}\right)^2 = 0.5024a^4$$

M.I of semi-circle about *AB*

$$I'_{x2} = I_{xG_2} + Ah_2^2 = 0.5024a^4 + \frac{\pi a^2}{2}\left[25 - \left(\frac{4a}{3\pi}\right)\right]^2$$

$$= 0.5024 \times 10^4 + \frac{\pi \times 10^2}{2}\left(25 - \frac{4 \times 10}{3\pi}\right)^2 = 72694.86 \text{cm}^4$$

$\therefore$ M.I of shaded area about *AB*

$$= I_{x1} - I'_{x2} = 104166.67 - 72694.86 = 31471.81 \text{cm}^4$$

EXAMPLE 48

Determine second moment of the shaded area about *X*-axis and *Y*-axis. Use the same differential element for both.

SOLUTION Second moment of rectangle *DE* = second moment of rectangle *DF* – second moment of rectangle *EF*.

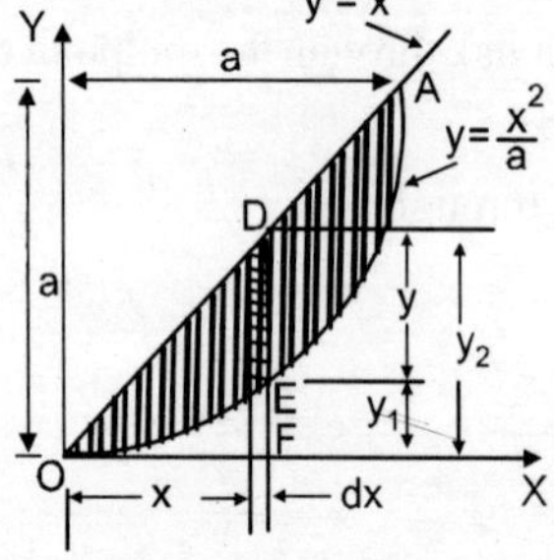

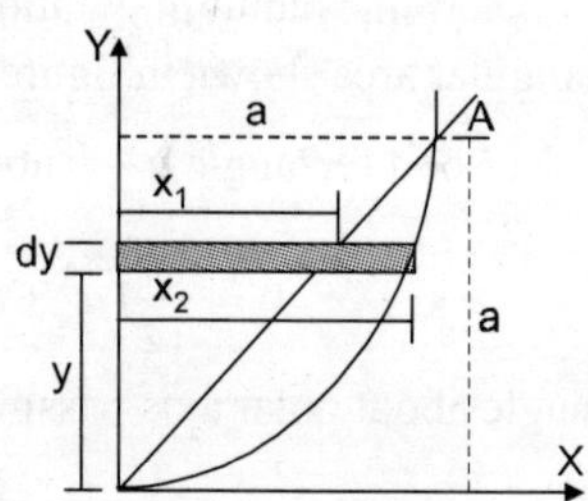

$$\therefore \quad dI_x = dI_{x2} - dI_{x1} = \frac{1}{3}y_2^3 dx - \frac{1}{3}y_1^3 dx = \frac{1}{3}\left[x^3 - \frac{x^6}{a^3}\right]dx$$

$$\therefore \quad I_x = \int dI_x = \frac{1}{3}\int_0^a \left(x^3 - \frac{x^6}{a^3}\right)dx = \frac{a^4}{28}$$

Again $\quad dI_y = dI_{y2} - dI_{y1} = \frac{1}{3}x_2^3 dx - \frac{1}{3}x_1^3 dx = \frac{1}{3}\left[(ay)^{\frac{3}{2}} - y^3\right]dy$

$$\therefore \quad I_y = \int dI_y = \frac{1}{3}\int_0^a \left((ay)^{\frac{3}{2}} - y^3\right)dy = \frac{a^4}{20}.$$

EXAMPLE 49

Figure shows a *T*-section of dimension 10 × 10 × 2cm. Determine the second moment of the section about the x_G and y_G axes passing through the centroid of the section.

SOLUTION For rectangle *ABCD*, $a_1 = 10 \times 2 = 20\text{cm}^2,\ x_1 = 0,\ y_1 = 8 + \dfrac{2}{2} = 9\text{cm}$

For rectangle *EFGH*, $a_2 = 8 \times 2 = 16\text{cm}^2,\ x_2 = 0,\ y_2 = \dfrac{8}{2} = 4\text{cm}$

As it is symmetrical about *y*-axis, $x_G = 0$

$$\therefore \quad y_G = \frac{a_1 y_1 + a_2 y_2}{a_1 + a_2} = \frac{20 \times 9 + 16 \times 4}{20 + 16} = 6.777\text{cm}$$

I_{x1} is second moment of the rectangle *ABCD* about centroidal x_G axis of composite section.

$$\therefore \quad I_{x1} = I_{xG1} + a_1 h_1^2 = \frac{1}{12} \times 10 \times 2^3 + 20(6.777 - 9)^2 = 105.501\text{cm}^4$$

I_{x2} is second moment of the rectangle *EFGH* about centroidal x_G axis of composite section.

$$\therefore \quad I_{x2} = I_{xG2} + a_2 h_2^2 = \frac{1}{12} \times 2 \times 8^3 + 16(6.777 - 4)^2$$

$$= 208.72\text{cm}^4$$

$$\therefore \quad I_{xG} = I_{x1} + I_{x2} = 105.501 + 208.72 = 314.221\text{cm}^4$$

And

$$I_{yG} = \frac{1}{12} \times 2 \times 10^3 + \frac{1}{12} \times 8 \times 2^3 = 172\text{cm}^4$$

EXAMPLE 50

Find the second moment of the *L*-section about reference axes *x* and *y*.

SOLUTION

$$I_{x1} = \frac{1}{3} \times 2 \times 10^3 = 666.67\text{cm}^4$$

and

$$I_{y1} = \frac{1}{3} \times 10 \times 2^3 = 26.67\text{cm}^4$$

$$I_{x2} = \frac{1}{3} \times (6-2) \times 2^3 = 10.67\text{cm}^4$$

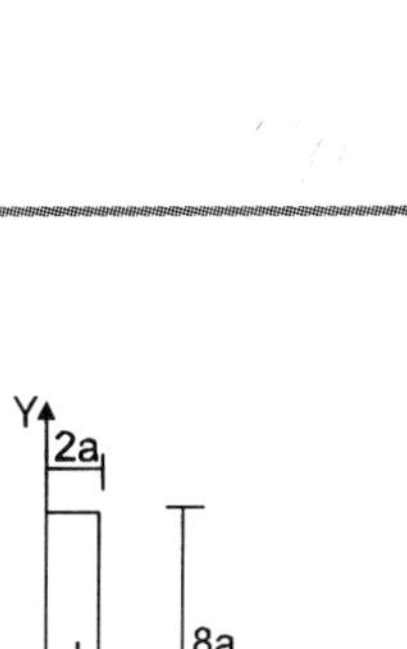

and

$$I_{y2} = \frac{1}{12} \times 2 \times (6-2)^3 + (6-2) \times 2 \times \left(2 + \frac{4}{2}\right)^2$$

$$= 138.67\text{cm}^4$$

Therefore, second moment of the entire section

$$I_x = I_{x1} + I_{x2} = 666.67 + 10.67 = 677.34\text{cm}^4$$

$$I_y = I_{y1} + I_{y2} = 26.67 + 138.67 = 165.34\text{cm}^4$$

EXAMPLE 51

Determine second moment of I.-Section about centroidal x_G axis.

SOLUTION Second moment of *ABLM* about x_G axis

$$= \frac{1}{12} \times 4a \times (4a)^3 = \frac{64a^4}{3}$$

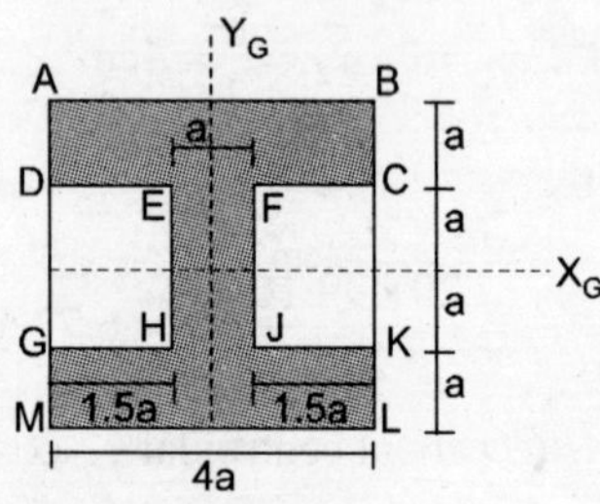

Second moment of $(DEHG + CFJK)$ about x_G axis

$$= \frac{1}{12} \times (4a - a) \times (2a)^3 = 2a^4$$

$\therefore$ second moment of I-section

$$= \frac{64a^4}{3} - 2a^4 = \frac{58a^4}{3}$$

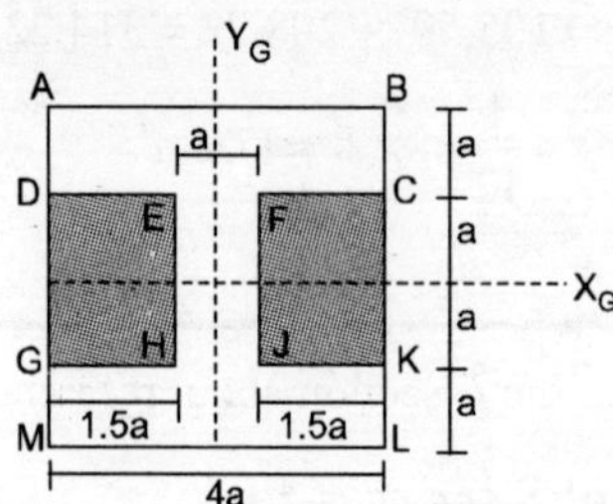

EXAMPLE 52

Determine second moment of shaded area about *X*-axis.

SOLUTION Here, $h = \sqrt{30^2 - 15^2} = 25.98\text{cm}$

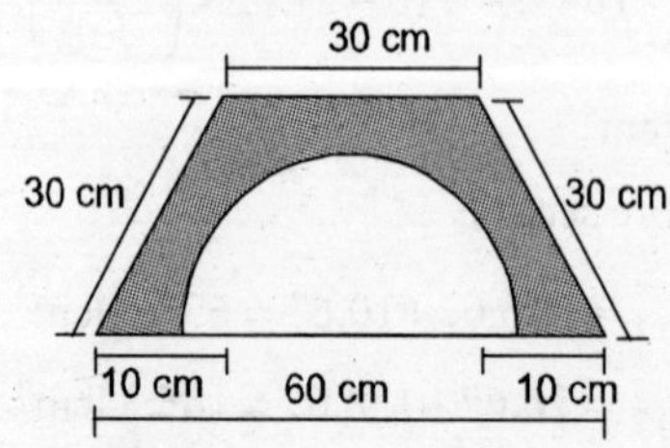

Second moment of trapezoidal section

$$I_{x1} = 2\times\frac{1}{12}\times 15(25.98)^3 + \frac{1}{3}\times 30(25.98)^3 = 219193.3919\text{cm}^4$$

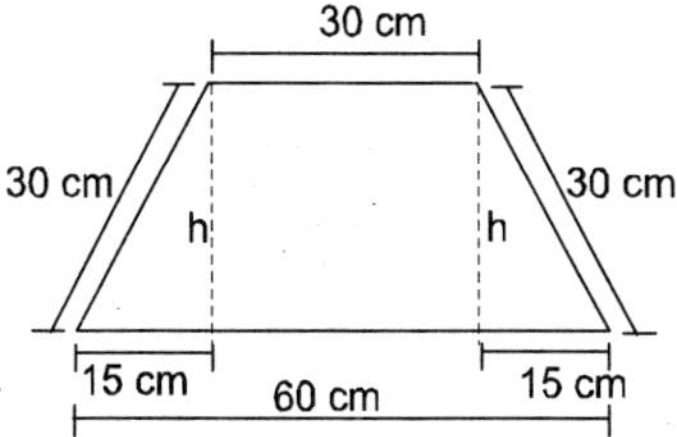

Second moment of semi-circle section

$$I_{x2} = \frac{1}{2}\times\frac{\pi}{4}\times 20^4 = 62831.853\text{cm}^4$$

Therefore, second moment of the shaded area

$$I_x = I_{x1} - I_{x2} = 219193.3919 - 62831.853 = 15.636\times 10^4\,\text{cm}^4$$

EXAMPLE 53

Determine Mass M.I of rectangular plate, about the axes shown.

SOLUTION

$$I_{YY',\ \text{mass}} = \rho\cdot t\cdot I_{YY',\ \text{area}} = \rho\cdot t\cdot\frac{ba^3}{12} = (\rho\, abt)\frac{a^2}{12} = \frac{1}{12}ma^2$$

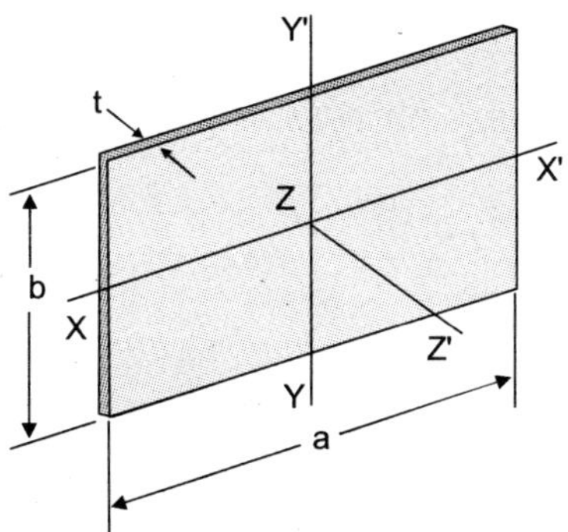

$$I_{XX',\ \text{mass}} = \rho\cdot t\cdot I_{XX',\ \text{area}} = \rho\cdot t\cdot\frac{ab^3}{12} = (\rho\, abt)\frac{b^2}{12} = \frac{1}{12}mb^2$$

$$I_{ZZ',\ \text{mass}} = I_{YY',\ \text{mass}} + I_{XX',\ \text{mass}} = \frac{1}{12}ma^2 + \frac{1}{12}mb^2.$$

EXAMPLE 54

Determine Mass M.I of circular plate, about the axes shown.

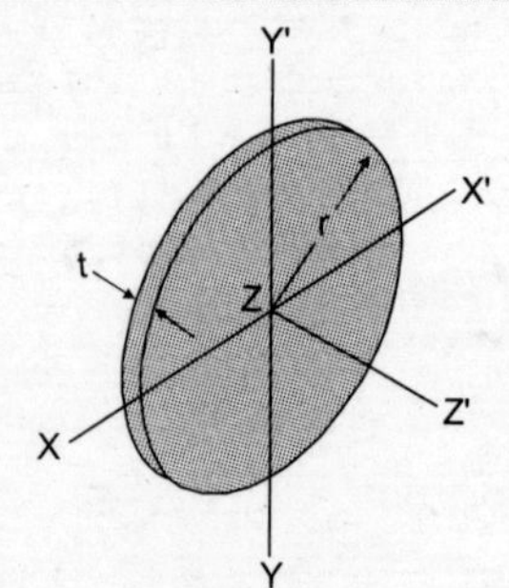

SOLUTION

$$I_{YY',\ \text{mass}} = \rho \cdot t \cdot I_{AA',\ \text{area}} = \rho \cdot t \cdot \frac{\pi r^4}{4}$$

$$= \left(\rho \pi r^2 t\right)\frac{r^2}{4} = \frac{1}{4}mr^2 = I_{XX',\ \text{mass}}$$

$$I_{ZZ',\ \text{mass}} = I_{YY',\ \text{mass}} + I_{XX',\ \text{mass}} = \frac{1}{4}mr^2 + \frac{1}{4}mr^2 = \frac{1}{2}mr^2.$$

EXAMPLE 55

Determine Mass M.I of homogeneous rectangular prism about $Z -$ axis

SOLUTION Considering a differential element of thin slab, elementary mass

$$dm = \rho bc\ dx \ \text{ and } \ dI_{z'} = \frac{1}{12}b^2 dm.$$

Applying parallel axis theorem

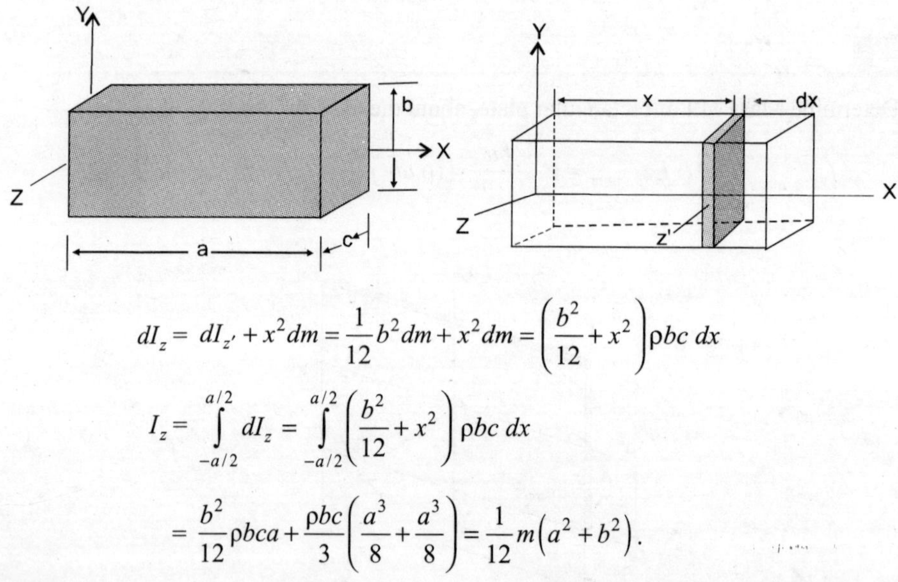

$$dI_z = dI_{z'} + x^2 dm = \frac{1}{12}b^2 dm + x^2 dm = \left(\frac{b^2}{12} + x^2\right)\rho bc\ dx$$

$$\therefore \quad I_z = \int_{-a/2}^{a/2} dI_z = \int_{-a/2}^{a/2}\left(\frac{b^2}{12} + x^2\right)\rho bc\ dx$$

$$= \frac{b^2}{12}\rho bca + \frac{\rho bc}{3}\left(\frac{a^3}{8} + \frac{a^3}{8}\right) = \frac{1}{12}m\left(a^2 + b^2\right).$$

EXAMPLE 56

Determine Mass M.I of slender rod, about Y-axis, Z-axis and X-axis.

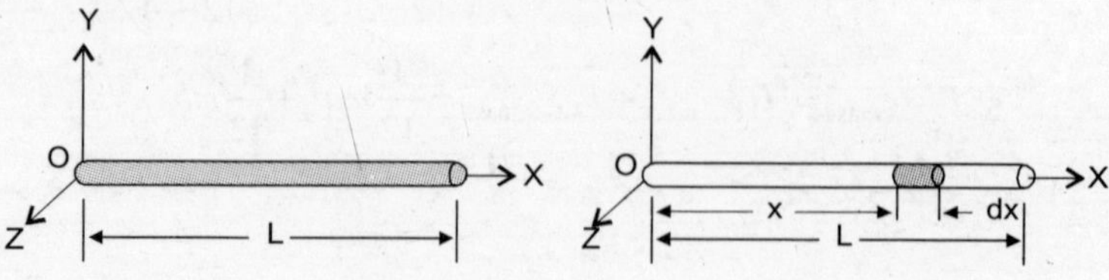

SOLUTION Mass of the elementary segment, $dm = \frac{m}{L}dx$

$$\therefore \quad I_y = \int x^2 dm = \int_{-L/2}^{L/2} x^2 \frac{m}{L} dx$$

$$= \frac{1}{12} mL^2 = I_z \text{ and } I_x = \frac{1}{3} mL^2 .$$

EXAMPLE 57

Determine Mass M.I of solid cylinder about the centroidal axes, as shown.

SOLUTION Applying parallel axis theorem, M.I of elemental disk

$$dI_x = \rho dz \frac{\pi a^4}{4} + \rho dz \cdot \pi a^2 \cdot z^2$$

So, for the entire cylinder

$$I_x = \int_{-l/2}^{l/2} dI_x = \int_{-l/2}^{l/2} \left(\rho \frac{\pi a^4}{4} + \rho \pi a^2 z^2 \right) dz$$

$$= \rho l \frac{\pi a^4}{4} + \rho \pi a^2 \frac{1}{3} \cdot \frac{l^3}{4}$$

$$= \pi a^2 l \rho \left(\frac{a^2}{4} + \frac{l^2}{12} \right) = m \left(\frac{a^2}{4} + \frac{l^2}{12} \right)$$

$$= \frac{m}{12} \left(3a^2 + l^2 \right) = I_y$$

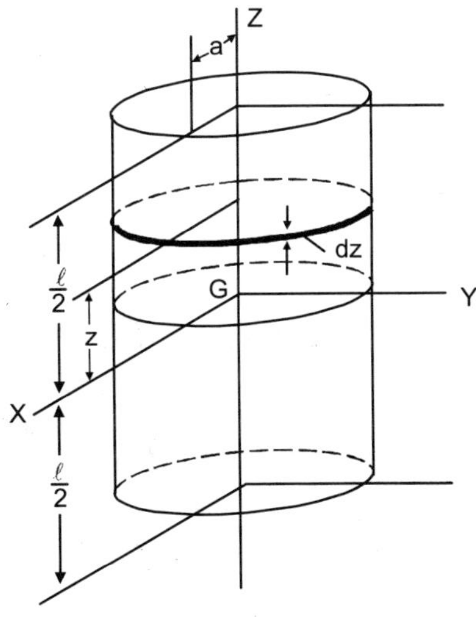

and just like circular disk $I_z = \frac{1}{2} ma^2$.

EXAMPLE 58

Determine Mass M.I of solid sphere of the centroidal axes.

SOLUTION If x, y, z be centroidal axes passing through the centre of gravity of a sphere, mass moment of inertia about centroidal axes will be

$$\left.\begin{aligned} I_x &= \int \left(y^2 + z^2\right) dm \\ I_y &= \int \left(x^2 + z^2\right) dm \\ I_z &= \int \left(x^2 + y^2\right) dm \end{aligned}\right\}$$

Adding above expressions together and observing that mass moments of inertia of sphere with respect to all diameters being equal,

$$I_x = I_y = I_z = \frac{2}{3}\int \left(x^2 + y^2 + z^2\right) dm = \frac{2}{3}\int r^2 dm$$

where r is the distance of the element dm from the centre of sphere. This elemental mass dm is the mass of a shell of thickness dr. So, $dm = 4\pi r^2 dr\rho$. Substituting the expression of dm we have,

$$I_x = I_y = I_z = \frac{2}{3}\int_0^a 4\pi r^2 \rho dr \cdot r^2 = \left(\frac{4}{3}\pi\rho a^3\right) \cdot \frac{2a^2}{5} = \frac{2}{5} m a^2 .$$

EXAMPLE 59

Determine Mass M.I of solid circular cone, about indicated X and Y axes and centroidal Y' axis.

SOLUTION Mass of the differential element as shown in the adjacent figure,

$$dm = \rho\pi r^2 dx = \rho\pi a^2 \frac{x^2}{h^2} dx \qquad \left[\because \frac{r}{a} = \frac{x}{h}\right]$$

Mass moment of inertia of this elemental mass,

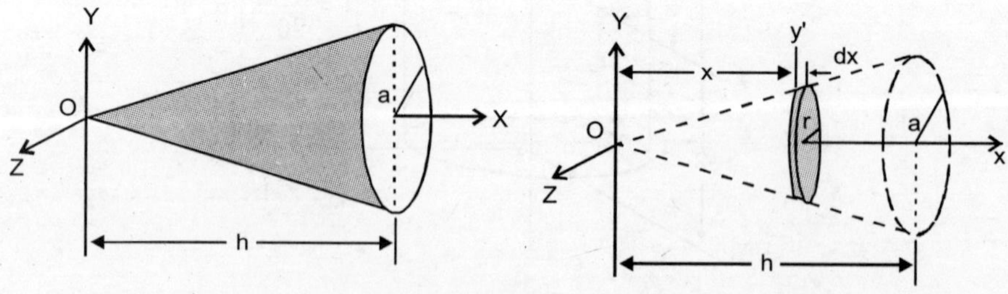

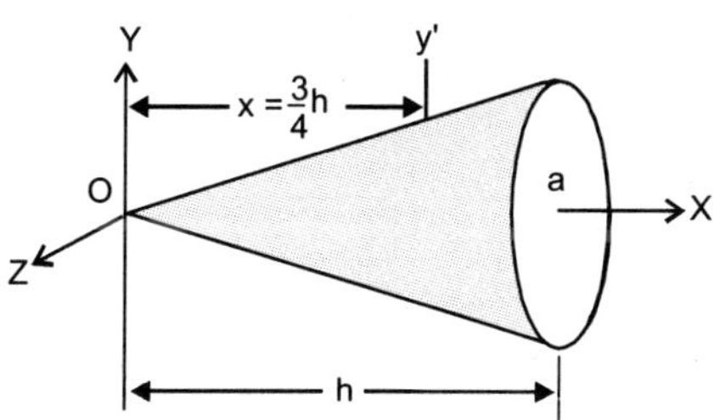

$$dI_x = \frac{1}{2}dm \cdot r^2 = \frac{1}{2}r^2 \cdot \rho\pi a^2 \frac{x^2}{h^2}dx = \frac{1}{2}\pi\rho\frac{a^4}{h^4}x^4 dx$$

$$\therefore \quad I_x = \int dI_x = \int_0^h \frac{1}{2}\pi\rho\frac{a^4}{h^4}x^4 dx = \left(\frac{1}{3}\rho\pi a^2 h\right)\frac{3}{10}a^2 = \frac{3}{10}ma^2$$

Using the same differential element,

$$dI_y = dI_{y'} + x^2 dm = \frac{1}{4}r^2 dm + x^2 dm = \rho\pi\frac{a^2}{h^2}\left(\frac{a^2}{4h^2}+1\right)x^4 dx$$

$$\therefore \quad I_y = \int_0^h dI_y = \rho\pi\frac{a^2}{h^2}\left(\frac{a^2}{4h^2}+1\right)\frac{h^5}{5}$$

$$= \left(\frac{1}{3}\rho\pi a^2 h\right)\frac{3}{5}\left(\frac{a^2}{4}+h^2\right) = \frac{3}{5}m\left(\frac{a^2}{4}+h^2\right)$$

In order to find out mass moment of inertia about centroidal y, axis, we can apply parallel axis theorem and obtain,

$$I_y = \bar{I}_{y''} + m\left(\bar{x}\right)^2$$

$$\bar{I}_{y''} = I_y - m\left(\bar{x}\right)^2 = \frac{3}{5}m\left(\frac{a^2}{4}+h^2\right) - m\left(\frac{3}{4}h\right)^2 = \frac{3}{20}m\left(a^2+\frac{h^2}{4}\right).$$

GEOMETRIC PROPERTIES OF PLANE FIGURE

FIGURE	*CENTROID*	*SECOND MOMENT OF AREA*
Arc Segment 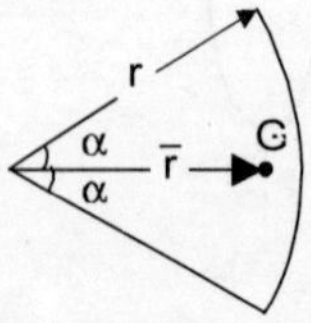 (r, α, α, $\bar{r}$, G)	$\bar{r} = \dfrac{r \sin \alpha}{\alpha}$	
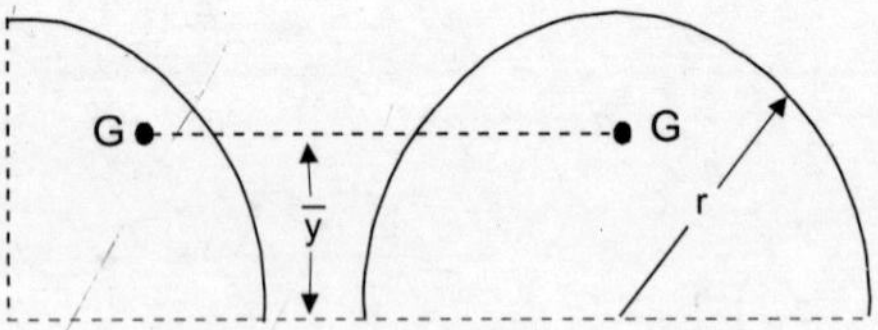 *Quarter and Semicircular Area* (G, G, $\bar{y}$, r)	$\bar{y} = \dfrac{2r}{\pi}$	
Circular Area 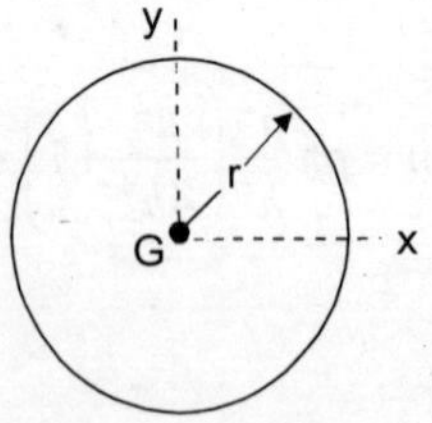 (y, r, G, x)		$I_x = I_y = \dfrac{\pi r^4}{4}$ $I_z = \dfrac{\pi r^4}{2}$
Semicircular Area 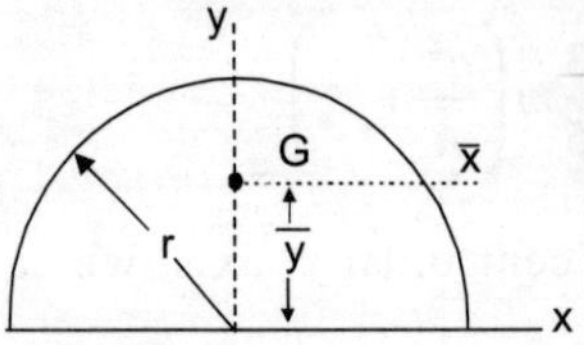 (y, G, $\bar{x}$, r, $\bar{y}$, x)	$\bar{y} = \dfrac{4r}{3\pi}$	$I_x = I_y = \dfrac{\pi r^4}{8} = I_z$ $\bar{I}_x = \left(\dfrac{\pi}{8} - \dfrac{8}{9\pi}\right) r^4$
Quarter-circular Area 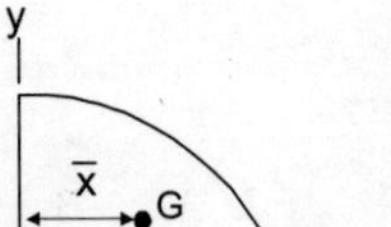 (y, $\bar{x}$, G, $\bar{y}$, x)	$\bar{x} = \bar{y} = \dfrac{4r}{3\pi}$	$I_x = I_y = \dfrac{\pi r^4}{16}$ $\bar{I}_x = \bar{I}_y \left(\dfrac{\pi}{16} - \dfrac{4}{9\pi}\right) r^4$ $I_z = \dfrac{\pi r^4}{8}$
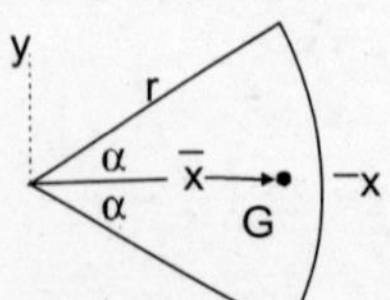 *Area of Circular Sector* (y, r, α, α, $\bar{x}$, G, x)	$\bar{x} = \dfrac{2}{3} \dfrac{r \sin \alpha}{\alpha}$	$I_x = \dfrac{r^4}{4}\left(\alpha - \dfrac{1}{2} \sin 2\alpha\right)$ $I_y = \dfrac{r^4}{4}\left(\alpha + \dfrac{1}{2} \sin 2\alpha\right)$ $I_z = \dfrac{1}{2} r^4 \alpha$

GEOMETRIC PROPERTIES OF PLANE FIGURE

FIGURE	***CENTROID***	***SECOND MOMENT OF AREA***
Rectangular Area 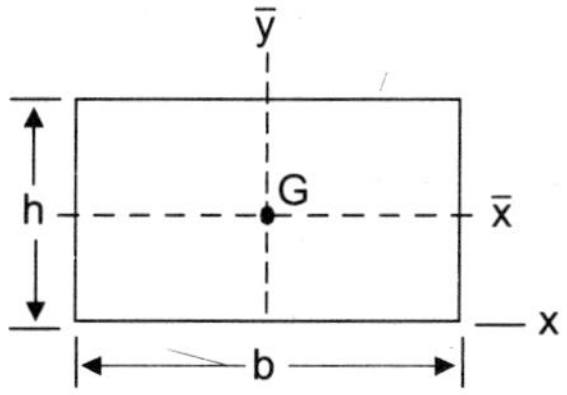		$I_x = \frac{bh^3}{3}$ $\bar{I}_x = \frac{bh^3}{12}, \bar{I}_y = \frac{hb^3}{12}$ $\bar{I}_z = \frac{bh}{12}\left(b^2 + h^2\right)$
Triangular Area 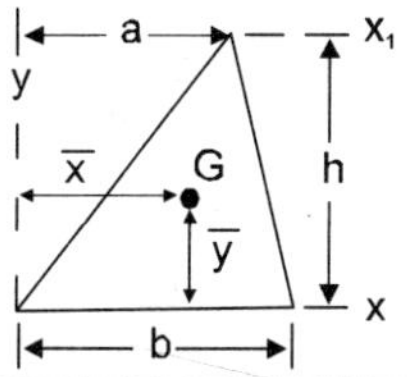	$\bar{x} = \frac{a+b}{3}$ $\bar{y} = \frac{h}{3}$	$I_x = \frac{bh^3}{12}, \bar{I}_x = \frac{bh^3}{36}$ $I_x = \frac{bh^3}{4}$
Area of Elliptical Quadrant 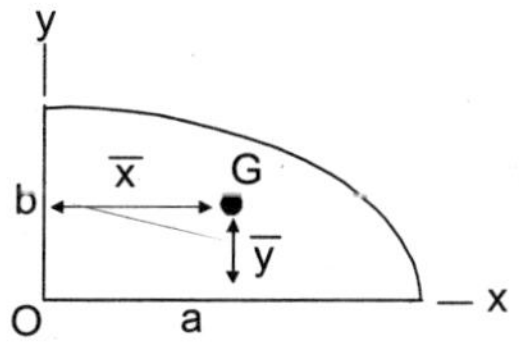	$\bar{x} = \frac{4a}{3\pi}$ $\bar{y} = \frac{4b}{3\pi}$	$I_x = \frac{\pi ab^3}{16}, \bar{I}_x = \left(\frac{\pi}{16} - \frac{4}{9\pi}\right)ab^3$ $I_y = \frac{\pi a^3 b}{16}, \bar{I}_y = \left(\frac{\pi}{16} - \frac{4}{9\pi}\right)a^3 b$ $I_z = \frac{\pi ab}{16}\left(a^2 + b^2\right)$
Subparabolic Area 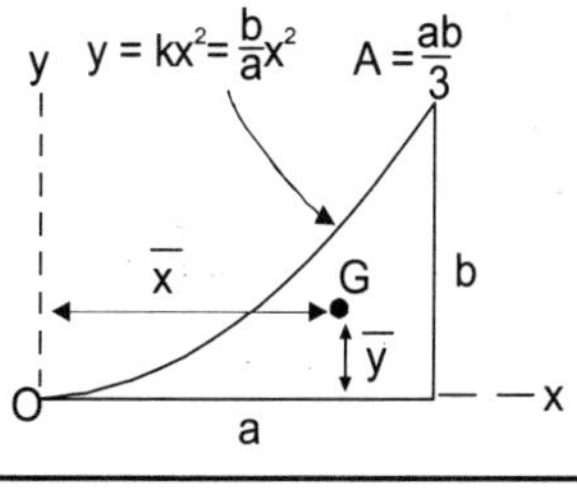	$\bar{x} = \frac{3a}{4}$ $\bar{y} = \frac{3b}{10}$	$I_x = \frac{ab^3}{21}$ $I_y = \frac{a^3 b}{5}$ $I_z = ab\left(\frac{a^3}{5} + \frac{b^2}{21}\right)$
Parabolic Area $A = \frac{2ab}{3}$ 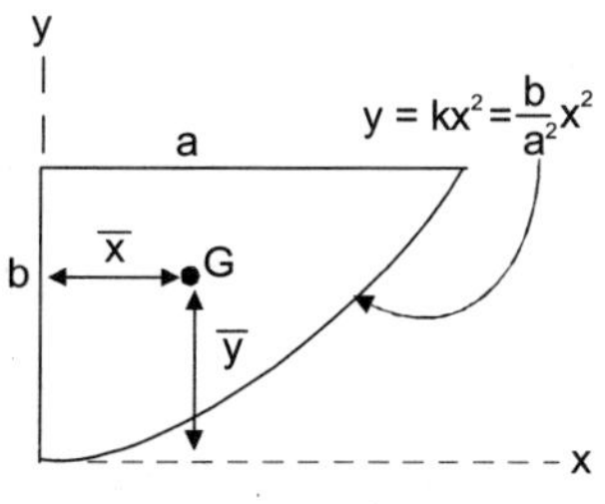	$\bar{x} = \frac{3a}{8}$ $\bar{y} = \frac{3b}{5}$	$I_x = \frac{2ab^3}{7}$ $I_y = \frac{2a^3 b}{15}$ $I_z = 2ab\left(\frac{a^3}{15} + \frac{b^2}{7}\right)$

PROPERTIES OF HOMOGENEOUS SOLIDS

BODY	*MASS CENTER*	*MASS MOMENTS OF INERTIA*
Circular Cylindrical Shell		$I_{xx} = \frac{1}{2}mr^2 + \frac{1}{12}ml^2$ $I_{x_1x_1} = \frac{1}{2}mr^2 + \frac{1}{3}ml^2$ $I_{zz} = mr^2$
Half Cylindrical Shell	$\bar{x} = \frac{2r}{\pi}$	$I_{xx} = I_{yy} = \frac{1}{2}mr^2 + \frac{1}{12}ml^2$ $I_{x_1x_1} = I_{y_1y_1} = \frac{1}{2}mr^2 + \frac{1}{3}ml^2$ $I_{zz} = mr^2$ $\bar{I}_{zz} = \left(1 - \frac{4}{\pi^2}\right)mr^2$
Circular Cylinder		$I_{xx} = \frac{1}{4}mr^2 + \frac{1}{12}ml^2$ $I_{x_1x_1} = \frac{1}{4}mr^2 + \frac{1}{3}ml^2$ $I_{zz} = \frac{1}{2}mr^2$
Semicylinder	$\bar{x} = \frac{4r}{3\pi}$	$I_{xx} = I_{yy} = \frac{1}{4}mr^2 + \frac{1}{12}ml^2$ $I_{x_1x_1} = I_{y_1y_1} = \frac{1}{4}mr^2 + \frac{1}{3}ml^2$ $I_{zz} = \frac{1}{2}mr^2$ $\bar{I}_{zz} = \left(\frac{1}{2} - \frac{16}{9\pi^2}\right)mr^2$
Rectangular Parallelepiped		$I_{xx} = \frac{1}{12}m\left(a^2 + l^2\right)$ $I_{yy} = \frac{1}{12}m\left(b^2 + l^2\right)$ $I_{zz} = \frac{1}{12}m\left(a^2 + b^2\right)$ $I_{y_1y_1} = \frac{1}{12}mb^2 + \frac{1}{3}ml^2$ $I_{y_2y_2} = \frac{1}{3}m\left(b^2 + l^2\right)$

PROPERTIES OF HOMOGENEOUS SOLIDS

BODY	*MASS CENTER*	*MASS MOMENTS OF INERTIA*
Spherical Shell 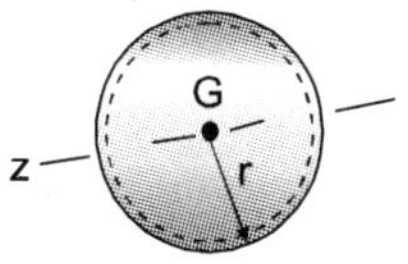		$I_{zz} = \frac{2}{3} mr^2$
Hemispherical Shell 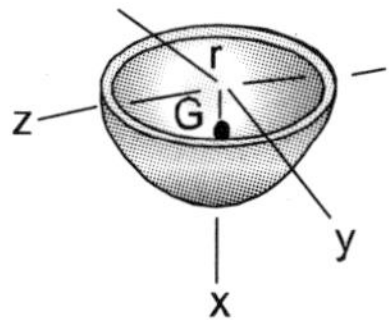	$\bar{x} = \frac{r}{2}$	$I_{xx} = I_{yy} = I_{zz} = \frac{2}{3} mr^2$ $\bar{I}_{yy} = \bar{I}_{zz} = \frac{5}{12} mr^2$
Sphere 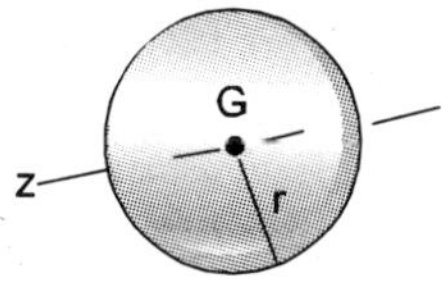		$I_{zz} = \frac{2}{5} mr^2$
Hemisphere 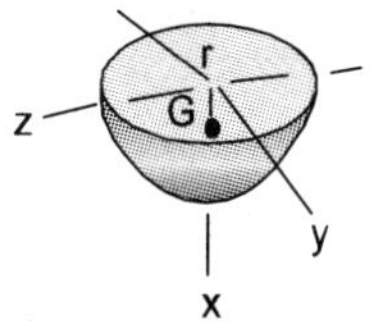	$\bar{x} = \frac{3r}{8}$	$I_{xx} = I_{yy} = I_{zz} = \frac{2}{5} mr^2$ $\bar{I}_{yy} = \bar{I}_{zz} = \frac{83}{320} mr^2$
Uniform Slender Rod 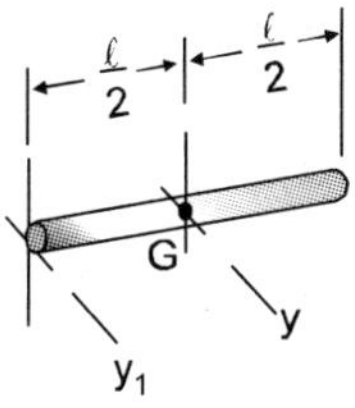		$I_{yy} = \frac{2}{12} ml^2$ $I_{y_1y_1} = \frac{1}{3} ml^2$

PROPERTIES OF HOMOGENEOUS SOLIDS

BODY	*MASS CENTER*	*MASS MOMENTS OF INERTIA*
Quarter - Circular Rod 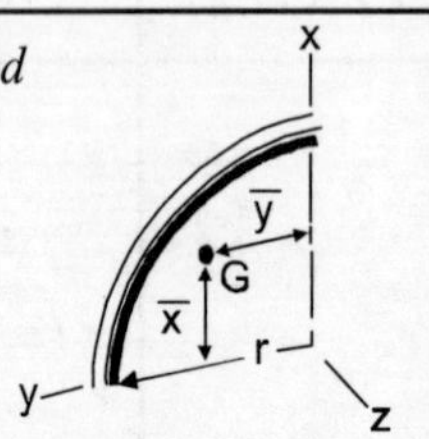	$\bar{x} = \bar{y} = \frac{2r}{\pi}$	$I_{xx} = I_{yy} = \frac{1}{2}mr^2$ $I_{zz} = mr^2$
Elliptical Cylinder 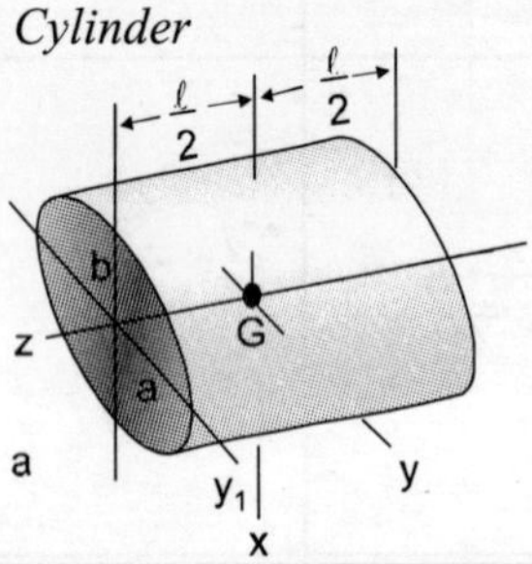		$I_{xx} = \frac{1}{4}ma^2 + \frac{1}{12}ml^2$ $I_{yy} = \frac{1}{4}mb^2 + \frac{1}{12}ml^2$ $I_{zz} = \frac{1}{4}m\left(a^2 + b^2\right)$ $I_{y_1y_1} = \frac{1}{4}mb^2 + \frac{1}{3}ml^2$
Conical Shell 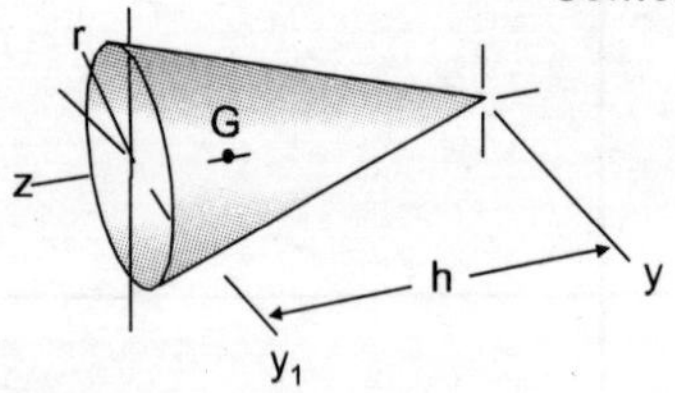	$\bar{z} = \frac{2h}{3}$	$I_{yy} = \frac{1}{4}mr^2 + \frac{1}{2}mh^2$ $I_{y_1y_1} = \frac{1}{4}mr^2 + \frac{1}{6}mh^2$ $I_{zz} = \frac{1}{2}mr^2$ $\bar{I}_{yy} = \frac{1}{4}mr^2 + \frac{1}{18}mh^2$
Half Conical Shell 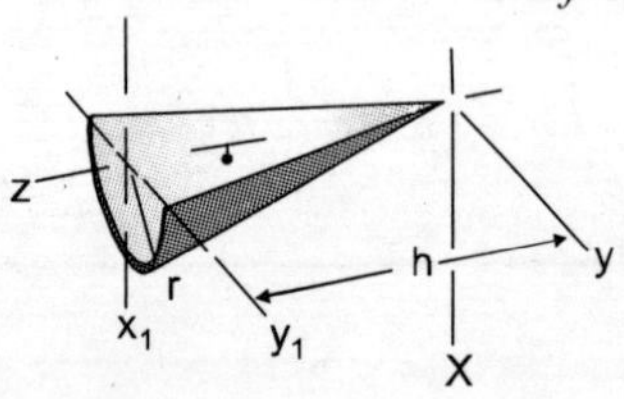	$\bar{x} = \frac{4r}{3\pi}$ $\bar{z} = \frac{2h}{3}$	$I_{xx} = I_{yy} = \frac{1}{4}mr^2 + \frac{1}{2}mh^2$ $I_{x_1x_1} = I_{y_1y_1} = \frac{1}{4}mr^2 + \frac{1}{6}mh^2$ $I_{zz} = \frac{1}{2}mr^2$ $\bar{I}_{zz} = \left(\frac{1}{2} - \frac{16}{9\pi^2}\right)mr^2$
Right Circular Cone 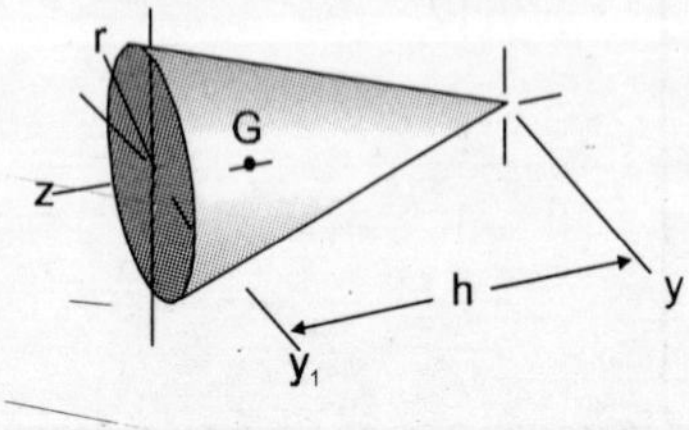	$\bar{z} = \frac{3h}{4}$	$I_{yy} = \frac{3}{20}mr^2 + \frac{3}{5}mh^2$ $I_{y_1y_1} = \frac{3}{20}mr^2 + \frac{1}{10}mh^2$ $I_{zz} = \frac{3}{10}mr^2$ $\bar{I}_{yy} = \frac{3}{20}mr^2 + \frac{3}{80}mh^2$

PROPERTIES OF HOMOGENEOUS SOLIDS

BODY	*MASS CENTER*	*MASS MOMENTS OF INERTIA*
Half Cone 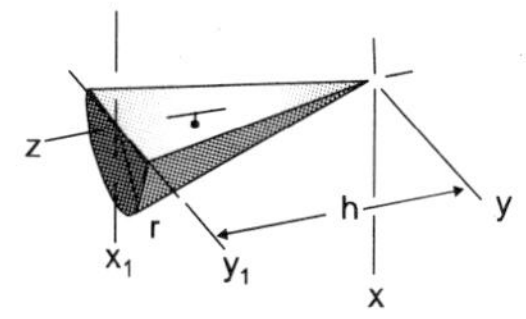	$\bar{x} = \frac{r}{\pi}$ $\bar{z} = \frac{3h}{4}$	$I_{xx} = I_{yy} = \frac{3}{20}mr^2 + \frac{3}{5}mh^2$ $I_{x_1x_1} = I_{y_1y_1} = \frac{3}{20}mr^2 + \frac{1}{10}mh^2$ $I_{zz} = \frac{3}{10}mr^2,\ \bar{I}_{zz} = \left(\frac{3}{10} - \frac{1}{\pi^2}\right)mr^2$
Semiellipsoid 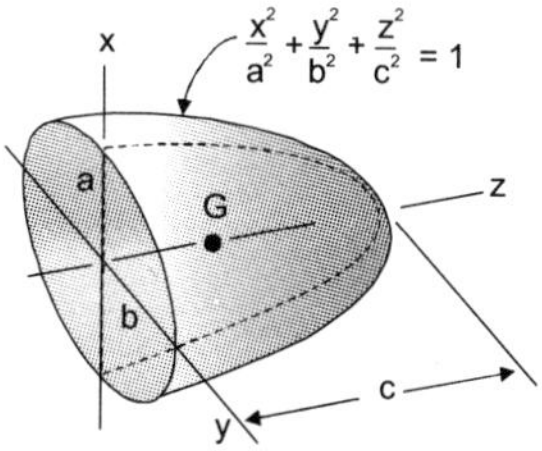 $\frac{x^2}{a^2} + \frac{y^2}{b^2} + \frac{z^2}{c^2} = 1$	$\bar{z} = \frac{3c}{8}$	$I_{xx} = \frac{1}{5}m\left(b^2 + c^2\right),\ I_{yy} = \frac{1}{5}m\left(a^2 + c^2\right)$ $I_{zz} = \frac{1}{5}m\left(a^2 + b^2\right),\ \bar{I}_{xx} = \frac{1}{5}m\left(b^2 + \frac{19}{64}c^2\right)$ $\bar{I}_{yy} = \frac{1}{5}m\left(a^2 + \frac{19}{64}c^2\right)$
Elliptic Paraboloid 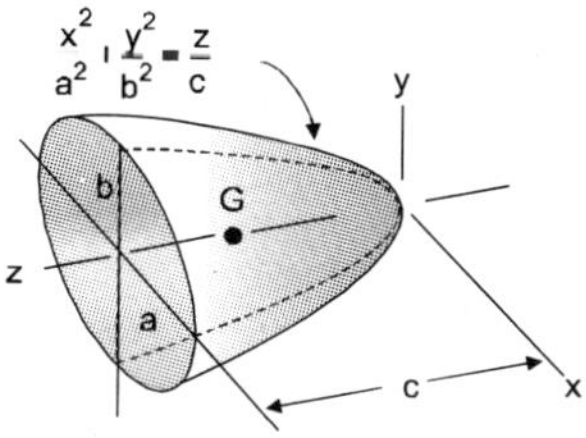 $\frac{x^2}{a^2} + \frac{y^2}{b^2} = \frac{z}{c}$	$\bar{z} = \frac{2c}{3}$	$I_{xx} = \frac{1}{6}mb^2 + \frac{1}{2}mc^2,\ I_{yy} = \frac{1}{6}ma^2 + \frac{1}{2}mc^2$ $I_{zz} = \frac{1}{6}m\left(a^2 + b^2\right),\ \bar{I}_{xx} = \frac{1}{6}m\left(b^2 + \frac{1}{3}c^2\right)$ $\bar{I}_{yy} = \frac{1}{6}m\left(a^2 + \frac{1}{3}c^2\right)$
Rectangular Tetrahedron 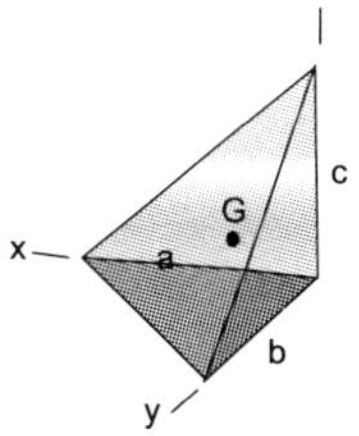	$\bar{x} = \frac{a}{4}$ $\bar{y} = \frac{b}{4}$ $\bar{z} = \frac{c}{4}$	$I_{xx} = \frac{1}{10}m\left(b^2 + c^2\right),\ I_{yy} = \frac{1}{10}m\left(a^2 + c^2\right)$ $I_{zz} = \frac{1}{10}m\left(a^2 + b^2\right),\ \bar{I}_{xx} = \frac{3}{80}m\left(b^2 + c^2\right)$ $\bar{I}_{yy} = \frac{3}{80}m\left(a^2 + c^2\right),\ \bar{I}_{zz} = \frac{3}{80}m\left(a^2 + b^2\right)$
Half Torus 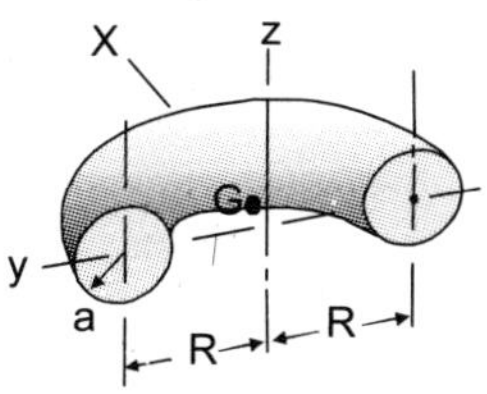	$\bar{x} = \frac{a^2 + 4R^2}{2\pi R}$	$I_{xx} = I_{yy} = \frac{1}{2}mR^2 + \frac{5}{8}ma^2$ $I_{zz} = mR^2 + \frac{3}{4}ma^2$

Exercise

1. Determine the x and y co-ordinates of the centroid of the shaded area.

Ans. [x_c=1.443k, y_c = 0.361k]

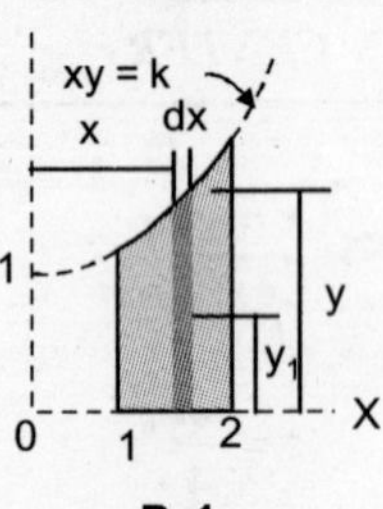

P-1

2. Determine the y co-ordinate of the centroid of the shaded area.

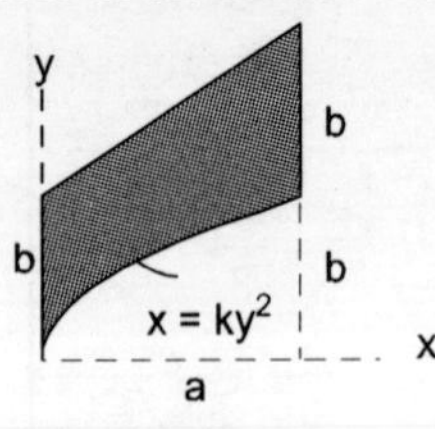

P-2

3. Determine, by integration, the co-ordinates x_c and y_c of centroid C of the quadrant AB of the arc of a circle of radius r.

Ans. [x_c = y_c= 2r/π]

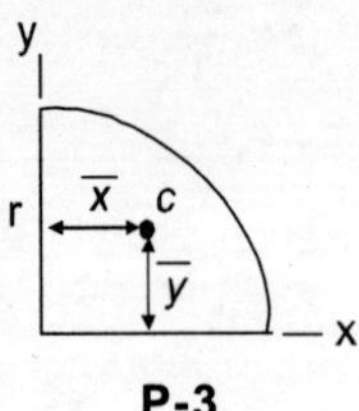

P-3

4. Determine the co-ordinate of the centroid with respect to the given axes of the shaded area shown in figure.

Ans. [x_c = 0.856cm, y_c = 1.143cm]

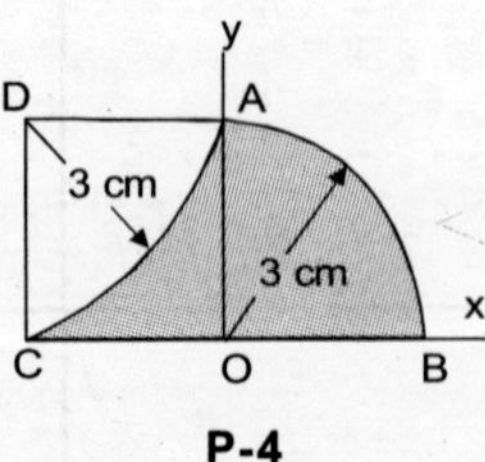

P-4

5. Determine the x and y co-ordinates of the centroid of (i) semicircle, (ii) semi-ellipse.

6. Determine the x and y co-ordinates of the centroid of the shaded area.

Ans. $\left[x_c = \dfrac{7a}{6(\pi-1)}, y_c = \dfrac{a}{\pi-1}\right]$

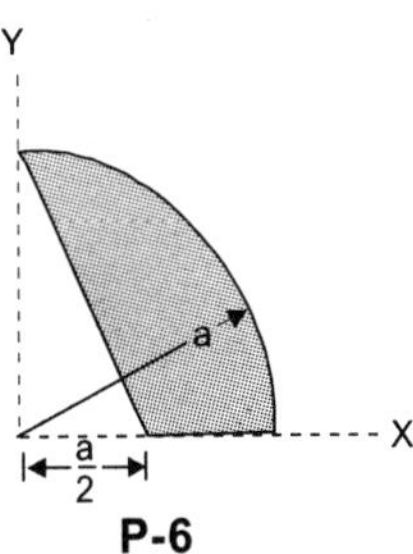

P-6

7. The rigidly connected unit consists of 2 kg circular disk, a 1.5 kg round shaft, and a 1kg square plate. Determine the y co-ordinate of the mass center of the unit.

Ans. [$y_c = 70\text{mm}$]

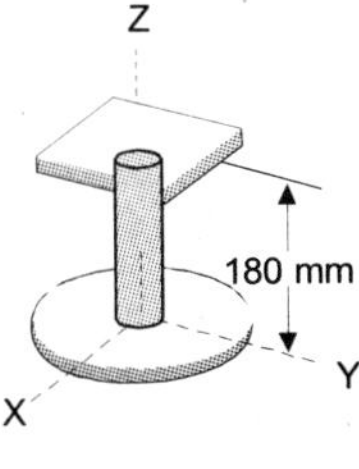

P-7

8. Determine the x co-ordinate of the mass center of the tapered steel rod of length L where the diameter at the larger end is twice the diameter at the small end.

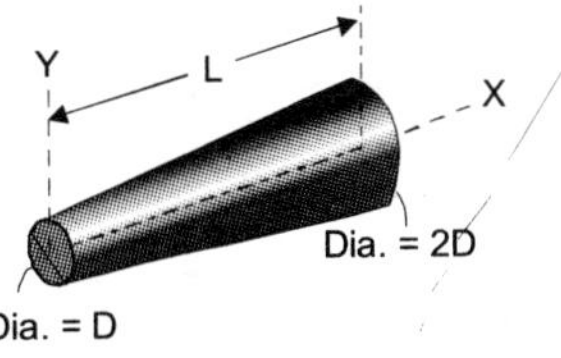

P-8

9. The homogeneous hemisphere with smaller hemispherical portion removed is shown. Determine the x co-ordinate of the mass center.

[**Ans.** $x_c = 45R/112$]

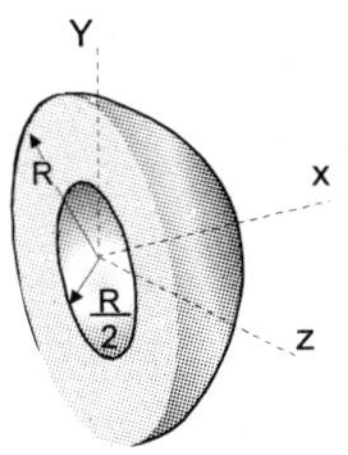

P-9

10. A homogenous wire ABC is bent and attached to the himge at A. Find the length of the portion BC, for which the portion BC of the wire is horizontal.

[**Ans.** 53.4cm]

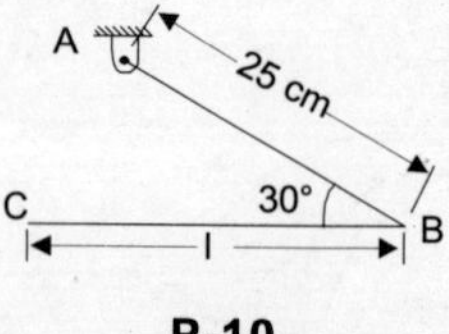

P-10

11. Determine the co-ordinates of the centroid of the shaded area as shown in figure.

[**Ans.** $x_c = 26.3a, y_c = 23.6a$]

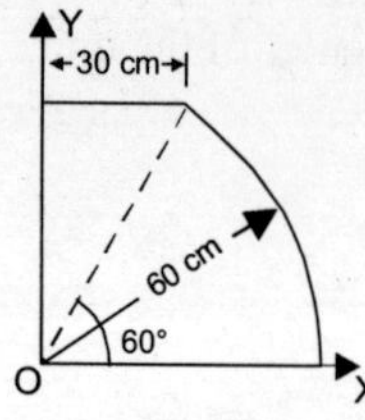

P-11

12. Find the co-ordinates of the centroid of the shaded area as shown in figure.

[**Ans.** $x_c = 6.98\text{cm}, \quad y_c = 12.72\text{cm}$]

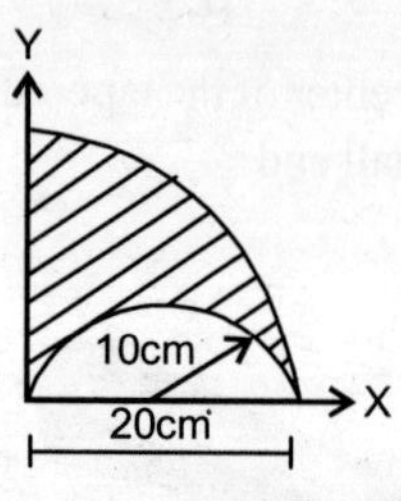

P-12

13. Determine the co-ordinate of the centroid of a L-section as shown in figure.

[**Ans.** $x_c = 2.61\text{cm}, \quad y_c = 3.015\text{cm}$]

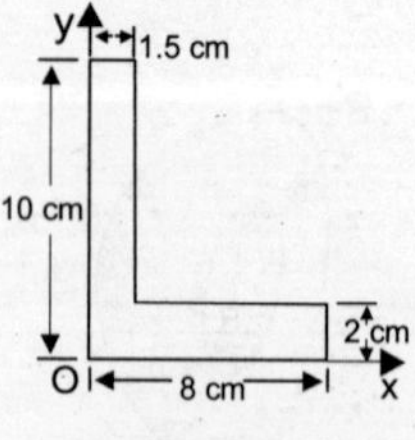

P-13

14. With respect to co-ordinate axes x and y locate the centroid of the shaded area shown in figure.

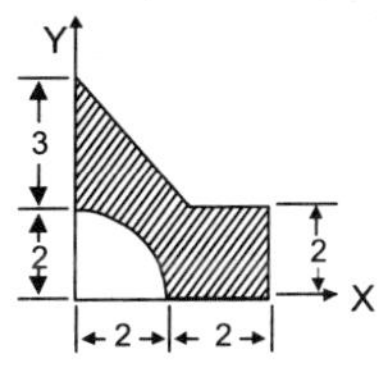

P-14

15. Calculate the moment of inertia of the shaded area about the x-axis. The equation of the curve OA is given by $x = ky^2$

Ans. $\left[I_x = \frac{a^4}{5}\right]$

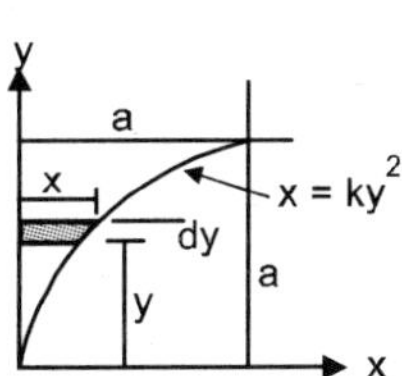

P-15

16. Determine the M.I about the x-axis of an area under the half sine curve, the equation for which is $y = b\ \sin\frac{\pi x}{a}$ shown in figure.

Ans. $\left[I_x = \frac{4ab^3}{9\pi}\right]$

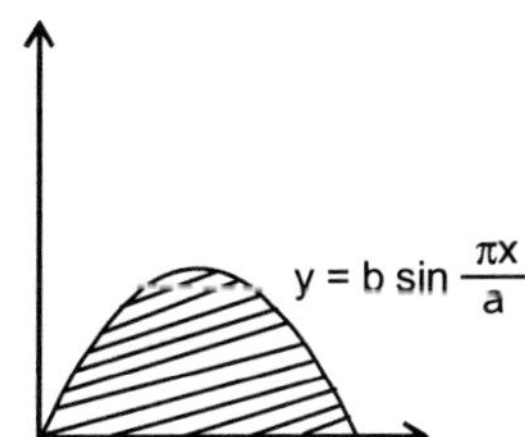

P-16

17. A T-section of dimensions $10 \times 10 \times 2$ cm. Determine the M.I of the section about the centroidal axes of the section.

Ans. [$I_{x_c} = 314.221\text{cm}^4,\ I_{y_c} = 172\text{cm}^4$]

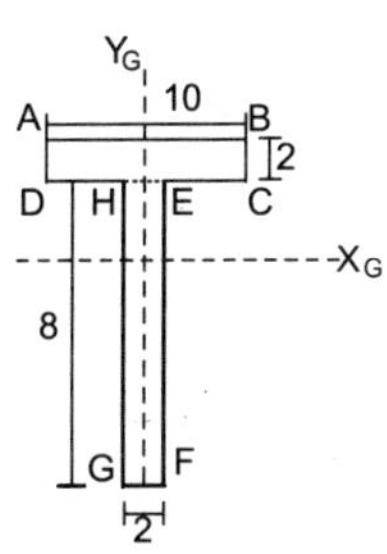

P-17

18. Find the M.I of the given figure about y-axis of a regular hexagon of side 'a'.

Ans. $\left[I_y = \frac{5\sqrt{3}}{16}a^4\right]$

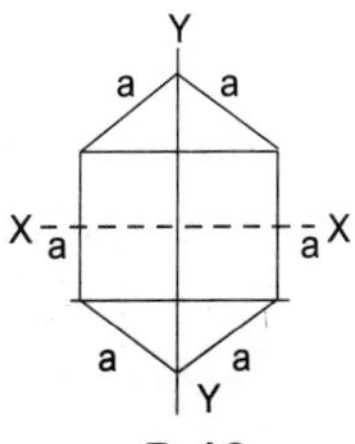

P-18

19. Find the M.I of the cross-section of the iron beam with respect to the centroidal axes.

Ans. [$I_{xG} = 288\text{cm}^4, I_{y_G} = 838\text{cm}^4$]

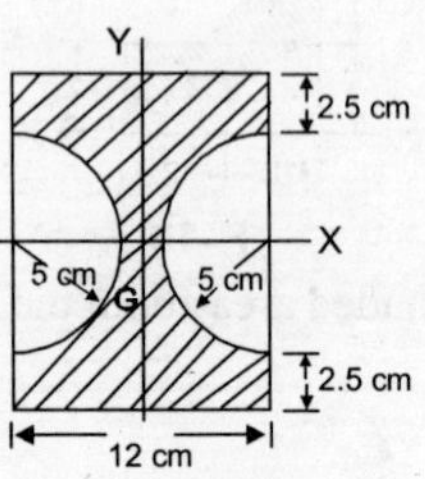

P-19

20. Find the M.I of the shaded area with respect to the centroidal axis parallel to *AB*.

Ans. [$I_{AB} = 429.3\text{cm}^4$]

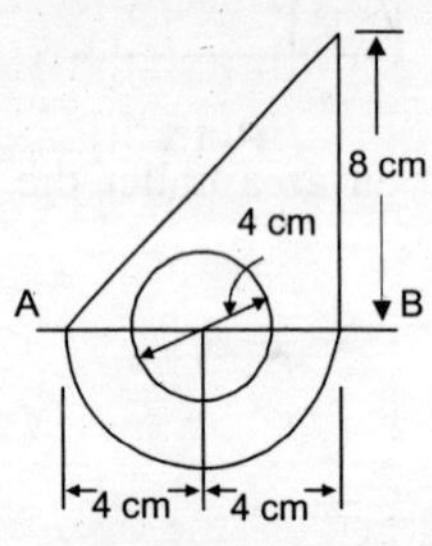

P-20

21. Find the M.I of a channel section as shown in figure about centroidal axes.

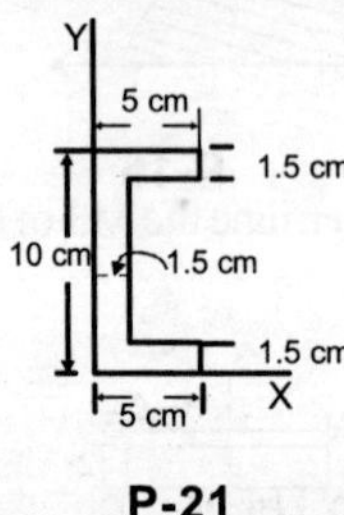

P-21

22. From a rectangular lamina *ABCD* 10 × 14cm a rectangular hole of 3 × 5cm is removed as shown in figure. Find the centroid of the remainder part of lamina.

Ans. [$x_C = 4.7\text{cm}, \ y_C = 6.444\text{cm}$]

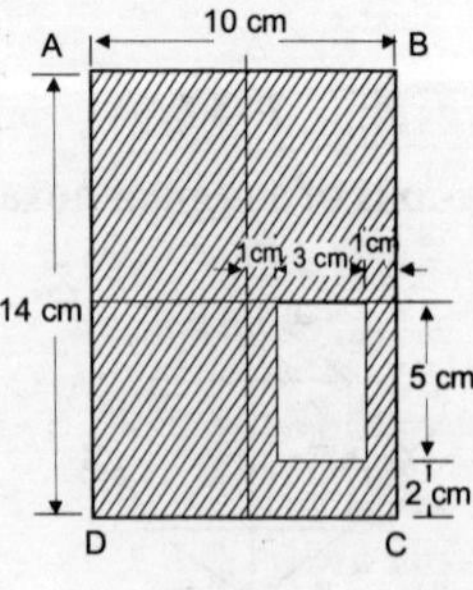

P-22

CHAPTER
5

PRINCIPLE OF VIRTUAL WORK

5.1 INTRODUCTION

In statics, we have studied till now, the procedure to isolate a body for exposing some unknown forces and then formulating equations of equilibrium that include all those unknown forces. Here an *alternative approach* of expressing equilibrium condition will be studied, where only some, not all, unknown forces acting on a body will be considered. This process has sometimes been found to be very useful for finding out simpler ways to obtain desired unknown quantities. This approach is designated as *Method of Virtual Work* or *Principle of Virtual Work.*

5.2 WORK AND VIRTUAL WORK

A differential amount of work dW of a force F acting on a particle equals the component of this force in the direction of movement of the particle multiplied by the differential displacement of the particle. So

$$dW = F \cdot dr \qquad \text{... (5.1)}$$

Hence if the particle moves from point 1 to point 2 as shown in Figure 5.1 following a particular path, total work done will be

$$W = \int_{r_1}^{r_2} F \cdot dr \qquad \text{...(5.2)}$$

Now, the word *virtual* means *imaginary* and not real. So virtual work can be defined as a *non-real* or *imaginary work*, done by a force on a body, causing *imaginary displacement* or *imaginary rotation.* This imaginary displacement or rotation is called *virtual displacement* or *virtual rotation*.

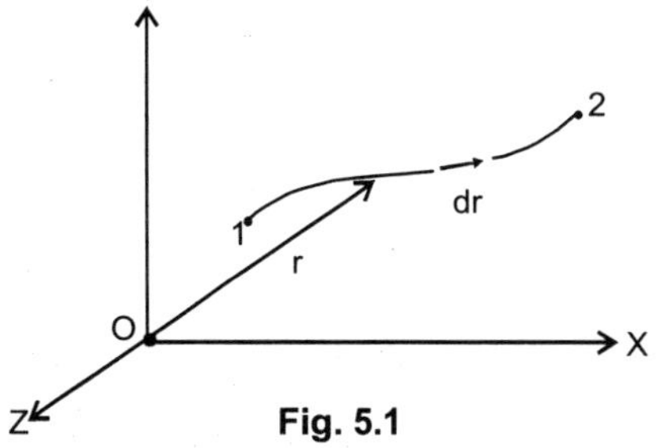

Y
2
dr
1
r
O
X
Z

Fig. 5.1

The principle of virtual work states that, *For a system of initially stationary rigid bodies, the algebraic summation of virtual works, done by all active forces causing virtual displacement consistent with geometrical conditions, will be zero.*

There exist several types of application of principle of virtual work on statics, which are discussed in the proceeding article. But before these illustrations, the effects of forces which are to be discarded during applying this principle are:

1. *Tension in inextensible string or axial forces in bars.*
2. *Weight of the body when its C.G moves horizontally.*
3. *Reaction between the body and smooth surfaces.*
4. *Reaction at frictionless hinge when the body rotates about it.*
5. *Friction force acting on wheel when it rolls without slip.*

5.3 VIRTUAL WORK APPLIED ON BEAM

Let us take the simply supported beam *ABCD*, loaded as shown in Figure 5.2 (*a*). Here the unknown are the support reactions at *A* and *B*, i.e. R_A and R_B. Now imagine, keeping point *A* fixed, R_B is allowed to push point *B* to *B′*. This purely imaginary movement as shown in Figure 5.2 (*b*) will cause virtual displacement at *B* $(BB') = \delta_{B_1}$, at *C* $(CC') = \delta_{C_1}$ and at *D* $(DD') = \delta_{D_1}$.

Applying principle of virtual work,

$$R_B \cdot \delta_{B_1} + P_1 \cdot \left(-\delta_{C_1}\right) + P_2 \cdot \left(-\delta_{D_1}\right) = 0 \qquad \text{....(5.3)}$$

Again, from similar triangles, $\Delta ABB'$, $\Delta ACC'$ and $\Delta ADD'$, we have

$$\frac{\delta_{C_1}}{\delta_{B_1}} = \frac{a}{l} \text{ and } \frac{\delta_{D_1}}{\delta_{B_1}} = \frac{b}{l} \qquad \text{.... (5.4)}$$

Substituting in equation (5.3), we obtain

$$R_B \cdot \delta_{B_1} - P_1 \cdot \left(\frac{a}{l}\right)\delta_{B_1} - P_2 \cdot \left(\frac{b}{l}\right)\delta_{B_1} = 0$$

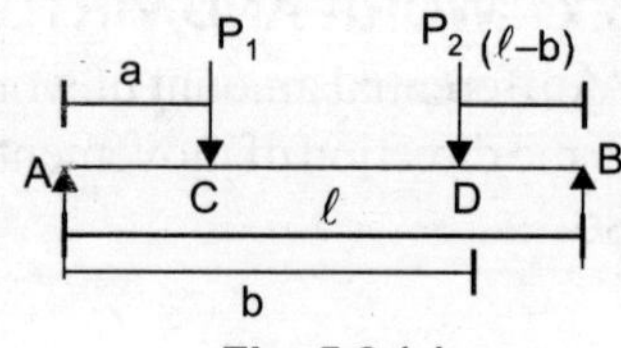

Fig. 5.2 (a)

$$\Rightarrow \quad R_B = \frac{1}{l}\left(P_1 \cdot a + P_2 \cdot b\right) \text{ as } \delta_{B_1} \neq 0 \qquad \text{...(5.5)}$$

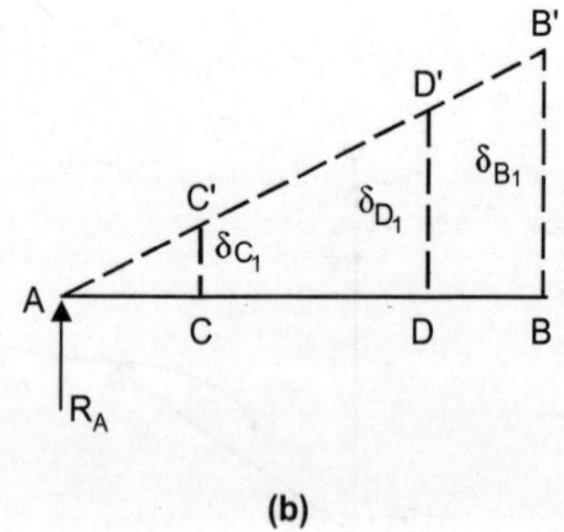

(b)

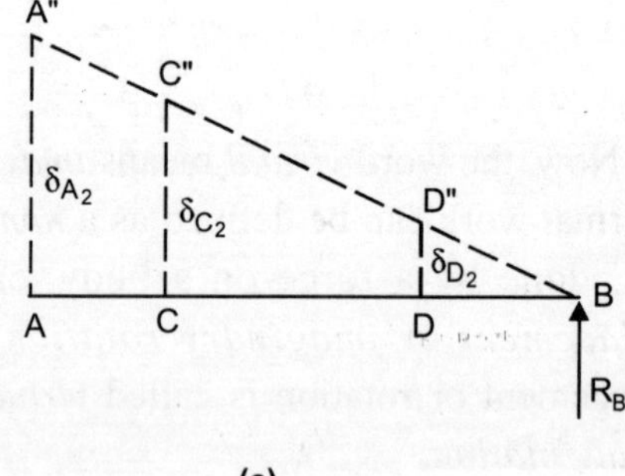

(c)

Fig. 5.2

Other way round, point *A* can be given virtual displacement to *A*", keeping point *B* fixed. Proceeding in the same fashion as above, we can write

$$R_A \cdot \delta_{A_2} - P_1 \cdot \left(\frac{l-a}{l}\right)\delta_{A_2} - P_2 \cdot \left(\frac{l-b}{l}\right)\delta_{A_2} = 0$$

$$\Rightarrow \qquad R_A = \frac{1}{l}\left[P_1 \cdot (l-a) + P_2 \cdot (l-b)\right] \text{ as } \delta_{A_2} \neq 0 \qquad \text{... (5.6)}$$

This basic approach can help us to solve problem on beams subjected to several loading conditions.

5.4 VIRTUAL WORK APPLIED ON THE BODY SUPPORTED ON TWO SURFACES

Let us assume one slender rod of self weight W supports on two mutually perpendicular surfaces, verticle surface being smooth. Figure 5.3 (a) shows the arrangement. Point A of the rod, resting on horizontal surface is allowed to slide at position A'. By virtue of this sliding, point B shifts to B' and C.G of the bar G to G'. Here we are interested to find out frictional force on horizontal surface to present sliding of the bar. From simple geometry, we can write,

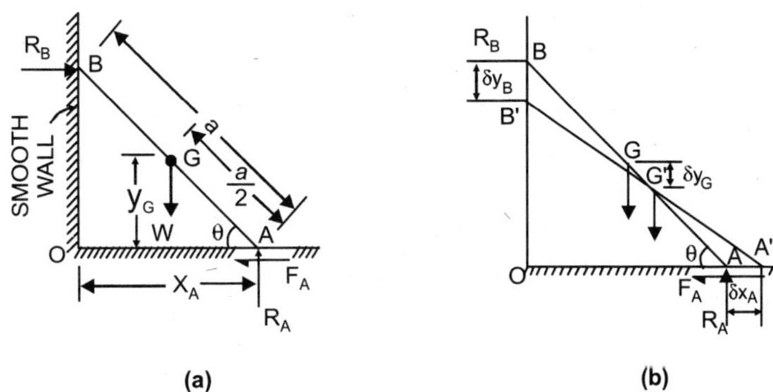

Fig. 5.3

$$\left.\begin{aligned} y_G &= AG\sin\theta = \frac{a}{2}\sin\theta \\ x_A &= AB\cos\theta = a\cos\theta \end{aligned}\right\} \qquad \text{...(5.7)}$$

The differential of equation (5.7) yield,

$$\left.\begin{aligned} dy_G &= \frac{a}{2}\cos\theta d\theta \\ dx_A &= -a\sin\theta d\theta \end{aligned}\right\} \qquad \text{... (5.8)}$$

Now, in terms virtual displacement as shown in Figure 5.3 (b), we can write,

$$\left.\begin{aligned} \delta y_G &= \frac{a}{2}\cos\theta\delta\theta \\ \delta x_A &= -a\sin\theta\delta\theta \end{aligned}\right\} \qquad \text{...(5.9)}$$

Here the effect of virtual work done R_B will be zero, as the component of R_B along δ_B is zero. In similar reason, virtual work done by R_A also is zero. Now applying principle of virtual work, the equation of equilibrium will be,

$$F_A \cdot \delta_{x_A} + W \cdot \delta_{y_G} = 0$$

$$\Rightarrow \quad F_A \cdot (-a \sin\theta \cdot \delta\theta) + W \cdot \left(\frac{a}{2} \cos\theta \cdot \delta\theta\right) = 0$$

$$\Rightarrow \quad F_A = \frac{W}{2} \cot\theta \quad \text{...(5.10)}$$

In the above equation, δ_{x_A} is not considered (–) ve, as it is already taken care in equation (5.9).

5.5 VIRTUAL WORK APPLIED ON LIFTING MACHINES

The basic principle of lifting machine is that load moves *upward* while effort acts *downward*. Considering the virtual work done by the effort and the load, the magnitude and direction of unknown load or effort can be determined. Stevinus (1548–1620) first made the application of virtual work in a pulley system in equilibrium. In the adjacent figure both the pulleys are frictionless. If the effort undergoes a virtual displacement of δy downward, load will move in upward direction by a virtual amount $\frac{\delta y}{2}$. Hence, applying the principle of virtual work, we obtain,

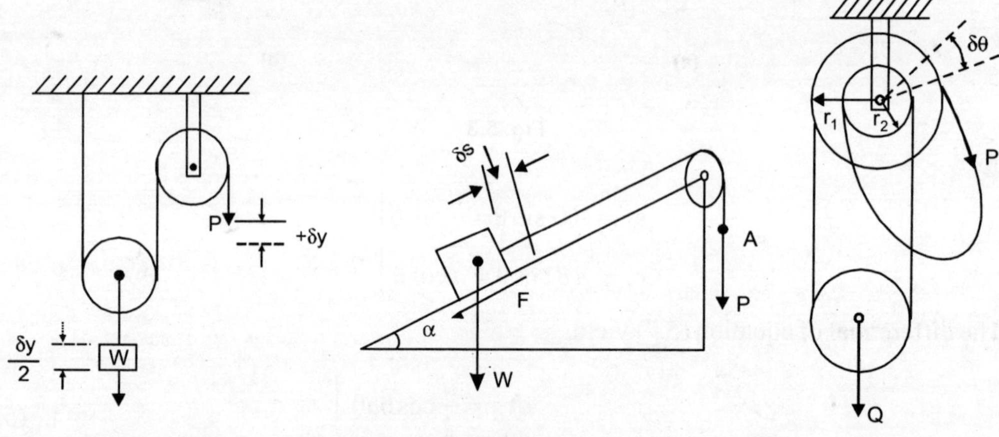

Fig. 5.4 Fig. 5.5 Fig. 5.6

$$P \cdot \delta y + W\left(-\frac{\delta y}{2}\right) = 0$$

$$\Rightarrow \quad P = \frac{W}{2} \quad \text{...(5.11)}$$

Another variety of this type of problem is pulling a body along inclined surface through a smooth pulley, as shown in Figure 5.5. If the effort be P and self weight of the body be W, we can write the equation of equilibrium as,

$$P \cdot \delta s + W \sin\alpha \cdot (-\delta s) + F \cdot (-\delta s) = 0$$

$$\Rightarrow \quad P \cdot \delta s + W \sin\alpha \cdot (-\delta s) + \mu W \cos\alpha \cdot (-\delta s) = 0$$

$$\Rightarrow \quad P = W(\mu \cos\alpha + \sin\alpha) \qquad ...(5.12)$$

In differential pulley, as shown in Figure 5.6, virtual rotation defined is $\delta\theta$. So, the virtual displacement for P is $r_1\delta\theta$ and that for Q is $\frac{1}{2}(r_1 - r_2)\delta\theta$. Then from principle of virtual work, we have,

$$P \cdot (r_1\delta\theta) + Q \cdot \left(-\frac{r_1 - r_2}{2}\delta\theta\right) = 0$$

$$P = \frac{r_1 - r_2}{2r_1}Q \qquad ... (5.13)$$

5.6 VIRTUAL WORK APPLIED ON INTERCONNECTED BODIES

In Figure 5.7 the body having weights P and Q respectively are connected with an inextensible string which passes through a smooth pulley. If we imagine the body of weight Q moves by an amount δs_1 along inclined plane, the equation of equlibrium will be,

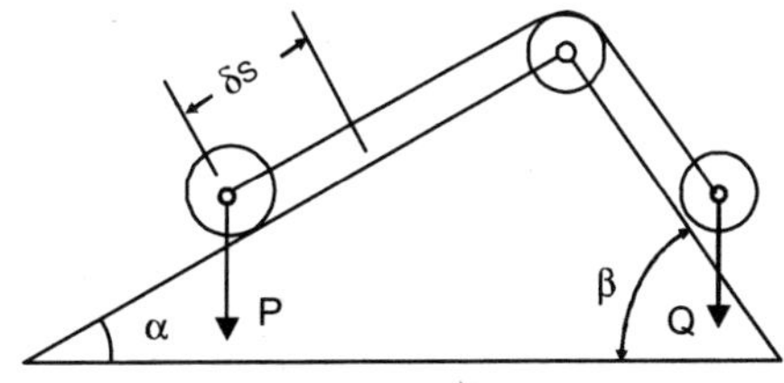

Fig. 5.7

$$P \sin\alpha \cdot (-\delta s) + Q \sin\beta \cdot (\delta s) = 0$$

$$\Rightarrow \quad \frac{P}{Q} = \frac{\sin\beta}{\sin\alpha} \qquad ...(5.14)$$

5.7 VIRTUAL WORK APPLIED ON MECHANISM

In *slider-crank* mechanism as shown in Figure 5.8 (*a*) cross head slides by an amount δx, while crank rotates by $\delta\theta$. From geometry of the configuration, we have,

$$x = OA = OC + CA = R\cos\theta + L\cos\phi$$

$$= R\cos\theta + L\sqrt{1-\left(\frac{R}{L}\right)^2 \sin^2\theta}$$

So,
$$\delta x = -R\sin\theta\cdot\delta\theta - \frac{R^2}{L}\cdot\frac{\sin\theta\cos\theta\delta\theta}{\sqrt{1-\left(\frac{R}{L}\right)^2\sin^2\theta}} \quad \text{... (5.15)}$$

From the principle of virtual work we can write,

$$M\cdot\delta\theta + F\cdot\delta x = 0 \quad \text{...(5.16)}$$

Substituting equation (5.13) and simplifying, we obtain,

$$M = FR\sin\theta\left[1+\frac{R\cos\theta}{L\sqrt{1-\left(\frac{R}{L}\right)^2\sin^2\theta}}\right] \quad \text{... (5.17)}$$

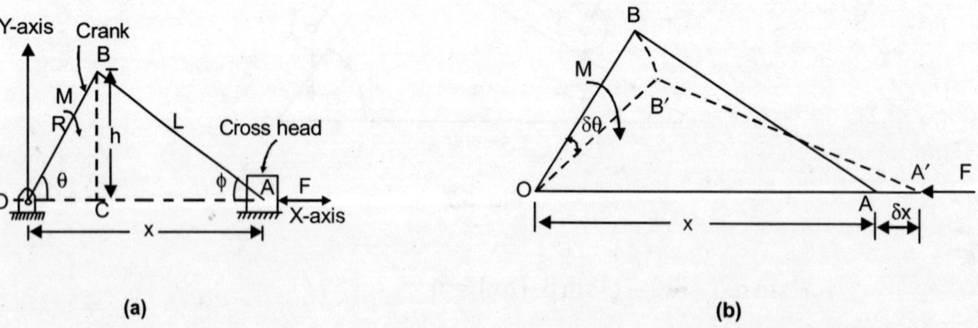

Fig. 5.8

5.8 VIRTUAL WORK APPLIED ON TRUSS STRUCTURE

This approach has also been found suitable for determination of support reaction of a truss as well as axial forces induced in different members of the truss. Figure 5.9 shows a truss *ABCDE* simply supported and loaded at joint *C*. In order to find out the reaction at *B*, we fix the truss at point *A* and allow R_B to provide an upward virtual displacement of the entire truss, by an amount δy. If the displacement at *B* be δy, from simply geometry we can say, the virtual displacement at *C* will be $\frac{\delta y}{2}$. Applying the principle of virtual work,

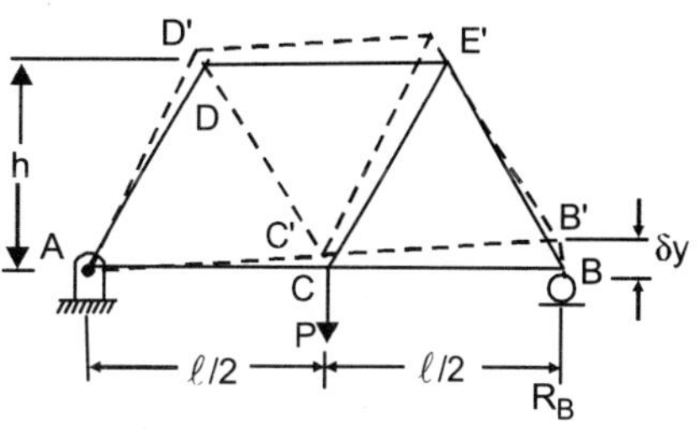

Fig. 5.9

$$R_B \cdot \delta y + P \cdot \left(-\frac{\delta y}{2} \right) = 0$$

$$\Rightarrow \qquad R_B = \frac{P}{2} \qquad \text{... (5.18)}$$

Multiple Choice Questions

Select the best alternative (s):

1. If the point of application of the force moves in a direction opposite to the force, the work done is
   (*a*) zero (*b*) positive (*c*) negative (*d*) none of the above
2. If the force is acting at an angle 90° with the direction of motion of the body, the work done will be
   (*a*) zero (*b*) positive (*c*) negative (*d*) none of the above
3. Virtual is the product of
   (*a*) displacement and virtual force (*b*) virtual displacement and virtual force
   (*c*) displacement and force (*d*) force and virtual displacement
4. In the equation of virtual work, following forces are neglected
   (*a*) reaction of any smooth surface with which the body is in contact
   (*b*) reaction of a rough surface of a body which rolls on it without slipping
   (*c*) reaction at a point or an axis, fixed in space, around which a body is constrained to turn
   (*d*) all of the above
   (*e*) none of the above

Answers

1. (*c*) 2. (*a*) 3. (*d*) 4. (*d*)

Numerical Examples

EXAMPLE 1

Find out the beam reaction by virtual work, where $l = 10\text{m}$, $a = 4\text{m}$, $b = 6\text{m}$, $P_1 = 15\text{kN}$ and $P_2 = 20\text{kN}$.

SOLUTION Proceeding the same manner as shown in article 5.3 the expression for R_B become, (equation 5.5)

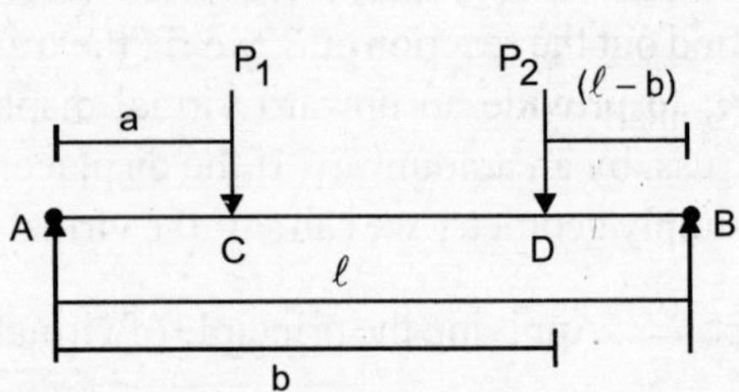

$$R_B = \frac{1}{l}(P_1 \cdot a + P_2 \cdot b)$$

$$= \frac{1}{10}(15 \times 4 + 20 \times 6) = 18\text{kN}$$

From the condition of virtual equilibrium, we have

$$R_A + R_B = 15 + 20 = 35$$

$$\Rightarrow \quad R_A = 17\text{kN}$$

EXAMPLE 2

Two beams AC and CD are hinged at C and simply supported at A and D. A hinged support is provided at B. The beam is loaded as shown in figure. Using the principle of virtual work, find reaction at B.

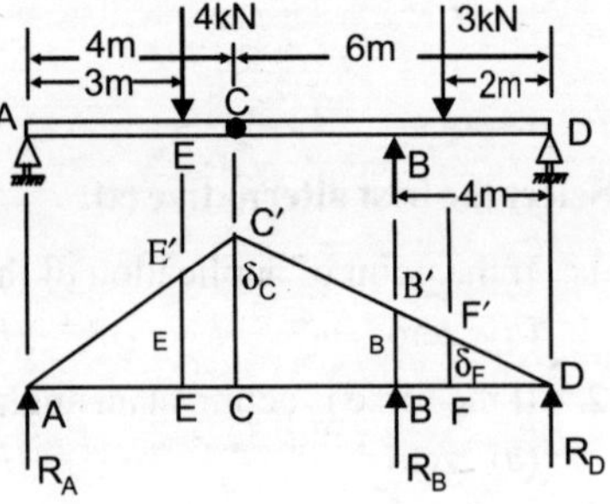

SOLUTION Keeping A and D fixed, a virtual displacement is provided at internal hinge C. From similar triangles DFF', DBB', DCC', we have

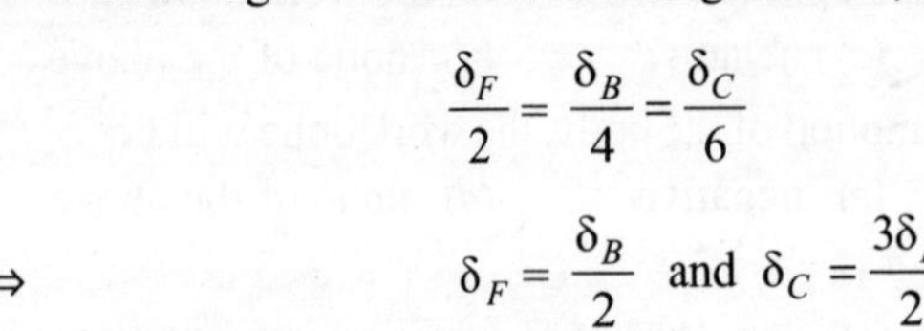

$$\frac{\delta_F}{2} = \frac{\delta_B}{4} = \frac{\delta_C}{6}$$

$$\Rightarrow \quad \delta_F = \frac{\delta_B}{2} \text{ and } \delta_C = \frac{3\delta_B}{2}$$

Again from similar triangles AEE' and ACC',

$$\frac{\delta_E}{3} = \frac{\delta_C}{4}$$

$$\Rightarrow \quad \delta_E = \frac{3}{4}\delta_C = \frac{3}{4} \times \frac{3}{2}\delta_B = \frac{9}{8}\delta_B$$

Now, applying principle of virtual work,

$$R_B \cdot \delta_B + 4 \cdot (-\delta_E) + 3 \cdot (-\delta_F) = 0$$

$$\Rightarrow \quad R_B \cdot \delta_B - 4 \cdot \frac{9}{8}\delta_B - 3 \cdot \frac{\delta_B}{2} = 0$$

$$\Rightarrow \quad R_B = 6\text{kN}$$

EXAMPLE 3

Find the reactions, by virtual work, at rollers B and D for the beam system as shown in figure.

SOLUTION Firstly consider beam DEF and the virtual displacement at D be δ. Following geometry of similar triangles, we have, $\delta_B = \frac{2}{3}\delta$, $\delta_E = \frac{\delta}{3}$

Similarly in beam ABC, $\delta_C = \delta_B \times \frac{6}{5} = \frac{4}{5}\delta$.

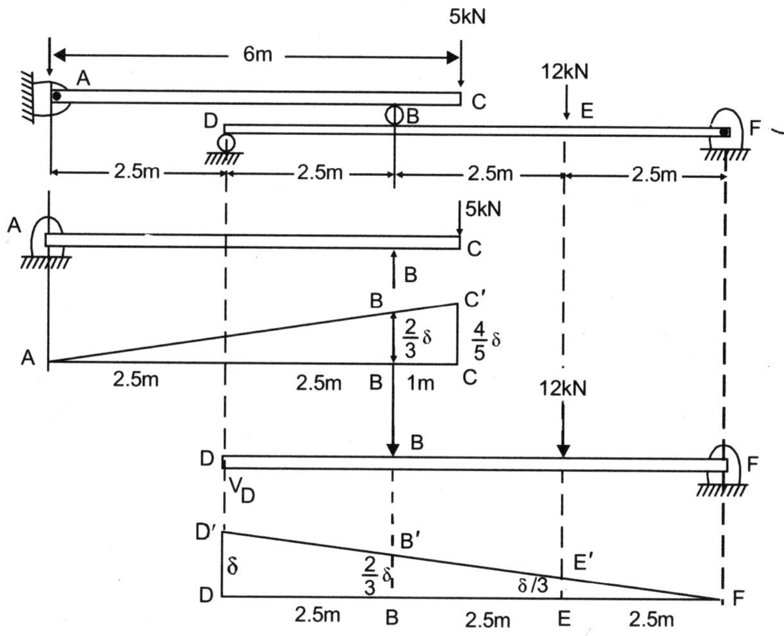

Applying principle of virtual work in beam ABC,

$$R_B \cdot \frac{2}{3}\delta + 5 \cdot \left(-\frac{4}{5}\delta\right) = 0$$

$$\Rightarrow \qquad R_B = 6\text{kN}$$

Again from beam DEF,

$$R_D \cdot \delta + R_B \cdot \left(-\frac{2}{3}\delta\right) + 12 \cdot \left(-\frac{\delta}{3}\right) = 0$$

$$\Rightarrow \qquad R_D \cdot \delta - 6 \cdot \frac{2}{3}\delta - 12 \cdot \frac{\delta}{3} = 0$$

$$\Rightarrow \qquad R_D = 8\text{kN}$$

EXAMPLE 4

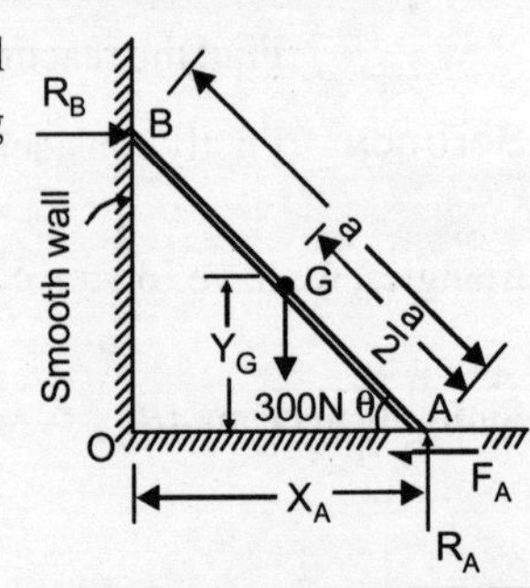

Following figure the weight of the ladder *AB* is 300 N and the vertical wall smooth while floor is rough. If $\theta = 60°$, find force of friction at the floor using method of virtual work.

SOLUTION Proceed in the identical fashion as worked out in article 5.4. Ultimately, the frictional resistance at *A* is obtained from equation (5.10) as

$$F_A = \frac{W}{2}\cot\theta = \frac{300}{2}\cot 60° = 86.6\text{kN}$$

EXAMPLE 5

A weightless ladder *AB* supports on two smooth surfaces as shown in figure and carries a load of 600N at *C*. Find the force *P* required at *A* to keep the ladder in equilibrium while $\theta = 60°$, using principle of virtual work.

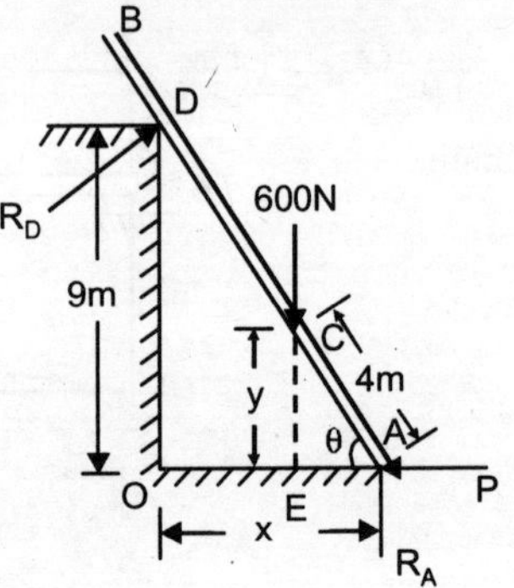

SOLUTION From similar triangles *AOD* and *AEC*,

$$x = 9\cot\theta, y = 4\sin\theta$$

Taking differentials,

$$\delta x = -9\,\text{cosec}^2\,\theta\cdot\delta\theta,$$

$$\delta y = 4\cos\theta\cdot\delta\theta$$

Where δx and δy are virtual displacements at *A* horizontally and at *C* vertically.

Now applying principle of virtual work,

$$P\cdot\delta x + W\cdot\delta y = 0$$

$$\Rightarrow \quad P\cdot\left(-9\,\text{cosec}^2\theta\cdot\delta\theta\right) + W\cdot\left(4\cos\theta\cdot\delta\theta\right) = 0$$

$$\Rightarrow \quad P = \frac{4W\cos\theta}{9\,\text{cosec}^2\theta} = 100\text{N}$$

EXAMPLE 6

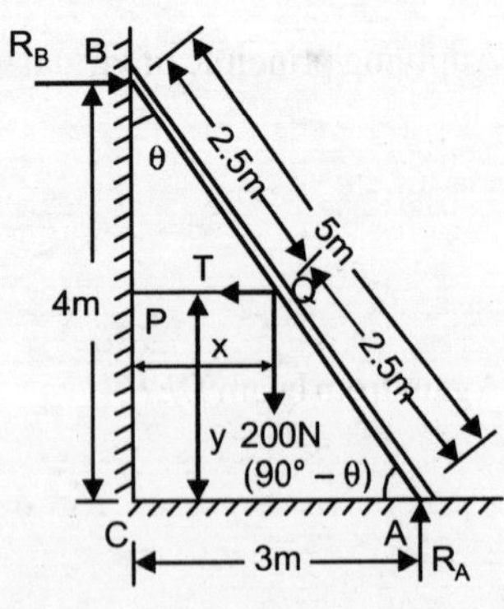

A uniform ladder, 5m long weighing 200N rests on a smooth floor at *A* and against a smooth wall at *B* as shown in figure. A horizontal rope *PQ* prevents the ladder from slipping. Using method of virtual work, determine tension *T* in the rope if $BQ = 2.5$m.

SOLUTION In the figure,

$$\tan\theta = \frac{3}{4} \text{ and}$$

$$x = 2.5\sin\theta, \quad y = 2.5\cos\theta$$

Taking differential,

$$\delta x = 2.5\cos\theta\cdot\delta\theta,$$

$$\delta y = -2.5\sin\theta\cdot\delta\theta$$

Applying principle of virtual work,

$$200\cdot\delta y + T\cdot\delta x = 0$$

$$\Rightarrow \quad 200 \cdot (-2.5 \sin\theta \cdot \delta\theta) + T \cdot (2.5 \cos\theta \cdot \delta\theta) = 0$$

$$\Rightarrow \quad T = \frac{200 \times 2.5}{2.5} \cdot \frac{\sin\theta}{\cos\theta} = 150\text{N}$$

EXAMPLE 7

Two balls D and E of weights P and Q can slide freely along the bars AC and BC. The balls are connected by an inextensible string DE. Find the value of the angle θ defining the position of equilibrium.

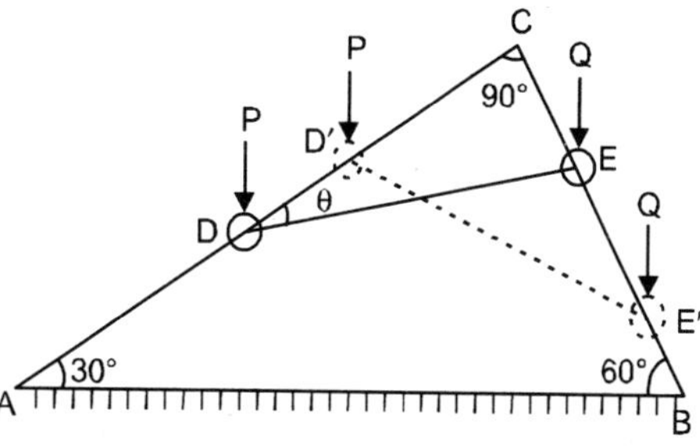

SOLUTION Let us assume the string DE undergoes a virtual rotation θ, and reaches $D'E'$. If length of DE be l,

$$\therefore \quad DC = l\cos\theta \text{ and } CE = l\sin\theta$$

Taking differential, we have,

$$\delta_{DC} = -l\sin\theta \cdot \delta\theta, \; \delta_{CE} = l\cos\theta \cdot \delta\theta$$

Now, applying principle of virtual work,

$$(P\sin 30°)\,\delta_{DC} + (Q\sin 60°)\,\delta_{CE} = 0$$

$$\Rightarrow \quad \frac{P}{2}(-l\sin\theta \cdot \delta\theta) + \frac{\sqrt{3}}{2}Q(l\cos\theta \cdot \delta\theta) = 0$$

$$\Rightarrow \quad \tan\theta = \frac{Q\sqrt{3}}{P}.$$

EXAMPLE 8

In the figure of pulley arrangement, the weight $W = 300$N, is to be raised through pully. Using principle of equilibrium, find the value of force P.

SOLUTION Approaching in the identical fashion as in the article (5.5), from equation (5.11), we can write,

$$P = \frac{W}{2} = \frac{300}{2} = 150\text{N}$$

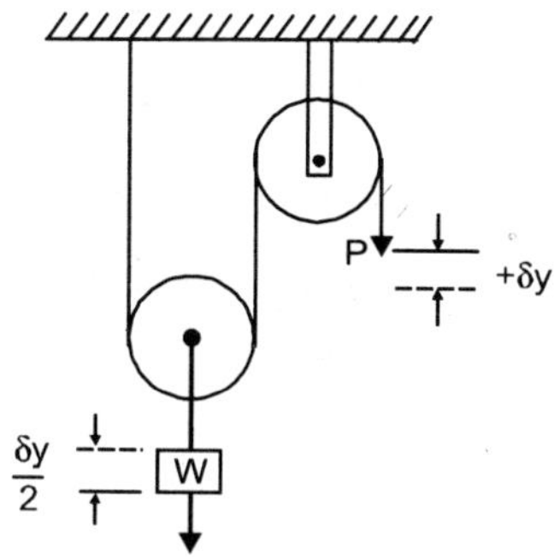

EXAMPLE 9

The diameters of two steps of the pully in differential pully block are 40cm and 30cm respectively. Using the principle of virtual work, determine value of the effort required to lift a load of 4kN.

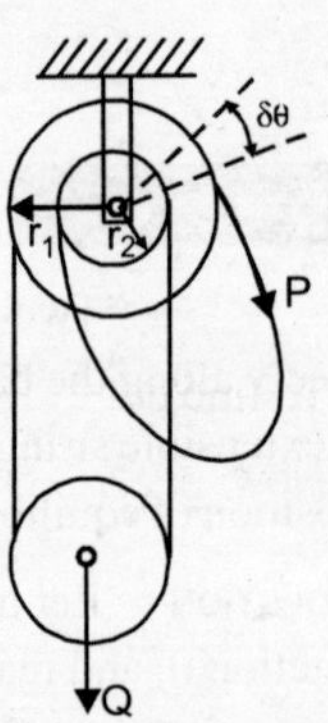

SOLUTION Proceeding in the same way as in the article 5.5, we obtain from equation (5.13)

$$P = \frac{r_1 - r_2}{2r_1} Q = \frac{20-15}{2 \times 20} \times 4 = 0.5\text{kN}$$

EXAMPLE 10

Find horizontal and vertical components of the reactions at A and B of the plane frame, loaded as shown in figure while $P = 500$N. The bars form three equal squares.

SOLUTION At first remove support B and replace it by reactive components, X_B and Y_B. Considering the origin of the co-ordinate system at A, as shown in figure we can write,

$$y_1 = y_2 = y = a \sin\theta,$$

and $$x = (a\cos\theta) \times 6 = 6a\cos\theta$$

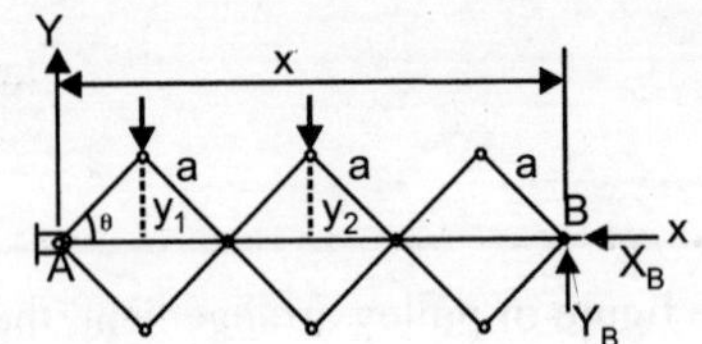

Here side of the square is considered a.

Now, $$\delta y = a\cos\cdot\delta\theta,\ \delta x = -6a\sin\theta\cdot\delta\theta$$

Let us give a virtual horizontal displacement δx of B, which causes a virtual vertical displacement of load points as δy.

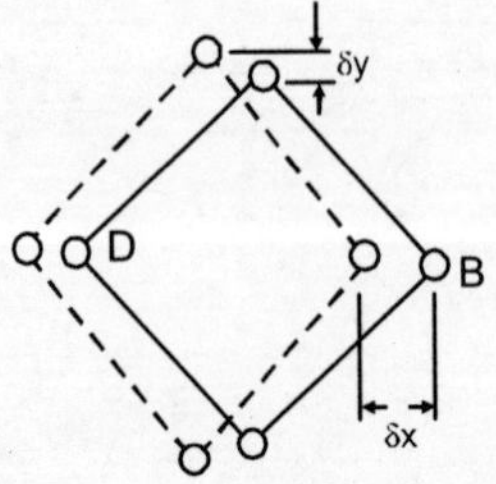

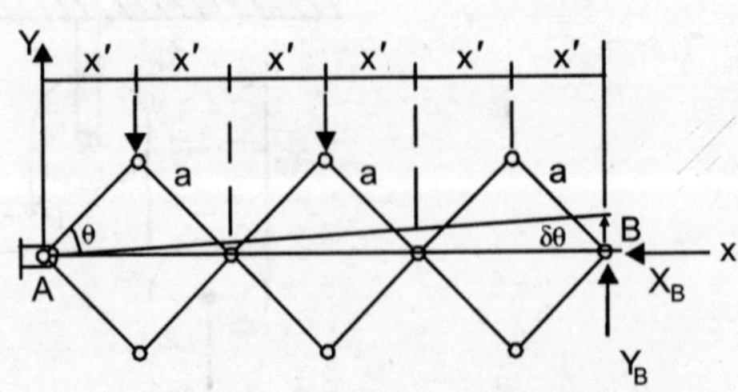

Applying principle of virtual work,

$$P\cdot\delta y + P\cdot\delta y + X_B\cdot\delta x = 0$$

$$\Rightarrow \quad 500(2a\cos\theta\cdot\delta\theta) - X_B\,(6a\sin\theta\cdot\delta\theta) = 0$$

$$\Rightarrow \quad X_B = \frac{2\times 500}{6}\times\frac{\cos\theta}{\sin\theta} = 166.67\text{N}$$

To find out Y_B keeping the frame fixed at A, rotate the entire frame by a virtual rotation $\delta\theta$, as shown in figure. Now

$$\text{vertical displacement of 1st loading point} = x'\cdot\delta\theta$$

$$\text{vertical displacement of 2nd loading point} = 3x'\cdot\delta\theta$$

$$\text{vertical displacement of } Y_B = 6x'\cdot\delta\theta$$

Applying principle of virtual work,

$$-500(x'\cdot\delta\theta) - 500(3x'\cdot\delta\theta) + Y_B(6x'\cdot\delta\theta) = 0$$

$$Y_B = \frac{2000}{6} = 333.34\text{N}$$

Following a similar approach, where A is to be made free and B kept fixed, we find reactions at A as, $X_A = 166.67\text{N}, Y_A = 666.67\text{N}$

EXAMPLE 11

Find the axial force S in the bar CD of the simple truss as shown in figure, by using virtual work, where $P = 700$N.

SOLUTION Firstly we cut the member CD and give a rotation at B by an amount $\delta\theta$ as shown in figure. From equation (5.18) of article (5.8)

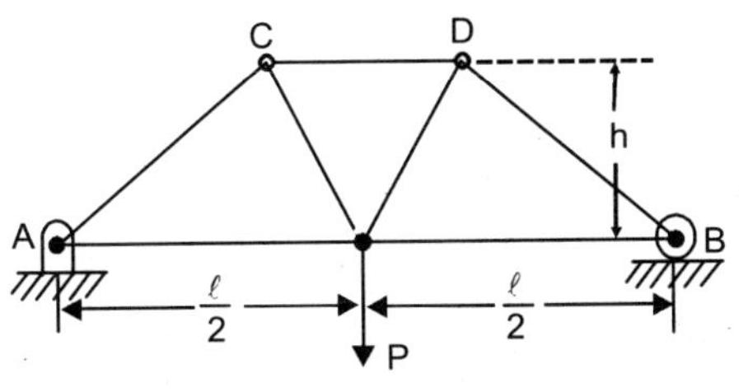

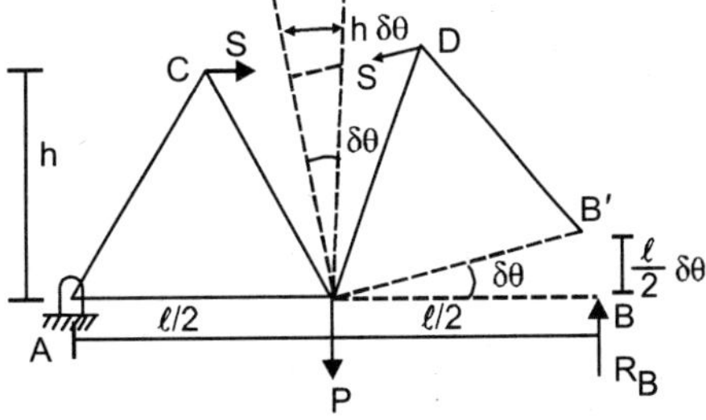

We obtain R_B as, $R_B = \frac{P}{2} = \frac{700}{2} = 350\text{N}$

Here the virtual work done by S will be $(h\cdot\delta\theta)$ and by R_B will be $\left(\frac{l}{2}\cdot\delta\theta\right)$.

Applying principle of virtual work,

$$-S \cdot (h \cdot \delta\theta) + R_B \cdot \left(\frac{l}{2} \cdot \delta\theta\right) = 0$$

$$S = R_B \cdot \frac{l}{2h} = 350 \times \frac{l}{2h} = 175\frac{l}{h}\text{ N}$$

EXAMPLE 12

Four bars of equal lengths l are hinged together at their ends in the form of a rhombus as shown in the figure. Using the principle of virtual work, find the relation between the active forces P and Q for equilibrium of the system in any configuration as defined by the angle θ, neglect the weight of the bars and friction.

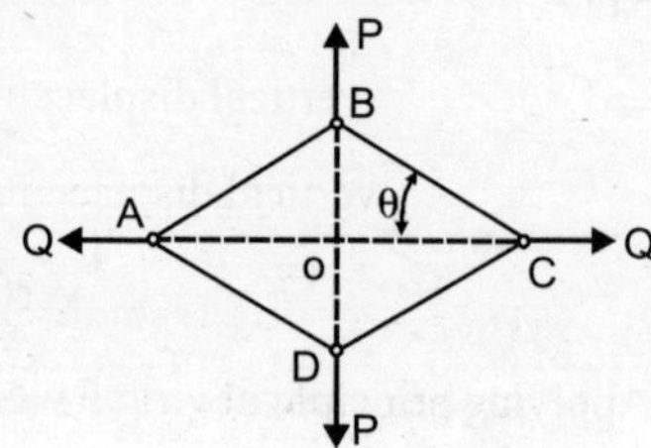

SOLUTION *Method I*: From figure we have,

$$OC^2 + OB^2 = BC^2$$

$$\Rightarrow \quad x^2 + y^2 = l^2$$

Suppose x changed by dx and y changes by dy and $x = l\cos\theta,\ y = l\sin\theta$

$$\therefore \quad 2xdx + 2ydy = 0$$

$$\Rightarrow \quad dy = -\frac{x}{y}dx$$

$$\Rightarrow \quad \frac{dx}{dy} = -\tan\theta$$

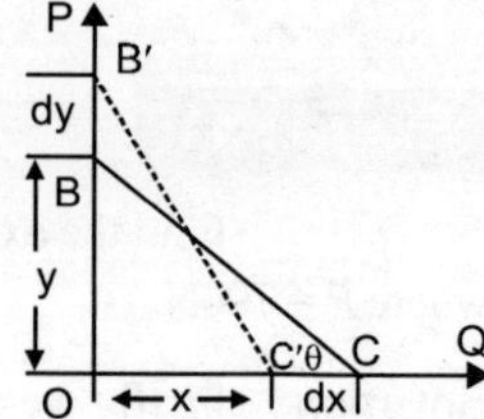

Now applying the principle of virtual work we have,

$$Pdy + Qdx = 0$$

$$\Rightarrow \quad \frac{P}{Q} = -\frac{dx}{dy} = \tan\theta$$

Method II: From figure we have,

$$OA = OC = x = l\cos\theta$$

$$\Rightarrow \quad dx = -l\sin\theta\, d\theta$$

$$OB = OD = y = l\sin\theta$$

$$\Rightarrow \quad dy = l\cos\theta\, d\theta$$

From principle of virtual work we have,

$$2P \times l\cos\theta d\theta - 2Q \times l\sin\theta d\theta = 0$$

$$\therefore \quad \frac{P}{Q} = \frac{\sin\theta}{\cos\theta} = \tan\theta$$

EXAMPLE 13

A rigid bar AB supported in a vertical plane by mutually perpendicular smooth surfaces OA and OB as shown in figure. Using the principle of virtual work, calculate the angle ϕ defining the configuration of equilibrium of the system.

SOLUTION

$$OA = x = (a+b)\cos\phi$$

$$\Rightarrow \quad dx = -(a+b)\sin\phi\, d\phi$$

$$OB = y = (a+b)\sin\phi$$

$$\Rightarrow \quad dy = (a+b)\cos\phi\, d\phi$$

$$\therefore \quad \frac{dx}{dy} = \frac{-(a+b)\sin\phi\, d\phi}{(a+b)\cos\phi\, d\phi} = -\tan\phi$$

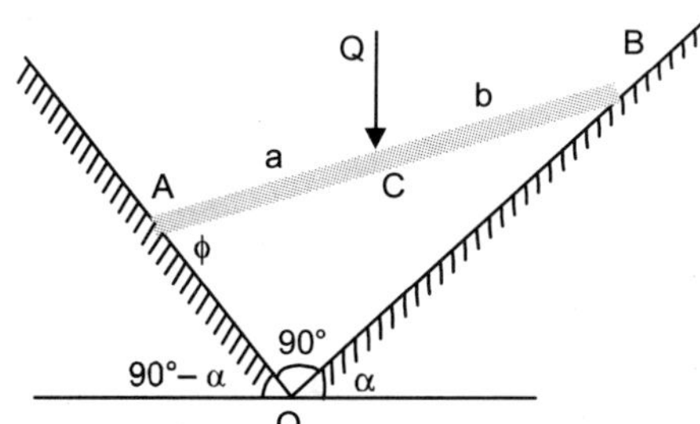

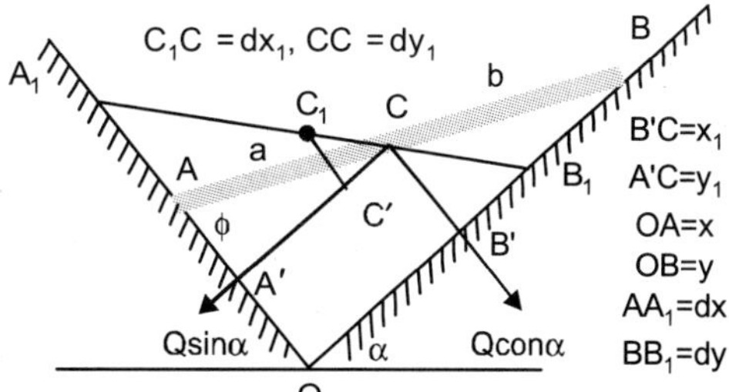

From similar ΔOAB and $\Delta B'CB$,

$$\frac{x_1}{x} = \frac{b}{a+b}$$

$$\Rightarrow \quad x_1 = \frac{b}{a+b}x$$

$$\Rightarrow \quad dx_1 = \frac{b}{a+b}dx$$

From similar ΔOAB and $\Delta A'AC$

$$\frac{y_1}{y} = \frac{a}{a+b}$$

$$\Rightarrow \quad y_1 = \frac{a}{a+b}y$$

$$\Rightarrow \quad dy_1 = \frac{a}{a+b}dy$$

$$\therefore \quad \frac{dx_1}{dy_1} = \frac{b}{a+b}\cdot\frac{a+b}{a}\cdot\frac{dx}{dy} = \frac{b}{a}\cdot\frac{dx}{dy} = \frac{-b}{a}\tan\phi$$

From the principle of virtual work we have,

$$Q\sin\alpha\cdot dy_1 + Q\cos\alpha\cdot dx_1 = 0$$

$$\therefore \quad \frac{dx_1}{dy_1} = \frac{\sin\alpha}{\cos\alpha}$$

$$\therefore \quad \frac{b}{a}\tan\phi = \tan\alpha$$

$$\therefore \quad \phi = \tan^{-1}\left(\frac{a}{b}\tan\alpha\right)$$

EXAMPLE 14

Two blocks weighing W_1 and W_2 resting on smooth inclined planes are connected by an inextensible string passing over a smooth pulley as shown in figure. Find the value of W_2 when $W_1 = 500\text{N}$ and $\alpha = 30^\circ, \beta = 60^\circ$.

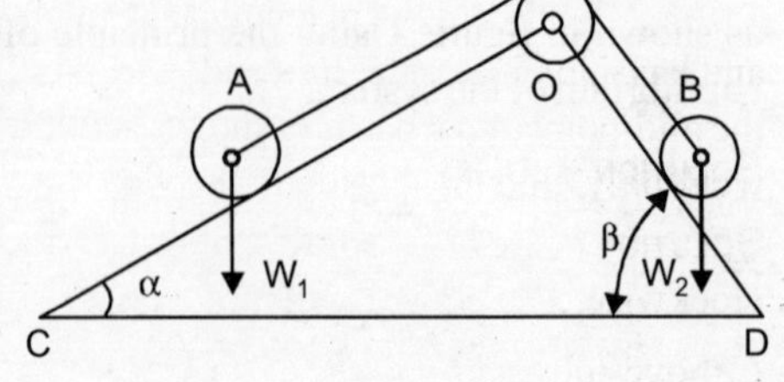

SOLUTION Let weight W_1 displaced a small distance dx along OC and W_2 will be displaced dy along OD. Equating the virtual work done to zero.

$$W_1 \sin\alpha \times dx - W_2 \sin\beta \times dy = 0$$

As the string AOB is inextensible, $dx = dy$

$$\therefore \quad W_2 \sin\beta = W_1 \sin\alpha$$

$$\therefore \quad W_2 = W_1 \frac{\sin\alpha}{\sin\beta} = 500 \times \frac{\sin 30^\circ}{\sin 60^\circ} = 288.67\text{N}$$

EXAMPLE 15

A structure with pin-connected members is shown in figure. Determine the value of θ for equilibrium using the principle of virtual work. Neglect friction and weight of the members. Take $P = \frac{W}{4}$.

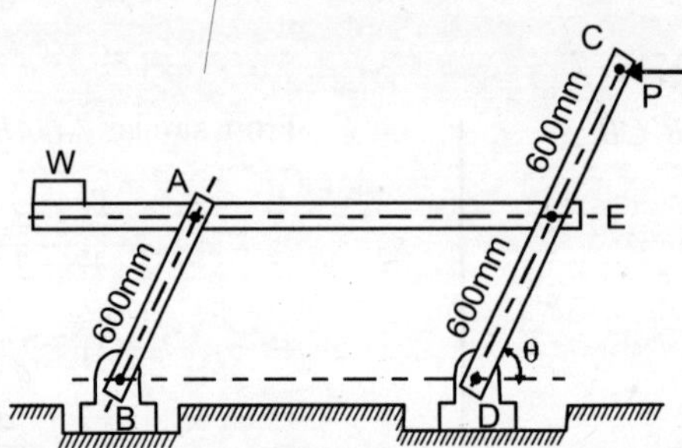

SOLUTION Let the system undergoes as small displacement such that θ will be $(\theta + d\theta)$.

$$DF = x = (0.6 + 0.6)\cos\theta = 1.2\cos\theta$$

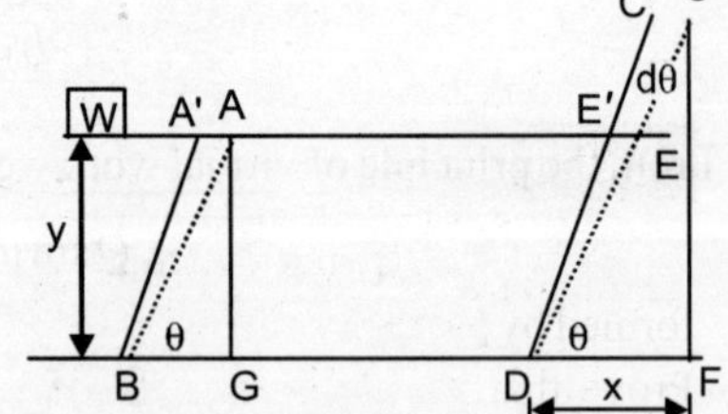

$$\Rightarrow \quad dx = -1.2\sin\theta d\theta$$

$$y = 0.6\sin\theta$$

$$\Rightarrow \quad dy = 0.6\cos\theta d\theta$$

From the principle of virtual work we have,

$$P \times dx + W \times dy = 0$$

$$\Rightarrow \quad \frac{W}{4} \times 1.2\sin\theta d\theta = W \times 0.6\cos\theta d\theta$$

$$\Rightarrow \quad \tan\theta = 2$$

$$\therefore \quad \theta = \tan^{-1}(2) = 63^\circ 26'$$

Example 16

Beam OA of length l and negligible weight is hinged at O and has a load W at the free end A. A wire BC supports this beam at the mid-point of the beam as shown in figure. By the method of virtual work only, find the tension in the wire.

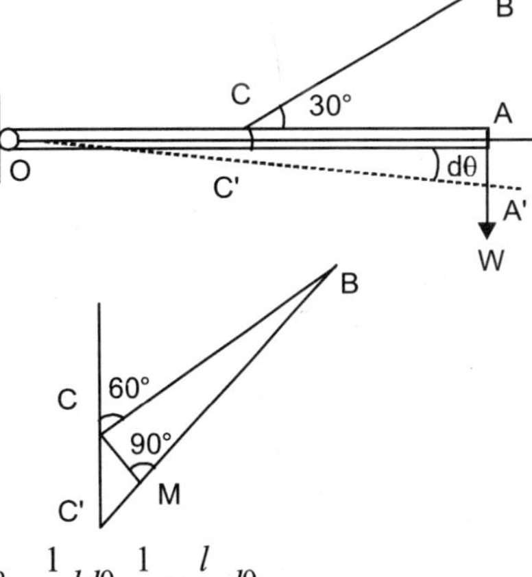

Solution Let OA rotates about O through a small angle $d\theta$ clockwise. The point A will move down a distance $AA' = dy = l d\theta$.

C moves down to C'.

$$\therefore \quad CC' = dy_1 = \frac{l}{2} d\theta$$

Draw CM perpendicular to BC'. BC makes with horizontal an angle of 30° and hence it makes with vertical an angle of 60°. We may take $\angle CC'M = 60°$ and $BM \approx BC$. MC' may be taken as increase in length of BC.

$$MC' = CC' \cos \angle CC'M = CC' \cos 60° = \frac{1}{2} l d\theta \cdot \frac{1}{2} = \frac{l}{4} d\theta$$

The equation of virtual work is

$$W l d\theta - T \times \frac{1}{4} l d\theta = 0$$

$$\therefore \quad T = 4W$$

Example 17

A beam supported on two identical rollers, each of radius R, is moved up an inclined plane by the application of a force F acting parallel to the plane. If the weight of the beam is W and the weight of each roller is W, find the force F needed for the equilibrium. Assume no slip between the rollers and the inclined plane or between the rollers and the beam.

Solution Let each roller moves up the plane by an amount x. Then the beam will move up the plane by $2x$.

The equation of virtual work is

$$F \times 2x - W \sin\alpha \times 2x - (W \sin\alpha + W \sin\alpha) x = 0$$

$$\Rightarrow \quad 2F = 4W \sin\alpha$$

$$\therefore \quad F = 2W \sin\alpha$$

Example 18

A string of length a of the shorter diagonal of a rhombus formed by four uniform rods each of length b and weight W hinged together. Prove that, if one of the rods is supported in a horizontal position, the tension in the string,

$$T = \frac{2W(2b^2 - a^2)}{b\sqrt{4b^2 - a^2}}.$$

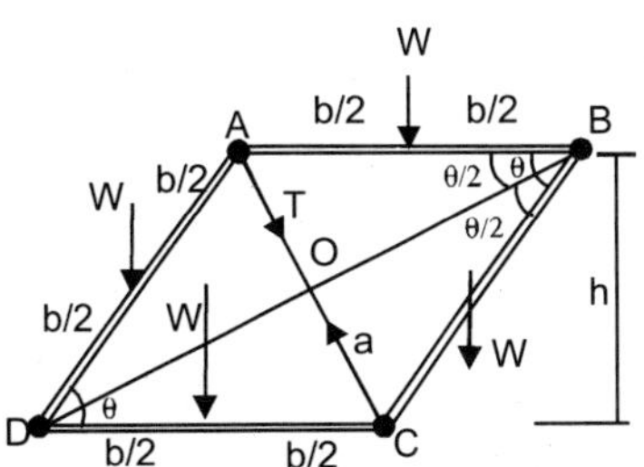

SOLUTION Let $ABCD$ be the rhombus formed by four uniform rods.

$\Rightarrow$ $$AB = BC = CD = DA = b \text{ and } AC = a, \angle ABC = \angle ADC = \theta$$

Here, $$a = 2b\sin(\theta/2) \text{ and } h = b\sin\theta$$

$\therefore$ $$\delta a = 2b\cos(\theta/2)\cdot\frac{1}{2}\delta\theta \text{ and } \delta h = b\cos\theta\delta\theta$$

Let T be the tension in the string. Let a small displacement be given so that angle θ is increased by $\delta\theta$. The rod CD will move down by δh and the middle points of AD and BC will move down by $\frac{\delta h}{2}$. Applying the principle of virtual work,

$$W\times\delta h + 2W\cdot\frac{\delta h}{2} - T\cdot\delta a = 0$$

$\Rightarrow$ $$T = 2W\frac{\delta h}{\delta a}$$

$$= 2W\frac{b\cos\theta\delta\theta}{b\cos(\theta/2)\delta\theta}$$

$$= 2W\frac{\cos\theta}{\cos(\theta/2)} = 2W\frac{2\cos^2(\theta/2)-1}{\cos(\theta/2)}$$

Again we know, $$\sin(\theta/2) = \frac{a}{2b}$$

$\Rightarrow$ $$\cos^2(\theta/2) = 1-\frac{a^2}{4b^2} = \frac{4b^2-a^2}{4b^2}$$

$\therefore$ $$T = 2W\frac{2\cdot\frac{4b^2-a^2}{4b^2}-1}{\sqrt{\frac{4b^2-a^2}{4b^2}}} = \frac{2W\left(2b^2-a^2\right)}{b\sqrt{4b^2-a^2}}$$

EXAMPLE 19

A uniform rod of length $2a$ rests in equilibrium against a smooth vertical wall and upon a smooth peg at a distance b from the wall. Show that, in the position of equilibrium, the beam is inclined to the wall at an angle $\theta = \left[\sin^{-1}(b/a)\right]^{1/3}$.

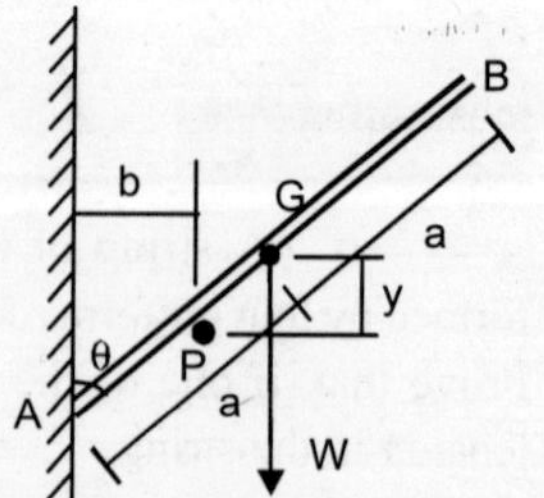

SOLUTION Let AB be the rod, G its C.G and P the position of peg.

$$y = AG\cos\theta - b\cot\theta = a\cos\theta - b\cot\theta$$

$\therefore$ $$dy = -a\sin\theta d\theta + b\,\text{cosec}^2\theta d\theta$$

Reaction at A and P do not perform any work. Hence equilibrium equation using virtual work is

$$W\left(-a\sin\theta d\theta + b\,\text{cosec}^2\theta d\theta\right) = 0$$

$$\Rightarrow \qquad b\,\text{cosec}^2\theta = a\sin\theta$$

$$\Rightarrow \qquad \frac{b}{\sin^2\theta} = a\sin\theta$$

$$\Rightarrow \qquad \sin^3\theta = \frac{b}{a}$$

$$\therefore \qquad \theta = \left[\sin^{-1}(b/a)\right]^{1/3}$$

EXAMPLE 20

A load Q is hoisted by the pulley system shown in figure. Determine the magnitude of the pull P required to raise the load Q, if $r = 14$cm, $d = 4$cm, coefficient of friction in the journals supporting the pulleys is $\mu = 0.25$, neglect friction in the movable pulley. What is the efficiency of this device?

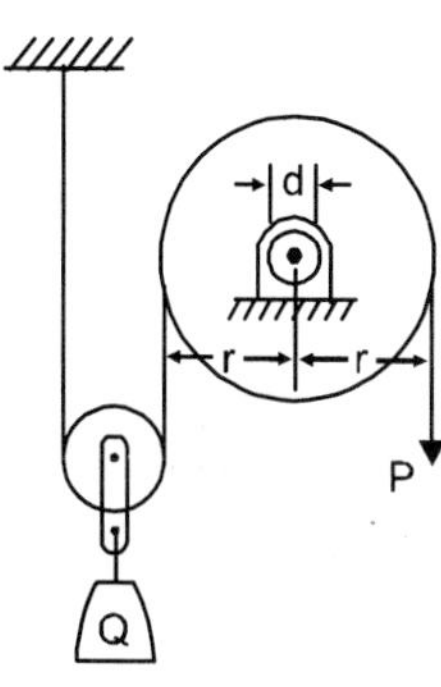

SOLUTION Let rotate the pulley through a small angle $\delta\theta$. It lifts centre of movable pulley δy_2 upward and pulls down P by δy_1 downward. From condition of compatibility

$$\Rightarrow \qquad 2\delta y_2 - \delta y_1 = 0$$

$$\therefore \qquad \delta y_2 = \frac{\delta y_1}{2} = \frac{r\delta\theta}{2}$$

Using principle of virtual work we have,

$$P \times \delta y_1 - F\frac{d}{2}\delta\theta - Q \times \delta y_2 = 0$$

$$\therefore \qquad Pr\delta\theta - F\frac{d}{2}\delta\theta - Q\frac{r\delta\theta}{2} = 0$$

From equilibrium of movable pulley $\qquad 2T = Q$

From equilibrium of fixed pulley $\qquad N = T + P = \dfrac{Q}{2} + P$

Again we know, $\qquad F = \mu N = \mu\left(\dfrac{Q}{2} + P\right)$

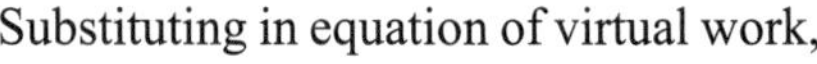

Substituting in equation of virtual work,

$$\Rightarrow \qquad P \times r = F\frac{d}{2} + Q\frac{r}{2} = \mu\frac{d}{2}\left(\frac{Q}{2} + P\right) + Q\frac{r}{2}$$

$$\Rightarrow \qquad P \times 14 = 0.25 \times \frac{4}{2}\left(\frac{Q}{2} + P\right) + Q\frac{14}{2}$$

$$P = 0.537Q$$

$$\therefore \qquad \text{efficiency} = \frac{\text{work output}}{\text{work input}} = \frac{Q \cdot r\delta\theta / 2}{P \cdot r\delta\theta} = \frac{Q}{2P} = \frac{1}{2 \times 0.537} = 0.931 \text{ or } 93.1\%$$

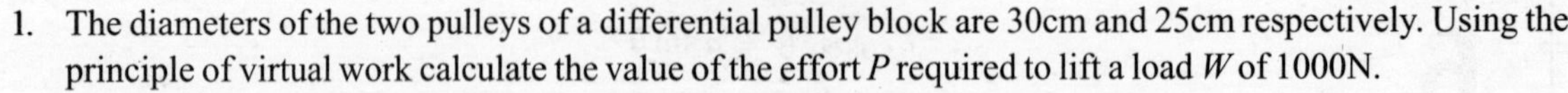

1. The diameters of the two pulleys of a differential pulley block are 30cm and 25cm respectively. Using the principle of virtual work calculate the value of the effort *P* required to lift a load *W* of 1000N.

Ans. [166.67N]

2. Using the principle of virtual work, determine the reaction of a beam *AB* of span 10m. The beam carries a point load of 6kN at a distance of 4m from *A*.

Ans. [R_A = 3.6kN, R_B = 2.4kN]

3. A beam *AB* of span 8m carries two point loads of 10kN and 15kN at 3m and 5m from the end *A* respectively. Determine the beam reactions by the principle of virtual work.

Ans. [R_A = 10.875kN, R_B = 13.125kN]

4. Two beams *AC* and *CD* of lengths 9m and 10m respectively are hinged at *C*. These are supported on rollers at the left and right ends (*A* and *D*). A hinged support is provided at *B*, 6m from *A*. Using the principle of virtual work, determine the reactions at the hinge *C* and at the support *B* when a load of 600N acts at a point 5m from *D*.

Ans. [R_C = 300N, R_B = 450N]

5. A uniform ladder of weight 200N rests with its upper end against a smooth vertical wall and its foot on a rough horizontal ground making an angle of 60° with the ground. Find the force of friction of the ground using the method of virtual work.

Ans. $\left[\frac{100}{\sqrt{3}}\text{N}\right]$

6. Four identical uniform rods, each of weight *W* are joined together to form a square *ABCD* and stiffened by a fifth rod *BD* also of weight *W*. The frame is suspended from *A* in the vertical. Find the thrust in the rod *BD*.

Ans. $\left[\frac{5W}{2}\right]$

7. A rigid beam *AB* of length 10m is hinged at the end *A* and supported on a roller at *C*, 6m away from end *A*. The beam is subjected to a point load of 900N at a distance of 6m from the end *B*. Determine the reaction at *C*, using virtual work principle.

Ans. [600N]

8. A uniform rectangular plank of height *h* and base width *b* rests on a cylindrical surface of radius *R*. Assuming sufficient friction to prevent slipping find the relation between *R* and *h* consistent with the stability.

Ans. [$R > h/2$]

9. A uniform body is made of a hemisphere of radius *R* and a right circular cylinder of height *H* which have a common base. Determine the largest value of the height of the cylinder consistent with the stability of the body in vertical position.

Ans. $\left[H = \frac{R}{\sqrt{2}}\right]$

CHAPTER

6

INTRODUCTION TO STRENGTH OF MATERIALS

6.1 INTRODUCTION

In the previous chapter of statics, the entire concepts were circulated around *rigid bodies.* But, in reality, not a single body is rigid at all. Better to say, all materials are deformable. When properties of deformation are taken into account to handle problem of statics, a branch of engineering science gets emerged, which is known as *Strength of Materials*. In other words, *Strength of Materials* can also be defined as *Statics of Deformable Bodies* or *Statics of Elastic Bodies*.

Here we have scope only to deal the elementary concept of deformation of elastic bodies. Elementary explanation of some basic parameters, basic law of elasticity, material properties and a handful of representative numerical problems are discussed as following :

6.2 GENERAL CONCEPT OF STRESS AND STRAIN

Every person, whether has conventional educational background or not, can understand very easily, the concept of stress and strain, in some form or other. Behind any effect, there must be some cause. In the same manners, if any kind of *generalised deformations* (lengthening, shortening, deflection, twist) be the cause of strain, *generalised forces* (pull, push, moment, torque, etc.) are the causes of stress.

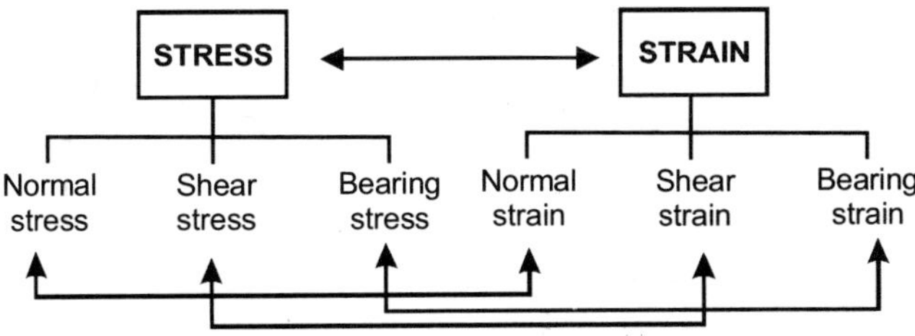

So let us know what is meant actually by stress. To be precise, stress or intensity of stress is force P per unit effective area A, i.e.

$$Stress = \frac{force}{effective\ \ area} = \frac{P}{A}$$

Similarly, strain is elongation or compression per unit length, i.e.,

$$Strain = \frac{elongation\ or\ compression}{original\ length}$$

Different types of stress and strain are shown in tree diagram.

6.3 NORMAL STRESS AND NORMAL STRAIN

When any force 'tensile or compressive' acts perpendicularly on any material surface, stress generated is called *normal stress*. Conventionally it is expressed by σ (sigma). If area of the shaded parts in the Figures 6.1 (*a*) and (*b*) be *A*, then

$$\sigma = \frac{P}{A} \qquad \text{... (6.1)}$$

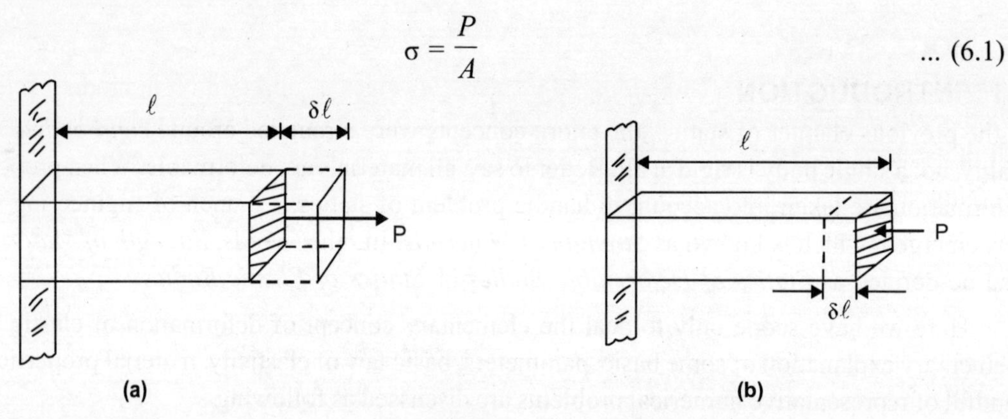

Fig. 6.1

Strain produced due to normal stress or better say, normal force is defined as *normal strain*. It is conventionally expressed by ε (epsilon). In the above figures, the original length of the members be and elongation or compression be δl in each case, so,

$$\varepsilon = \frac{\delta l}{l} \qquad \text{... (6.2)}$$

6.4 SHEAR STRESS AND SHEAR STRAIN

If the force, causing deformation of any members, acts along any material surface, stress induced is called *shear stress*. So, shear stress is produced in the effects of shear force, which may be either of vertical nature or of horizontal nature.

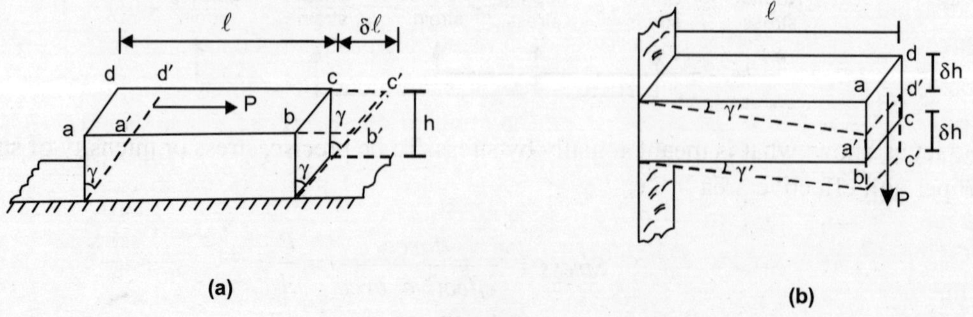

Fig. 6.2

Conventionally shear stress is represented by τ (tau). In the figure above, effective area on which shear force P is acting, let be A, in both the cases. In case of horizontal shear, as base is fixed, point a and b got shifted to a', and b', respectively, by virtue of application of force, and generating an angular deformation γ (gamma). Similarly in case of vertical shear, a shifted to a' and b shifted to b', and angular deformation is γ'. All these two cases angular deformation denote *shear strain.* The angle, γ' being very small

$$\gamma \approx \tan\gamma = \frac{\delta l}{h} \text{ or } \gamma \approx \tan\gamma = \frac{\delta l}{h} \qquad \text{... (6.3)}$$

6.5 HOOKE'S LAW

Robert Hooke (1635–1703) marked one of the most spectacular observation in modern science in 1678 that materials exhibit a certain relation between stress applied and strain induced, and came up with the statement "*ut tensio sic vis*" i.e., "*the extension is as large as the force.*"

Now accepting a modification of *force* to *stress* and *extension* to *strain*, the statement of *Hooke's law* appears as, "*stress is proportional to strain within elastic limit*". It is mathematically represented as

$$\sigma \propto \varepsilon$$

or, $$\sigma = E\cdot\varepsilon, \qquad \text{where } E = \text{constant of proportionality...(6.4)}$$

Now comes the question of *elastic limit* and thereby *elasticity.*

6.6 ELASTICITY

Elasticity is the property of the material by which it comes back or exhibits a tendency to come back to its orginal shape on removal of applied load.

Elastic limit is the point on the stress-strain curve or load-elongation curve up to which a load can be applied to a material without any permanent set.

Within the perview of this elementary theory of analysis, all materials are assumed to be perfectly elastic.

6.6.1 (a) Other Assumption of Material Property

In the present course of discussion, all materials are assumed to be perfectly *homogeneous* and *isotropic.*

If distribution of material property is uniform at entire part of the specimen, it is called *homogeneous.* Absence of this property is designated as *non-homogeneity.*

Isotropy is the presence of property at any point, same along all possible co-ordinate directions. And if property at a point differs from one direction to other direction, it is called *anisotropy.* If variation of material property at any point occurs in orthogonal (perpendicular) direction, it is called *orthotropy.*

6.7 YOUNG'S MODULUS OR MODULUS OF ELASTICITY

Now in the statement of his law, Robert Hooke, however, related total strain on a body to total force acting on it. He did not recognize a limit to this proportionality, as shown in equation (6.4).

The values of constant of proportionality, for Hooke's law had been experimentally determined by Thomas Young for a number of materials. He noticed that this constant holds good up to a range which is slightly lesser value to elastic stress. He designated this range as *limit of proportionality*. It was also noticed that this proportionality constant E, has a unique value for each material and is known as *Young's Modulus* or *Modulus of Elasticity*.

In fact there exists a fine difference between *elastic limit* and *limit of proportionality*. Limit of proportionality is the point on stress-strain plot or load-elongation plot beyond which the straight line starts to continue as curve. Elastic limit is slightly higher than limit of proportionality. So, we can say that beyond proportional limit, the material remains elastic up to elastic limit, but does not follow Hooke's Law.

Hence, the modified Hooke's Law states that '*stress is proportional to strain up to proportional limit*'.

6.8 MODULUS OF RIGIDITY

Hooke's Law, as discussed previously, does hold good for both normal stress and shear stress as well. Following this, it can be said that shear stress is proportional to shear strain.

$$\therefore \quad \tau \propto \gamma$$

$$\Rightarrow \quad t = G \cdot \gamma \qquad \text{...(6.5)}$$

Here G is the constant of proportionality, which is defined as *Modulus of Rigidity*.

6.9 POISSON'S RATIO

Any material subjected to strain in one direction, undergoes strain in its perpendicular direction also, because normal engineering material does not change volume under load.

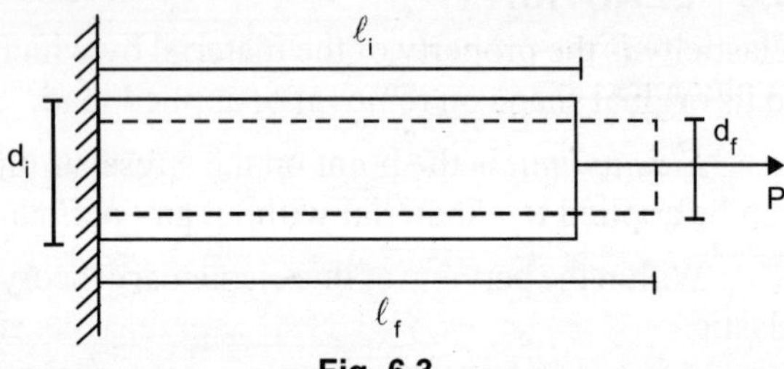

Fig. 6.3

In the above illustration, as the member is subjected to a tensile force P, strain occurs in longitudinal direction, changing l_i to l_f. At the same time diameter also gets reduced in transverse direction, making d_i to d_f, so that initial volume and final volume remain constant.

The ratio of lateral strain to longitudinal strain is defined by a ratio which is called *Poisson's Ratio*, which is symbolised by μ (*mu*)

$$\therefore \quad \frac{\text{Lateral strain}}{\text{Longitudinal strain}(\varepsilon)} = \text{Poisson's ratio}(\mu) \qquad \text{...(6.6)}$$

In normal engineering material, its value varies from 0 to 0.50, which is the maximum mathematically permissible value.

In the illustration we can write :

$$l_f = l_i(1+\varepsilon) \text{ and } d_f = d_i(1-\mu\varepsilon) \text{ as lateral strain} = \mu\,\varepsilon$$

So initial volume,

$$\forall_i = \frac{\pi}{4} d_i^2 \cdot l_i \qquad \text{... (6.7)}$$

and final volume, $$\forall_f = \frac{\pi}{4} d_f^2 \cdot l_f = \frac{\pi}{4} d_i^2 (1-\mu\varepsilon)^2 \cdot l_i (1+\varepsilon)$$

∴ Change in volume, $$\Delta\forall = \forall_f - \forall_i = \frac{\pi}{4} d_i^2 \cdot l_i \left[(1-\mu\varepsilon)^2 (1+\varepsilon) - 1\right]$$

$$= \frac{\pi}{4} d_i^2 \cdot l_i \, [1 - 2\mu\varepsilon + \mu^2\varepsilon^2 + \varepsilon - 2\mu\varepsilon^2 + \mu^2\varepsilon^3 - 1]$$

$$= \frac{\pi}{4} d_i^2 \cdot l_i (\varepsilon - 2\mu\varepsilon) \qquad ...(6.8)$$

∴ Rate change in volume

$$\frac{\Delta\forall}{\forall_i} = \varepsilon(1-2\mu). \qquad ... (6.9)$$

Now, as rate of change volume cannot be negative, $\varepsilon(1-2\mu) \geq 0$

$$\Rightarrow \qquad 1 - 2\mu \geq 0$$

$$\Rightarrow \qquad \mu \leq \frac{1}{2}$$

So, maximum mathematical value of *Poisson's Ratio* is 0.50. Poisson's Ratio is nearly 0 for cork and nearly 0.50 for rubber and paraffin. For steel it is roughly 0.30.

Modulus of Elasticity E, Modulus of Rigidity G, Poisson's Ratio μ, all are material constant. And all material constants are related to each other. Here E, G and μ are correlated as

$$E = 2G(1+\mu) \qquad ...(6.10)$$

For three-dimensional stress system, bulk modulus (K) is frequently used, which is the ratio of volumetric stress to volumetric strain and expressed as

$$E = 3K(1-2\mu)$$

Unfortunately the deduction of above relationship is not within the purview.

Units of Material Properties and Parametrs

Properties of Parameters	**Dimension**	**Unit (S.I)**
Stress	$\frac{MLT^{-2}}{L^2} = ML^{-1}T^{-2}$	$Pascal = \frac{N}{m^2}$
Strain	$\frac{L}{L} = 1$	Pure number
Modulus of Elasticity	$\frac{MLT^{-2}}{L^2} = ML^{-1}T^{-2}$	$Pascal = \frac{N}{m^2}$
Modulus of Rigidity	$\frac{MLT^{-2}}{L^2} = ML^{-1}T^{-2}$	$Pascal = \frac{N}{m^2}$
Poisson's Ratio	$\frac{(L/L)}{(L/L)} = 1$	Pure number

6.10 UNITS OF MATERIAL PROPERTIES AND PARAMETERS

Here Pascal being a small quantity, stresses or modulus of elasticity or rigidity are expressed in a larger unit as, following:

1 Mega Pascal (MPa) $= 1\text{MN/m}^2 = 1\text{N/mm}^2 = 10\text{kg/cm}^2$

1 Giga Pascal (GPa) $= 1\text{GN/m}^2 = 10^3\text{MPa} = 10^3\text{N/mm}^2$

6.11 BASIC GOVERNING EQUATION FOR DISPLACEMENT

Let us consider a bar of *homogeneous isotropic* material, of length l and cross-sectional area A hangs vertically fixed at top. It is subjected to a vertically downward pull P, by virtue of which an elongation of δl is occurred within elasticity limit. Modulus of elasticity of the material is E.

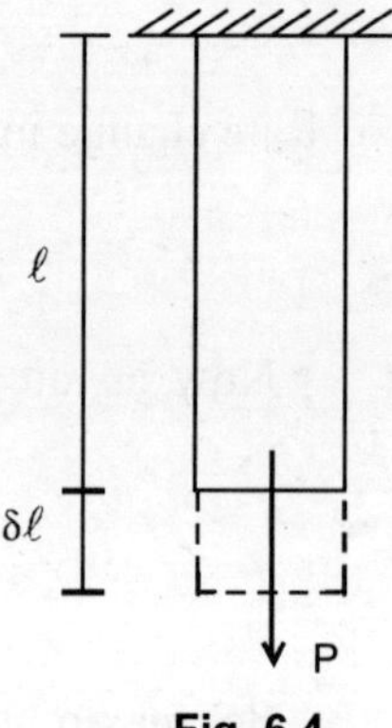

Fig. 6.4

Applying Hooke's Law, we can write,

$$\sigma = E \cdot \varepsilon$$

or

$$\frac{P}{A} = E \cdot \frac{\delta l}{l}$$

∴

$$\delta l = \frac{P \cdot l}{A \cdot E}$$

This is the basic relationship for computation of elongation or contraction. Now onwards, in order to express displacement, we will use only δ instead of δl.

Multiple Choice Questions

Select the best alternative (s):

1. Strain is defined as the ratio of
   (*a*) change in volume to original volume
   (*b*) change in length to orginal length
   (*c*) change in cross-sectional area to orginal cross-sectional area
   (*d*) none of the above
2. Hooke's Law is valid up to
   (*a*) yield point (*b*) limit of proportionality (*c*) breaking point
   (*d*) elastic limit (*e*) plastic limit
3. Young's modulus is defined as the ratio of
   (*a*) volumetric stress and volumetric strain (*b*) lateral stress and lateral strain
   (*c*) longitudinal stress and longitudinal strain (*d*) shear stress and shear strain
   (*e*) longitudinal stress and lateral strain
4. If equal and opposite force applied to a body tend to elongate it, the stress produced is
   (*a*) internal resistance (*b*) tensile stress
   (*c*) transverse stress (*d*) compressive stress
   (*e*) working stress
5. The material having same elastic properties in all directions are called
   (*a*) ideal material (*b*) uniform material (*c*) isotropic material
   (*d*) elastic material (*e*) none of the above

6. Modulus of rigidity is defined as the ratio of
   (*a*) longitudinal stress and longitudinal strain
   (*b*) volumetric stress and volumetric strain
   (*c*) lateral stress and lateral strain
   (*d*) shear stress and shear strain
   (*e*) linear stress and lateral strain
7. The property of material by virtue of which a body returns to its orginal shape after removal of the load is called
   (*a*) plastic
   (*b*) elasticity
   (*c*) ductility
   (*d*) malleability
   (*e*) resilience
8. Poisson's ratio is defined as
   (*a*) longitudinal stress and longitudinal strain
   (*b*) longitudinal stress and lateral strain
   (*c*) lateral stress and longitudinal stress
   (*d*) lateral stress and lateral strain
   (*e*) none of the above
9. The relationship between modulus of rigidity *G*, modulus of elasticity *E* is
   (*a*) $E = G(1+\mu)$
   (*b*) $G = E(2-\mu)$
   (*c*) $G = E/[2(1+\mu)]$
   (*d*) $G = E/(1+2\mu)$
   (*e*) $G = 2E/(1+2\mu)$

Answer

1. (*b*) 2. (*b*) 3. (*c*) 4. (*b*) 5. (*c*) 6. (*d*) 7. (*b*) 8. (*e*) 9. (*c*)

Numerical Examples

Example 1

A steel wire hangs vertically under its own weight. What is the greatest length, it can have, if the allowable tensile stress $\sigma_t = 200$MPa. The specific weight of steel is 80kN/m^3.

Solution Let the length of the wire is *l* metre and c/s area *A* m^2.

∴ Total weight of the wire (*W*) $= (l \times A \times 80)$ kN

∴ Stress induced in the wire $= \dfrac{W}{A} = (l \times 80)$ kN/m^2

According to the condition, $l \times 80 \times 10^3 = 200 \times 10^6$

$\Rightarrow \quad l = 2500$ m.

Example 2

The three pieces of glued wood as shown in figure has c/s of 3.75cm × 3.75cm. If $P = 37.5$kN, find average shear stress in each of glued joints.

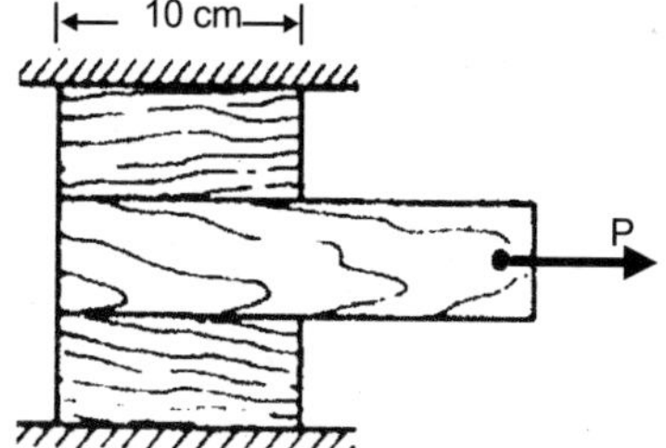

Solution Effective area for shear in each glued joint

$$= (3.75 \times 10)\text{cm}^2$$

∴ Total effective area for two interfaces

$$= (2 \times 3.75 \times 10)\text{cm}^2$$

∴ Average shear stress in each interface

$$= \frac{37.5\times10^3}{2\times3.75\times10\times100}\ \text{N/mm}^2 = 5\text{MPa}$$

EXAMPLE 3

Cross-section of the prismatic steel bar in figure is 3cm². Neglecting localized stress distribution, find net increase δ of the bar while $E = 2.1\times10^5\,\text{N/mm}^2$.

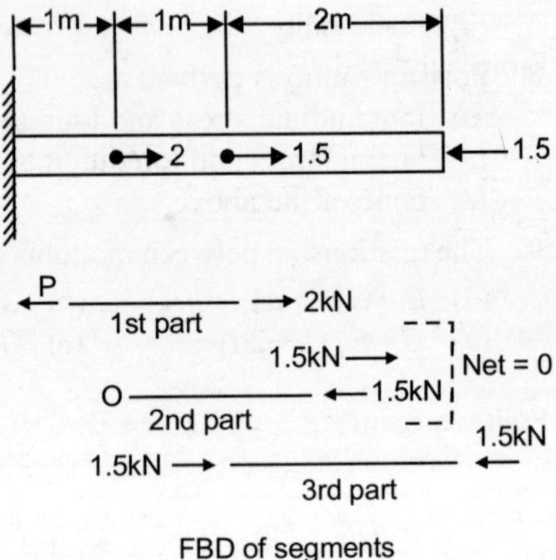

FBD of segments

SOLUTION Free-body diagrams of the three segments are as following and from that $P = 2\text{kN}$. From above, we observe that *3rd part* is subjected to 1.5kN compression force, *2nd part* subjected to zero force and *1st part* subjected to 2kN tensile force. Therefore, total displacement

$$\delta = \delta_1 + \delta_2 + \delta_3$$

$$= \frac{2000\times10^3\times1}{AE} + 0 + \left(-\frac{1.5\times10^3\times2000}{AE}\right)$$

$$= \frac{10^3\times10^3}{AE}(2-3) = -\frac{10^6}{AE}$$

$$= -\frac{10^6}{3\times100\times2.1\times10^5} = -0.01587\text{mm.}$$

EXAMPLE 4

In the adjoining figure, c/s of tie bar CD, 0.5cm² and allowable stress $\sigma_w = 150\text{MPa}$. Find out the magnitude of P and vertical deflection at B of the rod while $E = 2\times10^5\,\text{MPa}$.

SOLUTION $\tan\theta = \frac{1.5}{2} = \frac{3}{4}$

∴ $\sin\theta = \frac{3}{5}$

and $\cos\theta = \frac{4}{5}$

From the free-body diagram, taking moment at A,

$$P\times4 = S\sin\theta\times2$$

$$= S\left(\frac{3}{5}\right)\times2 = (0.5\times100)\times150\times\left(\frac{3}{5}\right)\times2$$

$$= 9000$$

⇒ $P = 2250\text{N}$

FBD of rod

So, here $S = \frac{2p}{\sin\theta}$

From deflected pattern of the system, we see that, B is shifted to B' and C to C'. Initial length of the tie CD becomes elongated by an amount $C_1C'(\delta)$.

Length $$CD = \sqrt{2^2 + 1.5^2} = 2.5\text{m}$$

Now, $$C_1C'(\delta) = \frac{S \times 2.5 \times 1000}{0.5 \times 10^2 \times 2 \times 10^5}$$

From right-angled $\Delta CC_1C'$,

$$CC' \sin\theta = CC_1$$

$$\Rightarrow \quad \delta_c \sin\theta = \delta$$

$$\Rightarrow \quad \delta_c = \frac{1}{\sin\theta} \cdot \frac{S \times 2.5 \times 1000}{0.5 \times 10^2 \times 2 \times 10^5}$$

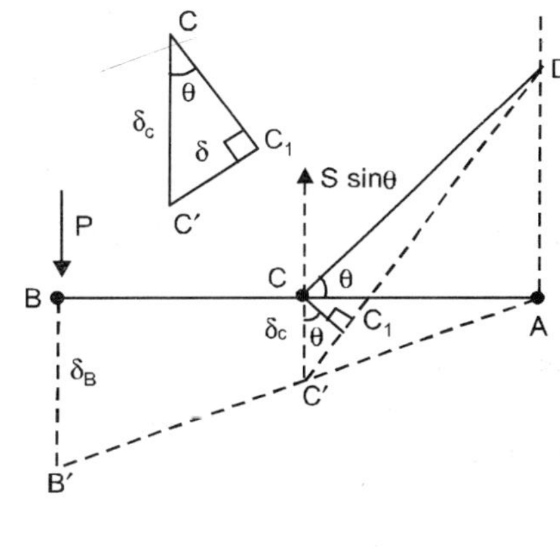

Following similar triangles ABB' and ACC'

$$\frac{BB'}{CC'} = \frac{AB}{AC}$$

$$\Rightarrow \quad \frac{\delta_B}{\delta_C} = \frac{4}{2}$$

$$\Rightarrow \quad \delta_B = 2\delta_c$$

$$= \frac{2}{\sin\theta} \cdot \frac{S \times 2.5 \times 1000}{0.5 \times 10^2 \times 2 \times 10^5}$$

$$= \frac{2}{\sin\theta} \cdot \frac{2P}{\sin\theta} \cdot \frac{2.5 \times 1000}{0.5 \times 10^2 \times 2 \times 10^5}$$

$$= \frac{2 \times 2 \times 2250 \times 2.5 \times 1000}{\left(\frac{3}{5}\right) \times \left(\frac{3}{5}\right) \times 0.5 \times 10^2 \times 2 \times 10^5} = 6.25\text{mm}$$

EXAMPLE 5

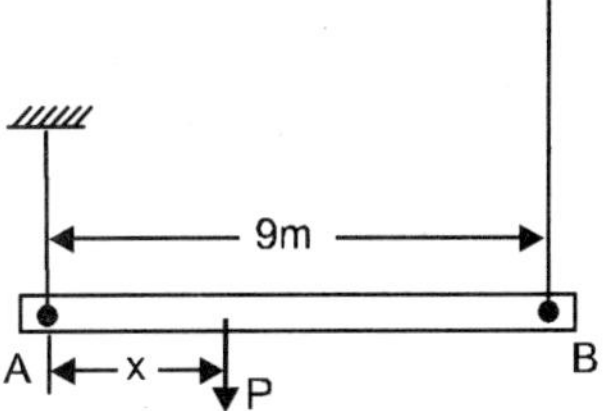

In the adjacent figure, the rod at A is of brass and at B is of steel, $L_a = 3\text{m}, A_a = 10\text{cm}^2, E_a = 1 \times 10^5$ MPa and $L_b = 5\text{m}, A_b = 4.45\text{cm}^2, E_b = 2.1 \times 10^5$ MP_a. At what distance x from A may a vertical load P be applied to make the bar remaining horizontal. In this condition, how much stress will be developed at those rods if $P = 3\text{kN}$.

SOLUTION Referring free-body diagram of the bar AB, reactive forces in the rods R_A and R_B at A and B respectively, can be expressed as,

$$R_B = P \cdot \frac{x}{9} \text{ and } R_A = P \cdot \frac{9 - x}{9}$$

To maintain static equilibrium, following the condition of compatibility, we can say that deflection at A and B must be equal.

$$\delta_A = \delta_B$$

$$\Rightarrow \qquad \frac{R_A \cdot L_a}{A_a \cdot E_a} = \frac{R_B \cdot L_b}{A_b \cdot E_b}$$

$$\Rightarrow \qquad P \cdot \frac{9-x}{9} \cdot \frac{3000}{10 \times 100 \times 1 \times 10^5} = P \cdot \frac{x}{9} \cdot \frac{5000}{4.45 \times 100 \times 2.1 \times 10^5}$$

$$\Rightarrow \qquad \frac{(9-x) \times 3}{10} = \frac{x \times 5}{4.45 \times 2.1}$$

$$\Rightarrow \qquad x = 3.23\text{m}$$

Now,

$$R_A = \frac{9-3.23}{9} P$$

$$= 0.6411 \times 3000 = 1923.3\text{N}$$

and

$$R_B = (3000 - 1923.3) = 1076.7\text{N}$$

So, stress induced at brass rod $= \dfrac{1923.3}{10 \times 100} = 1.923\text{MPa}$

and at steel rod $= \dfrac{1076.7}{4.45 \times 100} = 2.419\text{MPa}$

EXAMPLE 6

The hollow steel cylinder of length, $L = 30$ cm and inside diameter, $d = 15$cm and wall thickness, $t = 3$mm filled with concrete and compressed by parallel plates with $P = 500$kN. Calculate compressive force for each material and total shortening of cylinder while $A_S = 7.15\text{cm}^2$, $E_S = 2 \times 10^5$ MPa, $A_C = 176.5\text{cm}^2$, $E_C = 2 \times 10^4$ MPa,

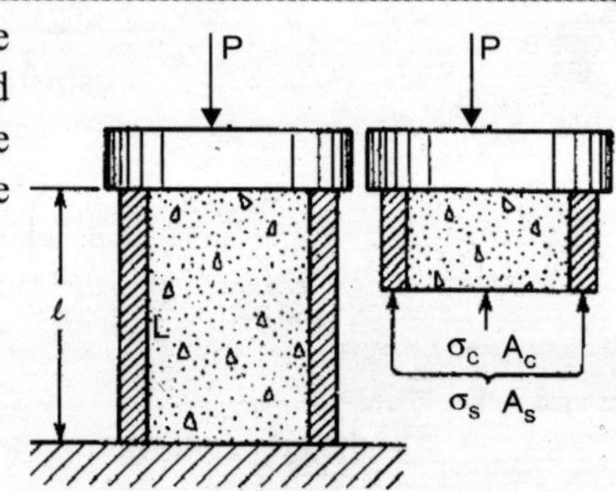

SOLUTION From free-body diagram,

$$\sigma_C A_C + \sigma_s A_s = P \qquad \text{... (1)}$$

From condition of compatibility, the strain in concrete and steel will be equal. Therefore,

$$\varepsilon_C = \varepsilon_S \qquad \text{... (2)}$$

$$\Rightarrow \qquad \frac{\sigma_C}{E_C} = \frac{\sigma_S}{E_S}$$

$$\Rightarrow \qquad \frac{\sigma_S}{\sigma_C} = \frac{E_S}{E_C} = 10$$

$\Rightarrow \qquad \sigma_S = 10\sigma_c$

Solving (1) and (2),

$$\sigma_C = \frac{P}{10A_S + A_C} = \frac{500 \times 10^3}{(10 \times 7.15 + 176.5) \times 100} = 20.16 \text{ MPa}$$

$$\sigma_S = \frac{10P}{10A_S + A_C} = \frac{10 \times 500 \times 10^3}{(10 \times 7.15 + 176.5) \times 100} = 201.6 \text{ MPa}$$

Shortening of the cylinder $= \varepsilon_C \times L$

or $= \varepsilon_S \times L$

$$= \frac{\sigma_C}{E_C} \times L$$

$$= \frac{20.16 \times 300}{2 \times 10^4} = 0.3024\text{mm}$$

EXAMPLE 7

Find the expression for tension induced in steel wires, which are identical in every sense and of length L_W.

SOLUTION Considering equilibrium of AB and taking moment about A,

$$S_1 a + S_2 b = Pl \qquad \text{...(1)}$$

Again, from equation of compatibility,

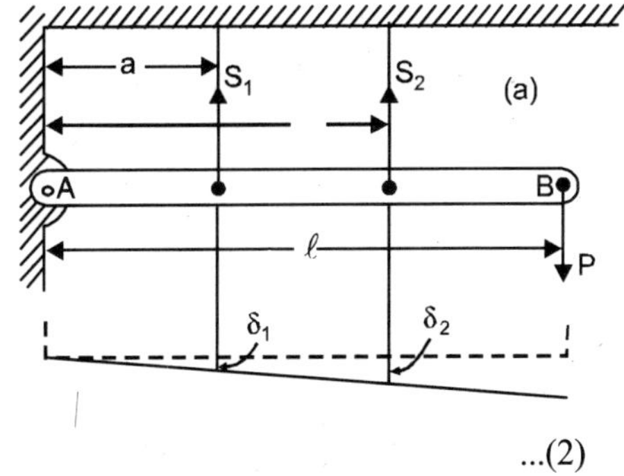

$$\frac{\delta_1}{\delta_2} = \frac{a}{b}$$

$$\Rightarrow \qquad \frac{\frac{S_1 L_W}{AE}}{\frac{S_2 L_W}{AE}} = \frac{a}{b}$$

$$\Rightarrow \qquad \frac{S_1}{S_2} = \frac{a}{b} \qquad \text{...(2)}$$

Solving equations (1) and (2), we find,

$$S_1 = \frac{Pal}{a^2 + b^2} \text{ and } S_2 = \frac{Pbl}{a^2 + b^2}.$$

EXAMPLE 8

In making prestressed beam, initial tensile stress of wire is σ_0. After prestressing, when external force P is removed, find residual stress in the materials. The ratio of Young's modulus of steel and concrete is 12:1 and that of c/s area is 1:15.

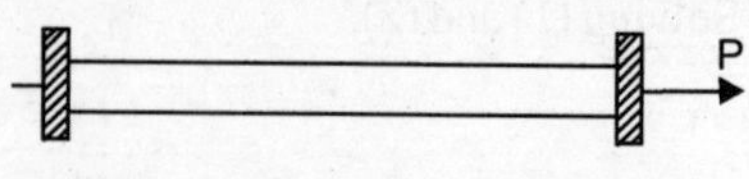

SOLUTION Here, $\dfrac{E_S}{E_C} = \dfrac{12}{1}$

and $\dfrac{A_S}{A_C} = \dfrac{1}{15}$

Now, the equation of force compatibility states that

$$\sigma_S A_S + \sigma_C A_C = P = \sigma_0 A_S$$

$$\Rightarrow \quad \sigma_S A_S + \sigma_C (15 A_S) = \sigma_0 A_S$$

$$\Rightarrow \quad \sigma_S + 15\sigma_C = \sigma_0 \quad \text{...(1)}$$

From strain compatibility, $\delta_S = \delta_C$

$$\Rightarrow \quad \frac{\sigma_S l}{E_S} = \frac{\sigma_C l}{E_C}$$

$$\Rightarrow \quad \sigma_S = \sigma_C \cdot \frac{E_S}{E_C} = 12\sigma_C \quad \text{...(2)}$$

Solving (1) and (2), we obtain, $\sigma_C = \dfrac{\sigma_0}{27}$ and $\sigma_S = \dfrac{5\sigma_0}{9}$

Initial strees in steel = σ_0, in concrete = 0

Final strees in steel = $4\sigma_0/9$, in concrete = $\sigma_0/27$

$\therefore$ Residual strees in steel = $\left(\sigma_0 - \dfrac{4\sigma_0}{9}\right) = \dfrac{5\sigma_0}{9}$, concrete $\left(0 - \dfrac{\sigma_0}{27}\right) = -\sigma_0/27$.

EXAMPLE 9

The steel guy wires in the adjacent structure are identical except for length and are just taut but free from stress before force P is applied. Find tensile forces S_1 and S_2.

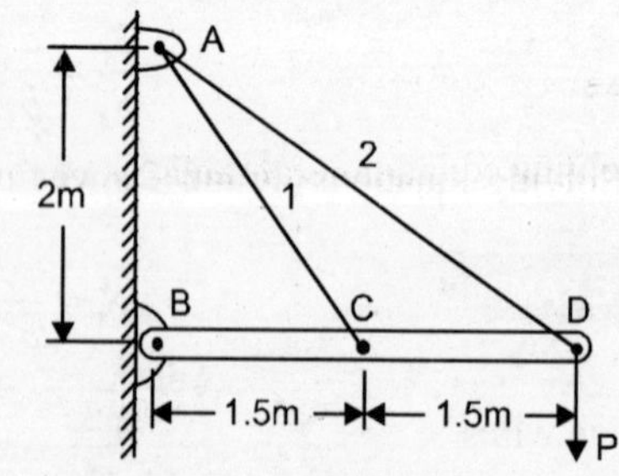

SOLUTION Here, $\tan\theta_1 = \dfrac{2}{1.5}$ and $\tan\theta_2 = \dfrac{2}{3}$

From free-body diagram, moment equilibrium equation at B

$$S_1 \sin\theta_1 \times 1.5 + S_2 \sin\theta_2 \times 3 = 3P$$

$$\Rightarrow \quad S_1\left(\frac{2}{2.5}\right) \times 1.5 + S_2\left(\frac{2}{\sqrt{13}}\right) \times 3 = 3P$$

$$\Rightarrow \qquad \frac{S_1}{2.5} + 2\frac{S_2}{\sqrt{13}} = P \qquad \text{....}(a)$$

From ΔCC_1C_2 and ΔDD_1D_2, we can write,

$$CC_1 \sin\theta_1 = C_1C_2 \text{ and } DD_1 \sin\theta_2 = D_1D_2$$

$$\Rightarrow \qquad CC_1 = \frac{C_1C_2}{\sin\theta_1} \qquad\qquad DD_1 = \frac{D_1D_2}{\sin\theta_2}$$

$$= \frac{S_1 \times \sqrt{2^2 + 1.5^2}}{AE \times \left(\frac{2}{2.5}\right)} \qquad\qquad = \frac{S_2 \times \sqrt{2^2 + 3^2}}{AE \times \left(\frac{2}{\sqrt{13}}\right)}$$

$$= S_1 \times \left(\frac{3.125}{AE}\right) \qquad\qquad = S_2 \times \left(\frac{6.5}{AE}\right)$$

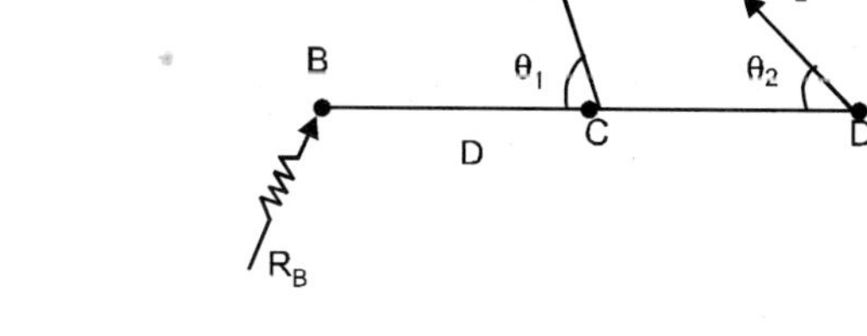

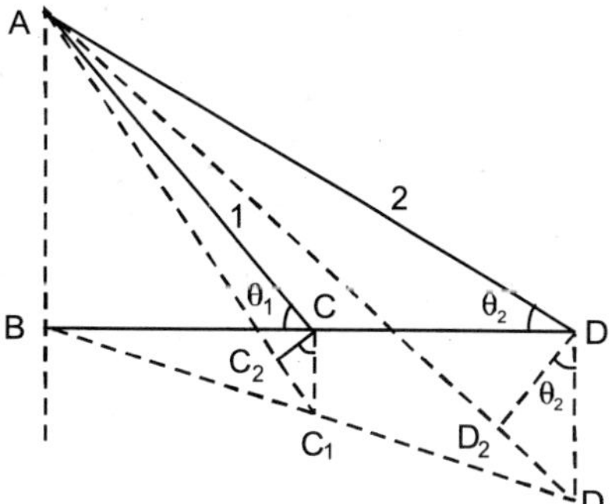

Considering equation of compatibility, from similar ΔBCC_1 and ΔBDD_1,

$$\frac{CC_1}{DD_1} = \frac{BC}{BD} = \frac{1.5}{3.0} = \frac{1}{2} \qquad \text{...(2)}$$

Substituting the expression, $\qquad \frac{S_1 \times 3.125}{S_2 \times 6.5} = \frac{1}{2}$

$$\Rightarrow \qquad S_1 = 1.04 S_2 \qquad \text{...(3)}$$

Solving equations (1) and (3), we have, $S_1 = 1.07P$, $S_2 = 1.03P$

EXAMPLE 10

The wires as shown, hold the bar horizontally. After P=200kN is applied, find out forces in each wires.

SOLUTION From the condition of equilibrium along vertical direction, we have,

$$P_1 + P_2 + P_3 = 200 \qquad \text{... (1)}$$

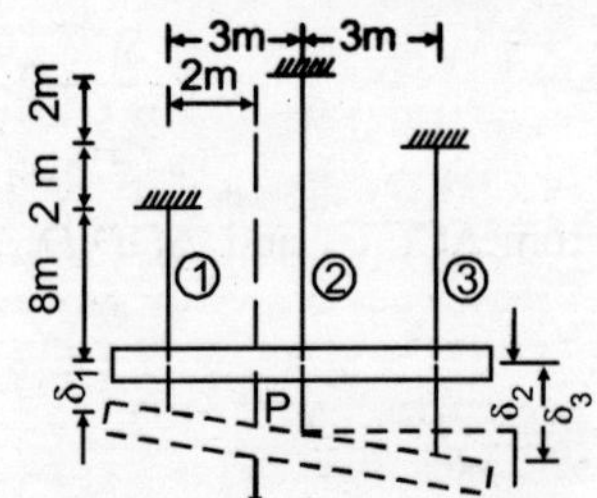

Now, taking moments of P_2 and P_3 about first wire,

$$3P_2 + 6P_3 = 2 \times 200$$

$$\Rightarrow \quad P_2 + 2P_3 = \frac{400}{3} \qquad ...(2)$$

From compatibility criterion on displacement we can say, from geometry,

$$\delta_2 = \frac{\delta_1 + \delta_1}{2}$$

$$\Rightarrow \quad \frac{P_2 L_2}{AE} = \frac{1}{2}\left[\frac{P_1 L_1}{AE} + \frac{P_3 L_3}{AE}\right]$$

$$\Rightarrow \quad P_2 \times 12 = \frac{1}{2}(P_1 \times 8 + P_3 \times 10)$$

$$\Rightarrow \quad 12P_2 = 4P_1 + 5P_3 \qquad ...(3)$$

Solving equations (1), (2), (3) for P_1, P_2 and P_3, we obtain

$$P_1 = 107.08\text{kN},\ P_2 = 52.52\text{kN},\ P_3 = 40.40\text{kN}.$$

EXAMPLE 11

A twisted clothesline wire of length L has c/s area A and modulus of elasticity E in tension. The wire is stretched horizontally between two fixed points, A and B, without appreciable initial tension. A load P is then suspended from the mid-point C of the line. Prove that vertical deflection Δ at point C will be $\Delta = \frac{L}{2}\sqrt[3]{\frac{P}{2AE}}$ assuming that Δ is small compared with the length L of the clothesline.

SOLUTION Considering equilibrium in vertical direction,

$$2S\sin\theta = P$$

$$\Rightarrow \quad S = \frac{P}{2\sin\theta} \qquad ...(1)$$

From ΔACC_1,

$$\sin\theta = \frac{\Delta}{\left[\Delta^2 + \frac{L^2}{4}\right]^{\frac{1}{2}}} = \frac{\Delta}{\left(\frac{L^2}{4}\right)^{\frac{1}{2}}} = \frac{2\Delta}{L} \qquad ...(2)$$

Here it is assumed that $\Delta^2 \ll \frac{L^2}{4}$

and

$$\Delta\sin\theta = C_1C_2 = S \cdot \frac{L/2}{AE} = \frac{PL}{4AE\sin\theta}$$

$$\Rightarrow \quad \Delta = \frac{PL}{4AE\sin^2\theta} = \frac{PL}{4AE} \times \frac{L^2}{4\Delta^2}$$

$$\Rightarrow \qquad \Delta^3 = \frac{PL^3}{16AE}$$

$$\Rightarrow \qquad \Delta = \frac{L}{2}\sqrt[3]{\frac{P}{2AE}} \qquad \text{(Proved).}$$

EXAMPLE 12

In the adjacent figure, a lever is attached to a spindle by means of a square key 6mm × 6mm by 3 cm long. If the average shear stress is not to exceed 65 MPa, what will be safe value of P?

SOLUTION Let us consider by virtue of application of P, a shear force F acts on the key on an effective area of 6mm × 3cm.

∴ Effective area on shear = 6 × 30 = 180mm²

∴ Effective shear force F = 180 × 65 = 11,700N

Taking moment about centre of spindle and applying condition of equilibrium, we have,

$$2.5 \times 10 \times F = 75 \times 10 \times P$$

$$\Rightarrow \qquad P = \frac{2.5 \times 11700}{75} = 390\text{N}$$

EXAMPLE 13

The frame in the figure is made up of 10cm × 10cm square wood posts. Find values of a, b, c if allowable stress in shear parallel to grain is 0.7 MPa and that in compression 2.8 MPa, while pull at top of post 1500N.

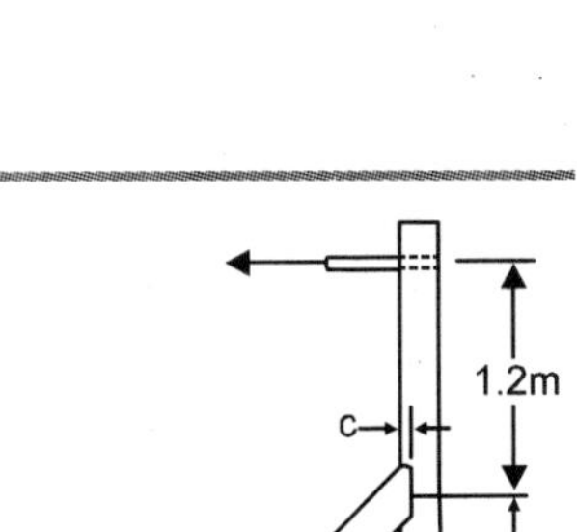

SOLUTION From free-body diagram of vertical posts, consider moment of all forces about hinge point.

$$S\cos\theta \times 0.9 = 1500 \times 2.1$$

$$\Rightarrow \qquad S\cos\theta = 3500\text{N}$$

$$\therefore \qquad S\sin\theta = \frac{3500}{\cos\theta} \times \sin\theta$$

$$= 3500\tan\theta$$

$$= 3500 \times \frac{0.9}{1.2} = 2625\text{N}$$

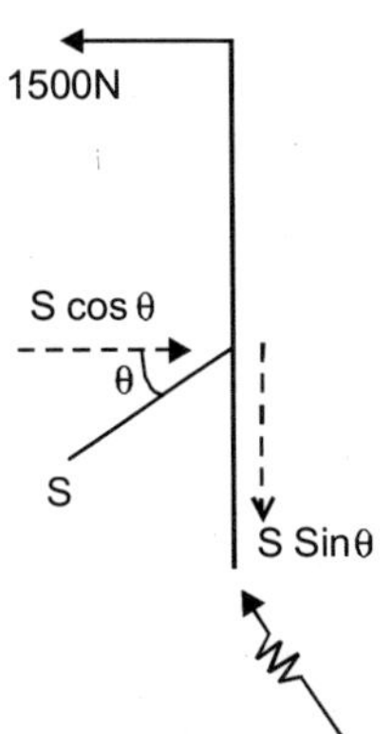

Now, 2625 N force acts as compression, parallel to grain on vertical post.

$$\therefore \qquad c \times 100 \times 2.8 = 2625$$

$$\Rightarrow \qquad c = 9.375\,\text{mm}$$

And 3500 N force acts compressively, parallel to grain on horizontal member,

$$\therefore \qquad b \times 100 \times 2.8 = 3500$$

$\Rightarrow$ $$b = 12.5\,\text{mm}$$

Again 3500 N force acts tangentially on horizontal member.

$\therefore$ $$a \times 100 \times 0.7 = 3500$$

$\Rightarrow$ $$a = 50\,\text{mm}$$

Exercise

1. A square concrete pedestal is to sustain a load of 500.0kN. Determine a suitable cross section for the pedestal, if the allowable stress is 15.0 MPa. What is the allowable load, if the deformation of the 2.0 m high pedestal should not exceed 1.0mm? Assume Young's modulus for concrete to be 25.0GPa.
2. A 2.0m long steel tie bar is subjected to a force of 150.0kN. Determine its cross-section so that (*i*) the stress does not exceed 140.0 MPa, and (*ii*) the extension is not more than 1.2mm. Assume a Young's modulus of 210.0 GPa; steel bars are available in increments of 5mm from 30mm diameter onwards.
3. Compute the stresses and deformations in the steel rope of a pulley supporting a load of 126.0kN. The rope is made of 12 wires of 2.5mm diameter. The rope is 12.6mm on one side of the pulley and 8.1m on the other side. Determine the elongation and the factor of safety of the system, if the yield stress for the wires is 1800.0MPa and the Young's modulus for the rope is 150.0GPa.
4. Determine the maximum load that can be supported by a 10.0m long steel rope comprising 12 wires of 2.0mm diameter, if the working stress is 230.0MPa and Young's modulus is 180.0GPa. The rope should not elongate more than 10mm under the load.
5. A rope comprising three 3.0mm diameter wires clamped together at the ends, lifts a load of 15.0kN. The lengths of the wires are 10.0m, 10.01m and 10.02m. Compute the stresses in each wire. What is the maximum load the rope can lift, if the working stress is limited to 800.0 MPa and the deformation of the rope to 40.0mm? Assume E=210.0GPa.
6. Design a steel rod to sustain a load of 80.0kN with a safety of 2.5. What is the maximum permissible length of the rod, if the allowable deformation is 0.5mm? Assume a yield stress of 230.0MPa and Young's modulus of 195.0GPa.
7. A bar of length l has its diameter increasing from d at one end to D at the other. Determine the deformation of the member subject to a tensile force of P.
8. A prismatic member of length l and unit weight γ is suspended freely from its end. Determine the elongation of the member under gravity.
9. Determine the profile of the member of the previous problem for uniform stress distribution along its length when a load P acts at its lower end. What is the corresponding elongation of the member?
10. A 5.0m long concrete column of size 300 × 400 mm is reinforced with eight bars of 25mm diameter. (*i*) Find the maximum load the 184 column can carry, (*ii*) Find the stresses in the steel bars when the column temperature is increased by 56.0K, and (*iii*) Determine the column deformations in both the cases.

Data	*Permissible stress* (MPa)	*Young's modulus* (GPa)	*Coefficient of thermal expansion* (K^{-1})
Steel	140	205	12×10^{-6}
Concrete	15	30	7×10^{-6}

11. A concrete column is reinforced with steel bars comprising 6 percent of the gross area of column section. What is the fraction of the compressive load sustained by steel bars, if the ratio of Young's moduli of steel and concrete is 12 : 5?

CHAPTER 7

STRESS-STRAIN DIAGRAM AND STRENGTH PARAMETERS

7.1 STRESS-STRAIN BEHAVIOUR OF MATERIAL

All engineering materials do not show same sort of behaviour when subjected to *tension* as well as *compression*. There exist some materials like metals, alloys etc., which are more or less *equally strong* in both tension and compression. And these materials are generally tested in tension again concrete, stones, bricks etc., are such type of materials which are *weaker in tension* and *stronger in compression*. Hence, these materials are tested in compression.

Now the stress-strain characteristics of *mild steel* are of specific importance to the community dealing with basic engineering science.

7.2 STRESS-STRAIN CHARACTERISTICS OF MILD STEEL (M.S)

In order to obtain stress-strain behaviour of M.S, a specimen of *uniform circular cross-section* is prepared following the specification laid in IS 1608:2005 identical to ISO 6892:1998. A specific length of maximum 4 inch or 100mm is generally selected in the well-middle part of the specimen and this length is designated as *gauge length*, over which the amount of elongation is studied.

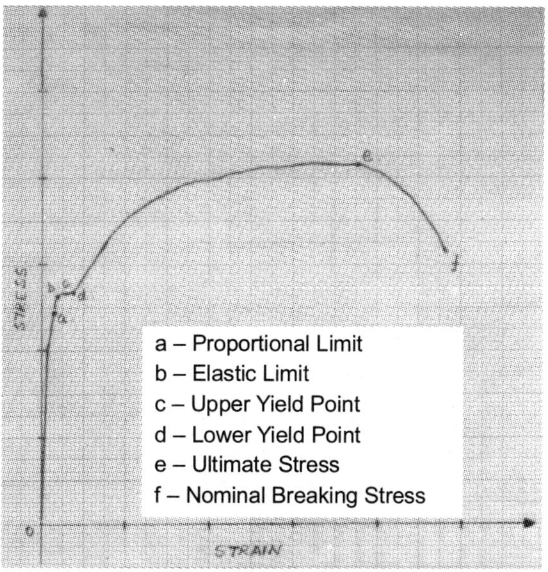

Fig. 7.1

Now the specimen, suitably fitted in *extensometer*, mounted on the machine where loading is started gradually from zero till failure. Following is a stress-strain curve of M.S specimen having gauge length 100mm, tested in *Amsler Universal Testing Machine* of capacity 20T . Various points on stress-strain curve are marked in Figure 7.1.

7.3 PROPORTIONAL LIMIT

It is the point on the stress-strain curve, up to which the plot is a *straight line* and stress is proportional to strain. Up to proportional limit, the material remains elastic and *strictly follows Hooke's Law*.

7.4 ELASTIC LIMIT

In the stress-strain curve, it is the point just beyond proportional limit. From proportional limit to elastic limit, the material *remains elastic* but *does not follow Hooke's Law* and so, stress and strain are not proportional.

7.5 YIELD POINT

When the specimen is loaded beyond elastic limit, it enters into *elasto-plastic zone*. In this region, elongation of specimen occurs by considerable amount without any perceivable amount of increase in load. Sometimes this yielding is accompanied by an abrupt reduction of load and thereby stress. In this case the upper and lower limits of stress are called *upper yield point or stress* and *lower yield point or stress*, respectively.

Lower yield stress is normally considered as yield stress σ_y of material, because upper yield stress is affected by speed of testing, form of specimen and shape of cross-section.

7.6 PROOF STRESS

Some materials like High Strength Deformed (HSD) steel, brass, duralumin etc., do not show any well defined yield point. For these materials, *proof stress* serves as analogous to *yield stress*.

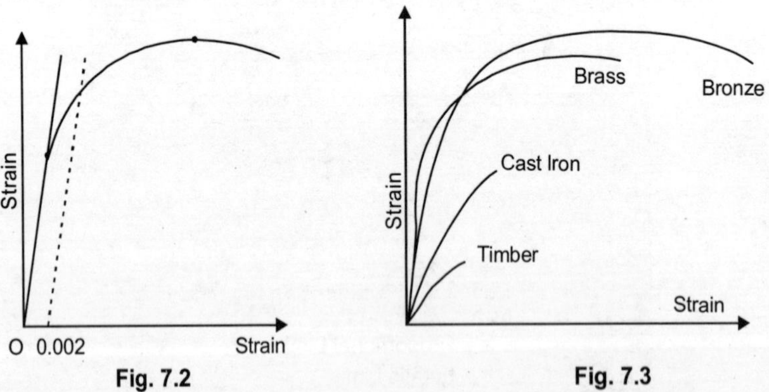

Fig. 7.2 Fig. 7.3

Proof stress is the stress that is just sufficient to produce under load, a *defined amount of permanent residual strain*, which a material can have without appreciable structural damage. This arbitrary value will be different for different material or different uses of same material.

It is determined from the stress-strain curve by drawing a line parallel to initial straight part or tangent of the curve and at a distance from the origin by an amount representing the defined residual strain (normally 0.1% or 0.2%) thus determining the stress at which the line cuts the curve.

In specifying proof stress, the amount of permanent strain considered, should be mentioned, i.e.,0.1% *proof stress*, 0.2% *proof stress* etc.

7.7 ULTIMATE STRESS

Yield point serves as the gateway to *plastic zone.* Beyond yield point, due to *sudden decrease in load,* material begins to strain-harden and recover some of the elastic property. And by virtue of that, gradual uprise of stress-strain curve occurs and terminates at a point, called *ultimate stress*. This is the maximum stress, the specimen can withstand, without any appreciable damage or permanent deformation.

7.8 BREAKING STRESS

While ultimate stress is the maximum stress with standing capacity prior to failure, further increase of ultimate stress leads to failure of the specimen and this occurs at breaking stress. Here the value of *breaking stress* lower than ultimate stress, as appearing in the stress-strain diagram obtained during experiment, of ductile material, is somehow misleading.

What happens *in reality* is that, beyond ultimate stress, there occurs a *reduction in area* of cross-section near at the middle of gauge length. This phenomenon is called *formation of neck* or *formation of waist*. As the grips of extensometer are attached at the end of gauge length, the effect of neck formation thereby the reduction in diameter of the specimen cannot be taken into account. By reason of which breaking stress exhibits value lower than ultimate stress. And this breaking stress is called *Nominal Breaking Stress*.

When the reduced cross-sectional area at neck is considered to compute actual stress, it is found that breaking stress is pretty higher than ultimate stress. And this is called *True Breaking Stress*.

In case of brittle material, ultimate stress is same as breaking stress.

7.9 WORKING STRESS AND FACTOR OF SAFETY

In practical design of structures, some *uncertainties* may be associated in terms of loading, material properties etc. Not only that, in some materials, like concrete, non-ferrous alloys etc., Hooke's Law does not hold good. To encompass all these aspects, it is essential to limit actual stress generated to a value comparatively lower than yield stress of the material. And this stress is considered as a safe one. This safe stress is designated at *Working Stress* (σ_w).

A pure number, *higher than* 1 (whole or fraction) that divides the yield stress to obtain working stress is called *Factor of Safety*.

$$\therefore \quad \sigma_w = \frac{\sigma_y}{n}, \text{ where } n = \text{factor of safety.} \qquad ...(7.1)$$

Sometimes working stress is computed deviding ultimate stress by factor of safety.

7.10 DUCTILITY

It is the property of a material which allows of its being drawn out by tension to a small section. Brittleness is the lack of ductility.

7.11 MALLEABILITY

It is the property of a material by virtue of which it can be turned to a very thin sheet by the application of pressure.

7.12 TOUGHNESS

It can be said, in general, resistance to deformation. This deformation may be due to impact, abrasive force, punch etc.

7.13 RESILIENCE

Let us consider a bar of length *l* and cross-sectional area *A* hanging vertically fixed at top, subjected to a normal pull *P*.

At zero-th instant of application of force, induced deflection is zero. Gradually load is increased within proportional limit. At position *D*, the amount of applied load is *P* which causes a displacement δ.

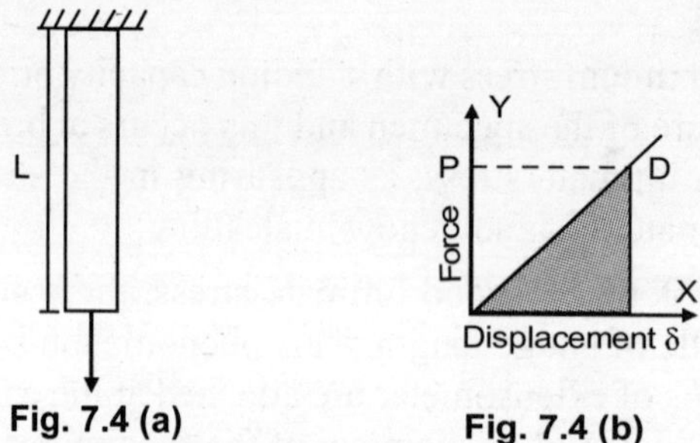

Fig. 7.4 (a) **Fig. 7.4 (b)**

Here, external work done by the force, will be represented by shaded area in Figure 7.4 (*b*).

$$\therefore \quad W_e = \frac{1}{2} P \cdot \delta \qquad \text{... (7.2)}$$

To resist the effect of external force, reactive work will be done by the force, generated internally due to deformation of body. This reactive work is defined as *Internal Work* or *Strain Energy* of the system, which is symbolised by *U*. Amount of strain energy is numerically equal to the external work done on the members.

As loading is done within proportional limit, as soon as load is released, the system also will lose energy and will come back to original position. This property of an elastic material to absorb and release energy with change in loading is called *Resilience*.

Here total strain energy of the system

$$U = \frac{1}{2} P \cdot \delta = \frac{1}{2} (\sigma A)(\varepsilon \cdot l) = \frac{1}{2} (\sigma \cdot \varepsilon)(A \cdot l) \qquad \text{...(7.3)}$$

Therefore strain energy per unit volume

$$u = \frac{U}{Al} = \frac{1}{2} \sigma \varepsilon = \frac{\sigma^2}{2E} \qquad \text{...(7.4)}$$

Another very importent parameter in this context is *Proof resilience*. It is the maximum strain energy stored in a body before yielding occurs. Strain energy will be maximum when the body will be stressed up to elastic limit. And proof resilience per unit volume is known as *Modulus of Resilience*.

$$\therefore \quad u_r = \frac{\sigma_p^2}{2E} \qquad \text{...(7.5)}$$

Sometimes σ_p stands for yield stress or proof stress, as the case may be.

7.14 THERMAL STRESS

Any engineering material, when subjected to change in temperature exhibits expansion in temperature rise and contraction in temperature fall. This change in temperature is often termed as *Thermal Loading*.

A structural member of length l if subjected to thermal loading of ΔT, can expand or contract by an amount, $\delta_t = l \alpha \Delta T$, where α = thermal coefficient of material, when free expansion or contraction is allowed. In *non-restrained* system no stress will be developed, though there presents thermal loading.

If free movement of the member is restricted partially or fully by somehow or other, some amount of reactive force will be generated within the member, which will give birth to a reactive stress, termed as *thermal stress*.

Multiple Choice Questions

Select the best alternative (s):

1. The impact strength of a material is an index of its
   (*a*) toughness (*b*) tensile strength (*c*) capability of being cold worked
   (*d*) hardness (*e*) fatigue strength
2. The property of a material which allows it to be drawn into a smaller section is called
   (*a*) plasticity (*b*) elasticity (*c*) ductility (*d*) malleability
3. The loss of strength in compression due to overloading is known as
   (*a*) hysteresis (*b*) relaxation (*c*) creep (*d*) resilience
   (*e*) bauschinger effect
4. The maximum strain energy that can be stored in a body is known as
   (*a*) impact energy (*b*) resilience (*c*) proof resilience
   (*d*) modulus of resilience (*e*) toughness
5. The total strain energy stored in a body is termed as
   (*a*) resilience (*b*) proof resilience (*c*) modulus of resilience
   (*d*) toughness
6. Proof resilience per unit volume of a material is known as
   (*a*) resilience (*b*) proof resilience (*c*) modulus of resilience
   (*d*) toughness
7. The Figure 7.1 shows the stress-strain diagram for mild steel. The elastic limit, upper yield point, lower yield point and proportional limit are represented by
   (*a*) A, B, C, D (*b*) A, C, D, B (*c*) B, C, D, A
   (*d*) C, B, D, A (*e*) B, C, A, D
8. Proof stress is
   (*a*) stress corresponding to proportional limit
   (*b*) stress causing materials to break
   (*c*) stress causing a specific permanent deformation usually 0.1% or 0.2%.
   (*d*) not related with engineering

9. Thermal strain caused in the material of a composite body due to change in temperature will be
(*a*) same nature (*b*) opposite nature (*c*) same magnitude (*d*) none of above

10. True stress ($\bar{\sigma}$) is related to simple stress (σ) and strain (ε) by__

(*a*) $\bar{\sigma} = \sigma(1 - \varepsilon)$ (*b*) $\bar{\sigma} = \sigma(1 + \varepsilon)$ (*c*) $\bar{\sigma} = \sigma\,\varepsilon$ (*d*) $\bar{\sigma} = \dfrac{\sigma}{\varepsilon}$

11. Bulk modulus *K* in terms of modulus of elasticity (*E*) and Poisson's ratio (μ) is given as equal to
(*a*) $E/3(1-2\mu)$ (*b*) $E(1-2\mu)$ (*c*) $3E(1-2\mu)$
(*d*) $E(1+2\mu)/3$ (*e*) $E(1-3\mu)/3$

12. The energy absorbed by a body, when it is strained within the elastic limit, is known as
(*a*) strain energy (*b*) resilience (*c*) proof resilience
(*d*) modulus of resilience (*e*) toughness

13. Value of factor of safety is
(*a*) greater than 1 (*b*) less than 1
(*c*) zero (*d*) none of above

14. Hooke's Law is truly valid up to
(*a*) elastic limit (*b*) proportional limit
(*c*) plastic limit (*d*) fatigue limit

15. In ductile material nominal breaking stress is
(*a*) lower than true breaking stress
(*b*) equals with true breaking stress
(*c*) higher than true breaking stress
(*d*) none of the above

16. In ductile material ultimate stress is
(*a*) higher than true breaking stress
(*b*) lower than nominal breaking stress
(*c*) higher than nominal breaking stress but lower than true breaking stress
(*d*) higher than true breaking stress but lower than nominal breaking stress

17. In a brass specimen subjected to tension, which of the following can be obtained in stress-strain diagram ?
(*a*) upper yield stress (*b*) lower yield stress (*c*) plastic stress (*d*) proof stress

18. Thermal change of length of a metal is related to its thermal coefficient
(*a*) inversely proportional (*b*) directly proportional (*c*) directly square proportional
(*d*) none of the above

19. Thermal strain of a body does not depend on
(*a*) length (*b*) thermal coefficient (*c*) change in temperature
(*d*) none of the above

20. In brittle material, normally, breaking stress is
(*a*) higher than ultimate stress (*b*) lower than ultimate stress
(*c*) equals with ultimate stress (*d*) none of the above

Answers

1. (*a*)	2. (*c*)	3. (*c*)	4. (*c*)	5. (*a*)	6. (*c*)
7. (*c*)	8. (*c*)	9. (*c*)	10. (*b*)	11. (*a*)	12. (*c*)
13. (*a*)	14. (*b*)	15. (*a*)	16. (*c*)	17. (*d*)	18. (*b*)
19. (*a*)	20. (*c*)				

Numerical Examples

Example 1

A steel bar of 25mm diameter was tested in tension and results were recorded as, limit of proportionality = 196.32kN, load at yield = 218.13kN, ultimate load = 278.20 kN. The elongation measured over a gauge length of 100mm was 0.189mm at proportionality limit, length of the bar between gauge marks after fracture was 112.62mm and minimum diameter was 23.64mm. Compute stress in the specimen at various stages, Young's modulus, % elongation and % contraction. Determine permissible stress in the material for a safety factor of 1.85.

Solution Initial c/s area of the bar $= \frac{\pi}{4} \times (25)^2 = 490.874 \text{ mm}^2$

$\therefore$ Strees at proportionality limit $= \frac{196.32 \times 10^3}{490.874} = 399.94$ MPa

$\therefore$ Strain at proportionality limit $= \frac{0.189}{100} = 1.89 \times 10^{-3}$

$\therefore$ Young's modulus $= \frac{399.94}{1.89 \times 10^{-3}} = 2.116 \times 10^5$ MPa

$\therefore$ Strees at yield point $= \frac{218.13 \times 10^3}{490.874} = 444.37$ MPa

$\therefore$ Ultimate strees $= \frac{278.20 \times 10^3}{490.874} = 566.74$ MPa

$\therefore$ Final c/s area at fracture $= \frac{\pi}{4} \times (23.64)^2 = 438.919 \text{mm}^2$

% elongation $= \frac{112.62 - 100}{100} \times 100 = 12.62$

% contraction in area $= \frac{490.874 - 438.919}{490.874} \times 100 = 10.584$

Permissible or working or allowable strees $= \frac{\text{yield stress}}{\text{factor of safety}}$

$$= \frac{444.37}{1.85} = 240.2 \text{ MPa}$$

EXAMPLE 2

Steel railroad, 10m long, is laid with a clearance of 3mm at 15ºC. At what temperature will the rails just touch? What stress will be induced in the rails at that temperature, if there were no initial clearance, while $a = 117$mm/mºC and $E = 200$ GPa.

SOLUTION Let the desired temperature = T°C

Now, free thermal elongation from 15°C to T°C $= 10 \times 10^3 \times 11.7 \times 10^{-6} \times (T - 15)$

$$= 0.117(T - 15)$$

As per condition provided, $0.117\,(T - 15) = 3$

$$\Rightarrow \quad T = 40.64\ °C$$

At no clearance condition, induced strain will be $= \dfrac{3}{10 \times 10^3} = 3 \times 10^{-4}$

and induced stress $= (3 \times 10^{-4} \times 200 \times 10^3)$ MPa = 60MPa

EXAMPLE 3

A steel rod 3ft long with c/s area of 0.25 inch^2 is stretched between two fixed points. The tensile force is 1200 lb at 40°F. Using $E = 29 \times 10^6$ psi and $a = 6.5 \times 10^{-6}$ inch/inch/ºF Calculate (*a*) the temperature at which the stress in the bar will be 10 ksi, (*b*) the temperature at which the stress will be zero.

SOLUTION Initial stress in the rod $= \dfrac{1200}{0.25} = 4{,}800$ psi

(*a*) Required strees is to be developed = 10,000 psi

∴ Additional stress to be developed = 10,000 – 4,800 = 5,200 psi

This additional stress will be generated due to rise in temperature, say at T° F

∴ Strain corresponding to additional stress $= \dfrac{5{,}200}{29 \times 10^6} = 1.793 \times 10^{-4}$

∴ Elongation due to above strain $= (1.793 \times 10^{-4})$inch

$$= 6.45 \times 10^{-3} \text{ inch}$$

Hence, $3 \times 12 \times 6.5 \times 10^{-6}\,(T - 40) = 6.45 \times 10^{-3}$

$$\Rightarrow \quad T = 67.564\ °F$$

(*b*) As stress is to be zero, temperature of the system will have to be reduced.

Let that temperature be T_1 °F.

Now, strain corresponding to initial stress $= \dfrac{4{,}800}{29 \times 10^6} = 1.655 \times 10^{-4}$

Thermal strain due to reduction of temperature $= 6.5 \times 10^{-6} \times (40 - T_1)$

Here, $6.5 \times 10^{-6} \times (40 - T_1) = 1.655 \times 10^{-4}$

$$\Rightarrow \quad T_1 = 14.538\ °F$$

EXAMPLE 4

A bronze bar, 3m long with a c/s area of 320mm^2 is placed between two rigid walls. At –20°C, the gap between bar and wall is 2.5mm. Find temperature at which compressive stress in the bar will be 35MPa. Take $a = 18 \times 10^{-6}$m/m/°C and E= 80GPa.

SOLUTION Strain corresponding to 35MPa stress $= \dfrac{35}{80 \times 10^3} = 4.375 \times 10^{-4}$

$\therefore$ Elongation due to above strain $= 4.375 \times 10^{-4} \times 3000 = 1.3125$mm

To generate above compressive stress, total elongation to be compensated by thermal rise will be $= (1.3125 + 2.5)$ mm $= 3.8125$ mm and let the final temperature be T°C.

Now, $$3 \times 10^3 \times 18 \times 10^{-6} \times (T + 20) = 3.8125$$

$$\Rightarrow \quad T = 50.6°\text{C}.$$

EXAMPLE 5

Calculate increase in stress for each segment of the compound bar, if temperature increases by 100°F. Assume unyielding supports and bar is suitably braced against buckling. If supports yield by 0.01inch, compute stresses.

SOLUTION If free thermal elongation would be allowed, it will be for aluminium

$$= (10 \times 12.8 \times 10^{-6} \times 100) = 0.0128 \text{ inch}$$

for steel $= (15 \times 6.5 \times 10^{-6} \times 100) = 9.75 \times 10^{-3}$ inch

and total $= (0.0128 + 9.75 \times 10^{-3}) = 0.02255$ inch ...(1)

As the supports are unyielding, to resist this elongation some mechanical stress will be developed and corresponding contraction will be:

$$\left.\begin{aligned} &\text{for aluminium} \quad = \frac{\sigma_a}{E_a} \times L_a = \left(\frac{\sigma_a}{10 \times 10^6} \times 10\right) \text{ inch} \\ &\text{for steel} \quad = \frac{\sigma_s}{E_s} \times L_s = \left(\frac{\sigma_s}{29 \times 10^6} \times 15\right) \text{ inch} \end{aligned}\right\} \quad ...(2)$$

To satisfy compatibility for deformation,

$$\left(\frac{\sigma_a}{10 \times 10^6} \times 10\right) + \left(\frac{\sigma_s}{29 \times 10^6} \times 15\right) = 0.02255$$

$$\Rightarrow \quad \sigma_a + 0.5172\sigma_s = 22550 \quad ...(3)$$

Another equation of compatibility,

$$\sigma_a A_a = \sigma_s A_s$$

$$\Rightarrow \quad \sigma_a \times 2.0 = \sigma_s \times 1.5$$

$$\Rightarrow \quad \sigma_a = 0.75\sigma_s \quad ...(4)$$

Solving equations (3) and (4), $\sigma_a = 13,346.35$ psi, $\sigma_s = 17,795.14$ psi

If the supports yield, equation of compatibility will be

$$\frac{\sigma_a}{10\times10^6}\times10+\frac{\sigma_s}{29\times10^6}\times15 = 0.02255-0.01$$

$$\Rightarrow \qquad \sigma_a + 0.5172\sigma_s = 12550 \qquad ...(5)$$

Solving equations (5) and (4), $\sigma_a = 7,427.79$ psi
$\sigma_s = 9,903.72$ psi

EXAMPLE 6

At 80ºC a steel tire, 12mm thick and 90mm wide is to be shrunk fit onto a locomative wheel, 2m in diameter, just fits over the wheel which is at a temperature of 25ºC. Determine contact pressure between tire and wheel at 25ºC. $\alpha = 11.76\times10^{-6}$ m/m/ºC and $E = 200$ GPa.

SOLUTION Let diameter of tire at 25ºC = d and of wheel = D

Considering condition of compatibility, $\pi\times D = \pi\times d[1+\alpha(80-25)]$

$$= \pi\times d[11.76\times10^{-6}\times55]$$

$$\Rightarrow \qquad \frac{D}{d} = 1.0006468 \qquad ...(1)$$

Circumferential strain in the wheel $\varepsilon = \dfrac{\pi D - \pi d}{\pi d} = \left(\dfrac{D}{d}-1\right)$

$$= 0.0006468$$

Corresponding strees in the wheel $= (0.0006468\times200\times10^3)$ MPa

$$= 129.36\text{MPa}$$

EXAMPLE 7

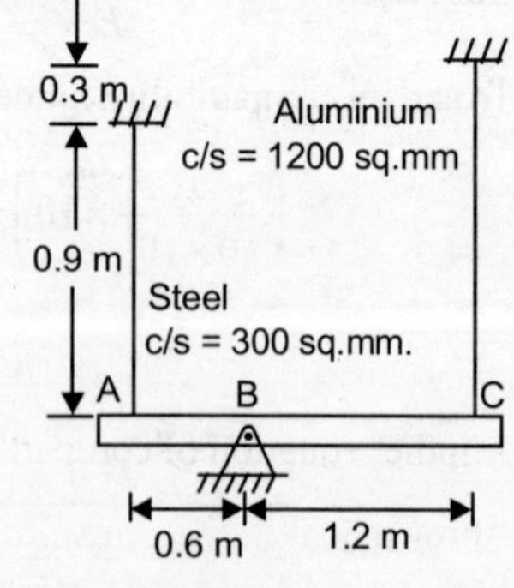

In the adjoining figure, bar ABC is initially horizontal and vertical rods are stress-free. Determine stress in the aluminium rod, if temperature of the steel rod is decreased by 40ºC.

$$E_s = 200\times10^9\text{ N/m}^2,\ E_a = 70\times10^9\text{ N/m}^2,$$

$$\alpha_s = 11.7\mu\text{m/m/ºC},\quad \alpha_a = 23\mu\text{m/m/ºC},$$

SOLUTION From free-body diagram of the bar, taking moment at B, the condition of equilibrium will be,

$$R_A\times0.6 = R_C\times1.2$$

$$\Rightarrow \qquad (\sigma_{st}\times A_{st})\times0.6 = (\sigma_{al}\times A_{al})\times1.2$$

R_A R_C B A C

$$\Rightarrow \qquad \sigma_{st} \times 300 \times 0.6 = \sigma_{al} \times 1200 \times 1.2$$

$$\Rightarrow \qquad \sigma_{st} = 8\sigma_{al} \qquad ...(1)$$

Due to reduction of temperature, strain in the steel rod

$$= 11.7 \times 10^{-6} \times 40$$

$$= 4.68 \times 10^{-4}$$

Corresponding stress steel rod (σ_{st}) $= (4.68 \times 10^{-4} \times 200 \times 10^{9} \times 10^{-6})$

$$= 93.6\text{MPa}$$

From (1), $$\sigma_{al} = \frac{93.6}{8} = 11.7\text{MPa}$$

EXAMPLE 8

A steel tube of 24mm external and 14mm internal diameters encloses a copper rod of 12 mm diameter. The assembly is held rigidly at both ends at 22°C. Compute (*i*) stresses at 122°C, (*ii*) the maximum temperature the assembly can withstand. Assume

$E_s = 220\text{GPa}$,

$\alpha_s = 11 \times 10^{-6} /°\text{C}$,

$\alpha_c = 18 \times 10^{-6} /°\text{C}$,

$(\sigma_s)_{\max} = 230\text{MPa}$, $(\sigma_c)_{\max} = 115\text{MPa}$,

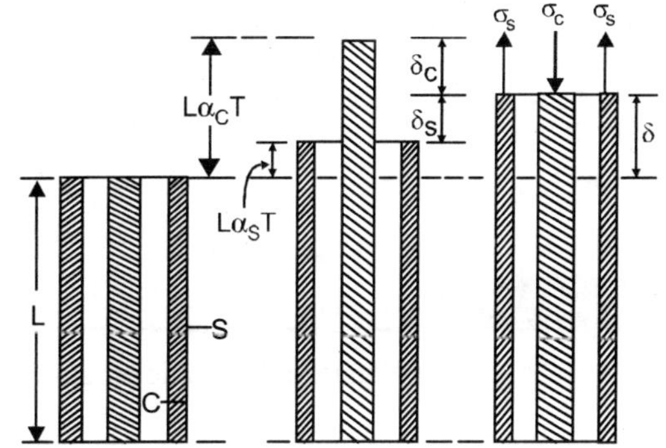

SOLUTION As $\alpha_c > \alpha_s$, copper rod will expand more than steel tube, if free expansion is allowed. As both ends of assembly are fastened regidly, free expansion will not be permitted in either rod or tube. A thermal compromise will be happened. By virtue of it, free expansion of copper rod will be reduced and free expansion of steel tube will be increased. At this level, a mechanically induced compressive force will act at copper rod and a tensile force will act at steel tube. Both of these forces are equal in magnitude.

Hence, $$L \cdot \alpha_c \cdot \Delta T - \frac{P_c \cdot L}{A_c \cdot E_c} = L \cdot \alpha_s \cdot \Delta T + \frac{P_s \cdot L}{A_s \cdot E_s} \qquad ...(1)$$

$$\Rightarrow \qquad (\alpha_c - \alpha_s)\Delta T = \frac{\sigma_c}{E_c} + \frac{\sigma_s}{E_s} \qquad ...(2)$$

$$7 \times 10^{-6} \times (122 - 22) = \left(\frac{\sigma_c}{110} + \frac{\sigma_s}{220}\right) \times \frac{1}{10^3}$$

$$\Rightarrow \qquad \sigma_c + 0.5\sigma_s = 77 \qquad ...(3)$$

From equation of compatibility,

$$P_c = P_s$$

$$\Rightarrow \qquad \sigma_c \cdot A_c = \sigma_s \cdot A_s$$

$$\sigma_c = \sigma_s \cdot \frac{A_s}{A_c} \qquad \text{... (4)}$$

$$= \sigma_s \cdot \frac{24^2 - 14^2}{12^2} = 2.639\sigma_s$$

Solving equations (3) and (4) $\left.\begin{aligned}\sigma_c &= 64.73 \text{ MPa} \\ \sigma_s &= 24.53 \text{ MPa}\end{aligned}\right\}$

To find maximum withstandable temperature T, we are to substitute maximum permissible stresses in equation (2)

$$7 \times 10^{-6} \times (T - 22) = \left[\frac{(\sigma_c)_{\max}}{110} + \frac{(\sigma_s)_{\max}}{220}\right] \times \frac{1}{10^3} \qquad \text{... (5)}$$

If $(\sigma_c)_{\max} = 230\text{MPa}$, according to equation (4), $\sigma_s = 87.154\text{MPa}$

If $(\sigma_s)_{\max} = 115\text{MPa}$, according to equation (4), $\sigma_c = 303.485\text{MPa}$, which is higher than $(\sigma_c)_{\max}$. So, maximum allowable stress in copper and steel will be 230MPa and 87.154MPa respectively.
Substituting in equation (5),

$$7 \times 10^{-6} \times (T - 22) = \left(\frac{87.154}{110} + \frac{230}{220}\right) \times \frac{1}{10^3}$$

$\Rightarrow$ $$T = 284.54°\text{C}$$

EXAMPLE 9

A bar of uniform c/s A and length L hangs vertically, subjected to its own weight. Prove that the strain energy stored within the bar is $U = \dfrac{A\rho^2 L^3}{6E}$.

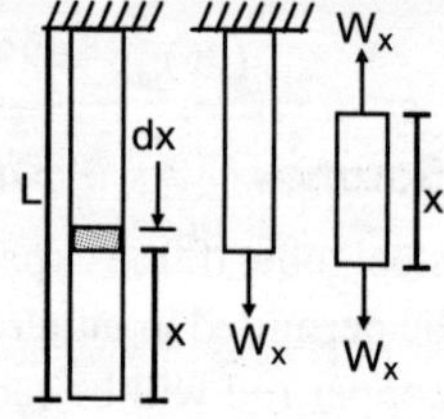

SOLUTION Let us take a section x distance from bottom, of thickness dx. The elongation of length dx be $d\delta$.

$\therefore$ Strain in length dx, $$\varepsilon_x = \frac{d\delta}{dx}$$

Stress in length dx, $$\sigma_x = \frac{W_x}{A} = \frac{\rho A x}{A} = \rho x$$

Now, $$E = \frac{\sigma_x}{\varepsilon_x} = \frac{\rho x dx}{d\delta}$$

$\Rightarrow$ $$d\delta = \frac{\rho x dx}{E}$$

Strain energy stored in dx

$$dU = \text{average weight} \times \text{elongation of } dx$$

$$= \left(\frac{1}{2} W_x\right) \times \frac{\rho x dx}{E}$$

$$= \frac{1}{2}(\rho A x)\left(\frac{\rho x dx}{E}\right) = \frac{A\rho^2}{2E} x^2 dx$$

$\therefore$ Total strain energy stored, $U \quad = \int_0^L dU = \int_0^L \frac{A\rho^2}{2E} x^2 dx$

$$= \frac{A\rho^2}{2E} \times \frac{L^3}{3} = \frac{A\rho^2 L^3}{6E}$$

EXAMPLE 10

In the adjacent figure, load is allowed to drop on the collar from a height h, find the expression of stress induced in the rod due to impact.

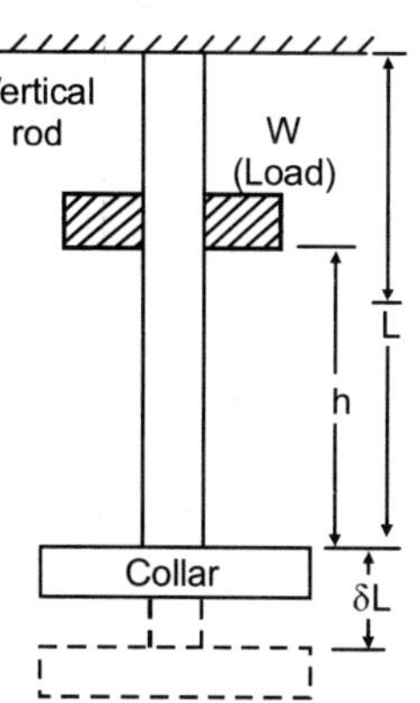

SOLUTION Strain in the bar $\quad \frac{\delta L}{L} = \frac{\sigma}{E}$

$$\Rightarrow \quad \delta L = \frac{\sigma}{E} \times L \quad \text{...(1)}$$

Work done by the load $\quad = W(h + \delta L) \quad$ ...(2)

Strain energy stored by the rod $\quad = \frac{\sigma^2}{2E}(AL) \quad$ ...(3)

Equating (2) and (3) following condition of equilibrium,

$$W(h + \delta L) = \frac{\sigma^2}{2E}(AL)$$

... (4)

Substituting the expression of (1) into (4), we have,

$$\sigma^2 - \left(\frac{2W}{A}\right)\sigma - \frac{2WEh}{AL} = 0$$

Solving, $\quad \sigma = \frac{W}{A}\left(1 + \sqrt{1 + \frac{2AEh}{WL}}\right) \quad$ ...(5)

Sometimes, $W\left(1 + \sqrt{1 + \frac{2AEh}{WL}}\right)$ term is designated as equivalent static load, i.e.,

$$W_e = W\left(1 + \sqrt{1 + \frac{2AEh}{NL}}\right)$$

(*i*) if $\delta L \ll h$, from equation (4),

$$Wh = \frac{\sigma^2}{2E}(AL)$$

$$\Rightarrow \quad \sigma = \sqrt{\frac{2EhW}{AL}} \quad \text{... (6)}$$

(*ii*) If $h \approx 0$, from equation (4)

$$\sigma = \frac{2W}{A} \qquad \text{...(7)}$$

EXAMPLE 11

Find the strain energy of these two members, loaded and shown in the figure.

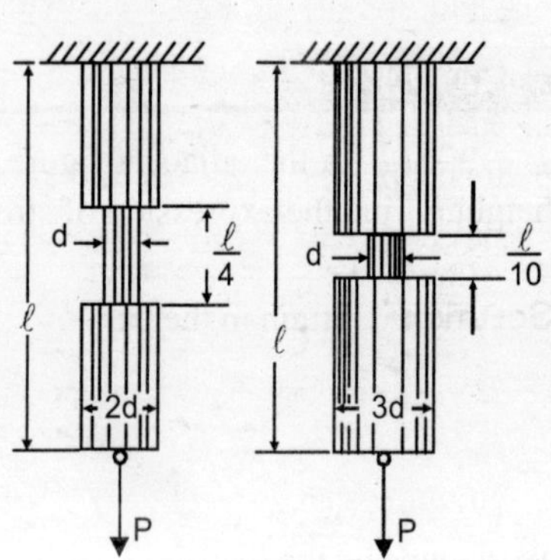

SOLUTION Strain energy (S.E) of a bar subjected to loading P

$$= \frac{1}{2} P \cdot \delta = \frac{1}{2}\frac{P^2 l}{AE}.$$

For the first member, S.E

$$= \frac{1}{2}\frac{P^2\left(\frac{3l}{8}\right)}{\frac{\pi}{4}(2d)^2 \cdot E} + \frac{1}{2}\frac{P^2\left(\frac{l}{4}\right)}{\frac{\pi}{4}(2d)^2 \cdot E} + \frac{1}{2}\frac{P^2\left(\frac{3l}{8}\right)}{\frac{\pi}{4}(2d)^2 \cdot E}$$

$$= \frac{1}{2} \times \frac{4}{\pi} \times \frac{P^2 l}{d^2 E}\left(\frac{3}{8\times 4} + \frac{1}{4} + \frac{3}{8\times 4}\right)$$

$$= \frac{2P^2 l}{\pi d^2 E} \times \frac{7}{16} = \frac{7}{8}\frac{P^2 l}{\pi d^2 E}$$

For the second member, S.E

$$= \frac{1}{2}\cdot\frac{P^2\left(\frac{9l}{20}\right)}{\frac{\pi}{4}(3d)^2 E} + \frac{1}{2}\cdot\frac{P^2\left(\frac{l}{10}\right)}{\frac{\pi}{4}(d)^2 E} + \frac{1}{2}\cdot\frac{P^2\left(\frac{9l}{20}\right)}{\frac{\pi}{4}(3d)^2 E}$$

$$= \frac{1}{2} \times \frac{4}{\pi} \times \frac{P^2 l}{d^2 E}\left(\frac{9}{20\times 9} + \frac{1}{10} + \frac{9}{20\times 9}\right)$$

$$= \frac{2P^2 l}{\pi d^2 E} \times \frac{1}{5} = \frac{2}{5}\frac{P^2 l}{\pi d^2 E}.$$

Exercise

1. A mass of 200 kg falls through a height of 500 mm on a concrete column of 300 × 400mm section. Determine the maximum stress and deformation in the 4.0m long column, considering modulus of elasticity of concrete 20.0GPa.

Ans. [–9.06MPa, –1.81mm]

2. A rod is 3m long at a temperature of 15°C. Find the expansion of the rod, when the temperature is raised to 95°C. If this expansion is prevented, find the stress induced in the material of the rod. Take $E = 1 \times 10^5$ N / mm^2 and $\alpha = 0.000012$ per degree centrigrade. **Ans.** [0.288cm, 96N/mm^2]

3. A steel rod 5cm diameter and 6m long is connected to two grips and the rod is maintained at a temperature of 100°C. Determine the stress and pull exerted when the temperature falls to 20°C if (*i*) the ends do not yield, and (*ii*) the ends yield by 0.15cm. Take $E = 200$GPa, $\alpha = 12\times10^{-6}$ /°C.

4. Compute the maximum force a 200mm long compound bar comprising a copper rod of 18 mm diameter enclosed in a mild steel tube of 200mm inner and 32 mm outer diameters can sustain. Assume Young's modulii for copper and steel to be 120GPa and 195GPa, respectively. What will be the reduction in the maximum load, if the bar temperature rises by 40K? Find the reduction in the strength when the temperature falls by 40K. Allowable stresses are 80MPa and 140MPa in copper and steel, respectively; $\alpha = 18 \times 10^{-6}$K^{-1} for copper and 12×10^{-6}K^{-1}for steel.

5. A 3.5m long steel column of cross-sectional area 5000mm^2 is subjected to a load of 1.6 MN. Determine the safety factor for the column, if the yield stress of steel is 550MPa. Determine the allowable load on the column, if the deformation of the column should not exceed 5mm.

6. A compound bar comprises a 12.5mm diameter aluminum rod and a copper tube of 14.5mm inner and 25mm outer diameters. If the Young's modulii of aluminium and copper are 80GP and 120GPa, respectively, determine the stress in the assembly when subject to (*i*) a temperature rise of 95K, and (*ii*) a temperature fall of 35K ; $\alpha = 14.6\times10^{-6}$ K^{-1} for aluminium and 16.8×10^{-6} K^{-1} for copper.

7. A steel rod of 20mm diameter passes centrally through a copper tube of 40mm external diameter and 30mm internal diameter. The tube is closed at each end by rigid plates of negligible thickness. The nuts are tightened lightly home on the projected parts of the rod. If the temperature of the assembly is raised by 60°C, calculate the stresses developed in copper and steel. Take E for steel and copper as 200GN/m^2 and 100GN/m^2 and α for steel and copper as 12×10^{-6} per °C and 18×10^{-6} per °C. **Ans.** [16.23, 28.4 N / mm^2]

CHAPTER

8

INTRODUCTION TO DYNAMICS

8.1 INTRODUCTION

While statics is a pretty old engineering science, dynamics, in contrast, is a comparatively new one. Normally it is believed that Galileo (1564–1642) was the first person who laid his footsetps in the fields of dynamics. Delayed growth in dynamics was mainly due to inconveniences associated with experimentation that were necessary to build up its foundation. After Galileo, many a stalwart like, Newton, Euler, D'Alembert, Lagrange, Laplace, Poinsot, Coriolis, Einstein etc. roamed and reigned the domain of dynamics in an involved fashion.

As we already have entered into an era of motion we find all machines or moving structures are being operated at high speeds and appreciable acceleration. So it is very much apt to make those calculations based on basic concepts of dynamics and not on statics. Hence concept of dynamic analysis has been found to serve as a basic to the analysis and design of moving structures, to fixed structures subjected to shock loads, to mechatronic and robotic devices, to automated control system, to rocket, missiles, spacecraft, to vehicular movent, to pumps, turbines etc.

8.2 KINEMATICS AND KINETICS

Conveniently dynamics is commonly seggregated into two parts, namely *kinematics* and *kinetics*.

Kinematics is that very part of dynamics which deals the motion of bodies without any reference to system of forces causing or resulting the motion. Sometimes kinematics is described as "*geometry of motion*." Calculation of trajectories, design of cams, gears are common examples of kinematic problem.

Kinetics is the study of imbalance of force causing motion. Analysis claiming laws of motion, work and energy principle, impulse and momentum methods are well within the purview of kinetics.

8.3 CLASSIFICATION OF DYNAMIC SYSTEMS

Dynamic systems or simply dynamics can conveniently classified as following :

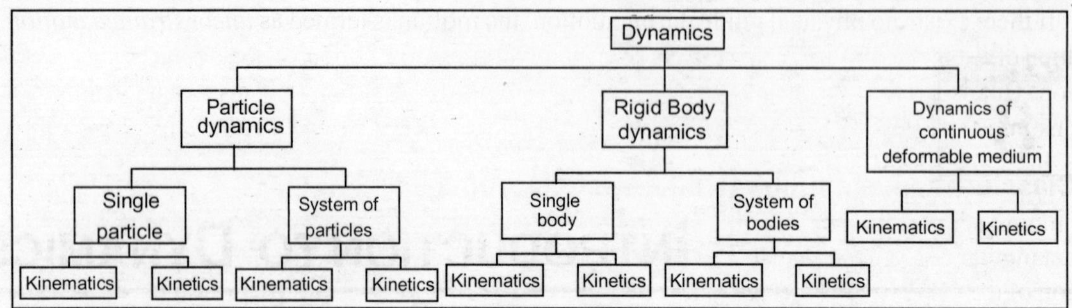

Fig. 8.1

8.4 PARTICLE AND RIGID BODY

A *particle* is defined as a material point bearing no dimensions but containing a definite quantity of matter. As a definite quantity of matter must occupy some space, there can exist no particle in the strictness of the term. Actually, when size of a body is extremely small, in comparison with its range of motion, it is considered as a particle, where the entire mass of the body is assumed to remain concentrated.

A *rigid body is* defined as a system of particles, for which the distances amongst particles remain unchanged. A rigid body undergoes no deformation or too small a deformation to be considered in the analysis.

8.5 PATH

When a particle moves through a space, it describes a curve which is called *path.*

If a path of a particle is described in a plane, it is called *plane path.*

If a path of a particle is described in space, it is called *tortuous path.*

8.6 MOTION

(a) Classification according to orientation of path

Depending upon choice of orientation of path, motion of a particle can be described as either *rectilinear motion* or *curvilinear motion.*

When a particle moves along a path which is a straight line, motion of the particle is defined as *rectilinear motion.*

When a particle traces a curved path during motion, the motion is called as *curvilinear motion.*

(b) Classification according to nature of path

Depending upon nature of path, motion of a particle can be classified as either *constrained motion* or *unconstrained motion.*

If a particle is confined to a specific path, the motion is designated as constrained motion. Sliding of beads along thread is an example of *constrained motion.*

If there exists no physical guide during motion, the motion is termed as *unconstrained motion.* Rotation of a stone tied with an inextensible string, around vertical or horizontal axis, depicts constrained motion. But as soon as the string will be torn, movement of the stone will serve as an example of unconstrained motion.

(c) Classification according to choice of co-ordinates

Any moving particle can be described in space, at any instant t, by specifying co-ordinates like, X, Y, Z (rectangular co-ordinates) or r, q, z (polar co-ordinates) or R, θ, ϕ (spherical co-ordinates). And motion of the particle can be classified depending on whether these co-ordinate systems are fixed or moving.

If motion of a particle is described is reference with a fixed reference axes, the motion is defined as an *absolute motion.*

If motion of a particle is decribed in reference with moving reference axes, the motion is called a *relative motion.*

8.7 DISPLACEMENT

Let us consider a particle P moving along a straight line. Considering a suitably fixed reference point O, the position of particle P at any instant of time t is defined at P_1 which is S distance away from O. At time instant $(t + \Delta t)$, the particle reaches at P_2 and its co-ordinate is $(S + \Delta S)$. This change in position-coordinate during the time interval Δt is defined as the displacement of particle, ΔS.

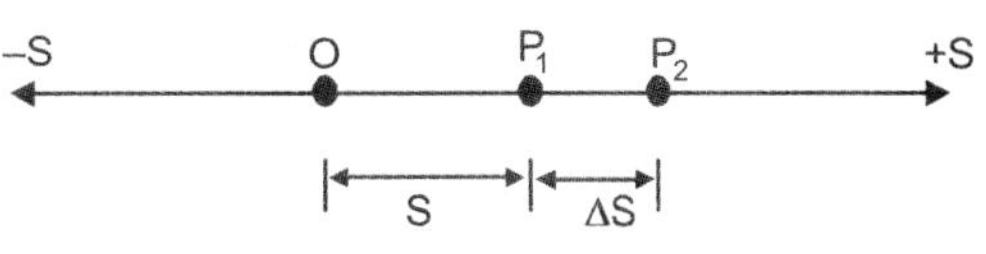

Fig. 8.2

Here it is observed that displacement is a function of time, i.e.,

$$S = f(t) \qquad \text{... (8.1}a\text{)}$$

Now the motion of a particle can be represented by various types of equation, depending on nature of movement.

(*i*) $S = c + bt$ ... (8.2*b*)

From this equation of motion it is suggested that at *zero-th* instant of time, initial displacement of the particle is c and rate of increment of displacement is b

(*ii*) $S = \dfrac{1}{2}gt^2$ ...(8.2*c*)

This is a parabolic relationship of displacement with time.

(*iii*) $S = be^{-kt}$ ... (8.1*d*)

This is an exponential or logarithemic representation of displacement-time relationship.

8.8 VELOCITY

Average velocity of the particle during time interval Δt is represented as

$$v_{av} = \frac{\Delta S}{\Delta t} \qquad \text{... (8.2)}$$

If Δt becomes smaller gradually and approaches to zero within limit, it will generate instantenous velocity. Symbolically it is expressed as v (without any suffix).

So, $$v = \lim_{\Delta t \to 0} \frac{\Delta S}{\Delta t} = \frac{dS}{dt} = \dot{S} \qquad \text{... (8.3)}$$

[*Only time derivative of any variable can be represented by the symbol of that variable with a dot on top of it. For double derivative will be represented by two dots on top of it.*]

Hence we can say that, position-velocity is the time rate change of position co-ordinate S. The nature of velocity i.e., whether positive or negative depends on the nature of displacement wheather it is positive or negative.

8.9 ACCELERATION

The average acceleration is obtained as dividing the change in velocity by the time interval Δt So,

$$a_{av} = \frac{\Delta v}{\Delta t}, \text{ where } \Delta v = \text{ change in velocity} \qquad \text{... (8.4)}$$

When Δv becomes very small and approaches to zero within limit, average acceleration changes into instantenous acceleration, which is symbolised by a only, without any suffix. Therefore

$$a = \lim_{\Delta t \to 0} \frac{\Delta v}{\Delta t} = \frac{dv}{dt} = \dot{v} \qquad \text{... (8.5)}$$

$$= \frac{d}{dt}\left(\frac{dS}{dt}\right) = \frac{d^2 S}{dt^2} = \ddot{S} \qquad \text{... (8.6)}$$

If velocity is increasing acceleration is positive, and if decreasing acceleration becomes negative.

By application of elimination of dt from equations (8.3) (8.5), we obtain,

$$\frac{v}{dS} = \frac{a}{dv}$$

$$\Rightarrow \qquad vdv = adS \qquad \text{... (8.7)}$$

8.10 GRAPHICAL REPRESENTATION IN ANALYTICAL BACKGROUND

Governing differential equations of rectilinear motion can be further clarified if those are represented graphically, in order to have some physical concept of those parameters.

Different equations of displacement as given in equations [8.1(*b*), (*c*), (*d*)] are plotted in Figs. 8.3 (*a*), (*b*), (*c*).

Equation 8.1 (*b*) is of a straight line having y-intercept c and slope b here. So slope of the line $\left(\frac{dS}{dt}\right)$ is the velocity.

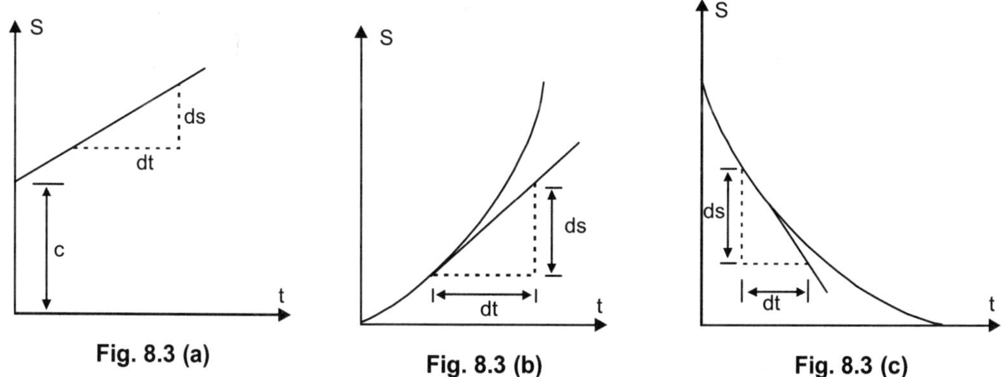

Fig. 8.3 (a) Fig. 8.3 (b) Fig. 8.3 (c)

In the similar fashion velocity can be represented graphically for other two equation [8.1 (*c*) and (*d*)] that are curvilinear by nature. Analytical values of velocity is these cases are as following:

$$v = \frac{d}{dt}(c + bt) = b \quad \text{... (8.8}a\text{)}$$

$$v = \frac{d}{dt}\left(\frac{1}{2}gt^2\right) = gt \quad \text{... (8.8}b\text{)}$$

$$v = \frac{d}{dt}\left(be^{-kt}\right) = -bke^{-kt} \quad \text{... (8.8}c\text{)}$$

Let us plot those equations in *v*-*t* axis in velocity-time plot.

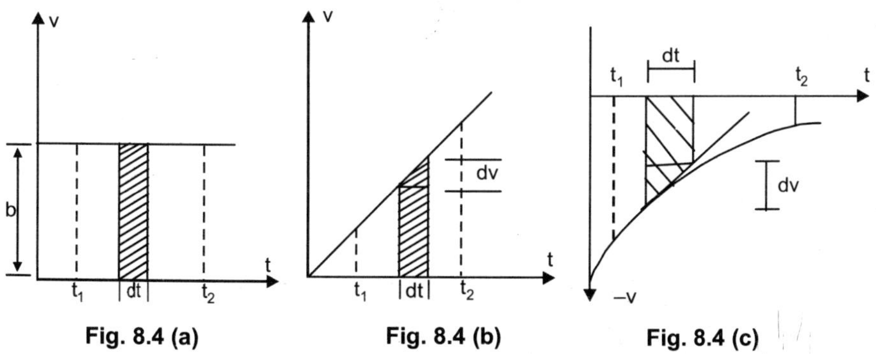

Fig. 8.4 (a) Fig. 8.4 (b) Fig. 8.4 (c)

As we know that $\left(\frac{dv}{dt}\right)$ represents the acceleration which is just slope of *v*-*t* curve, no acceleration will be obtained in Fig. 8.4 (*a*) as the plot becomes a straight line of constant value *b*. In the other two cases, numerical values of acceleration are possible to obtain.

From velocity-time diagram, entire distance traversed by the particle in some specific time *t* can be conceptualized. This distance is nothing but the area under the curve bound by *v*-axis and *t*-axis.

Analytically, the displacement can also be computed, from equation (8.3), we can write :

$$dS = vdt \qquad \text{... (8.9)}$$

Integrating it within limit,

$$\int_{S_1}^{S_2} dS = \int_{t_1}^{t_2} v\, dt = v\int_{t_1}^{t_2} dt$$

$$\Rightarrow \qquad S_2 - S_1 = v\left(t_2 - t_1\right) = \text{ area under } v - t \text{ curve.} \qquad \text{... (8.10)}$$

In analytical computation of acceleration from basic displacement equation as shown earlier, we obtain.

$$a = \frac{dv}{dt} = \frac{dv}{dt}\left(b\right) = 0 \qquad \text{... (8.11}a\text{)}$$

$$a = \frac{dv}{dt} = \frac{d}{dt}\left(gt\right) = g \qquad \text{... (8.11}b\text{)}$$

$$a = \frac{dv}{dt} = \frac{d}{dt}\left(-bke^{-kt}\right) = bk^2e^{-kt} \qquad \text{... (8.11}c\text{)}$$

The plot of $a - t$ curve on acceleration time diagram with be like this:

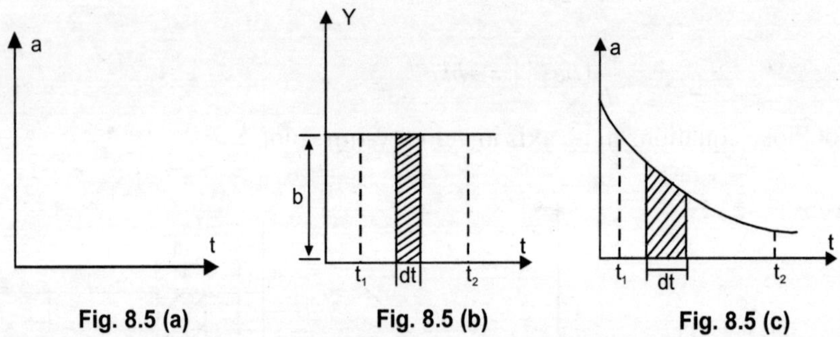

Fig. 8.5 (a) Fig. 8.5 (b) Fig. 8.5 (c)

Here the first plot represents no acceleration, second one represents a constant acceleration of amount g while the third plot depicts acceleration varying with time t.

The amount of change in velocity can be obtained from $a - t$ diagram. The area under $a - t$ curve will provide the change in velocity within a stipulated time limit.

From equation (8.5) the following relation can be written: $dv = a\, dt$ ... (8.12)

Integrating it within limit,

$$\int_{v_1}^{v_2} dv = \int_{t_1}^{t_2} a\, dt$$

$$\Rightarrow \qquad v_2 - v_1 = a\left(t_2 - t_1\right) \qquad \text{... (8.13)}$$

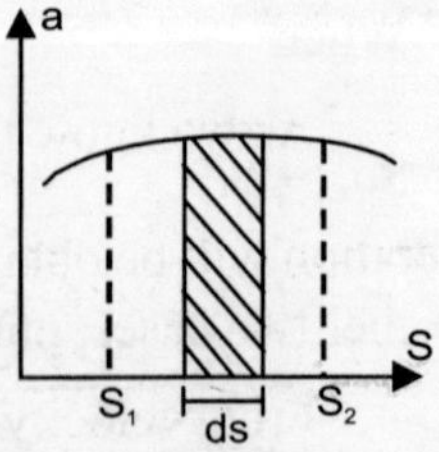

Fig. 8.6

So amount of change in velocity = area under $a - t$ curve.

Let us study an additional graphical presentation obeying equation (8.7) where acceleration is a function of displacement i.e., $a = f(s)$.

Integration of (8.7) within limits results.

$$\int_{v_1}^{v_2} v\,dv = \int_{S_1}^{S_2} a\,ds$$

$$\Rightarrow \qquad \frac{1}{2}\left(v_2^2 - v_1^2\right) = 2 \times \text{ area under } a\text{–}S \text{ curve.} \qquad \text{.... (8.14)}$$

8.11 MOTION WITH UNIFORM ACCELERATION

Considering the motion of the particle as with constant acceleration, various equations of motion can be derived. As we know.

$$a = \frac{dv}{dt}$$

$$\Rightarrow \qquad dv = a\,dt$$

Integrating both sides, we obtain $v = at + D$. When $t = 0$, v = initial velocity = u, So, $D = u$

$$\therefore \qquad v = u + at \qquad \text{... (8.15)}$$

Now again $\qquad v = \dfrac{dS}{dt}$

$$\Rightarrow \qquad dS = v\,dt = (u + at)\,dt$$

Integrating both sides, $\qquad S = ut + \dfrac{1}{2}at^2 + E$

When $\qquad t = 0, S = 0, \quad$ So, $\quad E = 0$

$$\therefore \qquad S = ut + \frac{1}{2}at^2 \qquad \text{... (8.16)}$$

Again we know $\qquad a\,dS = v\,dv$

Integrating on both sides, $\qquad aS = \dfrac{v^2}{2} + F$

When $S = 0$, v = initial velocity = u,

So, $\qquad F = \dfrac{-u^2}{2}$

$$\therefore \qquad aS = \frac{v^2}{2} - \frac{u^2}{2}$$

$$\Rightarrow \qquad v^2 = u^2 + 2\,aS \qquad \text{... (8.17)}$$

8.12 MOTION WITH NON-UNIFORM ACCELERATION

In all practical problems, acceleration may not be constant, and it may be a function of time or velocity or displacement, i.e.

$$\left.\begin{aligned} a &= f(t) \\ a &= f(u) \\ a &= f(S) \end{aligned}\right\} \qquad \text{... (8.18)}$$

In these three cases differential equation will have to be solved separately with case-specific boundary conditions.

8.13 RELATIVE MOTION

The entire concepts inculcated till now here revolve around the concept of measurement with respect to *fixed reference axes* system. But it is impractical to describe all the motions in this fashion. This necessity evolved the concept of analysing motion with reference to moving axes system, which is relative motion.

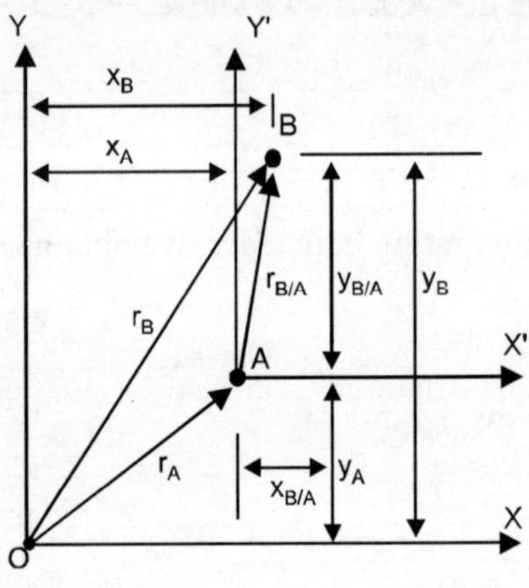

In the present case, our discussion will have to remain restricted within the scope of relative translation of particles only. When a particle moves in a straight line parallel to its previous position on the plane, it is said to undergo a translation.

Let us consider two particles A and B moving in the the plane. Fixed system of axes is represented by $X - Y$ having origin at O. And moving axes system is $X' - Y'$ with origin at A. Moving reference system can translate with reference to fixed axes system.

Position vectors r_A and r_B define the instantenous position of particle at A and B respectively.

The position vector of B as measured relative to A is symbolized as $r_{B/A}$, where the subscript notation B/A stands for B with respect to A and

$$r_{B/A} = x_i + y_j,$$

where i and j are unit vectors along reference axes. ... (8.19)

From triangular law of vector addition,

$$\overrightarrow{OA} + \overrightarrow{AB} = \overrightarrow{OB}$$

or, $$r_A + r_{B/A} = r_B \qquad \text{... (8.20)}$$

Separation of r into perpendicular components results,

$$\left.\begin{aligned} x_B &= x_A + x_{B/A} \\ y_B &= y_A + y_{B/A} \end{aligned}\right\} \qquad \text{...(8.21)}$$

Successive differentiation of vector equation (8.20) with respect to time, once and twice, will give expression for velocity and acceleration respectively as,

$$\left.\begin{aligned} r_B &= r_A + r_{B/A} \\ r_B &= r_A + r_{B/A} \end{aligned}\right\} \qquad \text{... (8.22)}$$

Separating into orthogonal components, we have

$$\left.\begin{aligned} \dot{x}_B &= \dot{x}_A + \dot{x}_{B/A} \\ \dot{y}_B &= \dot{y}_A + \dot{y}_{B/A} \end{aligned}\right\} \text{ and } \left.\begin{aligned} \ddot{x}_B &= \ddot{x}_A + \ddot{x}_{B/A} \\ \ddot{y}_B &= \ddot{y}_A + \ddot{y}_{B/A} \end{aligned}\right\} \qquad \text{...(8.23)}$$

Multiple Choice Questions

Select the best alternative (s):

1. A body weighing 100 kg falls vertically down on a cart weighing 200 kg moving at velocity V m/sec. The velocity of a cart after falling of weight would be
   (*a*) V m/sec (*b*) more than V m/sec (*c*) less than V m/sec
   (*d*) unpredictable (*e*) none of the above
2. A particle while sliding down a smooth plane of 19.86 $\sqrt{2}$ m length acquires a velocity of 19.86 m/sec. The inclination of plane is
   (*a*) 30º (*b*) 45º (*c*) 60º (*d*) 75º (*e*) none of the above
3. A body moves from rest with a constant acceleration of 5m/sec. The distance covered in 5 sec is most nearly
   (*a*) 38m (*b*) 625.5m (*c*) 96m (*d*) 124m (*e*) 240m
4. Equation of motion of a particle is $S = 2t^3 - t^2 - 2$ where S is metres and t in seconds. Acceleration of the particle after 1 sec will be.
   (*a*) 8 m/s^2 (*b*) 9 m/s^2 (*c*) 10 m/s^2 (*d*) 5 m/s^2

Answers

1. (*c*) 2. (*b*) 3. (*b*) 4. (*c*)

Numerical Examples

Example 1

The position co-ordinate of a particle which is confined to move along a straight line is given by $S = 2t^3 - 24t + 6$, where S is measured in metres from convenient origin and t is in seconds. Determine (*a*) the time required for the particle to reach a velocity of 72m/s from its initial condition at $t = 0$, (*b*) the acceleration of the particle when $v = 30$m/s and (*c*) the net displacement of the particle during the interval from $t = 1$s to $t = 4$s.

SOLUTION $v = \dfrac{dS}{dt} = (6t^2 - 24)\text{m/s}, \quad a = \dfrac{dv}{dt} = 12t \text{ m/s}^2$

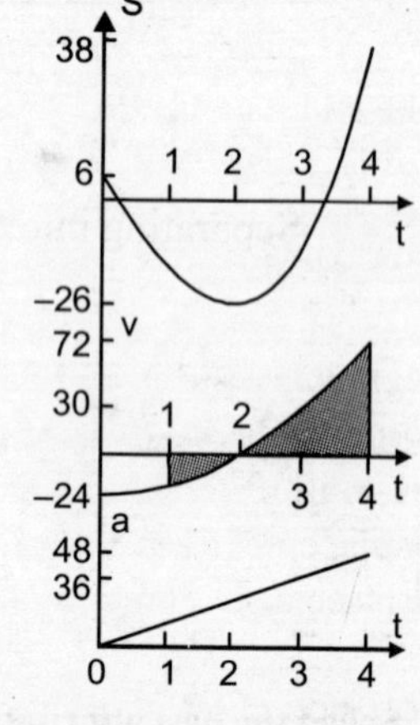

(*a*) When $v = 72\text{m/s}$, $72 = 6t^2 - 24$

$\Rightarrow$ $t = 4\text{s}$

(*b*) When $v = 30\text{m/s}$, $30 = 6t^2 - 24$

$\Rightarrow$ $t = 3\text{s}$

Here, $a = 12t = 12 \times 3 = 36\text{m/s}^2$

(*c*) Net displacement during specified interval,

$$\Delta S = S_4 - S_1 = [2(4)^3 - 24 \times 4 + 6] - [2(1)^3 - 24 \times 1 + 6] = 54\text{m}$$

Construction of $S - t$, $v - t$, $a - t$ graphs are shown.

EXAMPLE 2

A particle moves along the *x*-axis with an initial velocity $v_x = 60\text{m/s}$ at the origin when $t = 0$. For first 4 *seconds* it has no acceleration and thereafter it is acted on by a retarding force which gives it a constant acceleration $a_x = -15\text{m/s}^2$. Calculate the velocity and the *x*-coordinate of the particle for the condition of $t = 10s$ and $t = 15s$, and find the maximum positive *x*-coordinate reached t_y the particle.

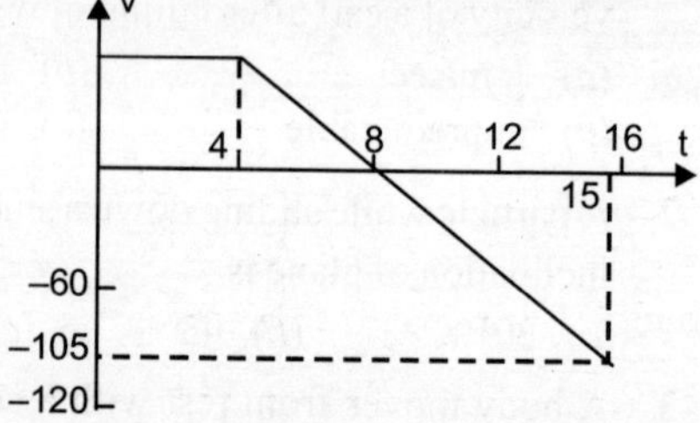

SOLUTION We know, $a = \dfrac{dv}{dt} \quad \Rightarrow \quad dv = adt$

So, $\displaystyle\int_{60}^{v_x} dv = -15\int_4^t dt \quad \Rightarrow \quad v_x = 120 - 15t$

When $t = 10$, $v_x = 120 - 15 \times 10 = -30\text{m/s}$

$t = 15$, $v_x = 120 - 15 \times 15 = -105\text{m/s}$

Again we know, $v = \dfrac{ds}{dt}$

So, $\displaystyle\int_{60\times 4}^{x} dS = \int_4^t vdt$

$$\Rightarrow \quad x - 240 = \left(120t - \frac{15}{2} \times t^2\right) - \left(120 \times 4 - \frac{15}{2} \times 4^2\right)$$

$$\Rightarrow \quad x = -\frac{15}{2}t^2 + 120t - 120$$

When, $t = 10s$, $x = 330\text{m}$ and $t = 15s$, $x = -7.5\text{m}$

When, $v_x = 0$, $t = \dfrac{120}{15} = 8\text{s}$.

So after $t = 8s$ motion in negative *x*-direction will start.

Maximum movement in positive x-direction $x_{max} = -\frac{15}{2}\times 8^2 + 120\times 8 - 120 = 360\text{m}$

EXAMPLE 3

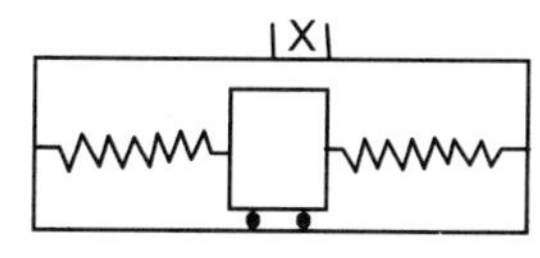

The spring-mounted slider moves in the horizontal guide with negligible friction and has a velocity v_0 in the x-direction as it crosses the mid position where $x = 0$ and $t = 0$. The two spring together exert a retarding force to the motion of the slider, which gives it an acceleration proportional to the displacement but oppositely and $a \propto -k^2x$, where k is constant. Determine the expressions for the displacement x and v as the function of time t.

SOLUTION Now,
$$a = \frac{dv}{dt} = \frac{dv}{dx}\cdot v$$

$\Rightarrow$
$$vdv = adx = -k^2xdx$$

So,
$$\int_{v_0}^{v} vdv = \int_0^x -k^2xdx$$

$\Rightarrow$
$$v = \sqrt{v_0^2 - k^2x^2}$$

Again,
$$\frac{dx}{dt} = v = \sqrt{v_0^2 - k^2x^2}$$

$\Rightarrow$
$$\frac{dx}{\sqrt{v_0^2 - k^2x^2}} = dt$$

So,
$$\int_0^x \frac{dx}{\sqrt{v_0^2 - k^2x^2}} = \int_0^t dt$$

$\Rightarrow$
$$\frac{1}{k}\sin^{-1}\frac{kx}{v_0} = t$$

$\Rightarrow$
$$x = \frac{v_0}{k}\sin kt$$

Again,
$$v = \frac{dx}{dt} = \frac{v_0}{k}k\cos kt = v_0\cos kt$$

EXAMPLE 4

The displacement of a particle which moves along the x-axis is given by $x = (-2 + 3t)e^{-0.5t}$, where x is in meters and t is in seconds. Plot the displacement, velocity, and acceleration versus time for the first 20 seconds of motion. Determine the time at which the acceleration is zero.

SOLUTION Here,
$$v = \frac{dx}{dt} = -2(-0.5)e^{-0.5t} + 3e^{-0.5t} + 3t(-0.5)e^{-0.5t} = (4 - 1.5t)e^{-0.5t}$$

$$a = \frac{dv}{dt} = 4(-0.5)e^{-0.5t} - 1.5e^{-0.5t} - 1.5t(-0.5)e^{-0.5t} = (-3.5 + 0.75t)e^{-0.5t}$$

For acceleration to be zero,

$$-3.5e^{-0.5t} + 0.75te^{-0.5t} = 0$$

$$\Rightarrow \qquad t = 4.67\text{s}$$

Following Table will help to plot $x - t$, $v - t$, $a - t$, curves:

$t =$	0s	2s	4s	4.67s	6s	8s	10s	16s	20s
x	–2	1.47	1.35	1.16	0.80	0.40	0.19	0.02	2.63×10^{-3}
v	4	0.37	–0.27	–0.29	–0.25	–0.15	–0.07	-6.7×10^{-3}	-1.8×10^{-3}
a	–3.5	–0.74	–0.07	0	0.05	0.04	0.03	2.9×10^{-3}	5.2×10^{-3}

EXAMPLE 5

The position of a particle is given by $x = 2t^3 - 6t^2 - 18t + 39$, where x is in meters and t is in seconds. Plot the position, velocity, and 12 seconds of the motion. Also determine the time at which the velocity is zero.

SOLUTION Here, $$v = \frac{dx}{dt} = (6t^2 - 12t - 18), \qquad a = \frac{dv}{dt} = 12\,(t-1)$$

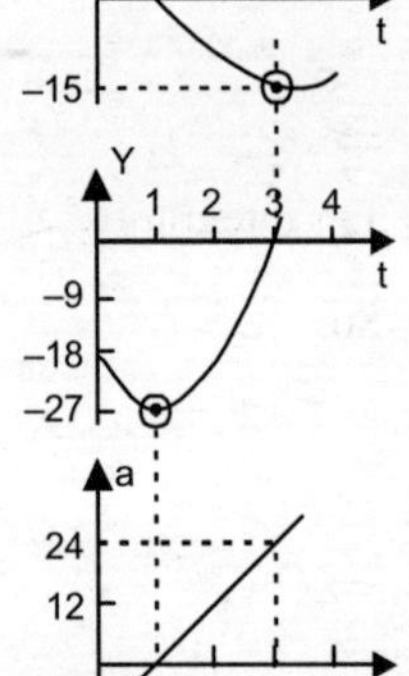

When velocity is zero, $6t^2 - 12t - 18 = 0$

$\therefore \qquad t = 3\text{s}$ [taking positive value]

Position of the particle at $t = 3$s,

$$x = 2(3)^3 - 18(3) + 39 = -15\text{m}$$

Acceleration of the particle at $t = 3s$, $a = 12(3 - 1) = 24\text{m/s}^2$.

Respective graphs are shown here. It may be noted that

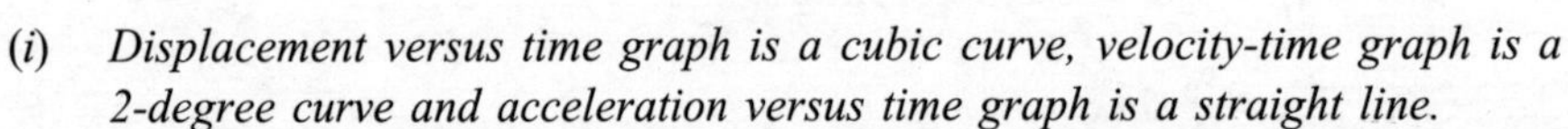

(*i*) *Displacement versus time graph is a cubic curve, velocity-time graph is a 2-degree curve and acceleration versus time graph is a straight line.*

(*ii*) *Particle does not move along any of these curves but along a straight line.*

(*iii*) *The point of inflexion on the displacement curve corresponds to zero velocity and the point of inflexion on the velocity curve corresponds to zero acceleration.*

EXAMPLE 6

A stone is thrown vertically upwards with a velocity of 19.6 m/s from the top of a tower 24.5 meters high. Calculate (*i*) time required for the stone to reach the ground, (*ii*) velocity of the stone, in its downward travel at the point in the same level as the point of projection, (*iii*) the maximum height to which the stone will rise in its flight.

SOLUTION (*i*) Velocity of stone at A upwards $u = 19.6$ m/s

Now, $$h = ut - \frac{1}{2}gt^2$$

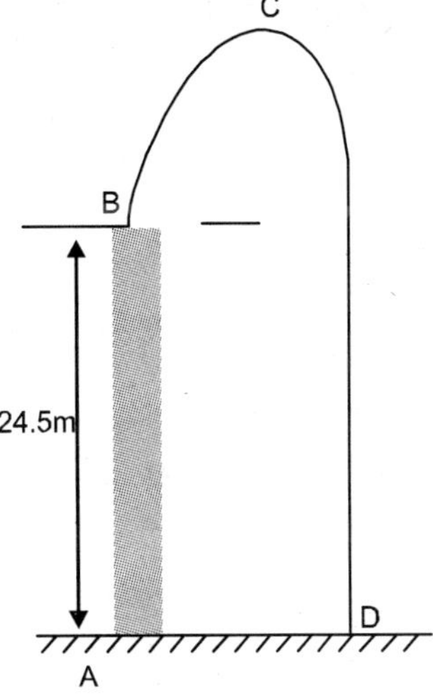

$$\Rightarrow \qquad -24.5 = 19.6t - \frac{1}{2} \times 9.8t^2$$

$$\Rightarrow \qquad t = 5, \quad \text{or } t \neq -1$$

So, time required to reach the ground is 5 sec.

(*ii*) Velocity of the stone in its downward motion at B = velocity of stone in its upwards motion with which it was thrown from A = 19.6 m/s.

(*iii*) Maximum height reached by the stone above A be

$$0 = u^2 - 2gh_m$$

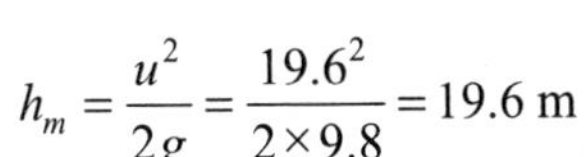

$$\Rightarrow \qquad h_m = \frac{u^2}{2g} = \frac{19.6^2}{2 \times 9.8} = 19.6 \text{ m}$$

EXAMPLE 7

A particle starts with velocity v_0. Its acceleration and velocity are related by equation $a = -kv$. Find the displacement time relation.

SOLUTION Here, $$\frac{dv}{dt} = a = -kv$$

$$\Rightarrow \qquad \frac{dv}{v} = -kdt$$

On integration,

$$\int_{v_0}^{v} \frac{dv}{v} = -\int_0^t kdt$$

$$\Rightarrow \qquad \ln\left(\frac{v}{v_0}\right) = -kt$$

$$\Rightarrow \qquad v = v_0 e^{-kt}$$

Again, $$\frac{dx}{dt} = v = v_0 e^{-kt}$$

$$\Rightarrow \qquad dx = v_0 e^{-kt} dt$$

On integration, $$\int_0^x dx = \int_0^t v_0 e^{-kt} dt$$

$$\Rightarrow \qquad x = -\frac{v_0}{k}(e^{-kt} - 1)$$

$$= \frac{v_0}{k}(1 - e^{-kt})$$

EXAMPLE 8

The crank of a reciprocating mechanism of an engine rotates at an angular velocity of ω rad/s. Find the displacement time relation for the motion of the piston P if the length of the connecting rod is l and radius of the crank is r. Assuming $l \gg r$, find the maximum velocity and acceleration of the piston and where it occurs. Also find the motion of the point B on the connecting rod between the point A and P.

SOLUTION Consider the position OA of the crank at any time t and θ be the angle turned with respect to $O - X$. So $\theta = \omega t$.

The co-ordinate of the piston P,

$x = r\cos\theta + l\cos\phi$. Again we have from triangle OAP,

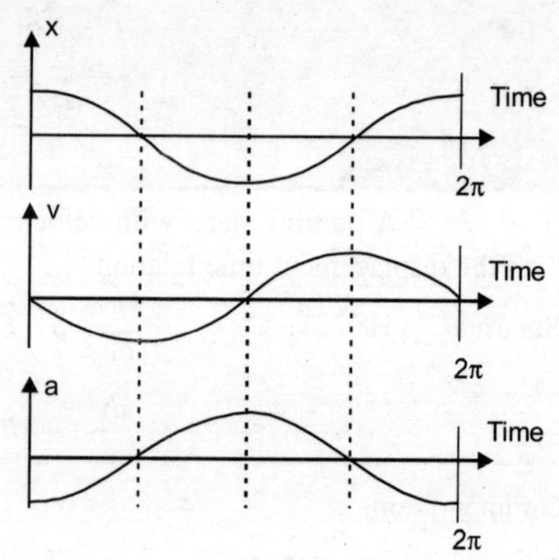

$$\frac{r}{\sin\phi} = \frac{l}{\sin\theta}$$

$$\Rightarrow \qquad \sin\phi = \frac{r}{l}\sin\theta = \frac{r}{l}\sin\omega t$$

$$\therefore \qquad \cos\phi = \sqrt{1 - \frac{r^2}{l^2}\sin^2\omega t}$$

$$\therefore \qquad x = r\cos\theta + l\sqrt{1 - \frac{r^2}{l^2}\sin^2\omega t}$$

As $\qquad l \gg r,$

$$\frac{r^2}{l^2} \simeq 0$$

$$\therefore \qquad x = r\cos\theta + l = l + r\cos\omega t \text{ and } v = \frac{dx}{dt} = -\omega r\sin\omega t$$

For velocity to be minimum, $\theta = 90°$, and maximum $\theta = 270°$

Acceleration $\qquad a = \frac{dv}{dt} = -\omega^2 r\cos\omega t.$

For acceleration to be minimum, $\theta = 0°$, and maximum, $\theta = 180°$. At any instant, the x and y co-ordinate of point B are given by

$$x = r\cos\theta + b\cos\phi, \text{ and } y = (l - b)\sin\phi.$$

Substituting for θ, $\cos\phi$, $\sin\phi$

$$x = r\cos\omega t + b\sqrt{1 - \frac{r^2}{l^2}\sin^2\omega t},$$

$$y = \frac{l-b}{l} r\sin\omega t$$

For the particular case when $l = r$,

$$x = r\cos\omega t + b\sqrt{1-\sin^2\omega t} = (r + b)\cos\omega t,$$
$$y = (r - b)\sin\omega t$$

$$\therefore \quad \frac{x^2}{(r+b)^2}+\frac{y^2}{(r-b)^2} = \cos^2\omega t+\sin^2\omega t = 1$$

This is the equation of an ellipse.

EXAMPLE 9

A system of two pulleys is shown in the figure. Find the relations connecting the displacements, velocities and accelerations of the blocks A and B.

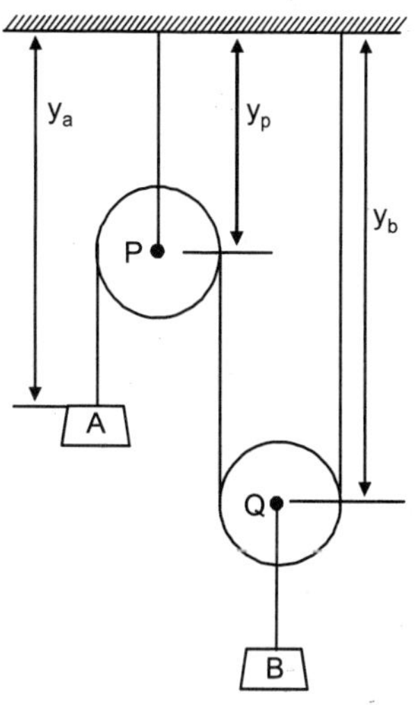

SOLUTION Distance of two blocks A and B with respect to the support is shown in the figure and total length of rope l.

$$\therefore \quad (y_a - y_p)+(y_b - y_p)+y_b = l$$

$$\Rightarrow \quad y_a + 2y_b = (l - 2y_p)$$

Differentiating above, $\quad \Delta y_a + 2\Delta y_b = 0$

$$\therefore \quad \Delta y_b = -\frac{\Delta y_a}{2}$$

Displacement of block B is half the displacement of block A and is in opposite direction.

Again, $\quad \dfrac{dy_a}{dt}+2\dfrac{dy_b}{dt} = 0$

$$\Rightarrow \quad v_b = -\frac{v_a}{2}$$

And, $\quad \dfrac{d^2y_a}{dt^2}+2\dfrac{d^2y_b}{dt^2} = 0$

$$\Rightarrow \quad a_b = -\frac{a_a}{2}$$

So the acceleration of the block B is half the acceleration of the block A in an opposite direction.

EXAMPLE 10

A trolley resting on a horizontal plane starts from rest and is moved to the right with a constant acceleration of 0.21 m/s^2. Determine (*i*) acceleration of the block B connected to the trolley, (*ii*) velocity of the trolley and the block after a time of 4 seconds and the distance moved by each of them.

SOLUTION The length of the string *a-b-c-d-e* connected the trolley T and block B is constant. From the system shown in the figure, we have,

$$x_T + 3x_B + cf = \text{constant.}$$

It may be noted that the motions of the centre of the pulley C and block B, are identical, hence expressed in terms of x_B, distance cf remains constant. Giving an increment Δx_T to the trolley,

$$\Delta x_T + 3\Delta x_B = 0$$

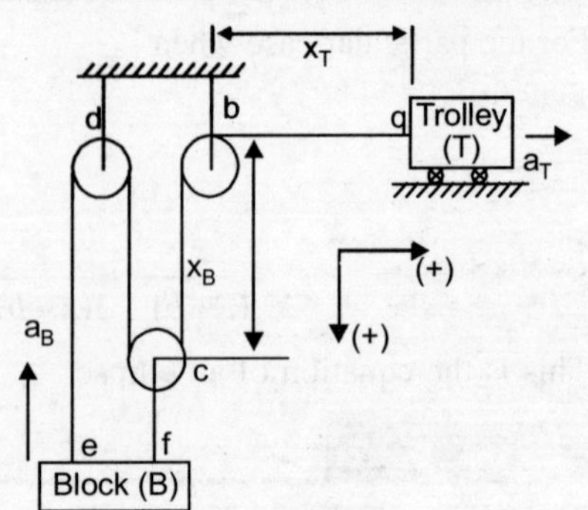

Differentiating with respect to time, we have,

$$\frac{dx_T}{dt} + 3\frac{dx_B}{dt} = 0$$

$$\Rightarrow \qquad v_T + 3v_B = 0$$

Again differentiating, $$\frac{dv_T}{dt} + 3\frac{dv_B}{dt} = 0$$

$$\Rightarrow \qquad a_T + 3a_B = 0$$

For $\qquad a_T = 0.21\ \text{m/s}^2, a_B = -\dfrac{a_T}{3} = -0.07\ \text{m/s}^2$

Velocity of the trolley after 4 sec, starting from rest is given by,

$$v_T = 0 + 0.21 \times 4 = 0.84\ \text{m/s}$$

Velocity of the block, $\qquad v_B = -\dfrac{0.84}{3} = -0.28\ \text{m/s}$

Distance moved by the trolley in 4 sec is given by,

$$S_T = 0 + \frac{1}{2} \times (0.21) \times 4^2 = 1.68\ \text{m}$$

Distance moved by the block, $\qquad S_B = -\dfrac{1.68}{2} = -0.56\ \text{m}$

EXAMPLE 11

Two blocks A and B are resting on a horizontal plane and connected by a cord as shown in the figure. The block A starts from rest and moves to the left with a constant acceleration. It was observed that the block B attains a velocity of 16 cm/s after moving a distance of 32 cm. Determine (*i*) acceleration of the block A and B, (*ii*) velocity and position of block A after 6 sec.

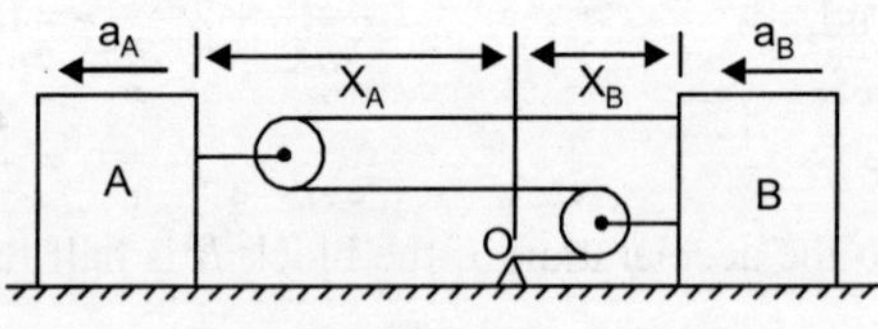

SOLUTION The length of the cord connecting the blocks A and B can be expressed in terms of the distance x_A and x_B of the block from a fixed point O as shown in figure.

Now we have, $\qquad 3x_B + 2x_A = \text{constant.}$

Giving an increment Δx_A to the block A, $3\Delta x_B + 2\Delta x_A = 0$

Differentiating we have, $$3\frac{dx_B}{dt} + 2\frac{dx_A}{dt} = 0$$

$$\therefore \qquad 3v_B + 2v_A = 0$$

Again differentiating, $3\dfrac{dv_B}{dt}+2\dfrac{dv_A}{dt}=0$

$\therefore$ $3a_B+2a_A=0$

Now for block B, $v_B = 16$ cm/s, $u_B = 0$, $S_B = 32$ cm

$\therefore$ $a_B=\dfrac{16^2-0^2}{2\times 32}=4\text{ cm/s}^2$

Acceleration of block A, $a_A=-\dfrac{3}{2}a_B=-\dfrac{3}{4}\times 4=-6\text{cm/s}^2$

Velocity of block A after 6 sec, $v_A = u_A + a_At = 0 + 6 \times 6 = 36$ cm/s

Position of block A after 6 sec, $S_A=u_At+\dfrac{1}{2}a_A\cdot t^2=0+\dfrac{1}{2}\times 6\times 6^2=108\text{ cm}=1.08\text{ m}$

EXAMPLE 12

A train starts from rest and increases its speed from zero to v m/s with a constant acceleration of a_1 m/s², runs at this speed for some time and finally comes to rest with a constant deceleration a_2 m/s². If the total distance travelled is x, find the total time t required for this journey.

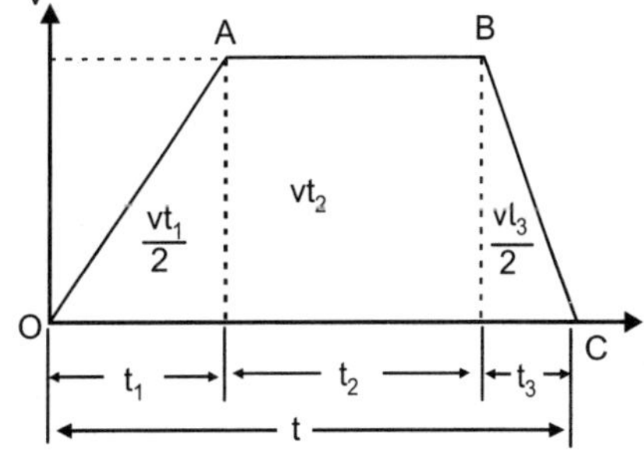

SOLUTION The figure represents the velocity-time graph for the journey. OA represents acceleration from zero to velocity v, AB of constant velocity and BC is deceleration from velocity v to zero. The time taken for the three portions of the journey be t_1, t_2, t_3 respectively.

For OA, we have, $v = u + a_1t_1$

$\therefore$ $t_1=\dfrac{v}{a_1}$

For BC, we have,

$$0 = v - a_2t_3$$

$\therefore$ $t_3=\dfrac{v}{a_2}$

The total distance travelled by the train, from velocity-time graph,

$$x=\frac{1}{2}vt_1+vt_2+\frac{1}{2}vt_3$$

$\therefore$ $t_2=\dfrac{x}{v}-\dfrac{t_1}{2}-\dfrac{t_3}{2}$

So, total time required for travel,

$$t = \frac{v}{a_1} + \left(\frac{x}{v} - \frac{t_1}{2} - \frac{t_3}{2}\right) + \frac{v}{a_2}$$

$$= \frac{v}{a_1} + \frac{x}{v} - \frac{v}{2a_1} - \frac{v}{2a_2} + \frac{v}{a_2} = \frac{x}{v} + \frac{v}{2}\left(\frac{1}{a_1} + \frac{1}{a_2}\right)$$

EXAMPLE 13

A train A passes a certain station at 72 km/hr and moves a distance of 20 km at this speed and then comes to rest at the next station 24 km from the first station. A train B starting from the first station reaches the second station in double the interval of time taken by the train A. Find the maximum speed reached by the train B which covers part of the distance with acceleration and rest of the distance with retardation.

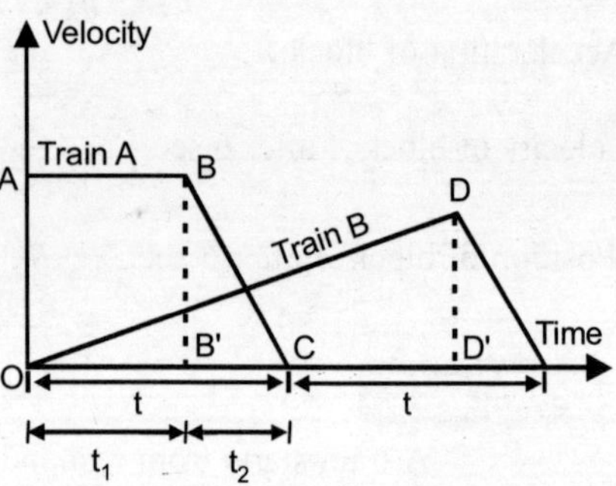

SOLUTION For the train A the distance moved at uniform motion is 20 km and time taken = t_1 hour.

Distance moved during uniform motion,

$$20 = t_1 \times 72$$

$$\therefore \quad t_1 = \frac{20}{72} = \frac{5}{18} \text{ hr}$$

So the distance moved by A during retarded motion $24 - 20 = 4$ km and corresponding time taken = t_2 hour.

Distance moved, $$4 = \frac{1}{2} t_2 \times 72$$

$$\therefore \quad t_2 = \frac{1}{9} \text{ hr}$$

So, total time taken by A, $$t = t_1 + t_2 = \frac{5}{18} + \frac{1}{9} = \frac{7}{18} \text{ hr}$$

For train B, total time taken $$2t = 2 \times \frac{7}{18} = \frac{7}{9} \text{ hr}$$

Distance travelled by B, $$24 = \frac{1}{2} \times 2t \times v_{max} = \frac{1}{2} \times \frac{7}{9} \times v_{max}$$

$$\therefore \quad v_{max} = 61.72 \text{ km/hr.}$$

EXAMPLE 14

Driver of a car travelling at 72 km/hr observes the light 250 m ahead of turning red. The traffic light is timed to remain red for 20 sec before it turns green. If the driver wishes to pass the light without stopping to wait for it to turn green, determine, (*i*) the required uniform acceleration of the car, (*ii*) the speed with which the driver crosses the traffic light.

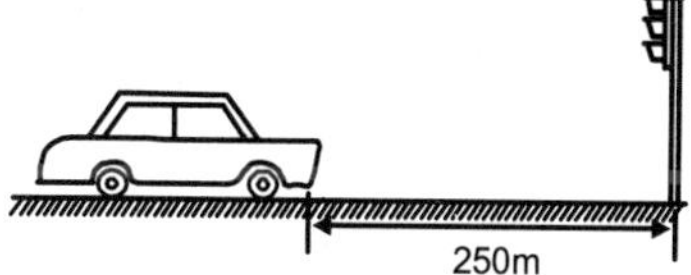

SOLUTION Initial speed $u = \frac{72 \times 1000}{60 \times 60} = 20 \text{ m/s}$

We know,
$$S = ut + \frac{1}{2}at^2$$

∴
$$250 = 20 \times 20 + \frac{1}{2}a \times 20^2$$

∴
$$a = -0.75 \text{ m/s}^2$$

Required velocity = $u + at = 20 - 0.75 \times 20 = 5$ m/s

EXAMPLE 15

An open platform elevator is moving down a mine shaft an acceleration of 0.5 m/s². After the elevator has travelled 25 m from the top of the shaft, a stone is dropped from the top of the shaft. Determine (*i*) the time taken the stone to hit the elevator (*ii*) the distance travelled by the elevator at the time of impact.

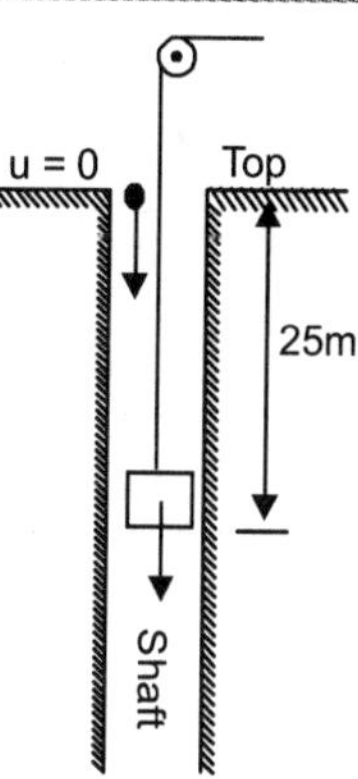

SOLUTION At the time the stone hits the elevator, both the elevator and the stone must be at the same distance from the top of the shaft. The stone hits the elevator after time t and distance travelled by this time $= ut + \frac{1}{2}gt^2 = \frac{1}{2} \times 9.81t^2$

The distance travelled by the elevator in time t be S_1. So the distance of the elevator from the top of the shaft after time t is $(S_1 + 25)$ m.

For the stone to hit the elevator, $\frac{1}{2} \times 9.81t^2 = S_1 + 25$

The elevator started from the rest, under and acceleration 0.5 m/s² and covered a distance of 25 m. So, at a distance of 25 m from the top, the velocity of elevator

$$= \sqrt{0^2 + 2 \times 0.5 \times 25} = 5 \text{ m/s}$$

So,
$$S_1 = 5t + \frac{1}{2}0.5t^2$$

By substitution, $5t + \frac{1}{2} \times 0.5t^2 + 25 = \frac{1}{2} \times 9.81t^2$

∴
$$t = 2.917 \text{ s}$$

EXAMPLE 16

A particle moves along a straight line so that the distance of the particle x from a point O on the straight line follows the law $x = 6t^2 - t^3$, where x is in metres and t in seconds. Determine the, (*a*) time at which the distance of the particle from O is zero, (*b*) time at which the distance of the particle from O is maximum, (*c*) the maximum distance of the particle from O, (*d*) time at which the velocity is zero, (*e*) time at which the velocity reaches a peak value, (*f*) time at which the acceleration is zero, also plot x–t, v–t and a–t graphs.

SOLUTION Here velocity, $$v = \frac{dx}{dt} = \frac{d}{dt}\left(6t^2 - t^3\right) = 12t - 3t^2$$

And acceleration, $$a = \frac{dv}{dt} = \frac{d}{dt}\left(12t - t^2\right) = 12 - 6t$$

Values of x, v, a for various values of t are tabulated below:

t	0s	1s	2s	3s	4s	5s	6s
x	0	5	16	27	32	25	0
v	0	9	12	9	0	–15	–36
a	12	6	0	–6	–12	–18	–24

For zero displacement,

$$0 = 6t^2 - t^3$$

$\Rightarrow$ $$t = 6\text{s}, 0\text{s}$$

For maximum displacement,

$$\frac{dx}{dt} = v = 0$$

$\Rightarrow$ $$12t - 3t^2 = 0$$

$\Rightarrow$ $$t = 4\text{ s}, 0\text{ s}$$

So, at $t = 0$, x is minimum and $t = 4$ s, x is maximum.

$\Rightarrow$ $$x_{max} = 6 \times 4^2 - 4^3 = 32\text{ m}$$

When velocity is zero, from previous calculation, we can say, it is at $t = 4$s, 0s.

For velocity to be maximum, $$\frac{dv}{dt} = a = 0$$

$\Rightarrow$ $$12 - 6t = 0$$

$\Rightarrow$ $$t = 2\text{s}$$

So, maximum velocity, $$v_{max} = 12 \times 2 - 3 \times 2^2 = 12\text{ m/s}$$

For acceleration to be zero, from above calculation, we can say, it is at $t = 2$s. Some important observations about the motion of the particle:

(1) *The particle starts from rest, as t increases the positive velocity increases. The particle moves in the positive direction from t = 0s to t = 2s, x, v and t are positive.*

(2) *At t = 2s, the acceleration reaches the zero value. The velocity reaches a maximum value. From t = 2s to t = 4s, v is positive but a is negative, i.e., the particle continues to move in the positive direction but the nature of motion is one of retardation.*

(3) *At t = 4s, the velocity of the particle reaches the zero value, when t > 4s, both v and a are negative. This means the particle moves in the negative direction with acceleration.*

(4) *At t = 6s, the particle reaches the point from where it started. For t > 6s, we find x, v and t are all negative i.e., the particle moves in the direction away from the starting point O with acceleration.*

EXAMPLE 17

A particle moving with an acceleration of $10 m/s^2$ travels a distance of 80m during the 6th seconds of its travel. Find its initial velocity.

SOLUTION The distance travelled by 6th second = distance travelled in 6 second – distance travelled in 5 second.

So,
$$S_6 = S_{0-6} - S_{0-5} = \left(6u + \frac{1}{2}\times 10.6^2\right) - \left(5u + \frac{1}{2}\times 10.5^2\right)$$

$$\therefore \quad 80 = u + 65$$

$$\Rightarrow \quad u = 15 \text{ m/s}$$

EXAMPLE 18

A particle moves in the $x - y$ plane such that the velocity components at any time t is given by, $v_x = 2t - 6$ and $v_y = 3t^2 - 18t + 27$. If the particle is in the position (9, –27) at $t = 0$, determine the equation of its path and also its linear acceleration at $t = 4s$.

SOLUTION We know
$$v_x = \frac{dx}{dt} = 2t - 6$$

$$\Rightarrow \quad dx = (2t - 6)dt$$

Integrating within limits,
$$\int_9^x dx = \int_0^t (2t - 6)\,dt$$

$$\Rightarrow \quad x = t^2 - 6t + 9$$

Again,
$$v_y = \frac{dy}{dt} = 3t^2 - 18t + 27$$

$$\Rightarrow \quad dy = (3t^2 - 18t + 27)dt$$

Integrating within limits,

$$\int_{-27}^{y} dy = \int_0^t (3t^2 - 18t + 27)dt$$

$$\therefore \quad y = t^3 - 9t^2 + 27t - 27$$

So the equation of the path of the particle in parametric form,

$$x = t^2 - 6t + 9,\ y = t^3 - 9t^2 + 27t - 27$$

The acceleration parallel to x-axis $\quad a_x = \frac{dv_x}{dt} = 2$

The acceleration parallel to y-axis $\quad a_y = \frac{dv_y}{dt} = 6t - 18$

At, $\quad t = 4\text{s},\ a_x = 2\ \text{m/s}^2$ and $a_y = 6 \times 4 - 18 = 6\ \text{m/s}^2$

So, resultant linear acceleration,

$$a = \sqrt{a_x^2 + a_y^2} = \sqrt{(2)^2 + 6^2} = 6.325\ \text{m/s}^2$$

EXAMPLE 19

The velocity of a body moving along a straight line follows the law $v = 1.25t - 0.125t^2$, where v is the velocity in m/s at time t sec. Find (a) the maximum acceleration, and (b) the distance moved in 10 sec.

SOLUTION For acceleration, $a = \frac{dv}{dt} = 1.25 - 0.25t$

So it is very much obvious from above equation that acceleration will be maximum when $t = 0$. So, $a_{max} = 1.25\ \text{m/s}^2$

Again, $\quad \frac{dx}{dt} = 1.25t - 0.125t^2$

$$\Rightarrow \quad dx = (1.25t - 0.125t^2)dt$$

Integrating within limits, $\int_0^x dx = \int_0^{10} (1.25t - 0.125t^2)dt$

$$\Rightarrow \quad x = \left[\frac{1.25}{2}t^2 - \frac{0.125}{3}t^3\right]_0^{10} = 20.833\ \text{m}$$

EXAMPLE 20

The acceleration of a particle moving along a straight line is given by the law $a = 25 - 3x^2$, where a in m/s², x is in metres. The particle starts from rest. Find (a) velocity when the displacement is 2 m and when velocity is again zero, (b) displacement at maximum velocity.

SOLUTION We know, $\quad a = \frac{dv}{dt} = \frac{dv}{dx} \times \frac{dx}{dt} = v\frac{dv}{dx}$

$$\Rightarrow \quad vdv = adx = (25 - 3x^2)dx$$

Integrating within limits, $\int_0^v vdv = \int_0^x (25 - 3x^2)dx$

$$\Rightarrow \qquad \frac{v^2}{2} = 25x - x^3$$

$$\Rightarrow \qquad v^2 = 50x - 2x^3$$

(*a*) When $x = 2$ m, $\qquad v = \sqrt{50 \times 2 - 2 \times 2^3} = 9.165$ m/s

When $\qquad v = 0 = 25x - x^3$

$$\Rightarrow \qquad x = 0 \text{ or } 5 \text{ m}$$

(*b*) For maximum velocity condition,

$$\frac{dv}{dt} = a = 0 = 25 - 3x^2$$

$$\Rightarrow \qquad x = 2.887 \text{ m}$$

EXAMPLE 21

An inextensible string *AB* is tied to a block *B* of negligible dimensions and passes over a small pulley *C* so that the free end *A* hangs h_1 units above the ground on which the block *B* rests. In this initial position shown in figure, the free end *A* is *h* units below *C*. If now the end *A* moves, horizontally with a velocity v_0, obtain an expression for the velocity of the block at any time *t*. If $h = 6$ m, $h_1 = 2$ m and $v_0 = 4$ m/s, find the time required for the block to reach the pulley.

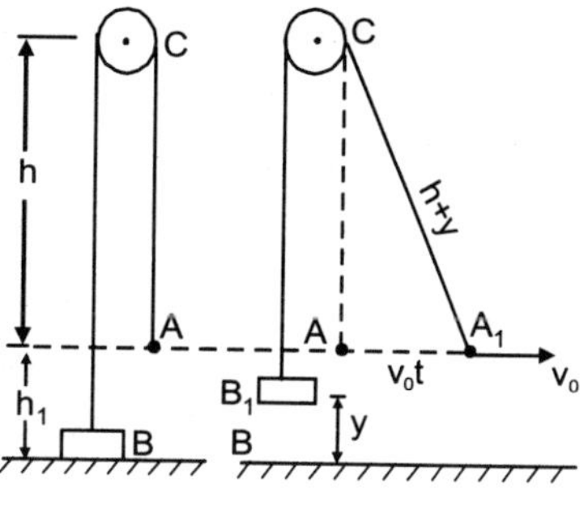

SOLUTION In a duration of *t* seconds the end *A* moves to the position A_1, so that $AA_1 = v_0t$. The block *B* moves upwards to the position B_1. Let $BB_1 = y$.

So the length $\qquad A_1C = h + y$

From the triangle ACA_1, $\qquad A_1C^2 = AC^2 + AA_1^2$

$$\Rightarrow \qquad (h+y)^2 = h^2 + (v_0t)^2$$

$$\Rightarrow \qquad y^2 + 2hy - v_0^2t^2 = 0$$

$$\Rightarrow \qquad y = -h + \sqrt{h^2 + v_0^2t^2}$$

This is the equation for displacement of the block.

Velocity of the block, $\qquad v = \frac{d}{dt}\left(-h + \sqrt{h^2 + v_0^2t^2}\right) = \frac{v_0t}{\sqrt{h^2 + v_0^2t^2}}$

When the block reaches at the pulley *C*, $h + h_1 = -h + \sqrt{h^2 + v_0^2t^2}$

$$\Rightarrow \qquad 2h + h_1 = \sqrt{h^2 + v_0^2t^2}$$

$$\Rightarrow \qquad 2\times 6+2=\sqrt{6^2+16t^2}$$

$$\Rightarrow \qquad t=3.16\text{s}$$

Exercise

1. Water drops from a faucet at a uniform rate of n drops per second. Find the distance x between any two adjacent drops as a function of the time t that the trailing drop has been in motion. Neglect air resistance.

Ans. $\left[x=\frac{gt}{n}+\frac{g}{2n^2}\right]$

2. When a stone has fallen a distance x after being dropped from the top of a tower, another stone is dropped from a point y below the top of the tower. If both of them reach the ground together, show that the top of the tower is $\frac{(x+y)^2}{4x}$.

3. The maximum possible acceleration or deceleration that a train may have is a while its maximum speed is v. Find the shortest interval of time in which the train can get from one station to the next if the distance between the station is S.

Ans. $\left[t=\frac{S}{v}+\frac{v}{a}\right]$

4. The distance between two stations is 2.50 km. A locomotive starting from one station gives the train an acceleration (reaching a speed of 36 km/hr. in 30 sec.) until the speed reaches 54 km/hr. This speed is maintained until the brakes are applied and the train is brought to rest at the second station under a retardation of 1m/s^2. Find the time taken to perform the journey, and the distances covered during the accelerated, uniform and retarded motion. **Ans.** [196.67 s, 337.5 m, 2050 m, 112.5 m]

5. An automobile starting from rest increases its speed from 0 to v with a constant acceleration a_1 runs at this speed for a time, and finally comes to rest with a constant deceleration a_2. If the total distance travelled is S find the total time t required.

Ans. $t=\left[\frac{S}{v}+\frac{v}{2}\left(\frac{1}{a_1}+\frac{1}{a_2}\right)\right]$

6. A man of height 1.80 m walks away from a lamp hanging at height 6m above the ground level. If the man walks with a speed 2.10 m/s determine the speed of the tip of the man's shadow. **Ans.**[3m/s]

7. A stone is dropped from the top of a tower of unkonwn height h. At the instance it has fallen C m a second stone is released from rest from a point b m below the top of the tower ($b>c$). The two stones strike the ground at the same time. Determine the height of the tower.

Ans. $\left[h=\frac{(b+c)^2}{4c}\right]$

8. Two balls are projected simultaneously with the same velocity from the top of the tower one vertically upwards and the other vertically downwards. If they reach the ground in times t_1 and t_2 respectively show that the time each will take to reach the ground if simply dropped from the top of the tower is given by $\sqrt{t_1 t_2}$.

9. A particle moves along a straight line according to the law $x = \frac{1}{k} v_0 \left(1 - e^{kt}\right)$ where x is the displacement at time t. Show that the acceleration at time t is given by $a = -kv$ where v is the velocity at time t.

10. A particle moves along a straight line. Its displacement x at time t follows the law $x = x_0\,(2e^{-kt} - e^{-2kt})$ where x_0 is the initial displacement and k is a constant and e is the natural logarithmic base. Find the maximum velocity reached by the particle and the time at which it is reached.

Ans. $\left[v_{\max} = -\frac{kx_0}{2}, t = \frac{\ln 2}{k}\right]$

11. A particle moving in a straight line is subjected to a resistance which produced a retardation of kv^3, where v is the velocity and k is a constant. Show that v and the time t are given in terms of S (the distance) by the equations.

$v = \dfrac{u}{1 + kSu}$ and $t = \dfrac{1}{2}kS^2 + \dfrac{S}{u}$, where u is the initial velocity.

12. The acceleration of a body starting from rest moving along a straight line follows the law $a = \frac{t}{30} + \frac{2}{3}$ where a is in m/s^2 and t is in seconds. Obtain expressions for velocity and displacement. Find also the velocity and displacement at $t = 10$sec.

13. A particle moves along a straight line according to the law $x = t^3 - 3t^2 - 9t + 10$ where x is the distance of the particle at t seconds, from a fixed point O on the straight line, in metres. Find (*i*) the time at which the velocity of the particle is zero, (*ii*) the position and the distance travelled by the particle at the time, (*iii*) the acceleration of the particle at that time, (*iv*) the distance travelled by the particle from $t = 2$s to $t = 4$s.

14. A stone is dropped from the top of a tower 50 m high. At the same time another stone is thrown up from the foot of the tower with a velocity of 25 m/s. At what distance from the top and after how much time the two stones cross each other? **Ans.** [19.6 m from the top; 2 second]

15. The velocity time diagram for the rectilinear motion of a particle is represented by a parabola as shown in the figure. Find the distance travelled by the particle in the time $T/2$.

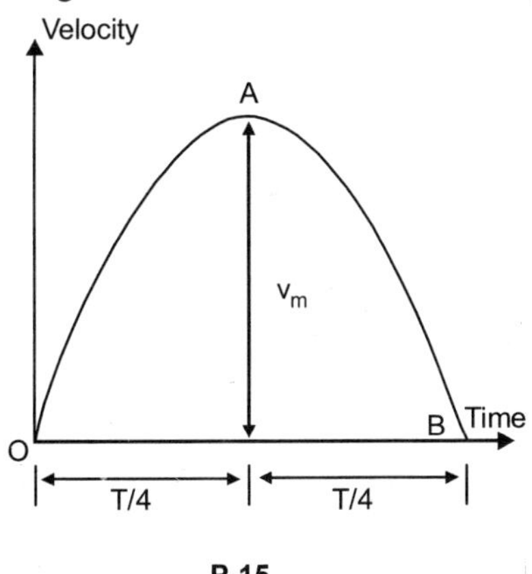

P-15

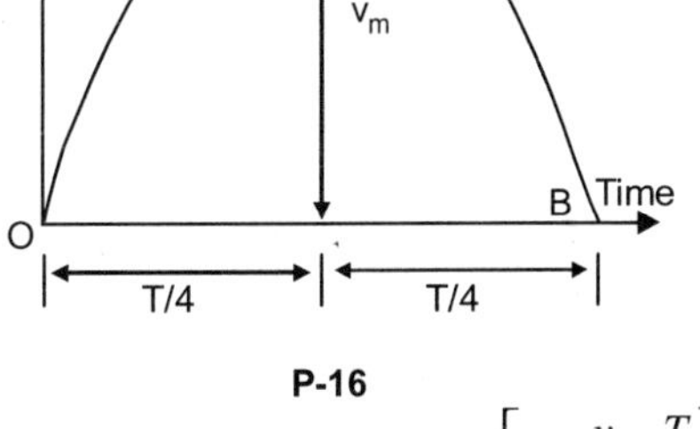

P-16

Ans. $\left[x = \frac{v_{\max} T}{3}\right]$

16. The velocity-time diagram for the rectilinear motion of a particle is represented by a portion of sine curve as shown in figure. Find the distance travelled by the particle is the time $T/2$. **Ans.** $\left[x = \frac{Tv_m}{\pi}\right]$

17. The position of a particle moving along a straight line is defined by the relation $x = t^3 - 9t^2 + 15t + 18$ where x is expressed in metres and t in seconds. Determine the time, position and acceleration of the particle when its velocity becomes zero. **Ans.** [$t = 1, x = 25\text{cm}, a = -12\text{m/s}^2,\ t = 5, x = -7\text{m}, a = 12\text{m/s}^2$]

18. The acceleration of a particle along a straight line is given by the equation, $a = 4 - t^2/9$. If the particle starts with zero velocity from a position $x = 0$ find (*a*) its velocity after 6 seconds, (*b*) distance travelled in 6 seconds. **Ans.** [16m/s; 60m]

CHAPTER
9

PLANE CURVILINEAR MOTION OF PARTICLE

9.1 INTRODUCTION

In this chapter we are going to analyse the *kinematics of a particle* moving along a *curved path* lying in a *single plane*.

Let us consider the motion of a particle with reference to x–y rectangular cartesian co-ordinate system. At any instant t, the position of the particle is at P, which is defined by the position vector $\overrightarrow{OP}$ or r, in terms of magnitude and direction. At time instant $(t + \Delta t)$, the particle moving along the path, shown in dotted line, reaches at P'. The new position of the particle is defined by position vector $\overrightarrow{OP'}$ or r'. Now joining P and P' and applying vector law of addition, we obtain

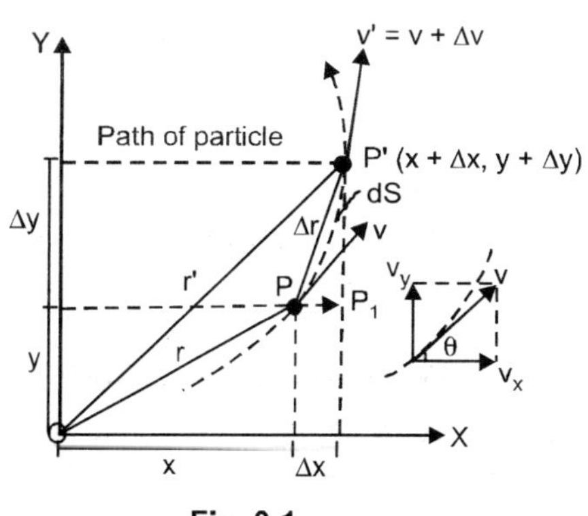

Fig. 9.1

$$r + \Delta r = r'$$

$$\Rightarrow \qquad \Delta r = r' - r \qquad \text{... (9.1)}$$

So, Δr represents the change in position vector, both in terms of magnitude as well as directions.

9.2 VELOCITY AND ACCELERATION

When the particle is at P, the tangent drawn on path of the particle at point P, we give the direction of velocity v at P. Similarly, the tangent drawn at P' will be the direction of velocity v'at P'.

Now average velocity of the particle in-between P and P' will be

$$v_{av} = \frac{\Delta r}{\Delta t} \qquad \text{...(9.2)}$$

This v_{av} is a vector whose direction is that of Δr and magnitude will be $\left|\frac{\Delta r}{\Delta t}\right|$. The instantaneous velocity will be the limiting value of the average velocity.

$$v = \lim_{t \to 0} \frac{\Delta r}{\Delta t} = \frac{dr}{dt} = \dot{r} \qquad \text{... (9.3)}$$

Now, magnitude of instantaneous velocity vector will be

$$|v| = \frac{dS}{dt}, \text{ while } dS = \text{length of line segment } PP' \qquad \text{...(9.4)}$$

The relationship between v and v' is as following

$$v' = v + \Delta v \qquad \text{... (9.5)}$$

The average acceleration of the particle in-between P and P' is

$$a_{av} = \frac{\Delta v}{\Delta t} \qquad \text{... (9.6)}$$

In the similar fashion, instantaneous acceleration will be the limiting value of average acceleration.

$$a = \lim_{t \to 0} \frac{\Delta v}{\Delta t} = \frac{dv}{dt} = \dot{v} \qquad \text{...(9.7)}$$

Generally, the acceleration of a particle in a curvilinear motion directs neither tangent nor normal to the path described.

9.3 RECTANGULAR COMPONENTS OF VELOCITY AND ACCELERATION

In the curvilinear motion of a particle, velocity and acceleration vector at any point of path at specific instant of time, can be resolved into two components, along the co-ordinate axes, x and y.

Referring Figure 9.1,

$$v = \lim_{\Delta t \to 0} \frac{\Delta r}{\Delta t} = \lim_{\Delta t \to 0} \frac{\left(\overrightarrow{PP_1} + \overrightarrow{P_1P'}\right)}{\Delta t}$$

$$= \lim_{\Delta t \to 0} \frac{\overrightarrow{PP_1}}{\Delta t} + \lim_{\Delta t \to 0} \frac{\overrightarrow{P_1P'}}{\Delta t}$$

$$= \lim_{\Delta t \to 0} \frac{\Delta x}{\Delta t} + \lim_{\Delta t \to 0} \frac{\Delta y}{\Delta t}$$

$$= \frac{dx}{dt} + \frac{dy}{dt}$$

$$= v_x + v_y \qquad \text{... (9.8)}$$

= velocity component along x axis + velocity component along y axis

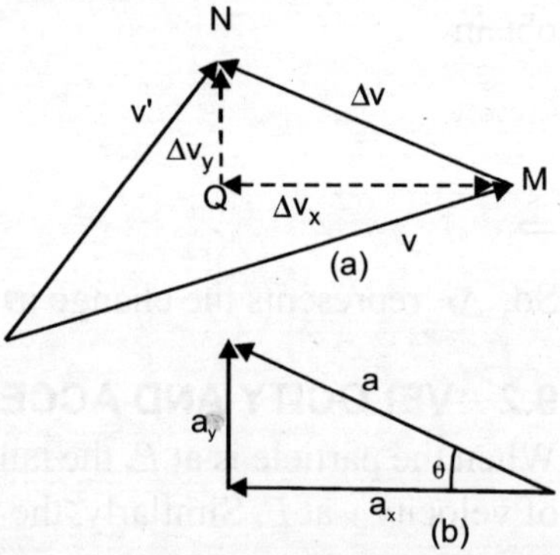

N
v'
Δv
Δv_y
Q
Δv_x
M
v
(a)
a
a_y
θ
a_x
(b)

Fig. 9.2

and $$|v| = \sqrt{v_x^{\,2} + v_y^{\,2}}, \quad \tan\theta = \left|\frac{v_y}{v_x}\right| \qquad \text{... (9.9)}$$

In case of acceleration, as shown in Figure 9.2,

$$a = \lim_{\Delta t \to 0} \frac{\Delta v}{\Delta t} = \lim_{\Delta t \to 0} \frac{\left(\overrightarrow{MQ} + \overrightarrow{QN}\right)}{\Delta t}$$

$$= \lim_{\Delta t \to 0} \frac{\overrightarrow{MQ}}{\Delta t} + \lim_{\Delta t \to 0} \frac{\overrightarrow{QN}}{\Delta t}$$

$$= \lim_{\Delta t \to 0} \frac{\Delta v_x}{\Delta t} + \lim_{\Delta t \to 0} \frac{\Delta v_y}{\Delta t}$$

$$= \frac{dv_x}{dt} + \frac{dv_y}{dt}$$

$$= a_x + a_y \qquad \text{....(9.10)}$$

= acceleration component along x axis

+ acceleration component along y axis

and $$|a| = \sqrt{a_x^{\,2} + a_y^{\,2}}, \tan\theta = \left|\frac{a_y}{a_x}\right| \qquad \text{... (9.11)}$$

9.4 VECTORIAL REPRESENTATION

To build up a concept of representation of kinematic parameters in terms of unit vector, a perpendicular co-ordinate system is assumed in Figure 9.3 where unit vectors i and j are represented along x and y axis respectively. With the help of unit vectors, r, v and a can be written as:

$$\left.\begin{aligned} r &= xi + yj \\ v &= \dot{r} = \dot{x}i + \dot{y}j \\ a &= \dot{v} = \ddot{r} = \ddot{x}i + \ddot{y}j \end{aligned}\right\} \qquad \text{... (9.12)}$$

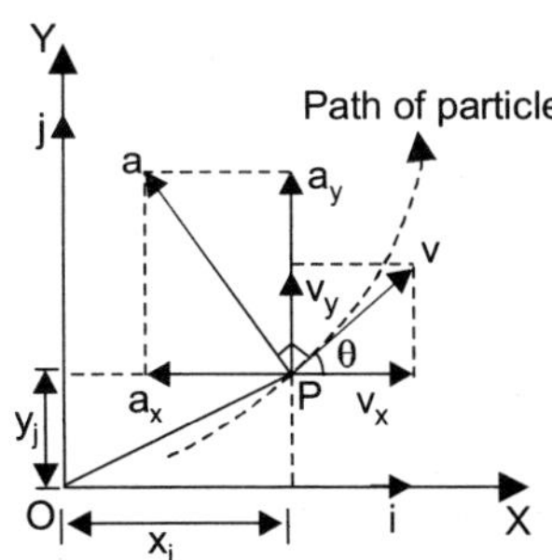

Fig. 9.3

9.5 PROJECTILE MOTION

A most practical application of two-dimensional kinematics in plane curvilinear motion is found in the analysis of trajectory or projectile motion. In doing so, following assumptions should be adhered :

(1) ***All kinds aerodynamic drags are neglected.***

(2) *The effects of curvature and rotation of the earth are neglected.*

(3) *The value of acceleration due to gravity is taken to be constant and does not get affected by the change in altitude.*

Here all the kinematic parameters are analysed with reference to rectangular co-ordinate system. At zero-th time instant i.e., at the start of the trajectory, referring to Figure 9.4, the velocity of projection is v_0, which makes an angle of projection θ with horizontal x-axis. Rectangular components of v_0 are

$$(v_x)_0 = v_0 \cos\theta, \text{ and } (v_y)_0 = v_0 \sin\theta \quad ...(9.13)$$

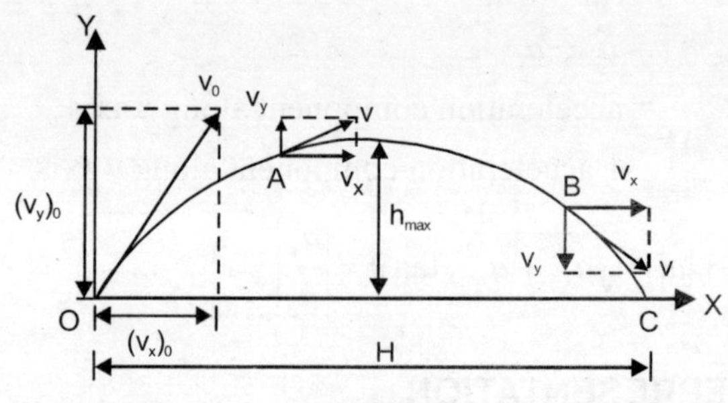

Fig. 9.4

Here horizontal component of velocity of projection remain constant. Throughout path of the trajectory $OABC$, the accelerations that come into play are

$$a_x = 0, \text{ and } a_y = -g \quad ...(9.14)$$

Here the equation of a trajectory in a cartesian co-ordinate system can be written as

$$y = x \tan\theta - \frac{gx^2}{2{v_0}^2 \cos^2\theta} \quad ...(9.15)$$

Here, (x, y) is the location of any point on the trajectory.

Now at any instant of time t, following equations of motion can be obtained :

$$\left.\begin{aligned} v_x &= (v_x)_0 = \text{constant} = v_0 \cdot \cos\theta \\ v_y &= (v_y)_0 - gt = v_0 \sin\theta - gt \\ x &= (v_x)_0 \cdot t = (v_0 \cdot \cos\theta)t \\ y &= (v_y)_0 \cdot t - \frac{1}{2}gt^2 = (v_0 \sin\theta)t - \frac{1}{2}gt^2 \\ {v_y}^2 &= (v_y)_0^2 - 2gy = {v_0}^2 \sin^2\theta - 2gy \end{aligned}\right\} \quad ....(9.16)$$

With the assistance of above equations, following parameters of a trajectory can be obtained as following :

(*i*) Maximum height attained, $h_{max} = \dfrac{v_0^2 \sin^2\theta}{2g}$ ...(9.17*a*)

(*ii*) Total time of flight, $T = \dfrac{2v_0 \sin\theta}{g}$ ...(9.17*b*)

(*iii*) Time to reach highest point, $T' = \dfrac{v_0 \sin\theta}{g} = \dfrac{T}{2}$ ...(9.17*c*)

(*iv*) Horizontal range, $H = \dfrac{v_0^2 \sin 2\theta}{g}$ ...(9.17*d*)

(*v*) Angle of projection for maximum horizontal range = 45° and $H_{max} = \dfrac{v_0^2}{g}$ ...(9.17*e*)

9.6 NORMAL AND TANGENTIAL COMPONENTS OF VELOCITY AND ACCELERATION

Another approach to express velocity and acceleration in curvilinear motion is to describe those kinematic parameters with reference to normal and tangential co-ordinate system. This *n–t* axes system as shown in Figure 9.5, sometimes found to provide very natural description of velocity and acceleration.

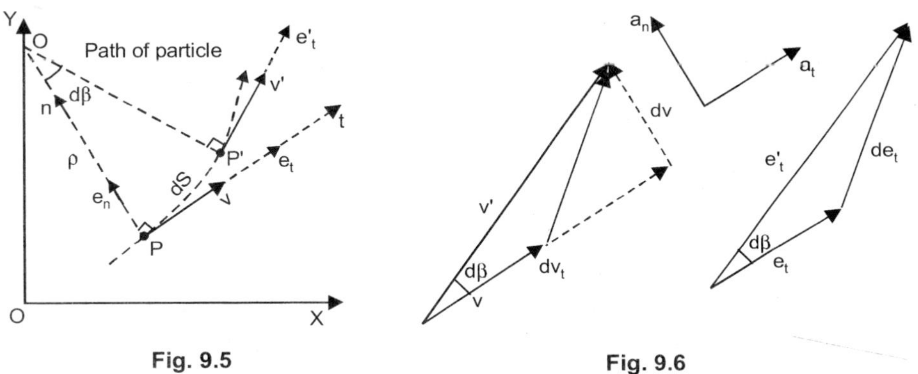

Fig. 9.5 **Fig. 9.6**

A separate idea of unit vector is brought into picture here. Unit vectors are assumed along tangential direction as e_t and along normal direction as e_n. The particle changes its position from *P* to *P′* and thereby makes a displacement *dS*, subtending an angle *d*β while ρ be radius of curvature of said path *PP′*.

So, $$dS = \rho \times d\beta \qquad ...(9.18)$$

∴ $$v = \frac{dS}{dt} = \rho \cdot \frac{d\beta}{dt} = \rho \cdot \dot{\beta}$$

∴ $$V = v \cdot e_t = \rho\dot{\beta}e_t \qquad ...(9.19)$$

When velocity vector is defined, acceleration a can also be enumerated. With the help of vector triangles in Figure 9.6, we can say

$$a = \frac{dV}{dt} = \frac{d}{dt}(ve_t) = \dot{v}e_t + v\dot{e}_t = \left(\frac{d^2S}{dt^2}\right)e_t + \frac{dS}{dt}\cdot\frac{de_t}{dt} \qquad ...(9.20)$$

Where the unit vector e_t has a nonzero derivative due to its change in direction.

$$\therefore \quad \frac{de_t}{dt} = \frac{de_t}{dS}\cdot\frac{dS}{dt} = v\cdot\frac{de_t}{dS} \qquad ...(9.21)$$

Here, if dS is sufficiently small, unit vectors e_t and e_t' are assumed to intersect and thus to form a plane, which is designated as *osculating plane.*

Here, $\quad de_t = |e_t|\,d\beta = d\beta = \frac{dS}{\rho}.$

Using e_n, again it can be written, $de_t = \frac{dS}{\rho}\cdot e_n$

$$\therefore \quad \frac{de_t}{dS} = \frac{e_n}{\rho}$$

$$\therefore \quad a = \left(\frac{d^2S}{dt^2}\right)e_t + \frac{dS}{dt}\cdot\frac{de_t}{dt} = \dot{v}e_t + v\cdot\left(v\frac{de_t}{dS}\right)$$

$$= \dot{v}e_t + v^2\frac{e_n}{\rho} = \dot{v}e_t + \frac{v^2}{\rho}e_n = a_t + a_n$$

Where, $a_t = \dot{v}e_t$ and $a_n = \frac{v^2}{\rho}e_n$

$$\therefore \quad |a_t| = \dot{v} = \frac{d^2S}{dt^2} \text{ and } |a_n| = \frac{v^2}{\rho}$$

9.7 RADIAL AND TRANSVERSE COMPONENTS OF VELOCITY AND ACCELERATION

Sometimes it is found in certain practical problems that representing kinetic parameters either with reference to rectangular system or *tangent – normal* system are not well convenient to work with. In those cases we have to opt for *radial – transverse* reference system. Sometimes it is called as *polar co-ordinate* system or $r - \theta$ system, as shown in Figure 9.7.

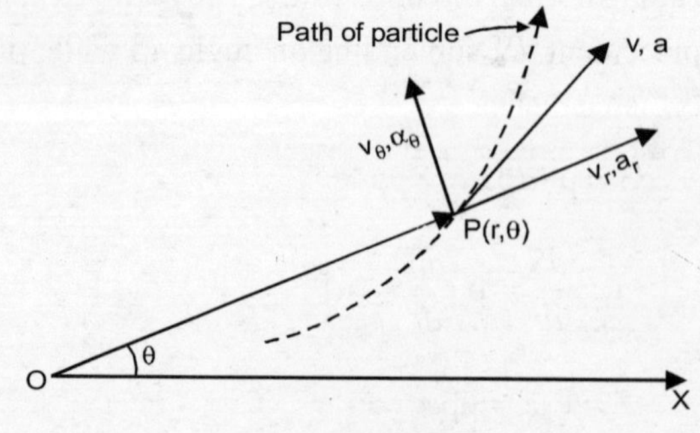

Fig. 9.7

Let us assume the position of the particle P which is r distance away from O and makes a positive angle θ with respect to x-axis. So the position at P is defined as (r, θ).

Considering the exact purview of the topic which will be truely effective to the student community, it has been decided here not to go through the tortuous analytical exercise, but to provide the ultimate results in brief.

So, $$v = v_r + v_\theta = \dot{r} + r\dot{\theta} \quad \text{and}$$

$$a = a_r + a_\theta = \left[\ddot{r} - r\left(\dot{\theta}\right)^2\right] + \left[r\ddot{\theta} + 2\dot{r}\dot{\theta}\right]$$

Another interdependent relationship of a and V are as following :

$$a_r = \dot{v}_r - v_\theta\dot{\theta}, \quad a_\theta = \dot{v}_\theta + v_r\dot{\theta}$$

Multiple Choice Questions

Select the best alternative (s):

1. The velocity of a body on reaching the ground from a height h is given by

   (*a*) $v = 2gh$ (*b*) $v = 2gh^2$ (*c*) $v = \sqrt{(2gh)}$

   (*d*) $v = 1/\sqrt{(2gh)}$ (*e*) $v = h^2/\sqrt{(2g)}$

2. Cartesian equation of a trajectory is

   (*a*) $y = x \sin\alpha - gx^2/(2u^2 \sin^2\alpha)$ (*b*) $y = x\tan\alpha - gx^2/(2u^2\tan^2\alpha)$
   (*c*) $y = x\tan\alpha - gx^2/(2u^2\cos^2\alpha)$ (*d*) $y = x\tan\alpha + gx^2/(2u^2\cos^2\alpha)$
   (*e*) $y = x\tan\alpha + gx^2/(2u^2\sin^2\alpha)$

3. A projectile is fired at an angle θ to the vertical, its horizontal range will be maximum when θ is

   (*a*) 0° (*b*) 30° (*c*) 45°
   (*d*) 60° (*e*) 90°

4. If the velocity of projection is um/sec and the angle of projection is α, the maximum height of the projection on a horizontal plane is

   (*a*) $u^2\cos^2\alpha\,(2g)$ (*b*) $u^2\sin^2\alpha/(2g)$ (*c*) $u^2\tan^2\alpha/2g$ (*d*) $u\sin\alpha/g$

5. A flywheel on a motor goes from rest to 1000 r.p.m in 6 sec. The number of revolutions made is nearly equal to

   (*a*) 25 (*b*) 50 (*c*) 100
   (*d*) 250 (*e*) 500

6. If a particle moves along the circumference of a circle of radius r with a uniform angular velocity ω radians/sec. The equation for the velocity of the particle is given by

   (*a*) $v = \omega\sqrt{(y^2 - r^2)}$ (*b*) $v = \omega\sqrt{(y - r)}$ (*c*) $v = \omega\sqrt{(r^2 - y^2)}$

   (*d*) $v = \omega\sqrt{(r^2 + y^2)}$ (*e*) none of the above

Answers

1. (*c*) 2. (*c*) 3. (*c*) 4. (*b*) 5. (*a*) 6. (*d*)

Numerical Examples

Example 1

A particle is projected with a velocity u. If the horizontal range is R show that there are two possible directions of projection to attain this range R.

Solution Let α be the angle of projection.

So, range $$R = \frac{u^2}{g}\sin 2\alpha$$

$$\Rightarrow \quad \sin 2\alpha = \frac{gR}{u^2}$$

Let α_1 and α_2 be the two possible solutions to the above equation. So, α_1 and α_2 are the two possible directions of projection.

So, $$2\alpha_1 = 180° - 2\alpha_2$$

$$\Rightarrow \quad \alpha_1 + \alpha_2 = 90°$$

In general if $\alpha_1 = 45° + \theta$ then $\alpha_2 = 45° - \theta$

But the direction of projection for maximum range is 45°.

Hence the two possible directions of projections are equally inclined to the direction of projection for maximum range.

Example 2

A particle is projected into the air at a given direction with a given velocity. Show that the path of the particle is parabola, neglecting the resistance of the air.

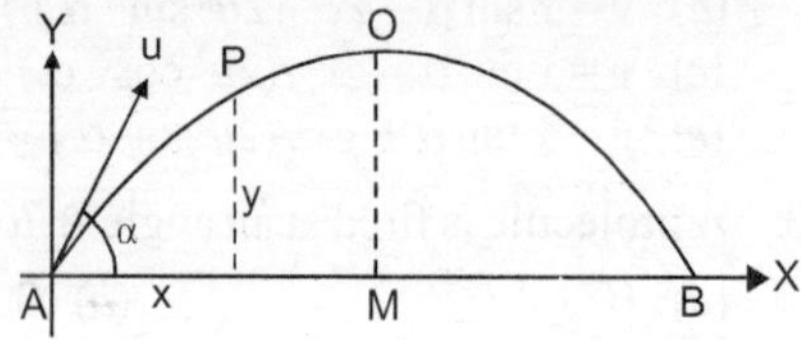

Solution Let A be point of projection. Let the particle be projected at a velocity u and let the angle of projection be α. Let after a time t the particle reaches P and attain the co-ordinates (x, y). So, vertical distance described in time t,

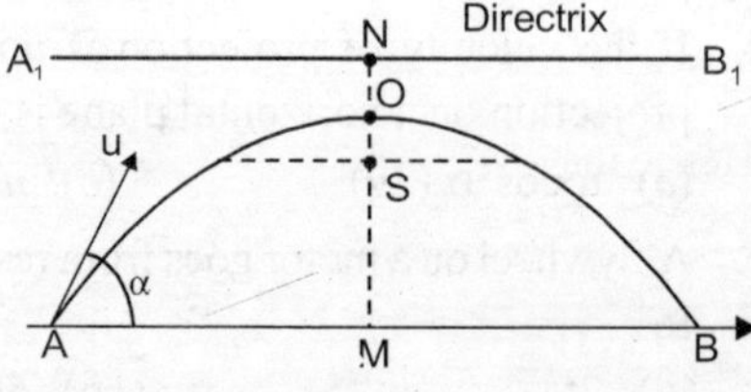

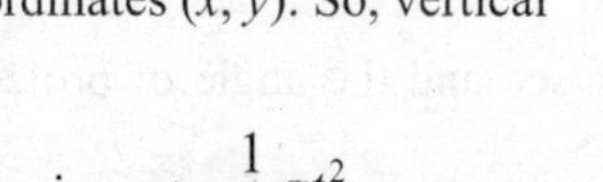

$$y = u\sin\alpha \cdot t - \frac{1}{2}gt^2$$

and horizontal distance described in time t, $x = u\cos\alpha \cdot t$

So, $$y = u\sin\alpha \cdot \frac{x}{u\cos\alpha} - \frac{g}{2}\frac{x^2}{\cos^2\alpha} = x\tan\alpha - \frac{gx^2}{2u^2}\sec^2\alpha$$

The above is the equation of a parabola. Let O be the highest point of the path. Let M be the middle point of the range AB.

So, $$OM = \frac{u^2\sin^2\alpha}{2g} \quad \text{and} \quad AM = \frac{u^2}{2g}\sin 2\alpha$$

Hence, co-ordinate of O with A as origin are $\left(\frac{u^2\sin 2\alpha}{2g}, \frac{u^2}{2g}\sin^2\alpha\right)$

Changing the origin to O, let (X, Y) be the co-ordinates of any point in the path.

So, $$x = X + \frac{u^2}{2g}\sin 2\alpha, \qquad y = Y + \frac{u^2}{2g}\sin^2\alpha$$

Substituting the above values of x and y, we have,

$$Y + \frac{u^2}{2g}\sin^2\alpha = \left(X + \frac{u^2}{2g}\sin 2\alpha\right)\tan\alpha - \frac{g}{2u^2}\left(X + \frac{u^2}{2g}\sin 2\alpha\right)^2 \sec\alpha$$

$$= X\tan\alpha + \frac{u^2}{g}\sin^2\alpha - \frac{g}{2u^2}\frac{X^2}{\cos^2\alpha} - X\tan\alpha - \frac{u^2}{2g}\sin^2\alpha$$

$$\Rightarrow \qquad X^2 = -\frac{2u^2\cos^2\alpha}{g}Y = -4aY$$

The above is the equation of a parabola whose *latus rectum*

$$4a = \frac{2u^2\cos^2\alpha}{g}.$$

Let S be the *focus* of the parabola

$$OS = a\frac{u^2\cos^2\alpha}{2g}, \quad OM = \frac{u^2\sin^2\alpha}{2g},$$

So, $$MS = \frac{u^2\sin^2\alpha}{2g} - \frac{u^2\cos^2\alpha}{2g} = -\frac{u^2}{2g}\cos 2\alpha$$

Let A_1B_1 be the *directrix*. Height of the directrix above the point of projection

$$= MN = MO + ON = MO + OS = \frac{u^2\sin^2\alpha}{2g} + \frac{u^2\cos^2\alpha}{2g} = \frac{u^2}{2g}$$

Hence the directrix of the parabola is a horizontal line at a height of $\frac{u^2}{2g}$ above the point of projection.

EXAMPLE 3

A particle is projected at an angle of elevation α and after t seconds it appears to have angle of elevation β as seen from the point of projection. Prove that the initial velocity is $\frac{gt\cos\beta}{2\sin(\alpha - \beta)}$.

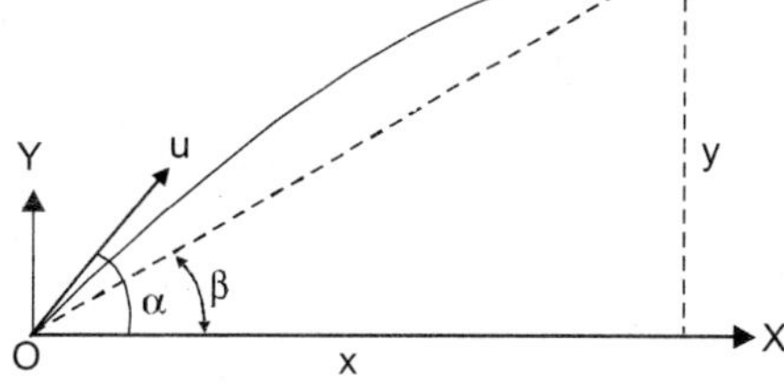

SOLUTION Let co-ordinates of the particle after t seconds be (x, y) referred to the point of projection as origin. Let u be the velocity of the projection. So,

$$x = (u \cos \alpha)t, \text{ and } y = (u \sin \alpha)t - \frac{1}{2}gt^2$$

$$\therefore \quad \tan\beta = \frac{y}{x} = \frac{(u\sin\alpha)t - \frac{1}{2}gt^2}{(u\cos\alpha)t} = \tan\alpha - \frac{gt}{2u\cos\alpha}$$

$$\Rightarrow \quad u = \frac{gt}{2\cos\alpha(\tan\alpha - \tan\beta)} = \frac{gt\cos\beta}{2\sin(\alpha-\beta)}$$

EXAMPLE 4

A particle is projected from a point O at an angle of elevation α. If after time t the particle is at P. Prove that $\tan\beta = \frac{\tan\alpha + \tan\theta}{2}$ where β and θ are the inclinations to the horizontal of OP and the direction of the motion of the particle when at P.

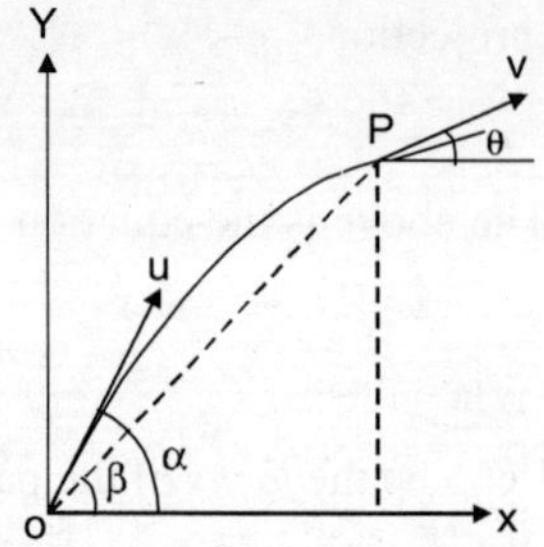

SOLUTION Let the co-ordinates of the projectile be (x, y) after time t. Let the velocity of projection be u and let the velocity after a time t be v.

So, $$x = u\cos\alpha \cdot t \text{ and } y = u\sin\alpha \cdot t - \frac{1}{2}gt^2$$

Hence $$\tan\beta = \frac{y}{x} = \frac{u\sin\alpha \cdot t - \frac{1}{2}gt^2}{u\cos\alpha \cdot t} = \tan\alpha - \frac{gt}{2u}\sec\alpha$$

Again, $v \cos\theta = u \cos\alpha$ and $v \sin\theta = u \sin\alpha - gt$

$$\therefore \quad \tan\theta = \tan\alpha - \frac{gt}{u}\sec\alpha$$

Solving the above two equations, we have,

$$2\tan\beta - \tan\theta = 2\tan\alpha - \frac{gt}{u}\sec\alpha - \tan\alpha + \frac{gt}{u}\sec\alpha$$

$$\Rightarrow \quad \tan\beta = \frac{\tan\alpha + \tan\theta}{2}$$

EXAMPLE 5

A gun fires a projectile with a muzzle velocity of 250m/s. Find the angle of projection so that it strikes a target at a horizontal distance of 4500 m from the gun and at a height of 300 m above it.

SOLUTION Here, $$y = x\tan\alpha - \frac{gx^2}{2u^2}\sec^2\alpha$$

$$\Rightarrow \quad 300 = 4500\tan\alpha - \frac{9.81 \times 4500^2}{2 \times 250^2}(1 + \tan^2\alpha)$$

$\Rightarrow$ $$\tan^2\alpha - 2.83158 \tan\alpha + 1.18877 = 0$$

$\Rightarrow$ $$\tan\alpha = 2.318945 \text{ and } 0.512635$$

$\Rightarrow$ $$\alpha = 66.673° \text{ and } 27.141°$$

EXAMPLE 6

A particle is projected so as just to clear two walls. The first one is of height a and at a distance b from the point of projection. The second one is of height b and at a distance a from the point of projection. Show that the angle of projection is $\tan^{-1}\left(\dfrac{a^2+ab+b^2}{ab}\right)$ and that the horizontal range is $\left(\dfrac{a^2+ab+b^2}{a+b}\right)$.

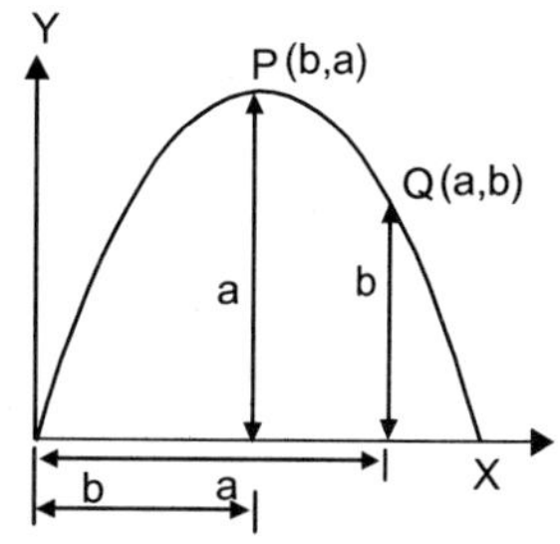

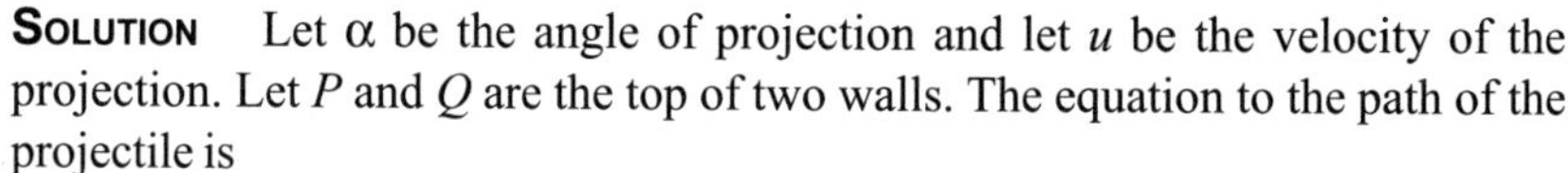

SOLUTION Let α be the angle of projection and let u be the velocity of the projection. Let P and Q are the top of two walls. The equation to the path of the projectile is

$$y = x\tan\alpha - \frac{gx^2}{2u^2}\sec^2\alpha$$

When the particle is at P, $$a = b\tan\alpha - \frac{gb^2}{2u^2}\sec^2\alpha \quad \text{...(1)}$$

When the particle is at Q, $$b = a\tan\alpha - \frac{ga^2}{2u^2}\sec^2\alpha \quad \text{...(2)}$$

Solving above two equations (1) and (2) we get,

$$a^3 - b^3 = ab\tan\alpha \cdot (a-b)$$

$\therefore$ $$\tan\alpha = \frac{a^3-b^3}{ab\times(a-b)} = \frac{(a-b)(a^2+ab+b^2)}{ab\times(a-b)} = \frac{(a^2+ab+b^2)}{ab}$$

$\therefore$ $$\alpha = \tan^{-1}\left(\frac{a^2+ab+b^2}{ab}\right)$$

And from equation (1)

$$\frac{gb^2}{2u^2\cos^2\alpha} = b\tan\alpha - a = \frac{a^2+ab+b^2}{a} - a = \frac{b(a+b)}{a}$$

$\therefore$ $$u^2\cos^2\alpha = \frac{ab}{2(a+b)}g$$

$\Rightarrow$ $$u\cos\alpha = \sqrt{\frac{ab}{2(a+b)}g}$$

$\Rightarrow$ $$u\sin\alpha = u\cos\alpha\cdot\tan\alpha = \sqrt{\frac{ab}{2(a+b)}g}\times\left(\frac{a^2+ab+b^2}{ab}\right)$$

So, horizontal range,

$$R = \frac{2u^2}{g}\sin\alpha\cdot\cos\alpha$$

$$= \sqrt{\frac{ab}{2(a+b)}g} \times \sqrt{\frac{ab}{2(a+b)}g} \times \left(\frac{a^2+ab+b^2}{ab}\right) = \frac{a^2+ab+b^2}{a+b}$$

Example 7

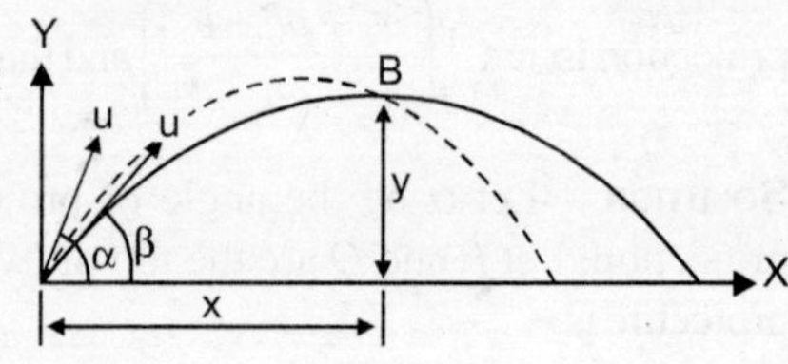

Two particles are projected in the same vertical plane, from the same point with the same velocity. If T and T' be the times to reach the respective maximum heights and t, t' be the time taken to reach the other common points of their paths, show that $Tt + T't'$ is the independent of directions of projection.

Solution Let the velocity of projection of each particle be u. Let the angles of projection be α and β.

So, $$T = \frac{u\sin\alpha}{g} \text{ and } T' = \frac{u\sin\beta}{g}.$$

Also, $$x = u\cos\alpha\cdot t = u\cos\beta\cdot t'$$

$\Rightarrow$ $$t = \frac{x}{u\cos\alpha} \text{ and } t' = \frac{x}{u\cos\beta}$$

So, $$Tt + T't' = \frac{u\sin\alpha}{g}\cdot\frac{x}{u\cos\alpha} + \frac{u\sin\beta}{g}\cdot\frac{x}{u\cos\beta} = (\tan\alpha + \tan\beta)\frac{x}{g}$$

But, $$y = x\tan\alpha - \frac{gx^2}{2u^2}\sec^2\alpha = x\tan\beta - \frac{gx^2}{2u^2}\sec^2\beta$$

$\Rightarrow$ $$x(\tan\alpha - \tan\beta) = \frac{gx^2}{2u^2}(\sec^2\alpha - \sec^2\beta) = \frac{gx^2}{2u^2}(\tan^2\alpha - \tan^2\beta)$$

$\Rightarrow$ $$x(\tan\alpha + \tan\beta) = \frac{2u^2}{g}$$

Hence, $$Tt + T't' = \frac{2u^2}{g}\cdot\frac{1}{g} = \frac{2u^2}{g^2}$$

Example 8

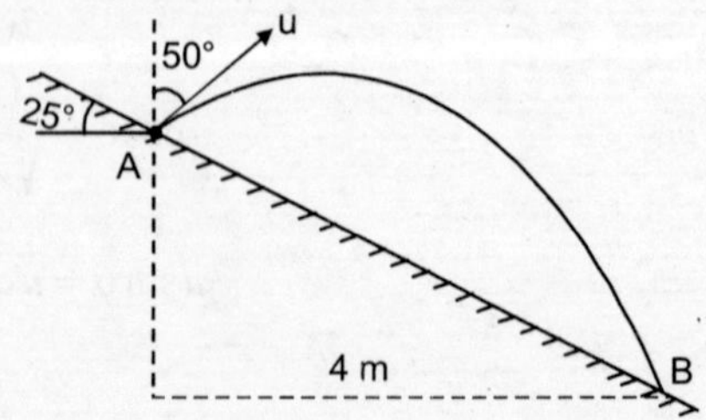

A ball is dropped vertically onto a 25° inclined plane at A as shown in figure. The direction of rebound is found to be at 50° with the vertical. After rebound, the ball strikes the plane at B at a horizontal distance of 4m from A. Find (*i*) the velocity of rebound at A, and (*ii*) the time interval for the ball to travel from A to B.

Solution Let the velocity of ball after rebound $= u$. Inclination of direction of projection with the horizontal $\alpha = 90° - 50° = 40°$. Inclination of the plane with the horizontal $\beta = 25°$.

Total time of flight

$$T = \frac{2u\sin(\alpha+\beta)}{g\cos\beta}$$

Horizontal displacement

$$u\cos\alpha \cdot T = 4$$

$$\Rightarrow \quad \frac{2u^2\sin(\alpha+\beta)}{g\cos\beta}\cos\alpha = 4$$

$$\Rightarrow \quad u^2 = \frac{4\times 9.81\cos 25^\circ}{2\sin(40^\circ+25^\circ)\cos 40^\circ}$$

$$\Rightarrow \quad u = 5.06 \text{ m/s}$$

Time interval required for travelling from A to B

$$T = \frac{2\times 5.06\times\sin 65^\circ}{9.81\times\cos 40^\circ} = 1.2204 \text{ s}$$

EXAMPLE 9

If a particle projected at an angle α with the horizon from the foot of an inclined plane whose inclination to the horizontal is β, strikes the plane at right angle show that, $\cot\beta = 2\tan(\alpha-\beta)$.

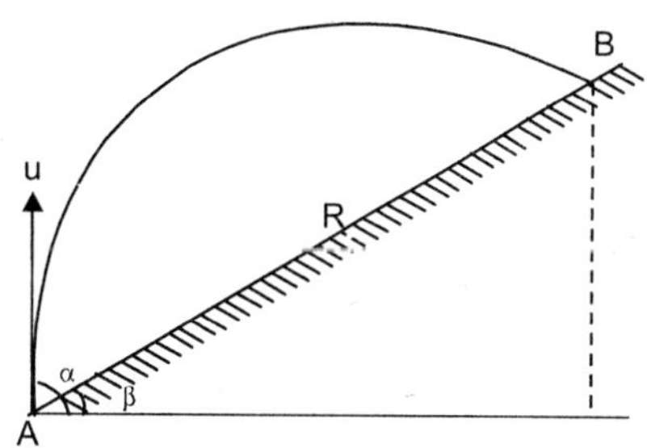

SOLUTION Let u be the velocity of projection. Initial velocity along the plane $- u\cos(\alpha-\beta)$ and that normal to the plane $= u\sin(\alpha-\beta)$. Acceleration due to gravity normal to the plane $= g\cos\beta$. Let T be the time of light to strike the plane. From the motion normal to the plane, since the net distance moved is zero, we have,

$$0 = u\sin(\alpha-\beta)\cdot T - \frac{1}{2}g\cos\beta\cdot T^2$$

$$\Rightarrow \quad T = 2u\frac{\sin(\alpha-\beta)}{g\cos\beta}$$

Now considering the motion along the plane, since the particle strikes the plane at B at right to it, the component of velocity along the plane at B is zero.

$$0 = u\cos(\alpha-\beta) - g\sin\beta\cdot T$$

$$= u\cos(\alpha-\beta) - g\sin\beta\cdot\frac{2u\sin(\alpha-\beta)}{g\cos\beta}$$

$$= u\cos(\alpha-\beta) - 2u\sin(\alpha-\beta)\tan\beta$$

$$\Rightarrow \quad 2u\sin(\alpha-\beta)\tan\beta = u\cos(\alpha-\beta)$$

$$\Rightarrow \quad 2\tan(\alpha-\beta) = \cot\beta$$

EXAMPLE 10

A motor cyclist wants to cross the ditch by jumping. Determine minimum velocity required at A. Also find direction and magnitude of velocity of the motor cyclist, at the instant of clearing the ditch.

SOLUTION Let time required to travel from A to B is t. So, horizontal displacement $ut = 5\text{m}$ and vertical displacement $\frac{1}{2}gt^2 = 3\text{m}$.

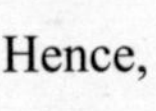

Hence, $$3 = \frac{1}{2}gt^2 = \frac{g}{2}\left(\frac{5}{u}\right)^2 = \frac{122.625}{u^2}$$

$$\Rightarrow \quad u = 6.39\text{m/s}$$

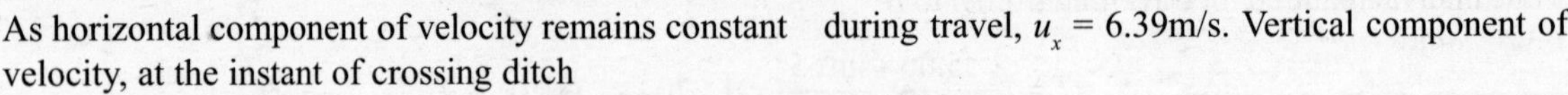

As horizontal component of velocity remains constant during travel, $u_x = 6.39\text{m/s}$. Vertical component of velocity, at the instant of crossing ditch

$$u_y = \sqrt{2gh} = \sqrt{2 \times 9.81 \times 3} = 7.67\text{m/s}$$

So, resultant velocity $$u = \sqrt{u_x^2 + u_y^2} = \sqrt{6.39^2 + 7.67^2} = 9.98 \text{ m/s}$$

And direction with horizontal $\theta = \tan^{-1}\left(\frac{7.67}{6.39}\right) = 50.2°$.

EXAMPLE 11

Compute the minimum speed at which a stuntman has to leave a 25° ramp at B, in order to clear the ditch.

SOLUTION As the movement from A to B describes a parabolic path, the related equation of motion is

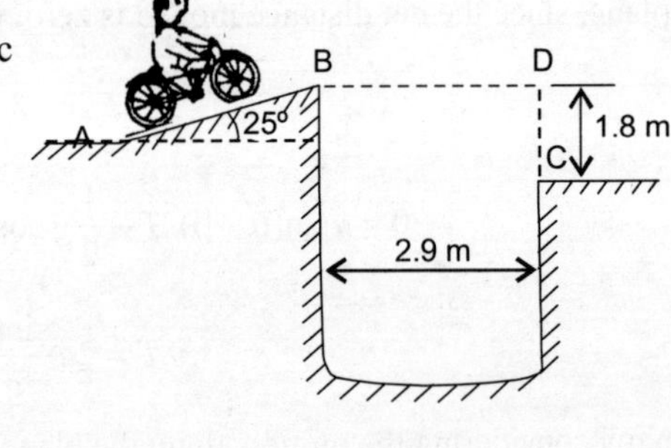

$$y = x\tan\alpha - \frac{gx^2}{2u^2\cos^2\alpha}$$

$$\Rightarrow \quad -1.8 = 2.9\tan 25° - \frac{9.81 \times 2.9^2}{2u^2\cos^2 25°}$$

$$\Rightarrow \quad u^2 = 15.9315$$

$$\Rightarrow \quad u = 3.99 \text{ m/s}$$

EXAMPLE 12

A Pakistani millitant positions himself at hill-top h m high at an elevation β. Prove that least velocity of projection to kill the millitant will be $\sqrt{gh(1+\text{cosec}\,\beta)}$.

SOLUTION Being u be the least velocity, maximum range of flight can be computed from ΔABC, $R = h\,\text{cosec}\,\beta$. Again we

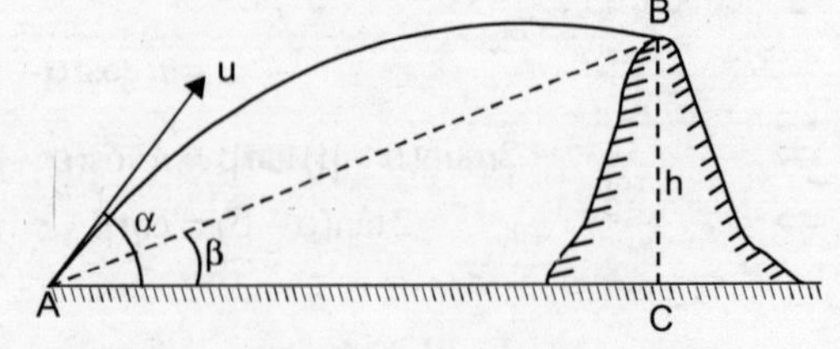

know, maximum range of flight $R = \dfrac{u^2}{g(1+\sin\beta)}$.

Hence, $h \operatorname{cosec} \beta = \dfrac{u^2}{g(1+\sin\beta)}$

$\Rightarrow$ $u^2 = gh \operatorname{cosec} \beta (1 + \sin \beta) = gh (1 + \operatorname{cosec} \beta)$

$\Rightarrow$ $u = \sqrt{gh(1+\operatorname{cosec}\beta)}$

EXAMPLE 13

A soldier fires from the top of a hill at angle 25° upward from horizontal at a velocity of 115 m/s. The enemy is 55 m below him. Find (*i*) maximum rise of bullet from horizontal, (*ii*) velocity of hitting the enemy (*iii*) time required to hit the enemy.

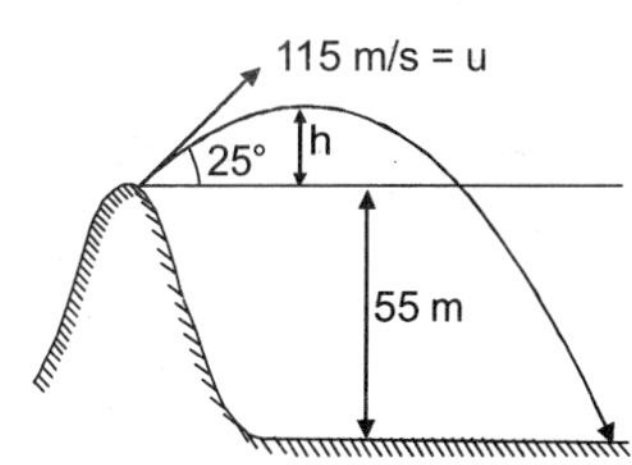

SOLUTION Initial horizontal velocity $u_x = 115 \cos 25° = 104.23$ m/s

Initial vertical velocity $u_y = 115 \sin 25° = 48.6$ m/s

After hitting, $u'_x = u_x = 104.23$ m/s

(*i*) Maximum rise of bullet

$$h = \frac{(0-48.6)^2}{-2\times 9.81} = 120.39 \text{ m}$$

(*ii*) Vertical velocity of bullet hitting enemy

$$(u'_y)^2 - (48.6)^2 = 2\times 9.81\times 55$$

$\Rightarrow$ $u'_y = \sqrt{981+48.6^2} = 58.66\text{m/s}$

So final velocity $u' = \sqrt{104.23^2 + 58.66^2} = 119.60$ m/s

(*iii*) If time required to hit enemy be t

$$t = \frac{48.6-(-58.66)}{9.81} = 10.93\text{s}$$

EXAMPLE 14

A shotput ball of mass 5kg starts sliding from A along a sloping roof having $\mu = 0.20$. Compute the distance of G where it strikes the ground.

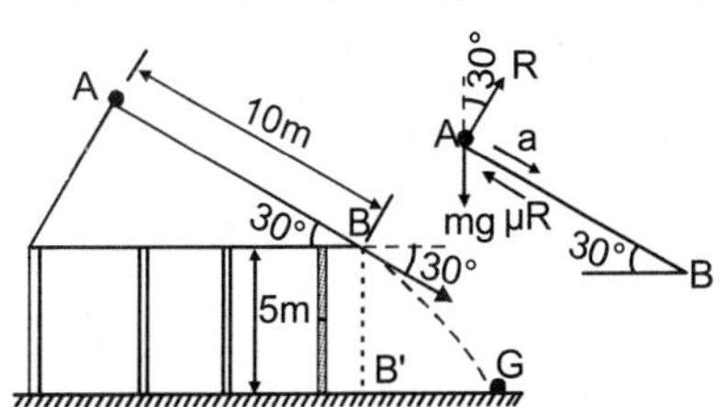

SOLUTION After sliding from A to B, the ball describes a projectile from B to G.

So, equations of motion along A to B,

$$\sum F_y = 0 \quad \Rightarrow \quad R - 5g\cos 30° = 0$$

$$\sum F_x = 0 \quad \Rightarrow \quad 5a = 5g\sin 30° - \mu R$$

$$= 5g\sin 30° - 0.2 \times 5g\cos 30°$$

$$\Rightarrow \qquad a = 3.206\text{m/s}^2$$

Therefore from A to B, the ball moves with constant acceleration a. So, the velocity of ball at B, parallel to the direction of roof slope $u = \sqrt{2 \times 3.206 \times 10 + 0^2} = 8\text{m/s}$. Now, initial velocity of ball along horizontal direction $u_x = 8 \cos 30° = 6.93$ m/s and that along vertical direction $u_y = 8 \sin 30° = 4$ m/s.

Considering vertical motion from B to B'

$$5 = u_y \cdot t + \frac{1}{2}gt^2 = 4t + 4.905t^2$$

$$\Rightarrow \qquad 4.905t^2 + 4t - 5 = 0$$

$$\Rightarrow \qquad t = 0.68\text{s}$$

So to move from B to G, the hammer will take 0.68 seconds. Considering horizontal motion, horizontal distance travelled

$$B'G = u_x \times t$$

$$= 6.93 \times 0.68 = 4.71\text{m}$$

EXAMPLE 15

A morter fired from a canon explode into two parts at a height 20 m, while horizontal distance is 1000 m. After exploding, first fragment hits the ground just below point of explosion while second fragment hits ground at C. Calculate (*i*) velocity of mortar just before explosion, (*ii*) velocity of first fragment just after explosion, (*iii*) velocity of second fragment just after explosion, (*iv*) the distance between point of firing to the point where second fragment touches ground.

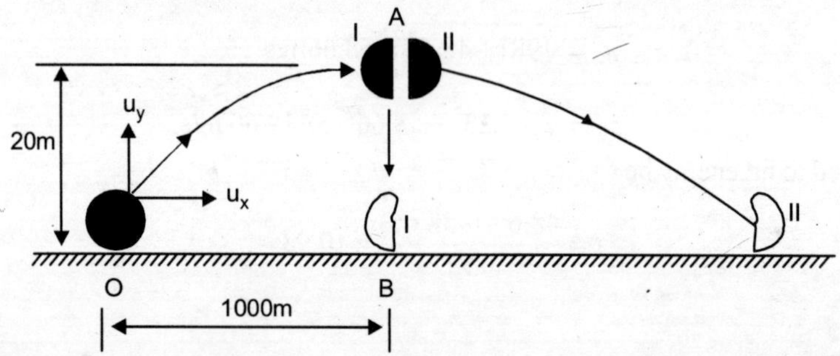

SOLUTION Considering motion from O to A, entire mortar attains maximum velocity at A. So at this point vertical velocity is zero.

Now, $$u_y^2 = 2gh + 0^2$$

$$\Rightarrow \qquad u_y = \sqrt{2 \times 9.81 \times 20} = 19.81 \text{ m/s}$$

Time taken by entire mortar to reach A, $$t = \frac{19.81 - 0}{9.81} = 2.02\text{s}$$

Again, horizontal velocity at time of firing $u_x = \dfrac{1000}{t} = \dfrac{1000}{2.02} = 495.05 \text{ m/s}$

So, at instant of explosion, vertical velocity is zero and horizontal velocity 495.05m/s.

Now applying conservation of momentum, momentum of entire mortar will be the algebraic summation of momenta of first and second fragments.

So, $$\text{m} \times 495.05 = 0.5\,\text{m} \times 0 + 0.5\,\text{m} \times u_{x_2}$$

$\Rightarrow$ $$u_{x_2} = 990.1 \text{m/s}$$

Distance travelled by second fragment from B to C

$$= u_{x_2} \times t = 990.1 \times 2.02 = 2000 \text{ m} = BC$$

So the distance from point of firing to point C will be $= 1000 + 2000 = 3000$ m.

EXAMPLE 16

On an inclined plane, a man at A can fire both upward and downward at a velocity 220 m/s. Determine the range BC.

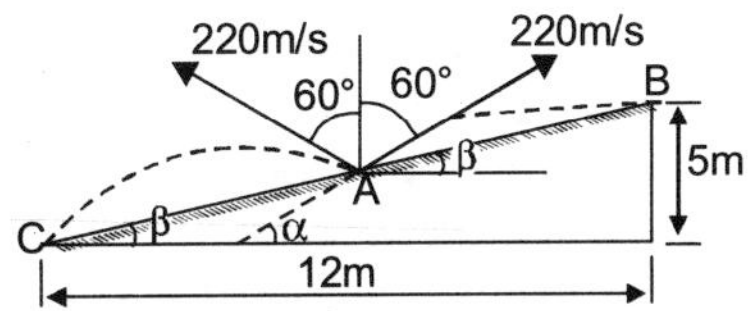

SOLUTION Inclination of plane $\beta = \tan^{-1}\left(\dfrac{5}{12}\right) = 22.62°$

Angle of projectile for upward firing, $\alpha = 90° - 60° = 30°$

For upward firing, range $AB = \dfrac{2u^2}{g\cos^2\beta}\sin(\alpha - \beta)\cos\alpha$

$$= \frac{2 \times 220^2}{9.81 \times \cos^2 22.62°}\sin(30° - 22.62°)\cos 30° = 1288.23 \text{ m}$$

For downward firing, range AC

$$= \frac{2u^2}{g\cos^2\beta}\sin(\alpha + \beta)\cos\alpha$$

$$= \frac{2 \times 220^2}{9.81 \times \cos^2 22.62°}\sin(30° - 22.62°)\cos 30°$$

$$= 7969.39 \text{ m}$$

So, total range, $BC = AB + AC = 1288.23 + 7969.39 = 9257.62$m

EXAMPLE 17

The motion of a particle is described by the following equations $x = 4(t+1)^2, y = 4(t+1)^2$. Show that the path travelled by the particle is a rectangular hyperbola. Find also, the velocity and the acceleration of the particle at $t = 0$.

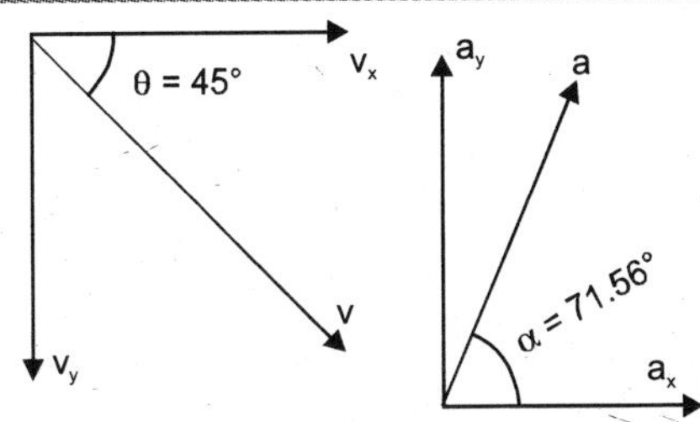

SOLUTION Here, we have, $x \cdot y = 4(t+1)^2 \times \dfrac{4}{(t+1)^2} = 16$.

So, the equation of the path of the particle is representing a rectangular hyperbola.

Now, along x direction, $v_x = \dfrac{dx}{dt} = 8(t+1)$, $a_x = \dfrac{dv_x}{dt}$

And along y direction, $v_y = \dfrac{dy}{dt} = -8(t+1)^{-3}$, $a_y = \dfrac{dv_y}{dt} = 24(t+1)^{-4}$

Substituting $t = 0$, we obtain, $v_x = 8\text{m/s}, a_x = 8\text{m/s}^2, v_y = -8\text{m/s}, a_y = 24\text{m/s}^2$

Finally $$v = \sqrt{v_x^2 + v_y^2} = \sqrt{64+64} = 11.31\text{m/s}, \quad \theta = \tan^{-1}\frac{8}{8} = 45°$$

And, $$a = \sqrt{a_x^2 + a_y^2} = \sqrt{64+576} = 25.298\text{m/s}^2, \quad \theta = \tan^{-1}\frac{24}{8} = 71.56°$$

EXAMPLE 18

A cyclist is travelling on a curved road of radius 100 m at a speed of 36 kmph. Find normal and tangential components of acceleration. If he applies brakes to slow down his car uniformly to a speed of 18 kmph in 10 seconds, find the normal and the tangential components of deceleration just after the brakes are applied.

36 kmph

18 kmph

100m

C

SOLUTION Here, $$v_1 = \frac{36 \times 1000}{3600} = 10 \text{ m/s},$$

$$v_2 = \frac{18 \times 1000}{3600} = 5\text{m/s}$$

So, normal acceleration, $$a_n = \frac{v_1^2}{r} = \frac{10^2}{100} = 1\text{m/s}^2$$

And tangential acceleration, $$a_t = \frac{dv_1}{dt} = 0$$

When the brakes are applied, the cycle has a tangential speed of 10 m/s and also there is a tangential deceleration.

So, $$a_t = \frac{v_1 - v_2}{dt} = \frac{10-5}{10} = 0.5\text{m/s}^2,$$

but normal deceleration remains constant.

EXAMPLE 19

A particle moves along the spiral shown in the figure. The motion of the particle is defined by relations $r = 10t$, $\theta = 2\pi t$, where r is expressed in centimetres, θ in radians and t in seconds. Determine the velocity and acceleration of the particle when (*i*) $t = 0$, and (*ii*) $t = 0.5$s.

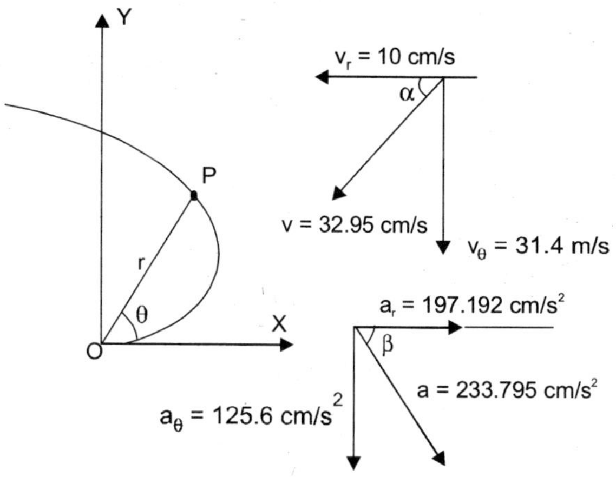

SOLUTION Here, $\dot{r} = 10$ and $\dot{\theta} = 2\pi$, $\ddot{r} = 0$ and $\ddot{\theta} = 0$

At $t = 0$, $r = 0$ and $\theta = 0$, $\dot{r} = 10$ and $\dot{\theta} = 2\pi$, $\ddot{r} = 0$ and $\ddot{\theta} = 0$

At $t = 0.5$, $r = 5$ and $\theta = \pi$, $\dot{r} = 10$ and $\dot{\theta} = 2\pi$, $\ddot{r} = 0$ and $\ddot{\theta} = 0$

At $t = 0$, $v = \sqrt{v_r^2 + (r\dot{\theta})^2} = \sqrt{10^2 + 0^2} = 10\text{cm/s}$ along v_r

So,

$$a_r = \ddot{r} - r(\dot{\theta})^2 = 0,$$

$$a_\theta = r\ddot{\theta} + 2\dot{r}\dot{\theta} = 2 \times 10 \times 2\pi = 125.6$$

$\Rightarrow$

$$a = \sqrt{a_r^2 + a_\theta^2} = \sqrt{125.6^2 + 0^2} = 125.6\text{cm/s}^2 \text{ along } a_\theta$$

At $t = 0.5$,

$$v = \sqrt{v_r^2 + (r\dot{\theta})^2} = \sqrt{10^2 + (5 \times 2\pi)^2} = 32.95\text{cm/s along } v_r$$

$$a_r = \ddot{r} - r(\dot{\theta})^2 = -5(2\pi)^2 = -197.19,$$

$$a_\theta = r\ddot{\theta} + 2\dot{r}\dot{\theta} = 2 \times 10 \times 2\pi = 125.6$$

$$a = \sqrt{a_r^2 + a_\theta^2} = \sqrt{(-197.19)^2 + 125.6^2} = 233.795\text{cm/s}^2 \text{ along } a_\theta$$

EXAMPLE 20

The rotation of rod *AB* is defined by the relation $\theta = \frac{\pi}{2}(4t - 3t^2)$. A collar *P* slides along this rod in such a way that its distance from *A* is given by $r = 1.25t^2 - 0.9t^3$. In these relations θ is expressed in radians, *r* in meters and *t* in seconds. Determine (*i*) the velocity of the collar, (*ii*) the total acceleration of the collar and, (*iii*) the acceleration of the collar relative to the rod when $t = 1\text{s}$.

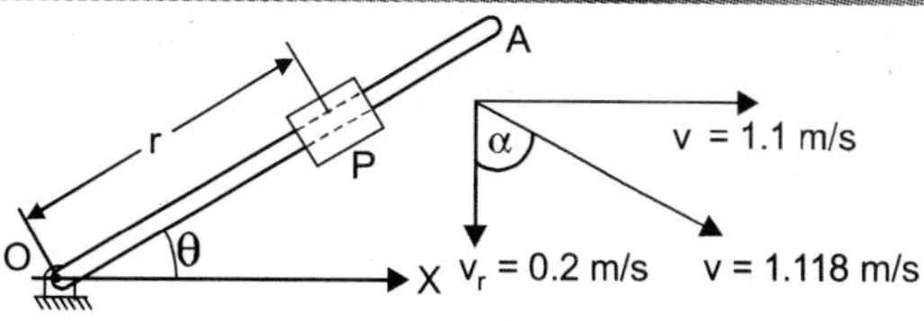

SOLUTION Here, $\dot{\theta} = \frac{\pi}{2}(4-6t)$, $\ddot{\theta} = -3\pi$, $\dot{r} = 2.5t - 2.7t^2$, $\ddot{r} = 2.5 - 5.4t$

At $t = 1$, $\theta = 0.5\pi$, $\dot{\theta} = -\pi$, $\ddot{\theta} = -3\pi$,

$r = 0.35$, $\dot{r} = -0.2$, $\ddot{r} = -2.9$

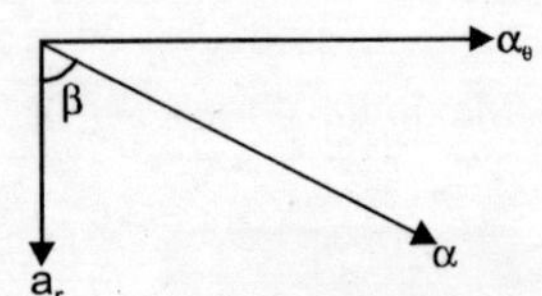

Here, velocity and acceleration of collar P,

$$v_r = \dot{r} = -0.2\text{m/s},$$

$$v_\theta = r\dot{\theta} = -1.1\text{m/s}$$

$$\therefore \quad v = \sqrt{v_r^2 + v_\theta^2} = \sqrt{(-0.2)^2 + (-1.1)^2} = 1.118 \text{ m/s}$$

$$\alpha = \tan^{-1}\frac{v_\theta}{v_r} = \tan^{-1}\frac{-1.1}{-0.2} = 79.7°$$

Now $$a_r = \ddot{r} - r(\dot{\theta})^2 = -6.35,$$

$$a_\theta = r\ddot{\theta} + 2\dot{r}\dot{\theta} = -2.04$$

$$\therefore \quad \alpha = \sqrt{a_r^2 + a_\theta^2} = \sqrt{(-6.35)^2 + (-2.04)^2} = 6.667\text{m/s}^2$$

$$\beta = \tan^{-1}\frac{a_\theta}{a_r} = \tan^{-1}\left(\frac{-2.04}{-6.35}\right) = 17.8°$$

EXAMPLE 21

The displacement time equation for the oscillations of a simple pendulum is given by $S = S\cos\left(\sqrt{\frac{g}{l}}\right)t$, where, S is the maximum displacement of oscillations.

Find (*i*) the maximum velocity, (*ii*) the maximum tangential and normal acceleration of the bob.

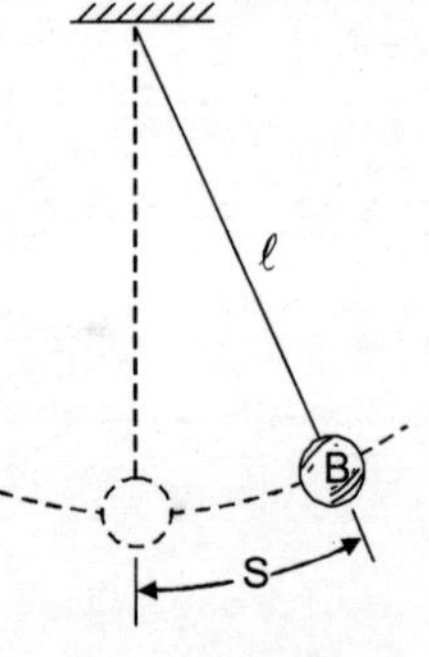

SOLUTION It may be recalled here, that to describe displacement-time relation for the curvilinear motion in cartesian co-ordinates it would require two equations.

A single equation of the curvilinear motion of the bob, $S = S\cos\left(\sqrt{\frac{g}{l}}\right)t$.

Given here, is expressed in the path co-ordinate, that is displacement is measured along the curve and time.

So, tangential velocity, $$v = \frac{dS}{dt} = -S\sqrt{\frac{g}{l}}\sin\left(\sqrt{\frac{g}{l}}\right)t$$

Hence, maximum tangential velocity, $v_{\max} = \pm S\sqrt{\frac{g}{l}}$

Tangential acceleration, $$a_t = \frac{dv}{dt} = -S\sqrt{\frac{g}{l}}\cos\left(\sqrt{\frac{g}{l}}\right)t$$

Maximum tangential acceleration $(a_t)_{\max} = \pm S\left(\sqrt{\frac{g}{l}}\right)$

Normal acceleration, $$a_n = \frac{v^2}{p} = S^2 \frac{g}{l^2} \sin\left(\sqrt{\frac{g}{l}}\right)t$$

Maximum normal acceleration, $$(a_n)_{max} = \frac{S^2 g}{l^2}$$

EXAMPLE 22

A car enters a curved portion of the road of radius 200m travelling at a constant speed of 36kmph. Determine the components of velocity and acceleration of the car in the x and y directions 15 seconds after it has entered the curved portion of the road. Also express the velocity and the acceleration of the car in terms of the normal and tangential components.

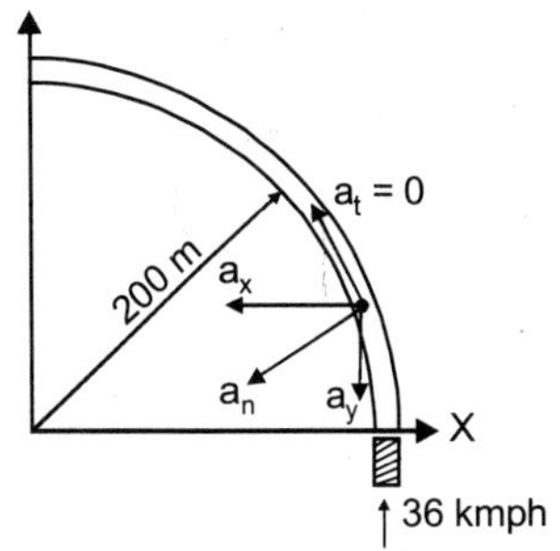

SOLUTION Linear velocity of the car, $$v = \frac{36 \times 1000}{3600} = 10\text{m/s}$$

Angular velocity of the car, $$\omega = \frac{v}{r} = \frac{10}{200} = \frac{1}{20}\text{rad/s}$$

After time t, if angular position of car be θ, $\theta = \omega t$

Displacement in x direction, $x = r\cos\theta = r\cos\omega t$

Displacement in y direction, $y = r\sin\theta = r\sin\omega t$

Velocity in x direction, $$v_x = \frac{dx}{dt} = \frac{d}{dt}(r\cos\omega t) = -\omega r\sin\omega t$$

Velocity in y direction, $$v_y = \frac{dy}{dt} = \frac{d}{dt}(r\sin\omega t) = \omega r\cos\omega t$$

Substituting $t = 15$s and $\omega = \frac{1}{20}$rad, we have,

$$v_x = 6.82\text{m/s(left)},\ v_y = 7.71\text{m/s (up)}$$

Acceleration in x direction,

$$a_x = \frac{d}{dt}(-\omega r\sin\omega t) = -\omega^2 r\cos\omega t$$

Acceleration in y direction,

$$a_y = \frac{d}{dt}(\omega r\cos\omega t) = -\omega^2 r\sin\omega t$$

Substituting $t = 15$s and $\omega = \frac{1}{20}$rad, we have,

$$a_x = 0.365\text{m/s}^2\text{(left)}, a_y = 0.341\text{m/s}^2\text{(down)}$$

$\therefore$ $$a = \sqrt{a_x^2 + a_y^2} = \sqrt{0.365^2 + 0.341^2} = 0.5\text{ m/s}^2$$

and $$\phi = \tan^{-1}\left(\frac{0.341}{0.365}\right) = 43.05^\circ$$

Tangential component of velocity, $v_t = 10\text{m/s} = v$ and normal component of velocity $v_n = 0$
Tangential component of acceleration,

$$a_t = \frac{dv}{dt} = 0$$

and normal component of acceleration,

$$a_n = \frac{v^2}{r} = \frac{10^2}{200} = 0.5\text{m/s}^2.$$

Angular position θ after 15 seconds,

$$\theta = \omega t = \frac{1}{2}\times\frac{180}{\pi}\times 15 = 42.97^\circ$$

EXAMPLE 23

Pivoted Link *OA* carries a pin *P* whose position is controlled by the horizontal slotted bar which can slide along a fixed vertical bar. What are the *x*-components of the velocity and acceleration of the pin *P* at the instant when $y = 5$cm and the slotted bar is moving upward at a constant velocity of 10cm/s?

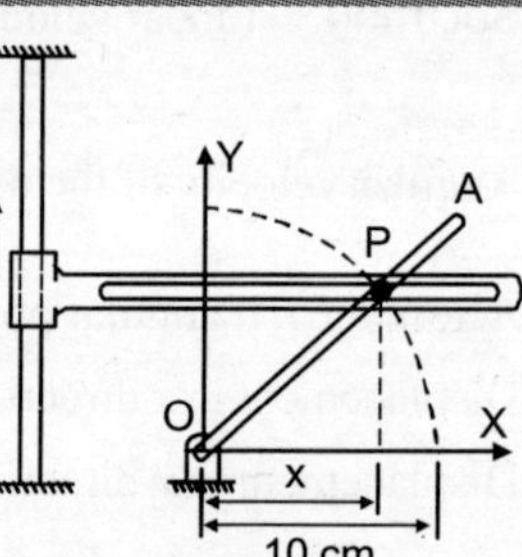

SOLUTION Equation of the path on which pin *P* moves is

$$x^2 + y^2 = r^2 = 10^2 = 100,$$

as it moves on a circular path guided by the rotating link *OA* and the slotted bar moving upward.

When the co-ordinate $y = 5$cm, at that instant *x*-coordinate will be,

$$x = \sqrt{100-25} = 8.66\text{cm}$$

As, $$\frac{dy}{dt} = \dot{y} = 10\text{cm/s}, \quad \frac{d\dot{y}}{dt} = \ddot{y} = 0$$

Differentiating the governing equation, $x^2 + y^2 = 100$, with respect to time,

$$2x\frac{dx}{dt} + 2y\frac{dy}{dt} = 0$$

$$\Rightarrow \quad x\dot{x} + y\dot{y} = 0 \quad ...(1)$$

Further differentiation results,

$$x\frac{d^2x}{dt^2} + \left(\frac{dx}{dt}\right)^2 + y\frac{d^2y}{dt^2} + \left(\frac{dy}{dt}\right)^2 = 0$$

$$\Rightarrow \quad x\ddot{x} + (\dot{x})^2 + y\ddot{y} + (\dot{y})^2 = 0 \quad ...(2)$$

Substituting, $x = 8.66$, $y = 5$, $\dot{y} = 10$, $\ddot{y} = 0$ we find, from equation (1)

$$8.66\dot{x} + 5\times 10 = 0$$

$$\Rightarrow \quad \dot{x} = -5.77\text{cm/s}$$

Further substitution in equation (2) results,

$$8.66\ddot{x} + (-5.77)^2 + 5\times 0 + 10^2 = 0$$

$$\Rightarrow \quad \ddot{x} = -15.4\text{cm/s}^2$$

EXAMPLE 24

A motorist of total weight $W = 500$ N travels in a vertical plane with a constant speed $v = 72$ kmph along the circular curve AB of radius $r = 1000$ m as shown in figure. Find the reaction N exerted on the motor cycle by the track as it passes the crest C of the curve and also find angle ϕ with the road reaction N become equal to zero.

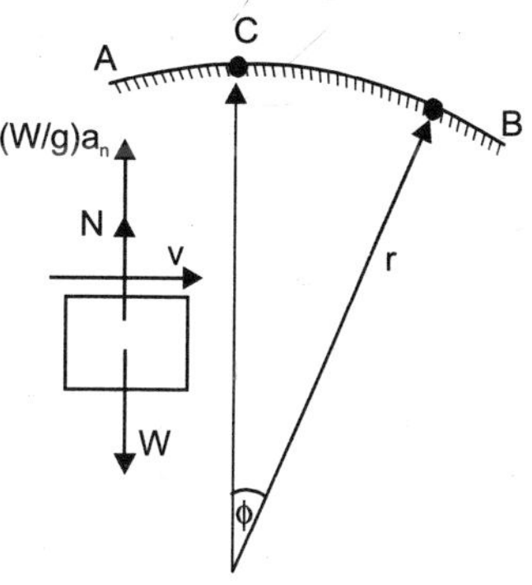

SOLUTION Here speed $v = \dfrac{72 \times 1000}{3600} = 20\text{m/s}$

From free body diagram of the motor cyclist,

$$\sum F_y = 0 \quad \Rightarrow \quad R + ma_n = W$$

$$\Rightarrow \quad N = W - ma_n = W - \frac{W}{g} \cdot \frac{v^2}{r} = 500 - \frac{500}{10} \cdot \frac{10^2}{1000} = 495\text{N}$$

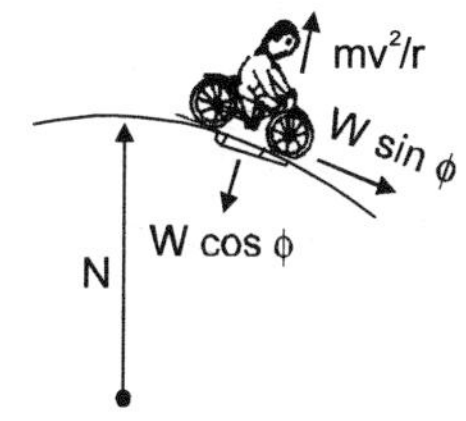

Consider the situation at ϕ inclination on the road.

Again, $\sum F_y = 0$

$$\Rightarrow \quad N + \frac{W}{g} \cdot \frac{v^2}{r} = W\cos\phi$$

$$\Rightarrow \quad 0 + \frac{500}{10} \cdot \frac{10^2}{1000} = 500\cos\phi$$

$$\Rightarrow \quad \cos\phi = 0.01$$

$$\Rightarrow \quad \phi = 89.5°$$

EXAMPLE 25

A small block of weight W rests on a horizontal turn table at a distance $r = 1$m from the centre of the turn table. Find the maximum uniform speed the block can have without slipping of the table. Assume the coefficient of friction between the block and the turn table to be $\mu = 0.5$.

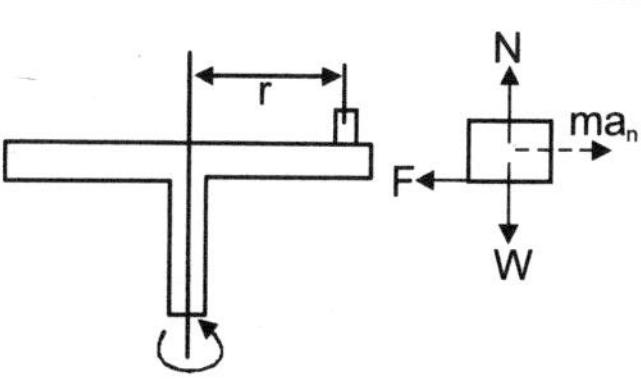

SOLUTION Equations of equilibrium of the block,

$$\sum F_y = 0$$

$$\Rightarrow \quad N = W$$

$$\sum F_x = 0$$

$$\Rightarrow \quad F = ma_n$$

$$\Rightarrow \quad \mu W = m\frac{v^2}{r}$$

$$\Rightarrow \quad v^2 = \mu g r$$

$$\Rightarrow \quad v = \sqrt{0.5 \times 9.81 \times 1} = 2.215\text{m/s}$$

EXAMPLE 26

A ball of weight 10N starts from rest from the point O of a smooth vertical track and rolls down under gravity along the track OAB. Find the reaction exerted on the ball at the point $A(l/2, h)$ if the curve OAB is defined by the equation $y = h\sin\left(\frac{\pi x}{l}\right)$ and $h = \frac{l}{3}$.

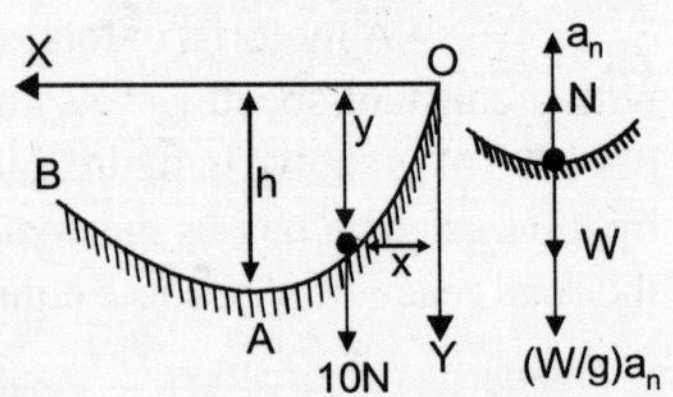

SOLUTION From free body diagram of the ball, $N = W\left(1+\frac{a_n}{g}\right) = W\left(1+\frac{v^2}{\rho g}\right)$,

Here, v is the velocity of the particle at A and ρ is the radius of curvature of the curve at A.

Now, $$\frac{dy}{dx} = \frac{\pi h}{l}\cos\left(\frac{\pi x}{l}\right) \quad \text{and} \quad \frac{d^2y}{dx^2} = -h\frac{\pi^2}{l^2}\sin\left(\frac{\pi x}{l}\right)$$

So, curvature $$\frac{1}{\rho} = \frac{\frac{d^2y}{dx^2}}{\left[1+\left(\frac{dy}{dx}\right)^2\right]^{\frac{3}{2}}} = \frac{-h\frac{\pi^2}{l^2}\sin\left(\frac{\pi x}{l}\right)}{\left[1+\frac{\pi^2h^2}{l^2}\cos^2\left(\frac{\pi x}{l}\right)\right]^{\frac{3}{2}}}$$

Hence, at A, $$\frac{1}{\rho} = \frac{\frac{-\pi^2 h}{l^2}}{1+0} = -\frac{\pi^2 h}{l^2} = \frac{\pi^2 h}{l^2} \text{(absolute value)}$$

Now velocity $$v = \sqrt{0^2 + 2gh} = \sqrt{2gh}$$

From governing equation, reaction

$$N = W\left(1+\frac{v^2}{\rho g}\right) = 10\left(1+\frac{2gh\times\pi^2 h}{g\times l^2}\right) = 10\left(1+\frac{2\pi^2h^2}{l^2}\right)$$

Substituting, $$h = \frac{l}{3},\ N = 31.9\text{N}$$

EXAMPLE 27

At what uniform speed of rotation N around the vertical axis AB, the balls P and Q of equal weight shall begin to lift the weight R of the governing device shown in figure. Given $P = Q = 50$ N, $R = 100$ N, $l = 30$ cm. Assume $\theta = 30°$ when the weight R begins to get lifted, when $r = 15$ cm.

SOLUTION Considering equilibrium of the weight,

$$\sum F_x = 0$$

$\Rightarrow$ $$T_2 \sin\theta = T_1 \sin\theta$$

$\Rightarrow$ $$T_1 = T_2 = T$$

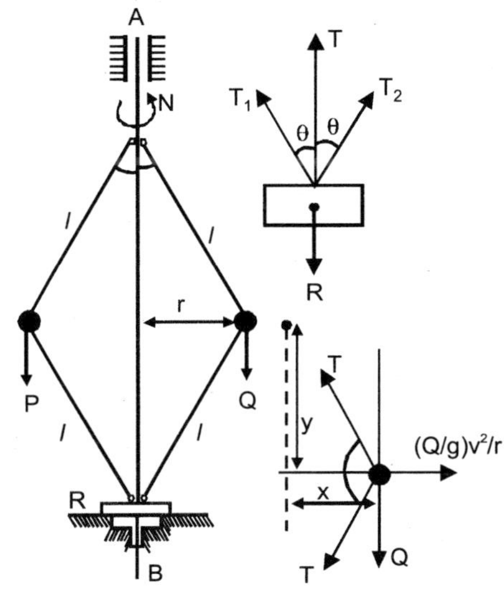

$$\sum F_y = 0$$

$\Rightarrow$ $$2T\cos\theta = R$$

$\Rightarrow$ $$T = \frac{R}{2\cos\theta}$$

Considering moment equilibrium

$$\sum M = 0$$

$\Rightarrow$ $$\frac{Q}{g}\cdot\frac{v^2}{r}y = y\cdot T\sin\theta + x(Q + T\cos\theta)$$

Substitute, $x = l\sin\theta$ and $y = l\cos\theta$

So, $$\frac{Qv^2}{rg}l\cos\theta = T\sin\theta\cdot l\cos\theta + (Q + T\cos\theta)l\sin\theta$$

$$= \frac{Rl\sin\theta}{2} + \left(Q + \frac{R}{2}\right)l\sin\theta$$

$\Rightarrow$ $$\frac{50v^2}{9.81\times 0.15}\cdot\frac{\sqrt{3}}{2} = \frac{100}{2\times 2} + \left(50 + \frac{100}{2}\right)\frac{1}{2}$$

$\Rightarrow$ $$v^2 = 2.5487$$

$\Rightarrow$ $$v = 1.596 \text{ m/s}$$

Again, we know,

$$N = \frac{60}{2\pi r}v = \frac{1.596\times 60}{2\pi\times 0.15} = 101.6 \text{ r.p.m}$$

EXAMPLE 28

Express the maximum speed with respect to skidding of a car travelling on a banked road in terms of the radius r of the curve, the banking angle θ and the friction angle ϕ between the tyres and the road.

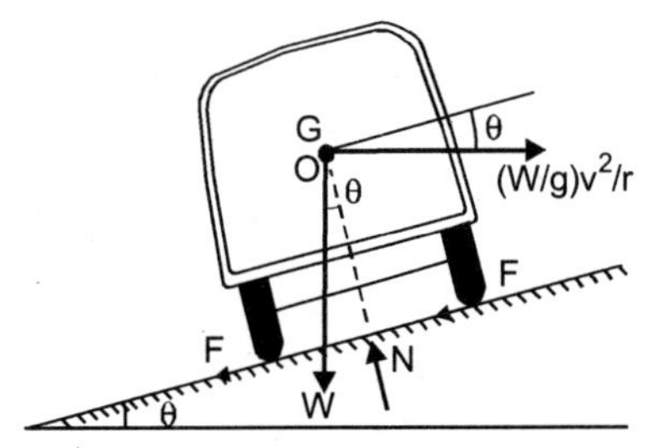

SOLUTION From equation of dynamic equilibrium, normal to incline

$$N - W\cos\theta - \frac{W}{g}\cdot\frac{v^2}{r}\sin\theta = 0$$

$\Rightarrow$ $$N = W\cos\theta + \frac{W}{g}\cdot\frac{v^2}{r}\sin\theta$$

From law of friction, frictional force

$$F = \mu N = \mu\left(W\cos\theta + \frac{W}{g}\cdot\frac{v^2}{r}\sin\theta\right) \quad ...(1)$$

From equilibrium equation along the incline,

So, $$F + W\sin\theta - \frac{Wv^2}{rg}\cos\theta = 0$$

$$\Rightarrow \quad F = \frac{Wv^2}{rg}\cos\theta - W\sin\theta \qquad ...(2)$$

Equating (1) and (2),

$$\mu\left(W\cos\theta + \frac{Wv^2}{rg}\sin\theta\right) = \frac{Wv^2}{rg}\cos\theta - W\sin\theta$$

$$\Rightarrow \quad v^2 = gr\left(\frac{\mu + \tan\theta}{1 - \mu\tan\theta}\right) = gr\left(\frac{\tan\phi + \tan\theta}{1 - \tan\phi\tan\theta}\right) = gr\tan(\phi + \theta)$$

$$\Rightarrow \quad v = \sqrt{gr\tan(\phi + \theta)}$$

EXAMPLE 29

A bob weighing 1N suspended by a cord from the ceiling of a railway carriage was found to make an angle $\theta = 8°$ with the vertical when the railway carriage was negotiating a curve of 100 m. Find the speed of the carriage and the tension in the cord.

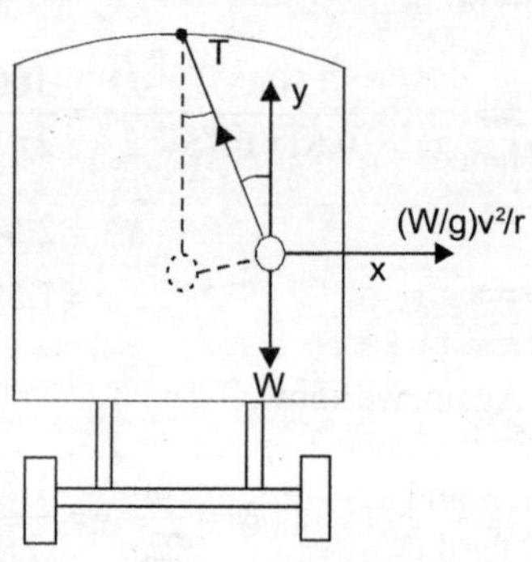

SOLUTION From free body diagram of the bob,

$$\sum F_x = 0 \quad \Rightarrow \quad T\sin\theta = \frac{W}{g}\cdot\frac{v^2}{r},$$

and $$\sum F_y = 0$$

$$\Rightarrow \quad T\cos\theta = W$$

So, $$\tan\theta = \frac{v^2}{rg}$$

$$\Rightarrow \quad v = \sqrt{rg\tan\theta} = \sqrt{100 \times 9.81 \times \tan 8°} = 11.74 \text{ m/s}$$

And, $$T = \frac{W}{\cos\theta} = \frac{1}{\cos 8°} = 1.01 \text{ N}$$

EXAMPLE 30

An automobile of weight W travels with uniform speed of $v = 100$ kmph, over a vertical curve ABC of parabolic shape as shown. Determine the total pressure exerted by the wheels of the car as it passes the point B.

SOLUTION Here, $$v = \frac{100 \times 1000}{3600} = 27.77 \text{m/s}$$

Dynamic equation of equilibrium of the automobile, along vertical

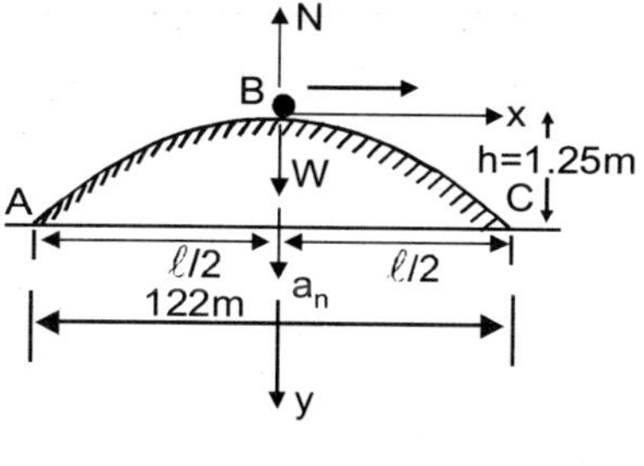

$$N = W - \frac{W}{g} a_n = W - \frac{W}{g} \cdot \frac{v^2}{\rho}$$

Let the equation of the path ABC be $x^2 = ky$.
At point B $(l/2, h)$,

$$(l/2)^2 = kh$$

$$\Rightarrow \qquad k = \frac{l^2}{4h}$$

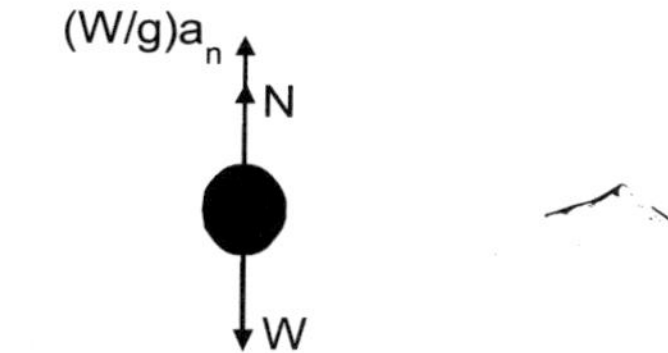

Now, $\dfrac{dy}{dx} = \dfrac{2x}{k}$ and $\dfrac{d^2y}{dx^2} = \dfrac{2}{k}$.

Hence, curvature of the path ABC

$$\frac{1}{\rho} = \frac{\dfrac{d^2y}{dx^2}}{\left[1+\left(\dfrac{dy}{dx}\right)^2\right]^{\frac{3}{2}}} = \frac{\dfrac{2}{k}}{\left[1+\left(\dfrac{2x}{k}\right)^2\right]^{\frac{3}{2}}}$$

At point B,

$$\frac{1}{\rho} = \frac{2/k}{\left(1+0^2\right)^{3/2}} = \frac{2}{k} = \frac{8h}{l^2} = \frac{8 \times 1.25}{122^2} = \frac{1}{1488.4}$$

Hence,

$$N = \mathrm{W}\left(1 \quad \frac{v^2}{\rho g}\right) = \mathrm{W}\left(1 - \frac{27.77^2}{1488.4 \times 9.81}\right) - 0.948\mathrm{W}$$

EXAMPLE 31

A portion ABC of a railway track consists of a spiral AB of length r and radius of curvature $\rho = r^2/S$ and the tangent BC. S is measured from B towards A. A diesel locomotive of weight 10 kN starts from rest at A and increases its speed along AB with a constant tangential acceleration $dv/dt = g/10$. Find the maximum lateral thrust, on the outer rail during the motion from A to B and where does it occur?

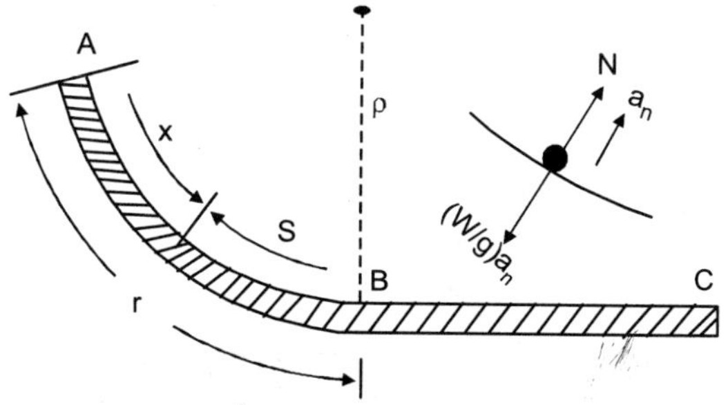

SOLUTION Equation of equilibrium of the locomotive $N = \dfrac{W}{g} a_n = \dfrac{W}{g} \cdot \dfrac{v^2}{\rho}$. When the locomotive is at a distance x from A, distance from B will be $S = r - x$ and so $\rho = \dfrac{r^2}{r-x}$.

The locomotive starts from rest and its final velocity

$$v^2 = 0^2 + 2 \cdot \frac{g}{10} \cdot x = \frac{gx}{5}$$

So, $$N = \frac{W}{g} \cdot \frac{v^2}{\rho} = \frac{Wx}{5}\left(\frac{r-x}{r^2}\right).$$

For maximum value of N,

$$\frac{dN}{dx} = 0$$

$$\Rightarrow \quad \frac{d}{dx}\left[\frac{Wx}{5}\left(\frac{r-x}{r^2}\right)\right] = 0$$

$$\Rightarrow \quad \frac{d}{dx}[x(r-x)] = 0$$

$$\Rightarrow \quad x = \frac{r}{2}$$

Hence, $$N_{\max} = \frac{Wr}{10}\left(\frac{r-0.5r}{r^2}\right) = \frac{W}{20} = \frac{10000}{20} = 500\text{N}$$

Exercise

1. A car starts from rest on a curved road of 250 m radius and accelerates at a constant tangential acceleration at 0.6 m/s². Determine the distance and the time for which that car will travel before the magnitude of the total acceleration attached by it becomes 0.75 m/s².
2. A car enters a curved section of the road of length equal to the quarter of a circle of radius 100 m at 18 kmph and leaves at 36 kmph. If the car is travelling with a constant tangential acceleration find the magnitude and direction when (*a*) it enters the curve, (*b*) it leaves the curve.

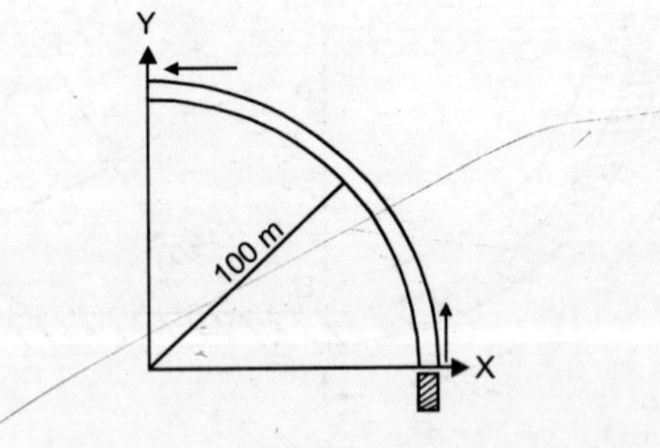

P-2

3. The telescopic rod forces the pin P to move along the fixed path $9y^2 = x^2$ where x and y are expressed in centimeters. At any instant t the x-coordinate of P is given by $x = t^3 - 14t$. Determine the components of the velocity and acceleration of P when $t = 15$s.

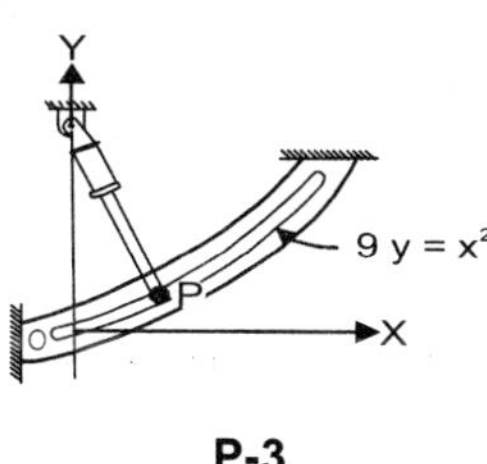

P-3

4. A cyclist riding on a level road has to turn a corner of radius 50 m. Find the maximum speed with which he can travel without the fear of skidding.

5. An automobile track is so designed that when a car travels at 100 kmph, the force between the car and track acts normal to the surface of the track. Find the angle of banking the track assuming it to be a circle of radius 250 m.

6. A roadway bridge is in the form of a circular arc of radius 200 m. Find the greatest speed with which a vehicle can cross the bridge without just loosing contact at the highest point of the roadway.

7. A small block rests on a turn-table which, starting from a rest, is rotated in such a way that the block undergoes at constant tangential acceleration $a_t = 2\text{m/ s}^2$. Determine (*a*) how long will it take for the block to start slipping on the turn-table, (*b*) the speed *v* of the block at that instant. Assume the coefficient of static friction between the block and the turn-table is 0.60.

8. A ball of weight *W* moves with uniform speed *v* along a vertical cosine curve *ABC*. Find the pressure exerted by the ball on the path as it passes the point *B*.

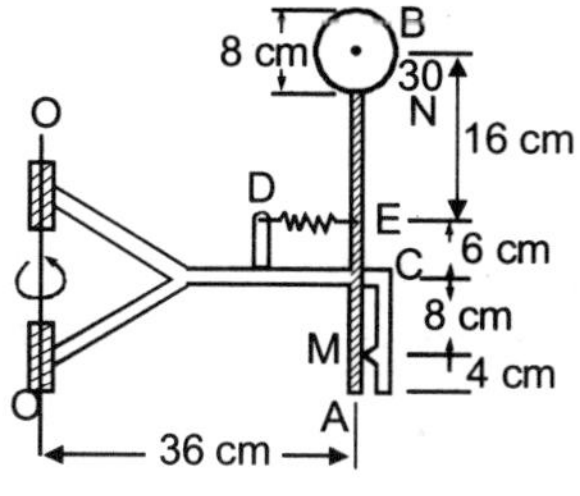

P-9

9. The arrangement shown in figure rotates about the vertical axis *OO* at a constant r.p.m. The weight of the vertical bar *AB* hinged at *C* is 15N and the weight of the ball at the top is 30 N. When the system is at rest, the initial tension in the spring *DE* is 100 N. At what speed in r.p.m will the contact at *M* be broken.

10. A railway track has a radius 400 m. The distance between the rails is 2 m and the superelevation of the outer rail is 5cm. If there is no lateral thrust on the rails find the speed of the train.

11. A body is projected at angle such that its horizontal range is five times the maximum height attained. Determine its angle of projection.

Ans. [$\alpha = 38.65°$]

12. An airplane flying at 600 kmph horizontaly, accidentally loses a rivet on its tail, when it is at a height of 6000 m above the ground. Determine the location, where the rivet would land if air resistance is neglected.

Ans. [5.83 km]

13. A cricket ball is thrown by a fielder from a height of 3 m at an angle of 40° with horizontal. The initial velocity gained by the ball is 30 m/s. The ball hits the wicket at a height 0.30 m with respect to the ground. Determine how far was the fielder from the wicket? **Ans.** [93.42 m]

14. Determine the position at which a ball thrown up to the right will strike the inclined surface shown in figure. The initial velocity of the ball is 30 m/s inclined at 60° with horizontal. **Ans.** [61.16 m]

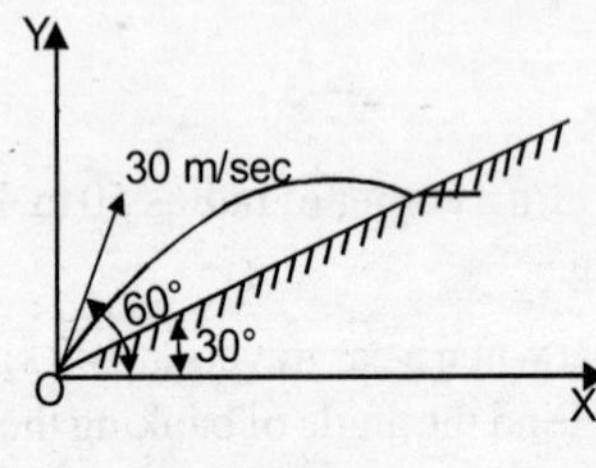

P-14

15. A ball thrown down an incline strikes it at a distance of $R = 76$ m. If the ball rises to a maximum height $h = 19$ m, above the point of release, compute its initial velocity and inclination of angle of projection to the horizontal. **Ans.**[u=24.23m/s, α = 52.85°]

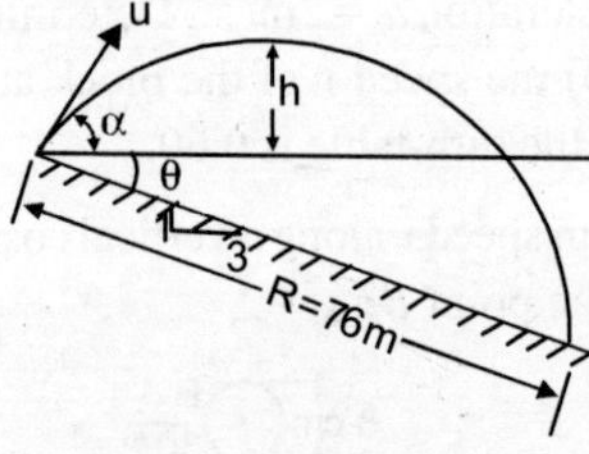

P-15

16. A ball falls vertically for 6 seconds on a plane inclined at 30° to the horizontal. If the coefficient of restitution is 0.8, when and where will the ball strike the plane again? **Ans.**[t = 9.6s and 56.5m]

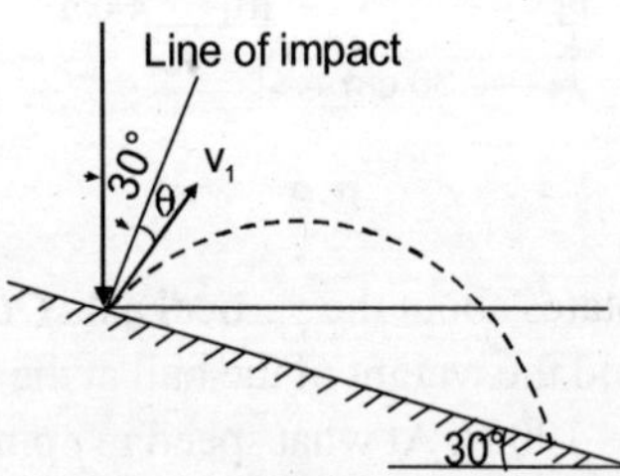

17. A ball is dropped from a height of 6 m upon an 25° incline. Find the resultant velocity of the ball after impact, if coefficient of restitution is 0.8. **Ans.** [9.1m/s]

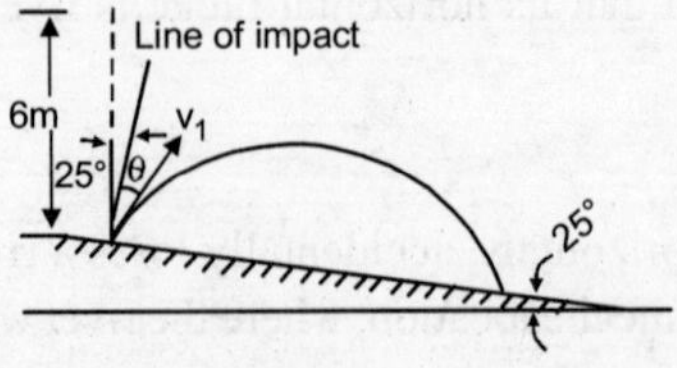

CHAPTER

10

PLANE KINEMATICS OF RIGID BODIES

10.1 INTRODUCTION

In the previous chapter on *dynamics*, we dealt about the *kinematics* of a particle for *rectilinear* and *curvilinear* motion, and developed governing relationships of displacement, velocity and acceleration. In *rigid body kinematics* we use those same relationship, in context with an additional parameter that is *rotation*. So it is evident that *rigid body kinematics* will involve displacements, velocities accelerations, both of *linear* as well as *angular* nature. Here all the *kinematic parameters* will be discussed in reference with a *single plane*.

10.2 PLANE MOTION

A rigid body is said to excute *plane motion* when each and every part of the body moves in *parallel planes*. And the planc of a motion is that specific plane that contains the centre of mass of the body while the body is treated as a thin slab with its motion confined to the plane of the slab. To encounter a large variety of rigid body motion, this idealization has been proved to be aptly adequate. Now the plane motion of a body may classified into following categories.

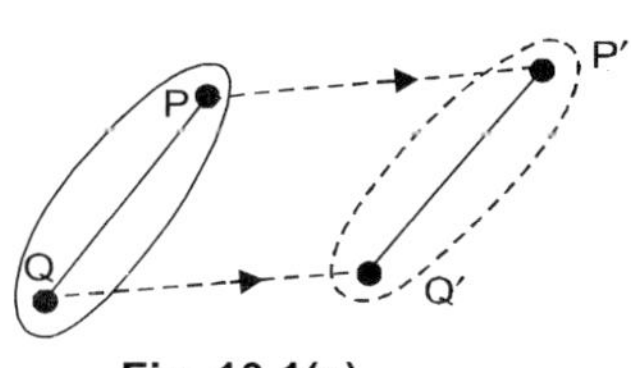

Fig. 10.1(a)

(a) Translation: It is a motion in which, every line in the body remains parallel to its original position. In fact, in rigid-body translation, all particles comprising a rigid body performs translation, separately. Rigid body translation can be separated as *rectilinear translation*, and *curvilinear translation*. In rectilinear translation, as shown in Figure 10.1(a) all points in a body move in *parallel straight lines*. In curvilinear translation, all points in a body move on *congruent curves*, as shown in Figure 10.1(b).

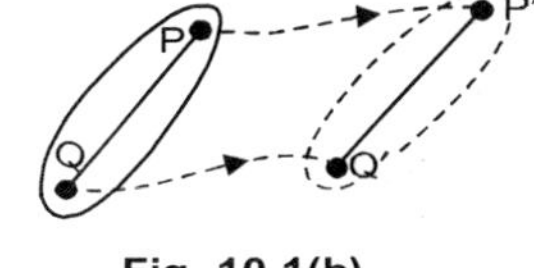

Fig. 10.1(b)

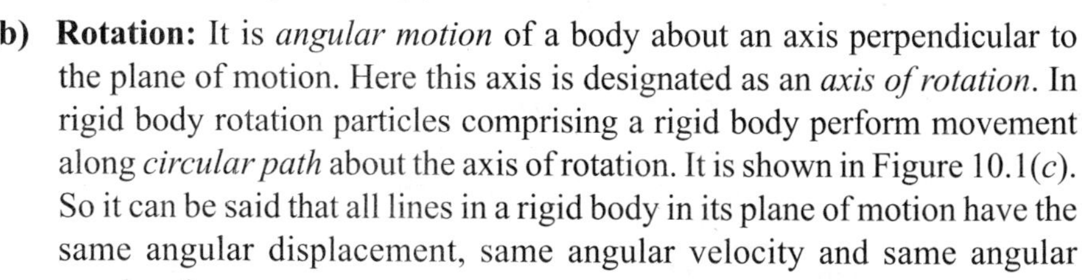

(b) Rotation: It is *angular motion* of a body about an axis perpendicular to the plane of motion. Here this axis is designated as an *axis of rotation*. In rigid body rotation particles comprising a rigid body perform movement along *circular path* about the axis of rotation. It is shown in Figure 10.1(c). So it can be said that all lines in a rigid body in its plane of motion have the same angular displacement, same angular velocity and same angular acceleration.

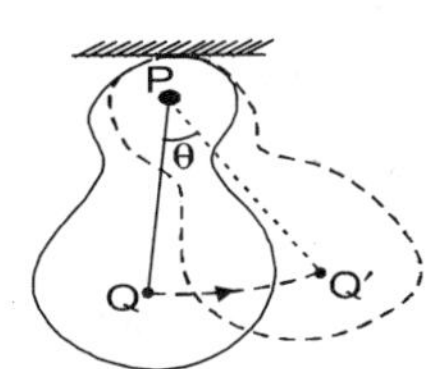

Fig. 10.1(c)

10.3 SOME KINEMATIC RELATIONS ASSOCIATED WITH ANGULAR MOTION

The angular velocity ω, angular acceleration α are related to angular displacement θ as following :

$$\omega = \frac{d\theta}{dt} = \dot{\theta},\ \alpha = \dot{\omega} = \frac{d\omega}{dt} = \ddot{\theta},\ \omega d\omega = \alpha d\theta$$

The equation of motion can be represented as

$$\left.\begin{aligned} \omega &= \omega_0 + \alpha t \\ \theta &= \omega_0 \cdot t + \frac{1}{2}\alpha t^2 \\ \omega^2 &= {\omega_0}^2 + 2\alpha\theta \end{aligned}\right\}$$

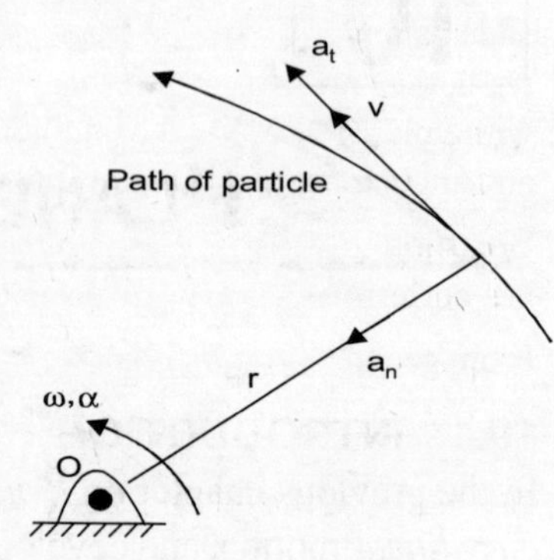

Fig. 10.1(d)

If the body rotates about a fixed axis perpendicular to plane of motion, as shown in Figure 10.1(*d*) the kinematic parameters are :

$$\left.\begin{aligned} v &= r\omega \\ a_n &= \omega^2 r = \frac{v^2}{r} = v\omega \\ a_t &= r\alpha \end{aligned}\right\}$$

Numerical Examples

EXAMPLE 1

The angle of rotation of a body is given as a function of time by the equation $\theta = \theta_0 + bt + ct^2$. Find general expressions for angular velocity and angular acceleration of the body. Determine also the values of constants *b* and *c* if initial angular velocity is 2π rad/s and 1 sec after it is 4π rad/s.

SOLUTION We have, $\theta = \theta_0 + bt + ct^2$

So, angular velocity $\omega = \dot{\theta} = \frac{d}{dt}(\theta_0 + bt + ct^2) = b + 2ct$

When $t = 0,\ \omega = 2\pi$ rad/s

Substituting this, $b = 2\pi$ and $\omega = 2\pi + 2ct$

Again when, $t = 1$s, $\omega = 4\pi$ rad/s

Substituting this, $c = \pi$ and $\omega = 2\pi + 2\pi m$

Angular acceleration, $\alpha = \ddot{\theta} = \frac{d}{dt}(2\pi + 2\pi t) = 2\pi$ rad/s

Hence, angular acceleration of a rigid body is constant.

Example 2

A slotted bar OA is hinged at O and supported by a pin B which can slide freely along the slot as the block to which the pin is attached slides along a horizontal plane shown in figure. If at the initial moment $t = 0$, the bar OA is in a vertical position and the block moves to the right with uniform velocity v_0, find general expressions for angular velocity ω and angular acceleration α of the bar OA.

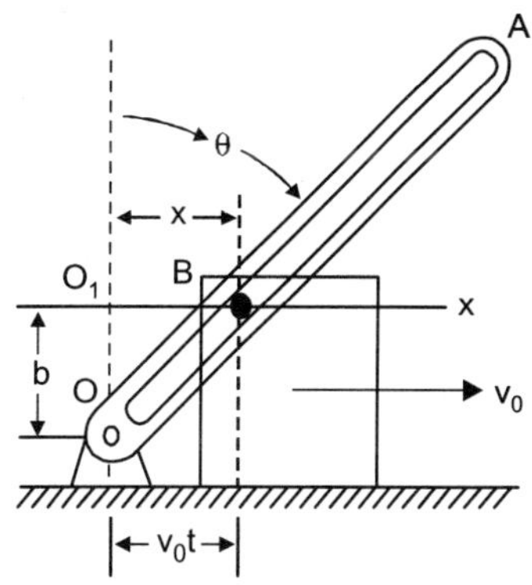

Solution After time t the block moves a distance $x = v_0 t$ while the bar has corresponding angular displacement θ.

From geometry of the figure we have,

$$\tan\theta = \frac{v_0 t}{b}$$

$$\Rightarrow \qquad \theta = \tan^{-1}\left(\frac{v_0 t}{b}\right)$$

So,

$$\omega = \dot{\theta} = \frac{d}{dt}\left(\tan^{-1}\frac{v_0 t}{b}\right) = \frac{v_0 t}{b^2 + v_0^2 t^2}$$

$$\alpha = \ddot{\theta} = \frac{d}{dt}\left(\frac{v_0 t}{b^2 + v_0^2 t^2}\right) = \frac{2v_0^3 bt}{(b^2 + v_0^2 t^2)^2}$$

Angular velocity will be maximum when, $t = 0$

So,

$$\omega_{max} = \frac{v_0}{b}.$$

Hence, initial angular velocity $\omega_0 = \omega_{max} = \dfrac{v_0}{b}$

Example 3

A slender but rigid semicircular wire of radius r is supported in its own vertical plane by a hinge at O and a smooth peg A as shown in figure. If the peg starts from O and moves with constant speed v_0 along the horizontal x-axis, find the angular velocity ω of the wire at the instant when $\theta = 60°$.

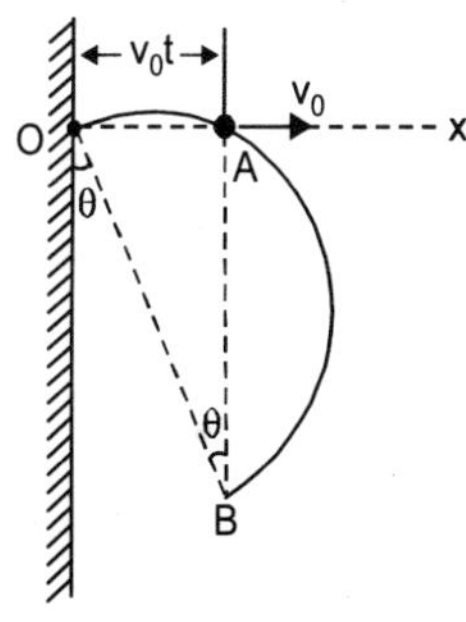

Solution After time t, the distance moved by the peg, $x = v_0 t$. From geometry of the figure, we have,

$$\sin\theta = \frac{OA}{OB} = \frac{v_0 t}{2r}$$

$$\Rightarrow \qquad \cos\theta = \sqrt{1 - \sin^2\theta} = \sqrt{1 - \frac{v_0^2 t^2}{4r^2}}$$

Now,

$$\frac{d}{dt}(\sin\theta) = \frac{d}{dt}\left(\frac{v_0 t}{2r}\right)$$

$$\Rightarrow \qquad \cos\theta\frac{d\theta}{dt} = \frac{v_0}{2r}$$

$$\Rightarrow \qquad \omega\sqrt{1-\frac{v_0^2t^2}{4r^2}} = \frac{v_0}{2r}$$

$$\Rightarrow \qquad \omega = \frac{v_0}{\sqrt{4r^2 - v_0^2t^2}}$$

Again for $\theta = 60°$ $\qquad \sin 60° = \frac{v_0 t}{2r}$

$$\Rightarrow \qquad t = \frac{r\sqrt{3}}{v_0}$$

Hence, $\qquad \omega = \frac{v_0}{\sqrt{4r^2 - v_0^2t^2}} = \frac{v_0}{\sqrt{4r^2 - \left(\frac{r\sqrt{3}}{v_0}\right)^2 v_0^2}} = \frac{v_0}{r}$

EXAMPLE 4

A slender prismatic bar *OA* of weight *W* and length *l* as shown in figure, can rotate freely about the fixed axis through *O* normal to the plane of the figure. By means of a horizontal bar *AB* and a crank shaft with crank radius *r* and crank pin *D* freely sliding in the slot *DC*, a simple harmonic motion is given to the end *A* of the bar *OA*. Determine the force *F* in the bar *AB*, assuming that its mass is negligible.

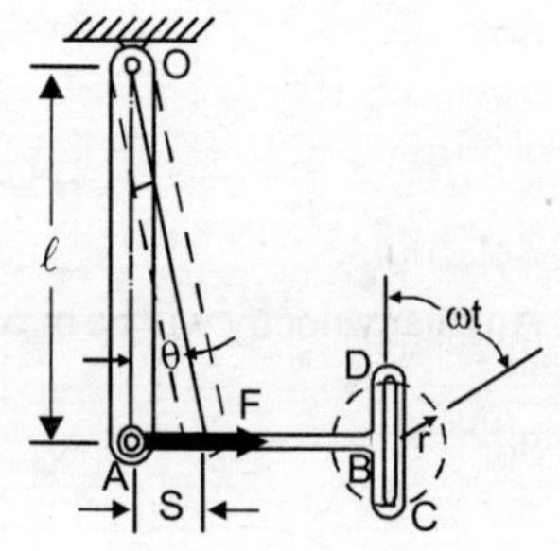

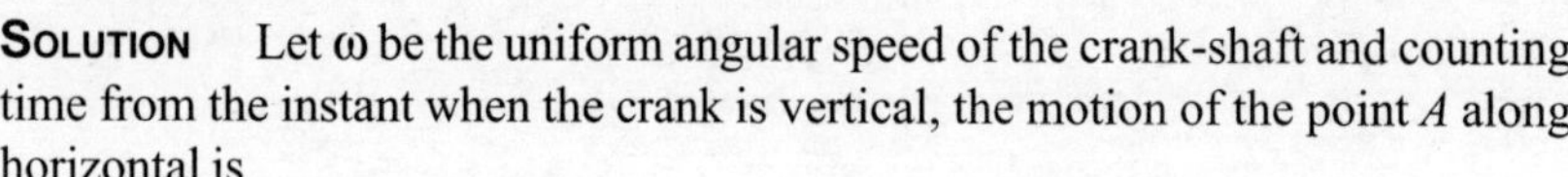

SOLUTION Let ω be the uniform angular speed of the crank-shaft and counting time from the instant when the crank is vertical, the motion of the point *A* along horizontal is

$$S = r \sin \omega t,$$

where the crank radius is *r* and it is small compared with the length *l* of the bar *OA*, vertical displacement is neglected. Then for small angles of oscillation of the bar, we have

$$\theta \approx \sin\theta = \frac{S}{l} = \frac{r \sin \omega t}{l}$$

Successive differentiation yields $\qquad \ddot{\theta} = -\frac{r}{l}\omega^2 \sin \omega t$

The moment of external force acting on the bar *OA* with respect to axis of rotation,

$$M = F \cdot l - W\frac{l}{2}\sin\theta = F \cdot l - W\frac{r}{2}\sin \omega t$$

Again we know, $\qquad M = I_0\ddot{\theta} = -I_0\frac{r}{l}\omega^2 \sin \omega t$

Equating these, we have, $\qquad -I_0\frac{r}{l}\omega^2 \sin \omega t = Fl - \frac{Wr}{2}\sin \omega t$

$$\Rightarrow \qquad F = \left(\frac{Wr}{2l} - I_0\frac{\omega^2 r}{l^2}\right)\sin \omega t$$

$$= \left[\frac{Wr}{2l} - \left(\frac{Wl^2}{12g} + \frac{Wl^2}{4g}\right)\frac{\omega^2 r}{l^2}\right]\sin\omega t = \frac{Wr}{l}\left(\frac{1}{2} - \frac{\omega^2 l}{3g}\right)\sin\omega t$$

For small value of ω, the term $\frac{\omega^2 l}{3g}$ can be neglected and hence we have,

$$F = \frac{Wr}{2l}\sin\omega t = \frac{WS}{2l}$$

For large value of ω, the force F decreases and becomes zero when $\frac{\omega^2 l}{3g} = \frac{1}{2}$.

So for this case, $\omega = \sqrt{\frac{3g}{2l}}$. And corresponding time for one revolution of the crank is $\tau = \frac{2\pi}{\omega} = 2\pi\sqrt{\frac{2l}{3g}}$.

Example 5

A string is wound around a pulley 0.3m in diameter. One of its end is fixed to the pulley while the other is fixed to a weight hanging freely. The weight describes a distance of 8m after starting from rest in 4*S*. Find the angular velocity of the pulley. Also determine the total distance moved by the weight to make the pulley rotate by 400 r.p.m.

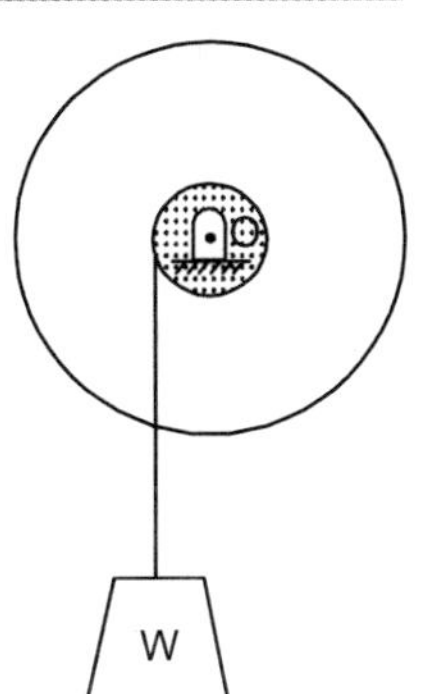

Solution Here, initial angular velocity, $\dot{\theta} = 0$

Initial linear velocity, $v_0 = 0$

Linear displacement of weight $S = 8$m

If a be the linear acceleration of the weight, we have,

$$S = ut + \frac{1}{2}at^2$$

$$\Rightarrow \quad 8 = 0 + \frac{1}{2}a \times 4^2$$

$$\Rightarrow \quad a = 1\text{m/s}^2$$

So, angular acceleration, $\ddot{\theta} = \frac{a}{r} = \frac{1}{0.15} = \frac{20}{3}\text{rad/s}^2$

Angular velocity after 4sec,

$$\dot{\theta} = \dot{\theta}_0 + \ddot{\theta}t = 0 + \frac{20}{3} \times 4 = \frac{80}{3}\text{rad/s}$$

Again we know $\omega = \dot{\theta} = \frac{2\pi N}{60} = 2\pi \times \frac{400}{60} = \frac{40}{3}\pi$

So, $t = \frac{\dot{\theta} - \dot{\theta}_0}{\ddot{\theta}} = \frac{\frac{40\pi}{3} - 0}{\frac{20}{3}} = 2\pi\,\text{sec}$

Angular displacement $\theta = \dot{\theta}_0 t + \frac{1}{2}\ddot{\theta}t^2 = 0 + \frac{1}{2}\cdot\frac{20}{3}\cdot(2\pi)^2 = \frac{40\pi^2}{3}\text{rad}$

Number of revolutions made $N = \dfrac{\theta}{2\pi} = \dfrac{40\pi^2}{3\times 2\pi} = \dfrac{20}{3}\pi$

Linear displacement of the weight $= \pi DN = \dfrac{20\pi}{3}\times\pi\times 0.3 = 19.74\text{m}$.

EXAMPLE 6

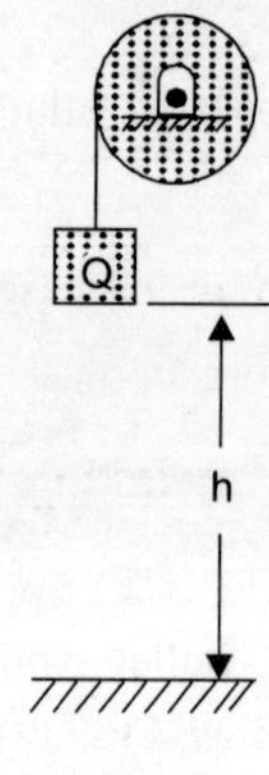

A solid right circular drum of radius r = 1 m and weight W= 32 N is free to rotate about its geometric axis as shown in figure. Wound around the circumference of the drum is a flexible cord carrying at its free end a weight Q = 10N. If the weight Q is released from rest,

(*a*) find the time t required for it to fall through the height h = 10m,
(*b*) with what velocity v will it strike the floor?

SOLUTION From equilibrium of vertical forces of the weight Q,

We can write, $$T + \frac{Q}{g}a - Q = 0$$

$$\Rightarrow \quad T = Q - \frac{Q}{g}r\ddot{\theta}$$

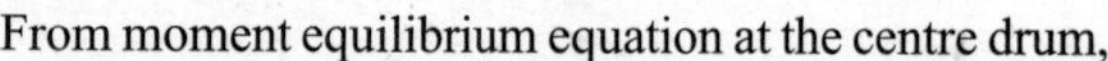
From moment equilibrium equation at the centre drum,

$$T\times r - I_0\ddot{\theta} = 0$$

$$\Rightarrow \quad r\left(Q - \frac{Q}{g}r\ddot{\theta}\right) = \frac{W}{g}\cdot\frac{r^2}{2}\cdot\ddot{\theta}$$

$$\Rightarrow \quad Q\times r - \frac{Q}{g}r^2\ddot{\theta} = \frac{W}{g}\cdot\frac{r^2}{2}\cdot\ddot{\theta}$$

$$\Rightarrow \quad Q = \ddot{\theta}\left(\frac{Qr}{g} + \frac{Wr}{2g}\right)$$

$$10 = \ddot{\theta}\left(\frac{10\times 1}{9.81} + \frac{32\times 1}{2\times 9.81}\right) = 2.65\ddot{\theta}$$

$$\Rightarrow \quad \ddot{\theta} = 3.774\text{rad/s}^2$$

$$\Rightarrow \quad a = r\ddot{\theta} = 3.774\text{m/s}^2$$

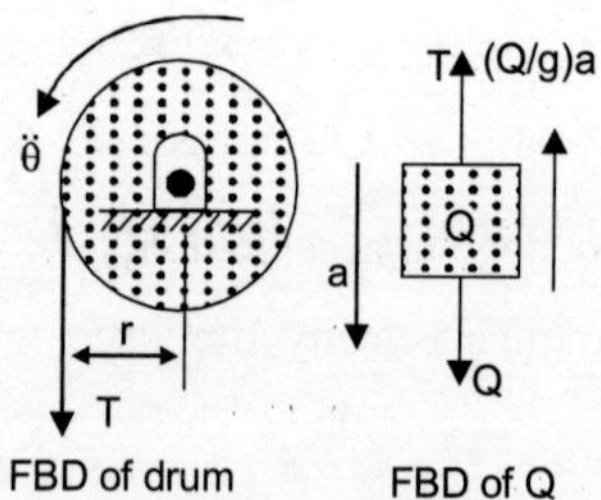

Again, $$t = \sqrt{\frac{h - ut}{\frac{1}{2}a}} = \sqrt{\frac{10\times 2}{3.774}} = 2.3\text{s}$$

and $$v = u + at = 0 + 3.774\times 2.3 = 8.68 \text{ m/s}$$

EXAMPLE 7

The rotor and shaft in figure weigh $W_1 = 500$ N and the radius of gyration with respect to the axis of rotation is $i = 10$ cm. Calculate the acceleration $\ddot{x}$ of the falling weight W= 100N. The shaft radius a = 5cm. Also calculate the torque M that must be applied to the shaft of the rotor to produce an upward acceleration $g/3$ of the attached weight W.

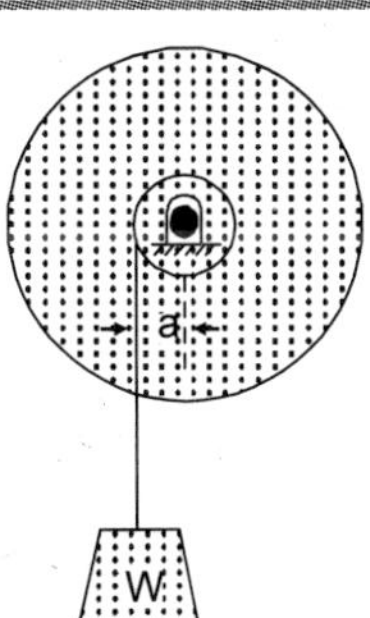

SOLUTION (*i*) From equilibrium of the vertical forces of the weight W,

$$T + \frac{W}{g}\ddot{x} = W$$

$$\Rightarrow \qquad T = W - \frac{W}{g}a\ddot{\theta}$$

From moment equilibrium at the centre of the rotor

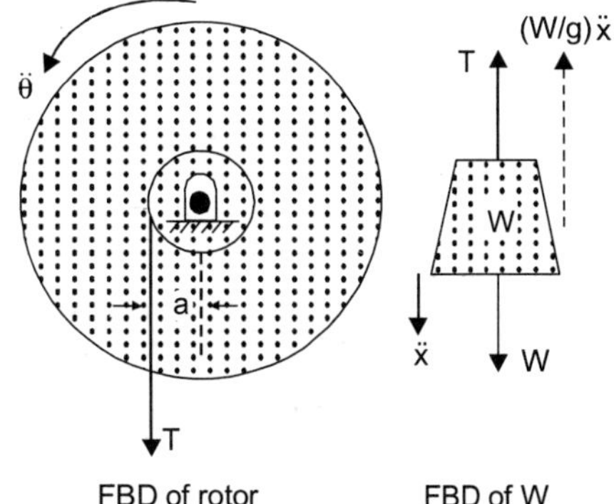

FBD of rotor FBD of W

$$I_0\ddot{\theta} = T \times a$$

$$\Rightarrow \qquad \frac{W_1}{g}\cdot i^2\ddot{\theta} = \left(W - \frac{W}{g}a\ddot{\theta}\right)a$$

$$\Rightarrow \qquad \frac{\ddot{\theta}}{g}(W_1 \cdot i^2 + W \cdot a^2) = W \cdot a$$

$$\Rightarrow \qquad \frac{\ddot{\theta}}{g}(500 \times 10^2 + 100 \times 5^2) = 100 \times 5$$

$$\Rightarrow \qquad \ddot{\theta} = \frac{g}{105}\text{rad/s}^2$$

$$\Rightarrow \qquad \ddot{x} = a\ddot{\theta} = \frac{g}{21}\text{m/s}^2$$

(*ii*) From equilibrium of the vertical forces of the weight W,

$$T = W + \frac{W}{g}\ddot{x} = W + \frac{W}{g}a\ddot{\theta}$$

From moment equilibrium of the pulley W,

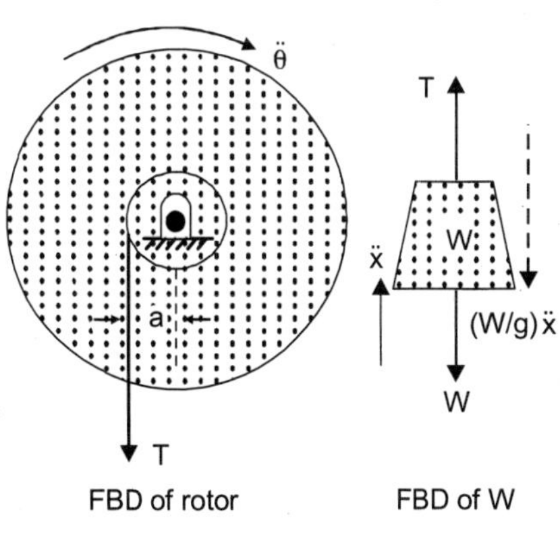

FBD of rotor FBD of W

$$M = I_0\ddot{\theta} + T \cdot a$$

$$= \frac{W_1}{g}i^2\ddot{\theta} + \left(W + \frac{W}{g}a\ddot{\theta}\right)a$$

$$= \frac{W_1}{g}i^2 \cdot \frac{\ddot{x}}{a}\left(W + \frac{W}{g}\ddot{x}\right)a$$

$$= \frac{500}{g}\cdot 10^2 \cdot \frac{g}{3}\cdot\frac{1}{5} + \left(100 + \frac{100}{g}\cdot\frac{g}{3}\right)5$$

$$= 4000\text{N·cm} = 40\text{N·m}$$

EXAMPLE 8

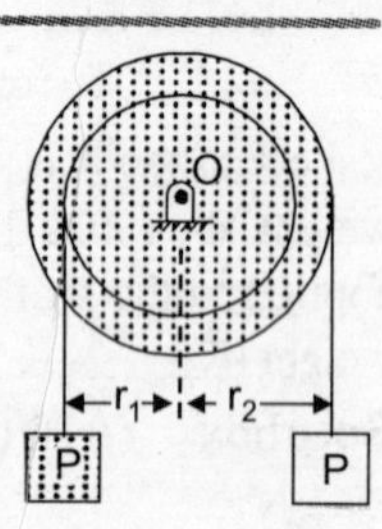

The two step pulley is shown in figure, has weight $W = 400\text{N}$ and radius of gyration $i_0 = 20\text{cm}$. Develop a formula for the downward acceleration of the following weight $P = 50\text{N}$ on the right. $r_1 = 10\text{cm}$ and $r_2 = 15\text{cm}$.

SOLUTION From vertical equilibrium of the weight at left,

$$T_1 = P + \frac{P}{g}\ddot{x}_1 = P + \frac{P}{g}r_1\ddot{\theta}$$

Vertical equilibrium of weight at right,

$$T_2 = P - \frac{P}{g}\ddot{x}_1 = P - \frac{P}{g}r_2\ddot{\theta}$$

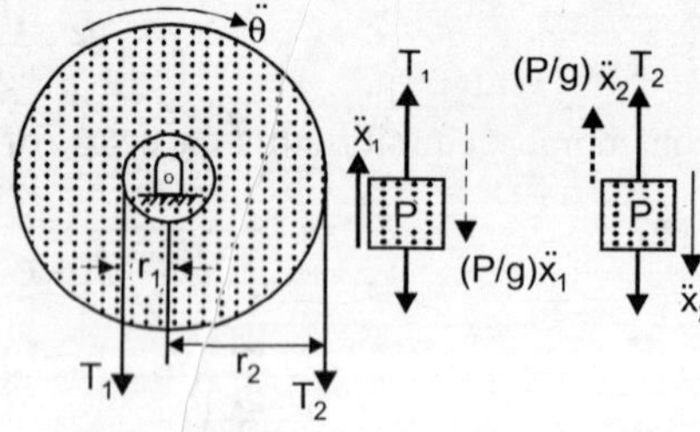

From moment equilibrium at centre of the pulley,

$$I_0\ddot{\theta} + T_1 r_1 = T_2 r_2$$

$$\Rightarrow \quad I_0\ddot{\theta} + \left(P + \frac{P}{g}r_1\ddot{\theta}\right)r_1 = \left(P - \frac{P}{g}r_2\ddot{\theta}\right)r_2$$

$$\Rightarrow \quad \frac{W}{g}\cdot i_0^2\ddot{\theta} + P\cdot r_1 + \frac{P}{g}r_1^2\ddot{\theta} = P\cdot r_2 - \frac{P}{g}r_2^2\ddot{\theta}$$

$$\Rightarrow \quad \ddot{\theta}\left(\frac{W}{g}\cdot i_0^2 + \frac{P}{g}r_1^2 + \frac{P}{g}r_2^2\right) = P\cdot r_2 - P\cdot r_1$$

$$\Rightarrow \quad \ddot{\theta}\left[400\times(0.2)^2 + 50\times(0.1)^2 + 50\times(0.15)^2\right] = 50(0.15 - 0.1)\times 9.81$$

$$\Rightarrow \quad \ddot{\theta} = 1.392 \text{ rad/s}^2$$

$$\Rightarrow \quad \ddot{x}_2 = 1.392\times 14 = 20.87\text{cm/s}^2$$

EXAMPLE 9

A flywheel rotating freely at 1800 r.p.m clockwise is subjected to a variable counterclockwise torque which is first applied at time $t = 0$. The torque produces a counterclockwise angular acceleration $\ddot{\theta} = 4t$ rad/s^2, where t is the time in seconds during which the torque is applied. Determine (*i*) the time required for the flywheel to reduce to clockwise angular speed to 900 r.p.m, (*ii*) the time required for the flywheel to reverse its direction of rotation, and (*iii*) the total number of revolutions, clockwise plus counterclockwise, turned by the flywheel during the first 14 seconds of torque application.

SOLUTION (*i*) Since angular acceleration $\ddot{\theta}$ is a function of the time t, we may integrate it to obtain angular velocity $\dot{\theta}$.

The initial angular velocity $\dot{\theta}_0 = -\dfrac{1800(2\pi)}{60} = -60\pi$ rad/s, the minus sign signifies clockwise direction.

We know,
$$\ddot{\theta} = \frac{d\dot{\theta}}{dt}$$

$$\Rightarrow \quad d\dot{\theta} = \ddot{\theta}dt$$

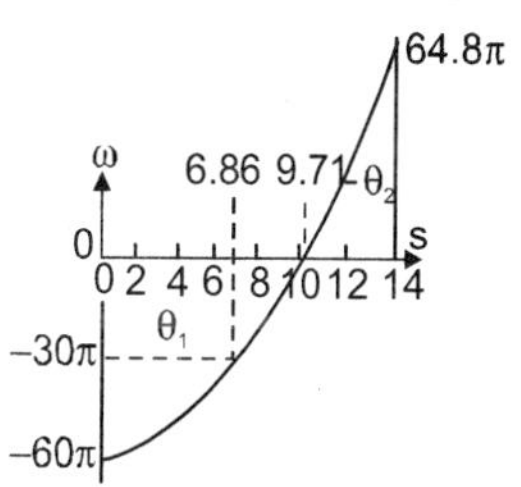

$$\Rightarrow \qquad \int_{-60\pi}^{\dot\theta} d\dot\theta = \int_0^t 4t\,dt$$

$$\Rightarrow \qquad \dot\theta = -60\pi + 2t^2$$

Substituting the clockwise angular speed of 900 r.p.m or

$$\dot\theta = -900(2\pi)/60 = -30\pi \text{ rad/s}, \; t = 6.86\text{s}$$

(*ii*) The flywheel changes direction when its angular velocity is momentarily zero, we have,

$$0 = -60\pi + 2t^2$$

$$\Rightarrow \qquad t = 9.71\text{s}$$

(*iii*) The total number of revolution through which the flywheel turns during 14 seconds is the number of clockwise turns N_1 during the first 9.71 seconds, plus the number of counterclockwise turns N_2 during the remainder of the interval. For first interval we have,

$$\dot\theta = \frac{d\theta}{dt}$$

$$\Rightarrow \qquad d\theta = \dot\theta dt$$

$$\Rightarrow \qquad \int_0^{\theta_1} d\theta = \int_0^{9.71} \dot\theta d\,t = \int_0^{9.71} \left(-60\pi + 2t^2\right) dt$$

$$\Rightarrow \qquad \theta_1 = -1220\text{rad}$$

So, $$N_1 = \frac{1220}{2\pi} = 194.2 \text{ revolutions clockwise}$$

For second interval $$\int_0^{\theta_2} d\theta = \int_{9.71}^{14} \left(-60\pi + 2t^2\right) dt$$

$$\Rightarrow \qquad \theta_2 = 410 \text{ rad}$$

So, $$N_2 = \frac{410}{2\pi} = 65.3 \text{ revolutions counter clockwise}$$

Hence total number of revolution,

$$N = N_1 + N_2 = 194.2 + 65.3 = 259.5 \simeq 259$$

EXAMPLE 10

The pinion A of the hoist motor drives gear B, which is attached to the hoisting drum is shown in figure. Load P is lifted from its rest position and acquires an upward velocity of 2m/s in a vertical rise of 0.8m with constant acceleration. As the load passes this position, compute (*i*) the acceleration of the point C on the cable in contact with drum, and (*ii*) the angular velocity and angular acceleration of the pinion A.

SOLUTION (*i*) The vertical velocity and acceleration of the load P are the same as the tangential velocity v and tangential acceleration a_t of the point C, if the cable does not slip on the drum. Normal acceleration a_n for rectilinear acceleration of load P is

$$a_n = \frac{v^2}{r} = \frac{2^2}{0.4} = 10\text{m/s}^2.$$

Tangential acceleration, $$a_t = a = \frac{v^2}{2s} = \frac{2^2}{2\times 0.8} = 2.5\text{m/s}^2.$$

So, total acceleration of the point C,

$$a_C = \sqrt{a_n^2 + a_t^2} = \sqrt{10^2 + 2.5^2} = 10.31\text{m/s}^2$$

(*ii*) Angular motion of gear A is determined from the angular motion of gear B by the velocity v_1 and tangential acceleration a_t of their common point of contact. Angular motion of gear B is determined from the motion of point C on the attached drum.

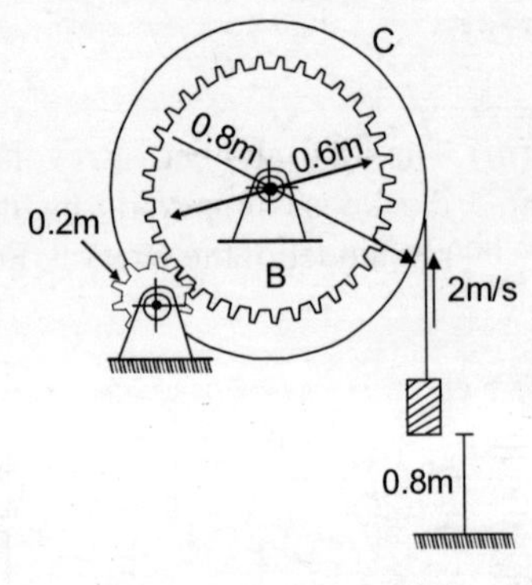

We know, $$v = \omega r$$

$$\Rightarrow \quad \omega_B = \frac{v}{r} = \frac{2}{0.4} = 5\text{rad/s}$$

And $$a_t = \alpha_B r$$

$$\Rightarrow \quad \alpha_B = \frac{a_t}{r} = \frac{2.5}{0.4} = 6.25\text{rad/s}^2$$

Then from $$v_1 = r_A\omega_A = r_B\omega_B \text{ and } a_1 = r_A\alpha_A = r_B\alpha_B$$

We have, $$\omega_A = \frac{r_B}{r_A}\omega_B = \frac{0.3}{0.1}\times 15 \text{ rad/s}$$

and $$\alpha_A = \frac{r_B}{r_A}\alpha_B = \frac{0.3}{0.1}\times 6.25 \text{ rad/s}^2$$

EXAMPLE 11

A wheel of radius r rolls on a flat surface without slipping. Determine the angular motion of the wheel in terms of the linear motion of its center O. Also determine the acceleration of a point on the rim of the wheel as the point comes into contact with surface on which the wheel rolls.

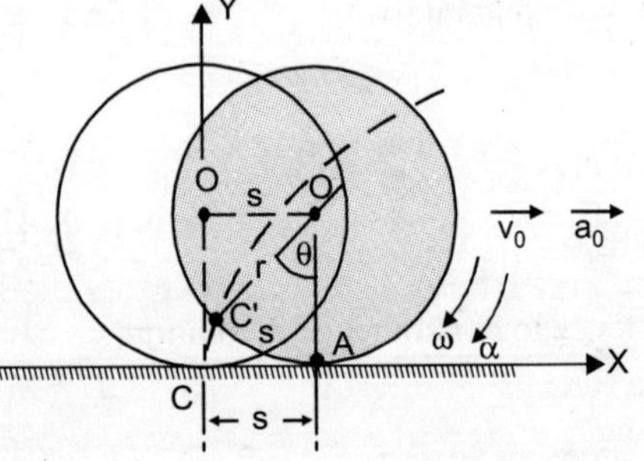

SOLUTION The figure shows the wheel rolling to the right from the dotted to the full position without slipping. The linear displacement of center O is $OO' = S$, as roll without slip arc length $AC' = S$. The radial line CO rotates to the new position $C'O'$ through the angle θ in radian.

From figure we have, $$S = r\cdot\theta$$

$$\Rightarrow \quad \frac{dS}{dt} = r\frac{d\theta}{dt}$$

$$\Rightarrow \quad v_0 = r\dot{\theta} = r\omega$$

Again, $$\frac{dv_0}{dt} = r\frac{d\omega}{dt}$$

$$\Rightarrow \quad a_0 = r\alpha$$

When point C has moved along its cycloidal path to C', its new co-ordinates (x, y) is taken at the point of contact between C on the rim of the wheel and the ground. So,

$x = S - r\sin\theta = r\theta - r\sin\theta$	$y = r - r\cos\theta = r(1-\cos\theta)$
$\Rightarrow \quad \dfrac{dx}{dt} = \dfrac{d}{dt}(r\theta - r\sin\theta)$	$\Rightarrow \quad \dfrac{dy}{dt} = \dfrac{dr}{dt}(1-\cos\theta)$
$\Rightarrow \quad \dot{x} = v_0(1-\cos\theta)$	$\Rightarrow \quad \dot{y} = v_0 \sin\theta$
$\Rightarrow \quad \dfrac{d\dot{x}}{dt} = \dfrac{dv_0}{dt}(1-\cos\theta)$	$\Rightarrow \quad \dfrac{d\dot{y}}{dt} = \dfrac{d}{dt}(v_0\sin\theta)$
$\Rightarrow \quad \ddot{x} = a_0(a - \cos\theta) + \omega^2 r\sin\theta$	$\Rightarrow \quad \ddot{y} = a_0\sin\theta + \omega^2 r\cos\theta$

When $\theta = 0°$ for the desired instant of contact, $\ddot{x} = 0$ and $\ddot{y} = \omega^2 r$

EXAMPLE 12

A cylinder of radius 2m rolls without slipping along a horizontal plane AB. Its centre has a uniform velocity of 30m/s. Find the velocity of the points E and F on the circumference of the cylinder shown in figure.

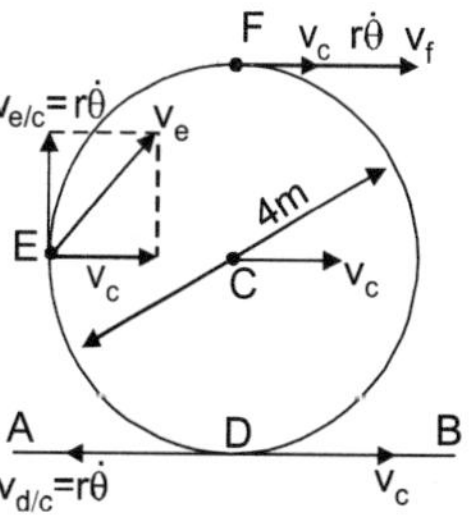

SOLUTION The cylinder rolls without slipping, so the velocity of the point D, in contact with the plane AB, must be zero, $v_d = 0$. The motion of the centre C of the cylinder is known, so we choose it as a pole and $v_c = 30$m/s.

Adding vectorially, $\quad \vec{v}_d = \vec{v}_c + \vec{v}_{d/c} = \vec{v}_c - r\dot{\theta}$

$$\Rightarrow \quad 0 = \vec{v}_c - r\dot{\theta}$$

$$\Rightarrow \quad \theta = \frac{30}{2} = 15\text{rad/s}$$

Similarly, $\quad \vec{v}_e = \vec{v}_c + \vec{v}_{e/c} = \vec{v}_c - r\dot{\theta}$

$$\Rightarrow \quad v_e = \sqrt{(v_c)^2 + (r\dot{\theta})^2} = \sqrt{(v_c)^2 + (v_c)^2} = v_c\sqrt{2} = 21.21\text{m/s}$$

And $\quad \theta = \tan^{-1}\left(\dfrac{r\dot{\theta}}{v_c}\right) = \tan^{-1}\left(\dfrac{v_c}{v_c}\right) = 45°$

Velocity of point F,

$$\vec{v}_f = \vec{v}_c + \vec{v}_{f/c} = \vec{v}_c + r\dot{\theta} = 2\vec{v}_c = 2\times 30 = 60\text{m/s}$$

EXAMPLE 13

A cylinderical roller is in contact at its top and bottom, with two conveyor belts PQ and RS as shown in figure. If the belts run at the uniform speeds of $v_1 = 6$m/s and $v_2 = 4$m/s, find the linear velocity and angular velocity of the roller. The diameter of roller may be assumed to be 60cm.

SOLUTION Choose C as the pole. Let V_c is the linear velocity of the centre of the roller and its angular velocity is $\dot{\theta}$.

Velocity of the point A, $\vec{v}_a = \vec{v}_c + \vec{v}_{a/c}$

$\Rightarrow$ $6 = v_c + r\dot{\theta}$

Velocity of the point B, $\vec{v}_b = \vec{v}_c - \vec{v}_{b/c}$

$\Rightarrow$ $4 = v_c - r\dot{\theta}$

Solving above equations, $v_c = 5\text{m/s},\ \dot{\theta} = 3.33\text{rad/s}$

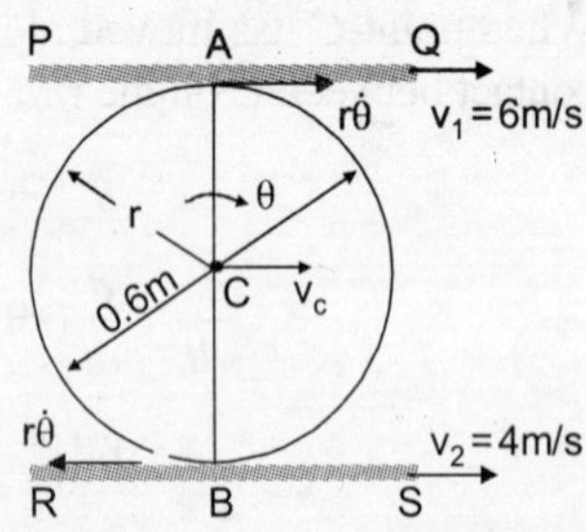

EXAMPLE 14

A reciprocating engine mechanism is shown in figure. The crank OA has a constant angular velocity of $N = 400$ r.p.m. The crank OA is of length $r = 12$cm and connecting rod AB is of length $l = 60$cm. Find (*i*) the angular velocity of connecting rod, (*ii*) the velocity of piston when crank makes an angle of with the horizontal. Shall the velocity of the piston be maximum when $\theta = 90°$?

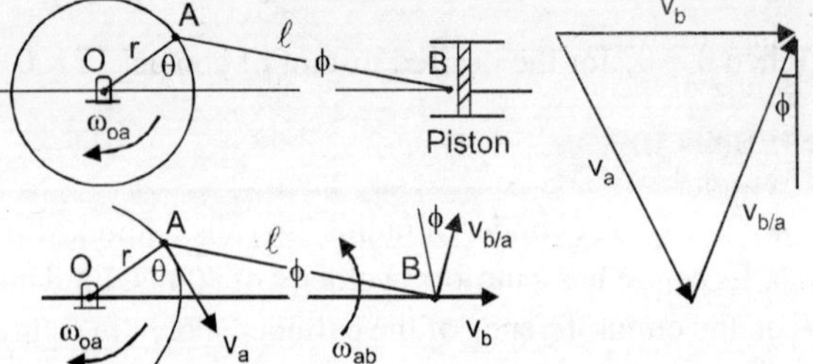

SOLUTION Here, $OA = r = 0.12$m,

$AB = l = 0.6\text{m}, \theta = 30°$

We have from triangle OAB,

$$\frac{\sin 30°}{l} = \frac{\sin\phi}{r}$$

$$\Rightarrow \quad \sin\phi = \frac{r}{l}\sin 30° = \frac{0.12}{0.6} \times \frac{1}{2}$$

$$\Rightarrow \quad \phi = 5.7°$$

Considering motion of the crank OA,

$$\omega_{oa} = \frac{2\pi N}{60} = \frac{2\pi \times 400}{60} = 41.87\text{rad/s},$$

$$v_a = OA \times \omega_{oa} = 0.12 \times 41.87 = 5.024\text{m/s (normal to } OA)$$

Choose A as pole which is the end of connecting rod. Velocity of the point B can be written as:

$$\vec{v}_b = \vec{v}_a + \vec{v}_{b/a} = \vec{v}_a + \vec{l} \times \vec{\omega}_{ab}$$

The velocity v_b of the end B must be horizontal and is equal to the velocity of the piston. The velocity $v_{b/a}$ is known by its direction and the velocity v_a is known both by magnitude and direction. The vector diagram of the velocities is shown.

Now we have, $$\frac{v_b}{\sin 35.7°} = \frac{v_a}{\sin 84.3°} = \frac{v_{b/a}}{\sin 60°}$$

$\Rightarrow \qquad \frac{v_b}{\sin 35.7^\circ} = \frac{5.024}{\sin 84.3^\circ} = \frac{v_{b/a}}{\sin 60^\circ}$

$\Rightarrow \qquad v_b = 2.95\text{m/s}, \; v_{b/a} = 4.37\text{m/s}$

So we have, $\qquad \omega_{ab} = \frac{v_{b/a}}{l} = \frac{4.37}{0.6} = 7.29$ rad

EXAMPLE 15

A cylinder of mass m and radius r resting on a inclined plane is released from rest and rolls down the inclined plane without slipping. Determine (*i*) the acceleration of its centre of mass, (*ii*) the maximum angle θ of the inclined plane for which the body will roll without slipping, (*iii*) the maximum velocity of centre of the cylinder after it has rolled a distance of 2m. Assume the coefficient of static friction $\mu = 0.2$. What would be the acceleration of the centre of the cylinder if it were to move down the inclined plane under the frictionless condition?

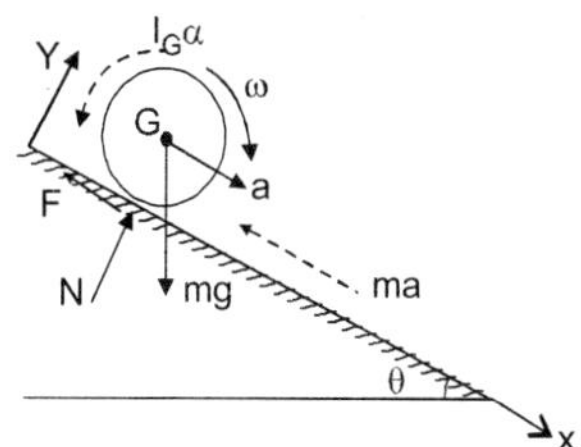

SOLUTION (*i*) Let the linear acceleration of the centre G be a and angular acceleration be a, From figure we have,

$$\left.\begin{aligned} \sum F_y &= 0 \Rightarrow & N &= mg\cos\theta \\ \sum F_x &= 0 \Rightarrow & F &= mg\sin\theta - ma \\ \sum M_G &= 0 \Rightarrow & F \times r &= I_G\alpha \end{aligned}\right\}$$

As the cylinder rolls without slipping, we have,

$$a = r\alpha$$

$$\Rightarrow \qquad \alpha = \frac{a}{r}$$

So, $\qquad F \times r = I_G \times \frac{a}{r}$

$$\Rightarrow \qquad (mg\sin\theta - ma)r = I_G \times \frac{a}{r}$$

$$\Rightarrow \qquad a = \frac{mg\sin\theta}{m + \frac{I_G}{r^2}} = \frac{mg\sin\theta}{m + \frac{mr^2}{2r^2}} = \frac{2}{3}g\sin\theta$$

(*ii*) We have the value of friction,

$$F = \frac{I_G}{r}\alpha = \frac{mr^2}{2r}\cdot\frac{a}{r} = \frac{ma}{2} = \frac{m}{2}\cdot\frac{2}{3}g\sin\theta = \frac{mg\sin\theta}{3}$$

The maximum frictional force possible for any value of θ, is when the body is just about to slide and is equal to $\mu N = \mu mg\cos\theta$.

So for the cylinder to roll without slipping, $F < \mu mg\cos\theta$

$$\Rightarrow \qquad \frac{mg\sin\theta}{3} < \mu mg\cos\theta$$

$$\Rightarrow \qquad \tan\theta < 3\mu$$

$$\Rightarrow \qquad \theta \le 30.963°$$

(*iii*) The maximum value of $\theta = 30.96°$.

So, $$a = \frac{2}{3} g \sin 30.96° = 3.36\text{m/s}^2$$

Again, $$v = \sqrt{0^2 + 2\times 3.36 \times 2} = 3.67\text{m/s}$$

If there is no friction, $F = 0$.

In this case, $mg \sin 30.96° = ma$

$$\Rightarrow \qquad a = 9.81 \times \sin 30.96° = 5.05\text{m/s}^2$$

EXAMPLE 16

A sphere, a cylinder and a hoop each having the same mass and radius are released from rest on an inclined plane of angle θ. If they roll down the inclined plane without slipping compare the acceleration of their centres.

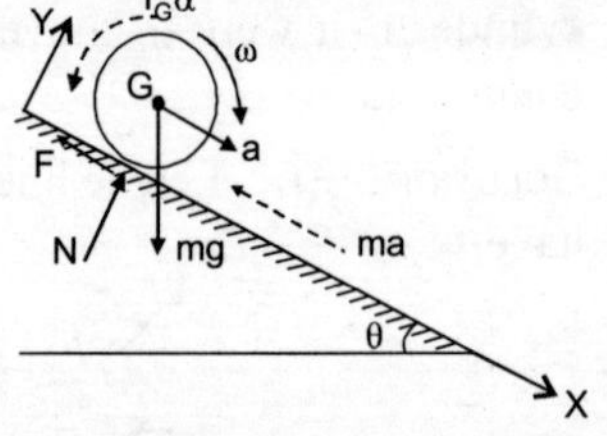

SOLUTION From the previous example we have,

$$a = \frac{mg \sin\theta}{m + \frac{I_G}{r^2}}$$

Mass moment of inertia of a sphere, a cylinder, a hoop are,

$$I_{GS} = \frac{2mr^2}{5}, \; I_{GC} = \frac{mr^2}{2}, \; I_{GH} = mr^2$$

Substituting these in above equation, we find the acceleration of a sphere, a cylinder, a hoop, $a_S = 0.714 g \sin\theta$, $a_C = 0.667 g \sin\theta$, $a_H = 0.5 g \sin\theta$

EXAMPLE 17

A right circular cylinder of mass m and radius r is suspended from a cord that is wound around its circumference. If the cylinder is allowed to fall freely, find the acceleration of its mass centre G and the tension in the cord.

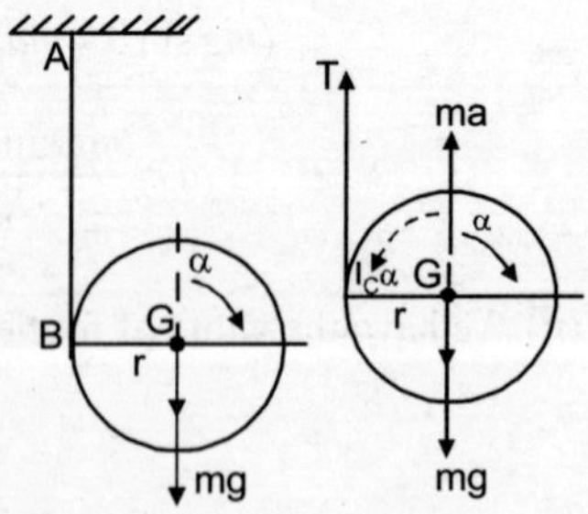

SOLUTION The forces acting on the cylinder are shown in figure. Acceleration of the mass centre G is a.

From figure we have,

$$\sum F_y = 0$$

$$\Rightarrow \qquad T = mg - ma$$

$$\sum M_c = 0$$

$$\Rightarrow \qquad T \times r = I_C \alpha = \frac{mr^2}{2} \cdot \frac{a}{r} = \frac{mar}{2}$$

$$\Rightarrow \qquad (mg - ma)r = \frac{mar}{2}$$

$$\Rightarrow \qquad a = \frac{2g}{3}$$

So, tension
$$T = mg - ma = mg - m\frac{2g}{3} = \frac{mg}{3}$$

EXAMPLE 18

A spool is pulled by a force P applied to the end of the cord wound around the axle of the spool. Find the acceleration of the spool if it rolls without slipping along a horizontal plane. Given $r_1 = 15\text{cm}$, $r_2 = 30\text{cm}$, $\theta = 30^\circ$, $P = 100\text{N}$, $W = 200\text{N}$, radius of gyration $i_0 = 10\text{cm}$.

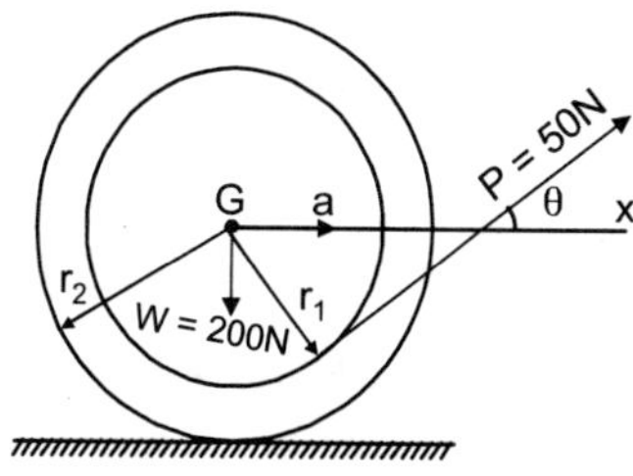

SOLUTION Forces acting on the spool are shown in figure. From figure we have,

$$\sum F_x = 0$$

$$\Rightarrow \qquad P\cos\theta - \frac{W}{g}a = F$$

$$\sum F_y = 0$$

$$\Rightarrow \qquad N = P\sin\theta$$

$$\sum M_G = 0$$

$$\Rightarrow \qquad Fr_2 - Pr_1 = \left(\frac{W}{g} i_0^2\right) \cdot \alpha$$

$$\Rightarrow \qquad \left(P\cos\theta - \frac{W}{g}a\right) r_2 - Pr_1 = \left(\frac{W}{g} i_0^2\right) \cdot \frac{a}{r_2}$$

FBD of spool

$$\Rightarrow \qquad a = \frac{Pg(r_2\cos\theta - r_1)}{W\left(\dfrac{i_0^2}{r_2} + r_2\right)}$$

$$= \frac{100 \times 9.81 \times (0.3\cos 30^\circ - 0.15)}{200\left(\dfrac{0.1^2}{0.3} + 0.3\right)} = 1.62\text{m/s}^2$$

EXAMPLE 19

A solid cylinder and a thin hoop of equal masses m and radii r are connected by a bar $C_2\,C_1$ as shown in figure. The system rolls down the inclined plane without slipping. Find the acceleration of the system and the tension in the bar.

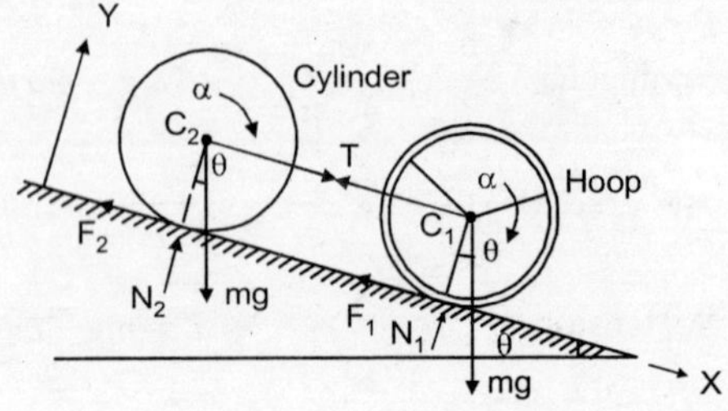

SOLUTION Forces acting are shown in figure. Equilibrium equations for hoop,

$$\left.\begin{aligned} \sum F_{xh} &= 0 && \Rightarrow F_1 = mg\sin\theta - T - ma \\ \sum F_{yh} &= 0 && \Rightarrow N_1 = mg\cos\theta \\ \sum M_{ch} &= 0 && \Rightarrow F_1 r = I_{c1}\alpha \end{aligned}\right\}$$

Equilibrium equations for cylinder

$$\left.\begin{aligned} \sum F_{xc} &= 0 && \Rightarrow F_2 = mg\sin\theta + T - ma \\ \sum F_{yc} &= 0 && \Rightarrow N_2 = mg\cos\theta \\ \sum M_{cc} &= 0 && \Rightarrow F_2 r = I_{c2}\alpha \end{aligned}\right\}$$

Now, $\sum M_{C_2} + \sum M_{C_1} = 0$ yields,

$$\Rightarrow \quad r(F_1 + F_2) = \alpha(I_{C1} + I_{C2})$$

$$\Rightarrow \quad a = \frac{4g}{7}\sin\theta$$

So,
$$\begin{aligned} T &= mg\sin\theta - F_1 - ma = mg\sin\theta - \frac{I_{C1}\alpha}{r} - ma \\ &= mg\sin\theta - \frac{mr^2}{r}\cdot\frac{a}{r} - ma \\ &= m(g\sin\theta - 2a) \\ &= m\left(g\sin\theta - \frac{2\times 4}{7}g\sin\theta\right) \\ &= -\frac{mg}{7}\sin\theta \\ &= \frac{mg}{7}\sin\theta \text{ (compressive)} \end{aligned}$$

Exercise

1. A solid right circular rotor of radius r and weight W tied to a horizontal plane by a rod AC has initial angular velocity ω_0 as shown in figure. If the rotor is suddenly allowed to rest its full weight on the plane, what time t will elapse before it comes to rest? The coefficient of friction at B is μ. Numerical data are given as follows: **Ans.** [t =3.64sec]

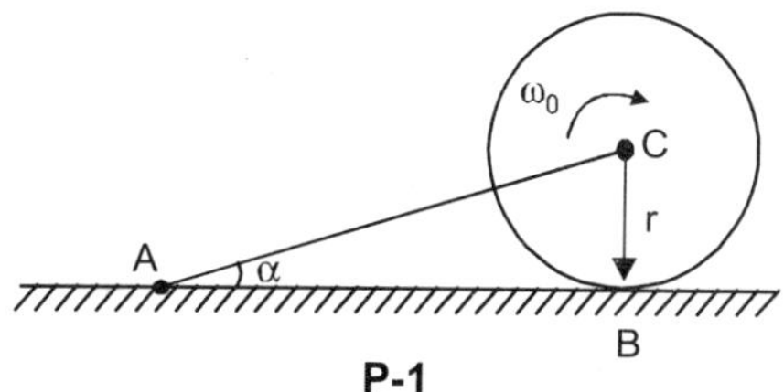

P-1

2. It is a system consisting of a block A of weight $W_a = 866$N connected to a block B of weight $W_b = 644$N by a flexible cord which runs over a pulley of radius $r = 12$cm and weight $W_c = 322$N. The block A slides along a plane inclined to the horizontal by the angle $\alpha = 30°$ and for which the coefficient of friction is $\mu = 0.1$. Friction on the axle of the pulley is assumed to supply a constant resisting torque of 10. Assuming that no slipping occurs between the cord and the pulley. Determine (a) the acceleration of the blocks, and (b) the maximum tensile force is the cord. **Ans.** [(a) $a = 2.43$ ft / sec^2 (b) $S_{max} = 395$N]

3. The circular disk suspended by a slender rod, as shown in figure performs torsional oscillations such that its angle of rotation measured from the equilibrium position is given by the equation $\theta = \theta_0 \cos pt$. Find general expressions for the angular velocity and angular acceleration of the disk and also general expressions for the radial and tangential components of the acceleration of a point on the periphery of the disk. Determine the maximum values of these components if the disk makes 2 oscillations/sec, has the radius $r = 5$cm, and maximum angular velocity 2π sec^{-1}.

Ans. [$(a_t)_{max} = 40\pi^2$ cm /sec^2, $(a_n)_{max} = 20\pi^2$ cm/sec^{-1}]

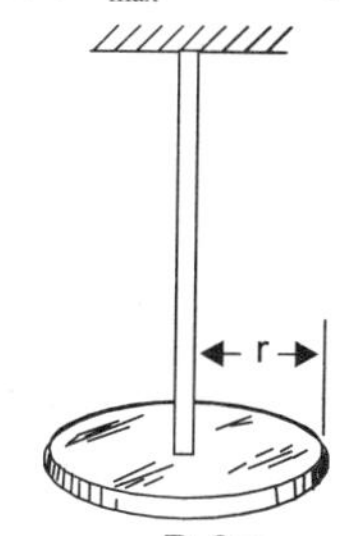

P-3

4. A homogeneous sphere, of radius $a = 10$cm and weight $W = 100$N, can rotate freely about a diameter. If it starts from rest and gains, with constant angular acceleration, an angular speed n = 180 r.p.m in 12 revolutions. Find the acting moment M. **Ans.**[M= 24.4cm – N]

5. A solid circular rotor of radius r and weight W which can rotate about its geometric axis is braked by the device shown in figure. If the coefficient of friction between rotor and brake shoe is μ and the rotor has an initial angular velocity ω before the brake is applied, how many revolutions will it make before coming to rest? Neglect friction in the bearings. **Ans.** [118 revolutions]

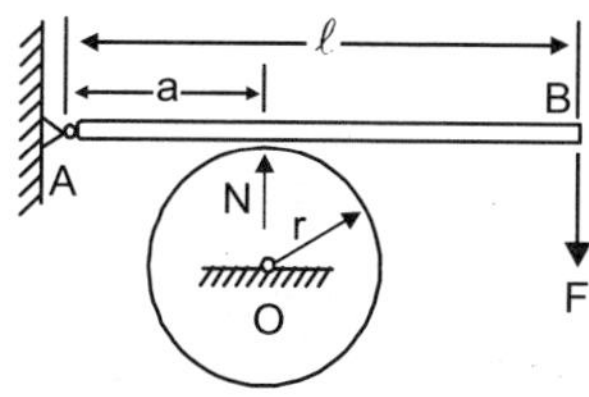

P-5

6. The square plate rotates about the fixed point O. At the instant represented, its angular velocity is $\omega = 6$ rad/s and its angular acceleration is $\alpha = 4$ rad /s^2, in the directions shown in the figure. Determine the velocity and acceleration of (a) point A, and (b) point B.

Ans. [(a) $v_A = 270i$ mm/s, $a_A = -180i - 1620j$ mm/s^2, (b) $v_B = 270i + 180j$ mm/s, $a_B = 900i - 1740j$ mm/s^2]

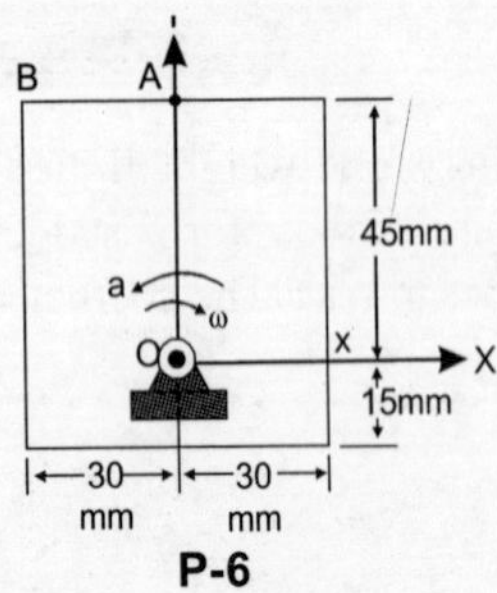

P-6

7. The angular acceleration of a body which is rotating about a fixed axis is given by $\alpha = -k\omega^2$, where the constant $k = 0.1$ (no units). Determine the angular displacement and time elapsed when the angular velocity has been reduced to one-third its initial value $\omega_0 = 12$ rad/s.

8. The flywheel is rotating with an angular velocity ω_0 at time $t = 0$ when a torque is applied to increase its angular velocity. If the torque is controlled so that the angle θ between the total acceleration of point A remains constant, determine the angular velocity ω and the angular acceleration α as functions of the time t.

9. A wheel 3m diameter rolls to the right on a horizontal plane with an angular velocity of 8 rad/s (clockwise) and an angular acceleration counterclockwise of 8 rad/s^2 as shown in figure. The latter merely indicates that the angular velocity of the wheel is decreasing. Determine the linear velocity and acceleration of the top point B on the wheel.

10. The angular velocity of a gear is controlled according to $\omega = 12 - 3t^2$ where ω is in radians per second and is positive in the clockwise sense and where t is the time in seconds. Find the net angular displacement $\Delta\theta$ from the time $t = 0$ to $t = 3$s. Also find the total number of revolutions N through which the gear turns during 3 secs.

11. A wheel of diameter 1m rolls without slipping on a flat surface. The center of the wheel is moving with a velocity of 10m/s. Find the velocity of the points A, B and C.

Ans. [$v_a = 10$m/s, $v_b = 20$m/s, $v_c = 10$m/s]

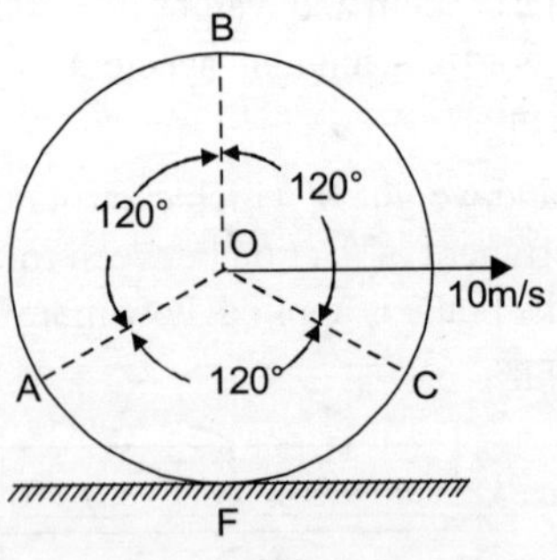

P-11

12. A reciprocating engine mechanism is shown in figure. The crank OA is of length 15cm and rotating at 600 r.p.m. The connecting rod AB is 70cm long. Find (a) the angular velocity of the connecting rod, (b) the velocity of the piston B, (c) the velocity of a point C on the connecting rod at a distance 20cm from A, when $\theta = 45°$.

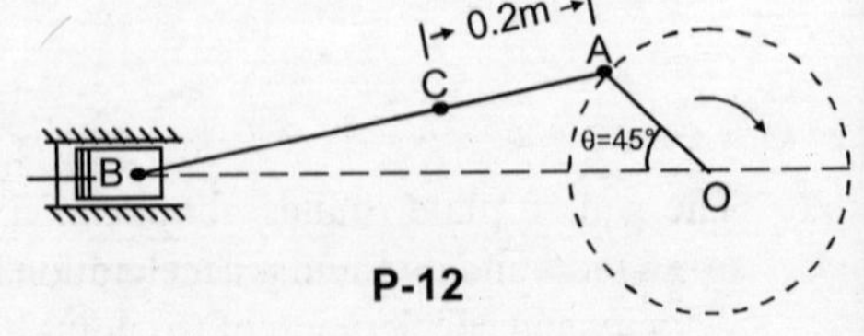

P-12

Ans.[$\omega_{ab} = 9.64$rad/s, $v_b = 7.67$m/s, $v_c = 8.4$m/s]

KINETICS OF PARTICLES

11.1 INTRODUCTION

In the previous chapter of dynamics discussions have been done on *kinetics of motion of a particle*, both *rectilinear* and *curvilinear* without regards to its cause. Here we are going to concentrate on *kinetics of rectilinear motion*, where the relationship of particle motion is set up in the context of forcing function. At the very onset, the law's of motion which govern principles of kinetics will be discussed. The complete set of principles and their final formulation are the work of *Newton* and are commonly known as *Newton's Law of Motion.*

11.2 LAWS OF MOTION

(a) First Law: Every body continues in its state of rest or of uniform motion in a straight line except in so far as it may be compelled by externally impressed forced to change that state. This law was first discovered by Galileo and sometimes it is called as Law of Inertia.

(b) Second Law: The acceleration of a given particle is proportional to the forced applied to it and takes place in the direction of the straight line in which the force acts. This law was also first discovered by Galileo, and later verified by Newton through several experiments.

(c) Third Law: To every action there is always an equal and opposite reaction.

11.3 EQUATION OF MOTION OF A PARTICLE

If a particle is subjected to a gravity force W, from the second law of motion, we can write,

$$g \propto W \quad \text{... (11.1)}$$

where g is the acceleration due to gravity. If a force F, instead of gravity force W, acts on the same particle and generates an acceleration a, in the direction of the force, then

$$a \propto F \quad \text{... (11.2)}$$

From the above two equations,

$$\frac{g}{a} = \frac{W}{F}$$

$$\Rightarrow \quad F = \frac{W}{g} a \quad \text{... (11.3)}$$

This is the general equation of motion of a particle. It can be inferred from equation (11.3) that, for a specific magnitude of force F, produces acceleration a, is inversely proportional to (W/g) factor. This factor is termed as *factor of sluggishness*, which is a measure of *inertia.* It is called the mass of the particle and denoted by m,

$$\frac{W}{g} = m \qquad \text{... (11.4)}$$

So the general equation of motion changes to

$$ma = F \qquad \text{... (11.5)}$$

The left hand side terms of the above equation i.e., ma is defined as the *inertia force.* As it tries or do oppose the motion of the particle, *inertia force acts oppositely in the direction of motion.* This inertia force is not *visible* but the effect is *perceivable.* Hence, this force is sometimes characterised as a *fictitious force.*

Often the acceleration is expressed in differential form, and subsequently, the equation of motion changes to,

$$m\ddot{x} = F$$

$$\Rightarrow \quad \frac{W}{g}\ddot{x} = F \qquad \text{... (11.6)}$$

11.4 VARIETY OF FORCING FUNCTION

The right hand side of the equation (11.6), denotes the externally applied force on the particle. This force F, can be a constant, a function of time, a function of speed, a function of position or any combination of these. A description of forcing function here is presented, in brief.

(a) Force as a function of time: A particle of mass m is subjected to a time-dependent forcing function $F(t)$, causing motion, as shown in the figure. Now from the law of motion, we write

$$m\ddot{x} = F(t)$$

$$\Rightarrow \quad m\frac{d^2x}{dt^2} = F(t)$$

$$\Rightarrow \quad \frac{d}{dt}\left(\frac{dx}{dt}\right) = \frac{F(t)}{m}$$

$$\Rightarrow \quad d\left(\frac{dx}{dt}\right) = \frac{F(t)}{m}\cdot dt$$

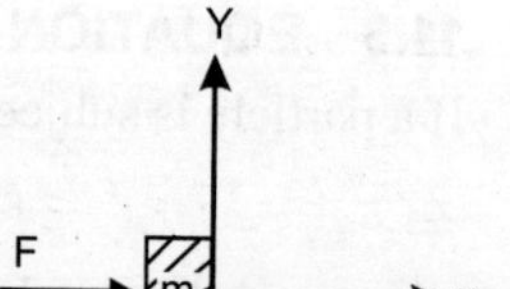

Fig. 11.1

Integrating once,

$$\frac{dx}{dt} = \int \frac{F(t)}{m}\,dt + C_1 \qquad \text{... (11.7)}$$

Again integrating,

$$x = \int\left[\int \frac{F(t)}{m}dt\right]dt + C_1 t + C_2 \qquad \text{... (11.8)}$$

Applying proper boundary conditions the values of constants of integration C_1 and C_2 can be obtained.

(b) Force as a function of speed: Aerodynamic drag force on air plane or missile is such a type of force which is a function of speed. Now, equation of motion appears as,

$$\frac{dV}{dt} = \frac{F(V)}{m}$$

$$\Rightarrow \qquad \frac{dV}{F(V)} = \frac{1}{m}\cdot dt$$

Considering seperation of variables and integrating once we have,

$$V = G(t, C_1) \qquad \text{... (11.9)}$$

Again, a second integration yields,

$$x = \int G(t, C_1)dt + C_2 \qquad \text{... (11.10)}$$

The constants of integration will be obtained from suitable boundary conditions.

(c) Force as a function of position: Frictionless *mass-spring* system as shown in the figure, is the simplest example of position-dependent forcing function. Now, equation of motion comes,

$$m\frac{dV}{dt} = F(x)$$

$$\Rightarrow \qquad m\frac{dV}{dx}\cdot\frac{dx}{dt} = F(x)$$

$$\Rightarrow \qquad m\cdot V\cdot\frac{dV}{dx} = F(x)$$

$$\Rightarrow \qquad m\cdot VdV = F(x)dx$$

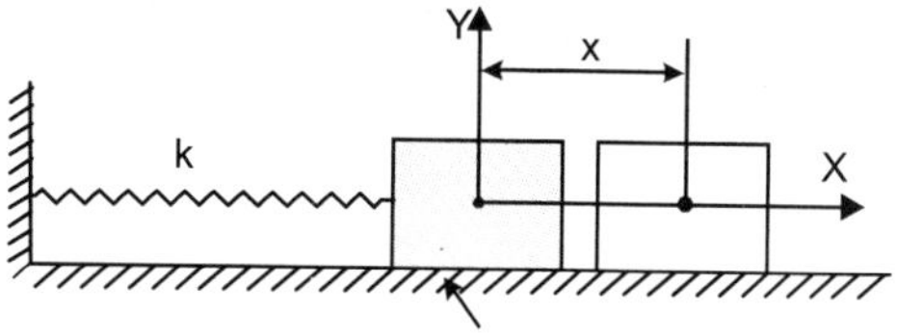

Fig. 11.2

Integrating once,

$$\frac{mV^2}{2} = \int F(x)dx + C_1 \qquad \text{... (11.11)}$$

Solving for V we have,

$$\frac{dx}{dt} = V = \left[\frac{2}{m}\int F(x)dx + C_1\right]^{1/2}$$

Separating variables and integrating again,

$$t = \int \frac{dx}{\left[\frac{2}{m}\int F(x)dx + C_1\right]^{1/2}} + C_2$$

A common force, we often encounter is *restoring force*, linear or non-linear in nature. *Linear restoring force* occurs when a particle of weight W and mass m is constrained by linear spring as shown in the figure. Here the force F is proportional to x measured from a position of W corresponding to undeformed configuration of the system i.e.,

$$F_{\text{spring}} \propto x$$

$$\Rightarrow \quad F_{\text{spring}} = -kx \qquad \text{... (11.12)}$$

This constant of proportionality k is defined as *spring constant* and *minus sign* indicates restoring effect.

Thus for a spring-mass system, the differential equation of motion is

$$\frac{d^2x}{dt^2} + \frac{k}{m}x = 0 \qquad \text{...(11.13)}$$

Solution of this differential equation yields,

$$x = C_1 \cos\sqrt{\frac{k}{m}}\cdot t + C_2 \sin\sqrt{\frac{k}{m}}\cdot t \qquad \text{...(11.14)}$$

This is the solution of the equation of free vibration of a particle. In studying this type of harmonic motion, following parameters need to be discussed.

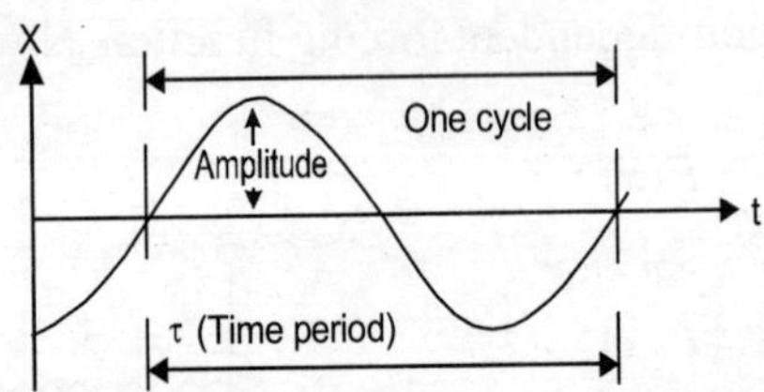

Fig. 11.3

(i) Cycle: *The cycle is that portion of a motion or series of events which, when repeated forms the motion.*

(ii) Frequency: *The number of cycles per unit time is the frequency.*

So,
$$f = \frac{1}{2\pi}\sqrt{\frac{k}{m}} \qquad \text{... (11.15a)}$$

Often $\sqrt{\frac{k}{m}}$ *is termed as the natural frequency of the system in radians per unit time.*

(iii) Period: *The period* (τ), *is the time of one cycle and thus the reciprocal of frequency*

$$\tau = 2\pi\sqrt{\frac{m}{k}} \qquad \text{...(11.15b)}$$

(iv) Amplitude: *The largest displacement attained by the particle during a cycle is the amplitude.*

In spring-mass system, sometimes more than one spring is used, either in *parallel* or in *series* condition, as shown in Figures 11.4 (*a*) and (*b*). Hence an equivalent spring constant k_e will come in the picture. In case of parallel connection, and in series connection,

$$k_e = k_1 + k_2 \quad \text{... (11.16a)}$$

$$\frac{1}{k_e} = \frac{1}{k_1} + \frac{1}{k_2} \quad \text{... (11.16b)}$$

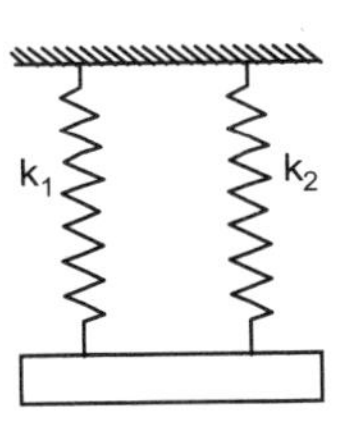

Parallel connection

Series connection

Fig. 11.4

11.5 D' ALEMBERT'S PRINCIPLE

The differential equation of rectilinear motion of a particle can be written in the form

$$F - m\ddot{x} = 0 \quad \text{...(11.17)}$$

This equation of motion of a particle is of the same form as an equation of static equilibrium and may be consider as an equation of *dynamic equilibrium.*

D' Alembert was first to point out that equations of motions could be written as equilibrium equations simply by introducing inertia forces in addition to the real forces acting on the system. And this idea is known as *D'Alembert's Principle.*

Multiple Choice Questions

Select the best alternative (s):

1. D'Alembert's principle is used for
   (*a*) reducing the problem of kinetics to equivalent statics problem
   (*b*) determining stresses in the truss
   (*c*) stability of floating bodies
   (*d*) designing safe structures
   (*e*) solving kinematic problems
2. An elevator weighing 1000 kg attains an upward velocity of 4m/s in two second with uniform acceleration. The tension in the supporting cables will be
   (*a*) 1000 kg
   (*b*) 800 kg
   (*c*) 1200 kg
   (*d*) 2000kg
   (*e*) not possible to determine
3. When a body slides down an inclined surface the acceleration (f) of the body is given by
   (*a*) $f = g$
   (*b*) $f = g \sin \theta$
   (*c*) $f = g \cos \theta$
   (*d*) $f = g \tan \theta$
   (*c*) $f = g \cos \theta$

4. As the acceleration of a vibrating particle executing simple harmonic motion increases, its time period
   (*a*) increases (*b*) decreases (*c*) remains unchanged
   (*d*) first decreases and then increases (*e*) first increases and then decreases
5. A boy is swinging on a swing. If another boy sits along with him without disturbing his motion, then the time period of swing will
   (*a*) increase (*b*) decrease (*c*) be doubled
   (*d*) remain the same (*e*) is halved
6. A body in S.H.M will have maximum velocity when its amplitude is
   (*a*) maximum (*b*) –ve maximum (*c*) zero
   (*d*) average (*e*) at mid value
7. In order to double the period of simple pendulum
   (*a*) the mass of its bob should be doubled (*b*) the mass of its bob should be quadrupled
   (*c*) the length should be doubled (*d*) the length should be quadrupled
   (*e*) the length should be halved

Answers

1. (*a*) 2. (*c*) 3. (*b*) 4. (*b*) 5. (*a*) 6. (*c*) 7. (*d*)

Numerical Examples

Example 1

Weights *W* and 2*W* are supported in a vertical plane by a string and pulley arrangement shown in figure. Find magnitude of additional weight *Q* which will give a downward acceleration $a = 0.1g$.

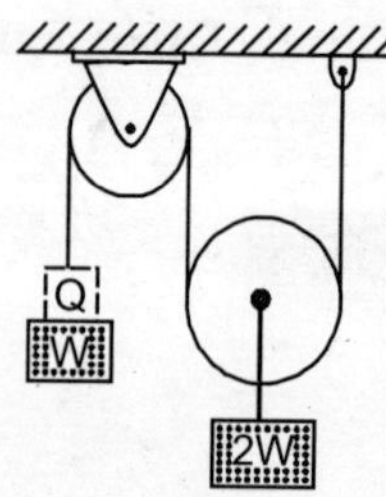

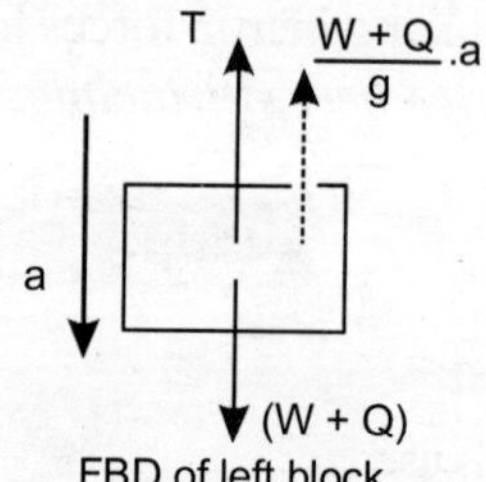

FBD of left block

Solution From free body diagram the equation of motion along vertical direction becomes,

$$T + \left(\frac{W+Q}{g}\right)a = (W+Q)$$

$$\Rightarrow \quad T = 0.9(W+Q)$$

$$\Rightarrow \quad 2T = 1.8(W+Q)$$

Again from pulley,

$$T_1 = 2T$$

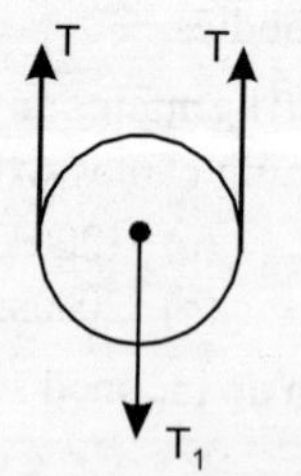

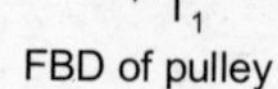
FBD of pulley

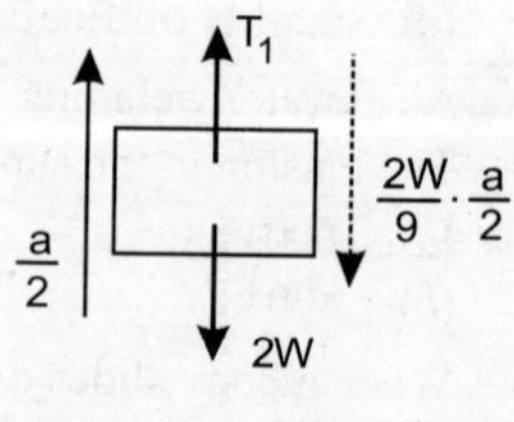

FBD of right block

And, from vertical force equilibrium of right block

$$T_1 = 2W + \frac{2W}{g} \cdot \frac{a}{2}$$

$\Rightarrow$ $$2T = 2W + 0.1W$$

$\Rightarrow$ $$1.8(W + Q) = 2.1W$$

$\Rightarrow$ $$Q = \frac{W}{6}$$

EXAMPLE 2

A prismatic bar AB of weight W is placed horizontally on the top edges of two identical disks rotating with equal angular speed in opposite direction. Show that, if centre of gravity G of the bar is displaced from the middle plane $O - O'$, then the bar is released, it will perform S.H.M and find out the period of oscillation.

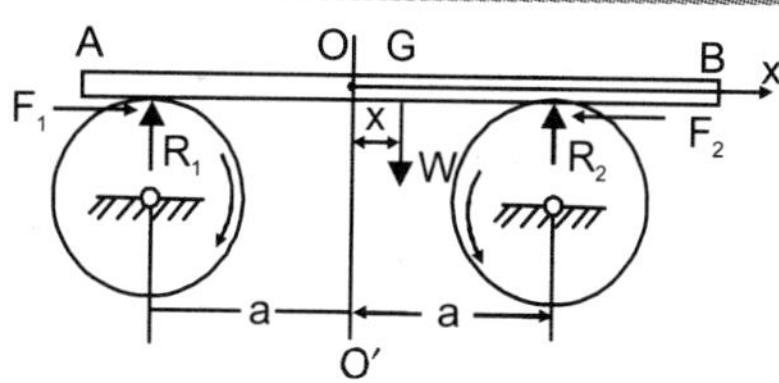

SOLUTION From statics, the vertical reactions can be computed as

$$R_1 = W\left(\frac{a - x}{2a}\right) \quad \text{and} \quad R_2 = W\left(\frac{a + x}{2a}\right)$$

Corresponding frictional forces, $F_1 = \mu R_1$ and $F_2 = \mu R_2$

As, $R_2 > R_1$, So, $F_2 > F_1$

Now the differential equation of motion becomes,

$$\frac{W}{g}\ddot{x} = F_1 - F_2 = \mu W\left(\frac{a - x}{2a}\right) \quad \mu W\left(\frac{a + x}{2a}\right) = \quad 2\mu W \frac{x}{2a} = -\mu \frac{W}{a} x$$

This is the equation of simple harmonic motion.

Following the equation [11.15(*b*)], the period oscillation of the system can be written as, $\tau = 2\pi\sqrt{\frac{a}{\mu g}}$.

EXAMPLE 3

Find the maximum acceleration along a level road that the rear-wheel drive automobile as shown in figure can attain, if the coefficient of friction between tires and pavement is μ.

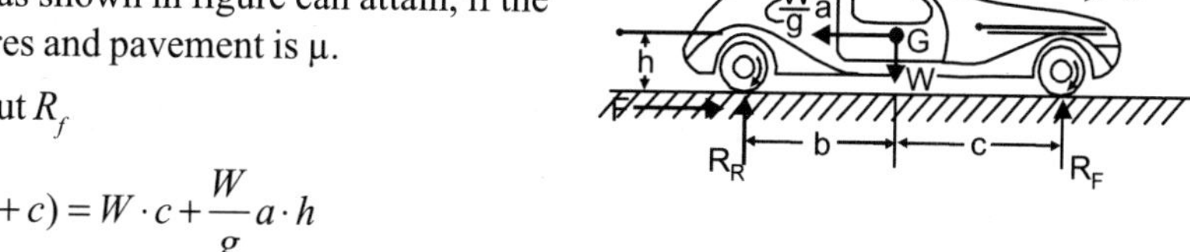

SOLUTION Taking moment about R_f

$$R_R(b + c) = W \cdot c + \frac{W}{g} a \cdot h$$

$\Rightarrow$ $$R_R = \frac{W}{b + c}\left(c + \frac{ah}{g}\right)$$

Now the equation of dynamic equilibrium along horizontal direction will be

$$F = \frac{W}{g} a$$

$\Rightarrow$ $$\mu \frac{W}{b + c}\left(c + \frac{ah}{g}\right) = \frac{W}{g} a$$

$$\Rightarrow \qquad \frac{W}{g}a\left(1-\frac{\mu h}{b+c}\right)=\mu\frac{Wc}{b+c}$$

$$\Rightarrow \qquad a=\frac{\mu cg}{b+c-\mu h}.$$

EXAMPLE 4

In the two-step pulley, find acceleration of following weight P. Assume the pulley as free of friction and inertia, and $P = 80\text{N}$, $Q = 120\text{N}$, $r_2 = 2r_1$.

SOLUTION From free body diagram, we can say, for a rotation of θ of the pulley,

$$\theta=\frac{S_1}{r_1}=\frac{S_2}{r_2}$$

$$\Rightarrow \qquad S_1r_2 = S_2r_1$$

Differentiating twice with respect to time, $a_1r_2 = a_2r_1$

$$\Rightarrow \qquad a_2 = a_1\frac{r_2}{r_1} = 2a_1$$

Again, from equation of vertical equilibrium of blocks, we have

$$T_1=\frac{Q}{g}a_1+Q \quad \text{and} \quad T_2=P-\frac{P}{g}a_2=P-\frac{2P}{g}a_1$$

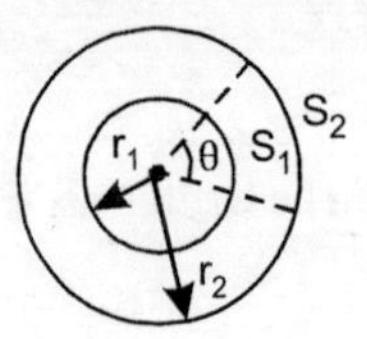

FBD of pulleys

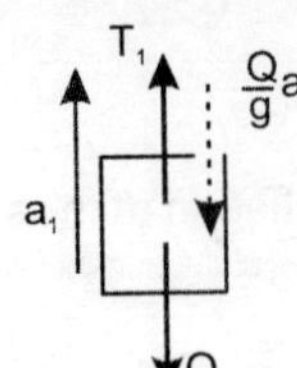

FBD of left block

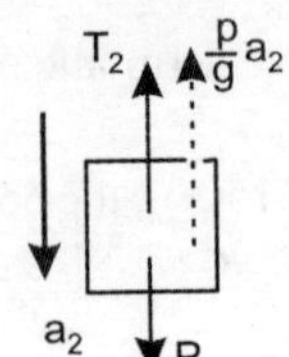

FBD of right block

Taking moment about centre of pulley, $T_1r_1 = T_2r_2$

$$\Rightarrow \qquad Q\left(1+\frac{a_1}{g}\right)r_1 = P\left(1-\frac{2a_1}{g}\right)2r_1$$

Substituting values we have, $\frac{3}{2}\left(1+\frac{a_1}{g}\right)=2\left(1-\frac{2a_1}{g}\right)$

$$\Rightarrow \qquad a_1=\frac{g}{11}\text{m/s}^2$$

$$\Rightarrow \qquad a_2=2\frac{g}{11}\text{ m/s}^2$$

EXAMPLE 5

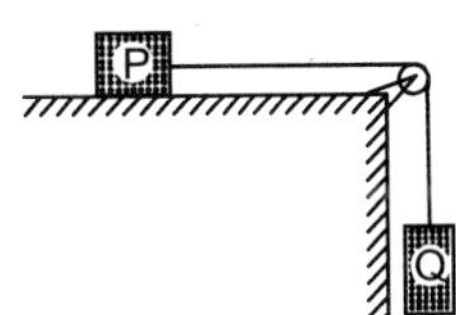

The cord connecting P and Q is inextensible. If coefficient of friction on horizontal surface be μ, find acceleration of the system and tensile force in the cord, while $P = 10\text{kg}$, $Q = 5\text{kg}$ and $\mu = 1/3$.

SOLUTION From free body diagram of block P, $R_p = 10g$.

So equilibrium equation of motion of block P,

$$T = 10a + F_p = 10a + 10\mu g$$

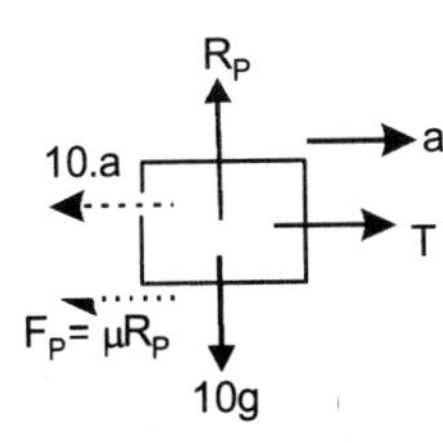

FBD of block P

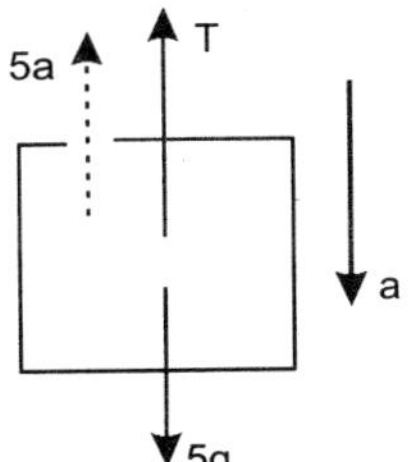

FBD of block Q

Now, from vertical force equilibrium of block Q,

$$T + 5a = 5g$$

$\Rightarrow$ $$10a + 10\mu g + 5a = 5g$$

$\Rightarrow$ $$15a = 5 \times 9.81 - \frac{1}{3} \times 10 \times 9.81$$

$\Rightarrow$ $$a = 1.09\text{m/s}^2$$

$$T = 10 \times 1.09 + \frac{1}{3} \times 10 \times 9.81 = 43.6\text{N}$$

EXAMPLE 6

In the figure the coefficient of friction on incline is μ. Find acceleration of the system and tension in the string, while $M_1 = 160\text{kg}$, $M_2 = 90\text{kg}$, $\theta = 15°$ and $\mu = 0.25$.

SOLUTION From free body diagram of M_1 equation of force equilibrium along specified X and Y direction are

$$R_1 = M_1 g \cos\theta$$

and $$T = M_1 a + M_1 g \sin\theta + \mu R_1$$

$$= M_1 a + M_1 g (\sin\theta + \mu \cos\theta)$$

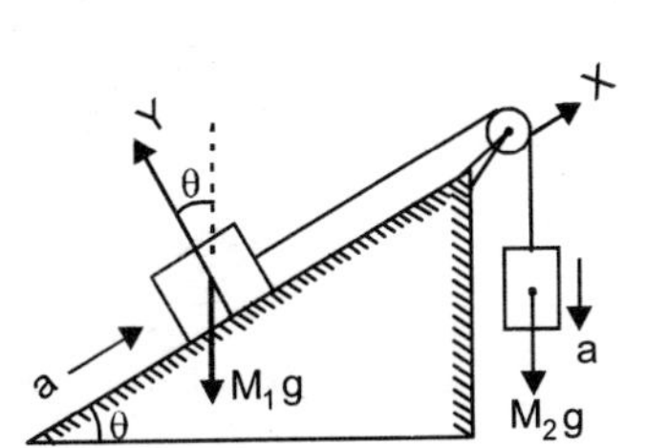

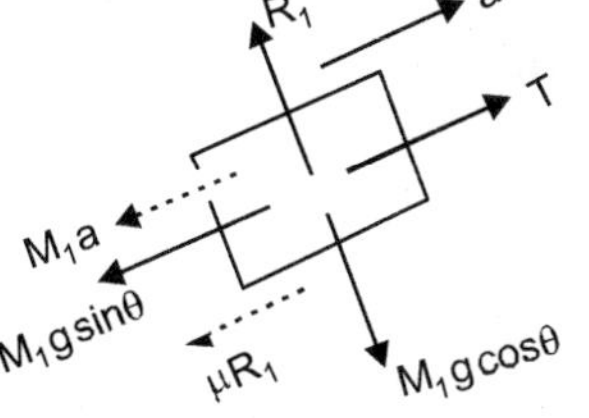

FBD of block M_1

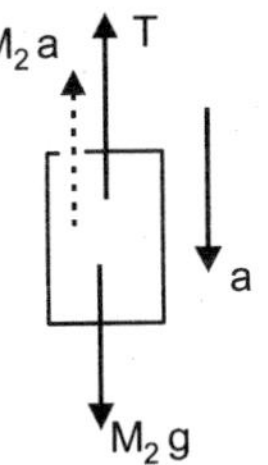

FBD of block M_2

From free body diagram of M_2, force equilibrium equation along vertical is

$$T + M_2 a = M_2 g$$

$$\Rightarrow \qquad T = M_2(g - a)$$

$$\Rightarrow \qquad M_1 a + M_1 g(\sin\theta + \mu\cos\theta) = M_2(g - a)$$

$$\Rightarrow \qquad (M_1 + M_2)a = [M_2 - M_1(\sin\theta + \mu\cos\theta)]g$$

Substituting values, we obtain, $\qquad a = 0.39 \text{ m/s}^2, T = 847.8\text{N}$

EXAMPLE 7

In the arrangement of connected bodies, as shown in figure, find the acceleration of the system and tension in the string.

SOLUTION From free body diagram of M_2, equation of force equilibrium along incline and along its normal are,

$$R_2 = M_2 g\cos\theta_2 \text{ and,}$$

$$T + M_2 a + \mu M_2 g\cos\theta_2 = M_2 g\sin\theta_2$$

$$\Rightarrow \qquad T + M_2 a = M_2 g\sin\theta_2 - \mu M_2 g\cos\theta_2$$

From free body diagram of M_1, equations of force equilibrium along incline and along its normal are,

$$R_1 = M_1 g\cos\theta_1 \text{ and,}$$

$$T = M_1 a + \mu M_1 g\cos\theta_1 + M_1 g\sin\theta_1$$

$$\Rightarrow \qquad T - M_1 a = M_1 g\sin\theta_1 + \mu M_1 g\cos\theta_1$$

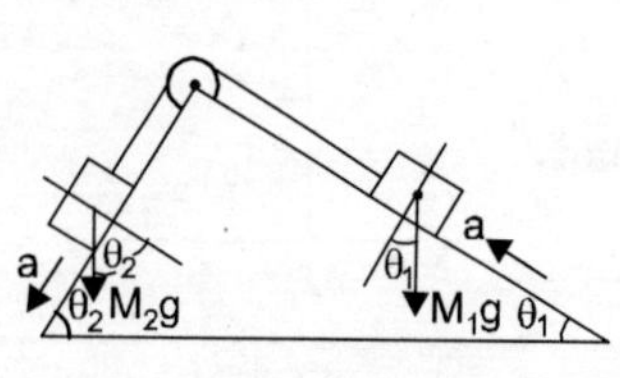

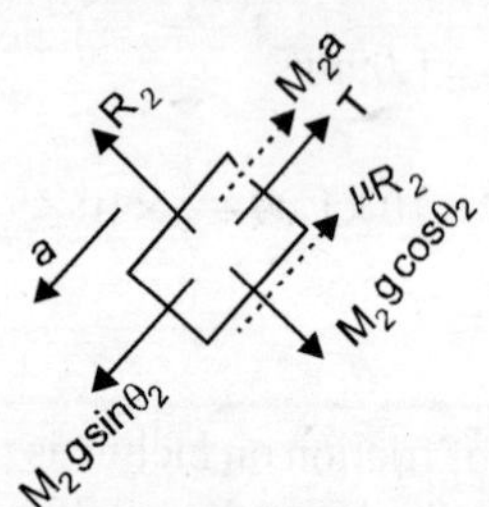

FBD of block M_2

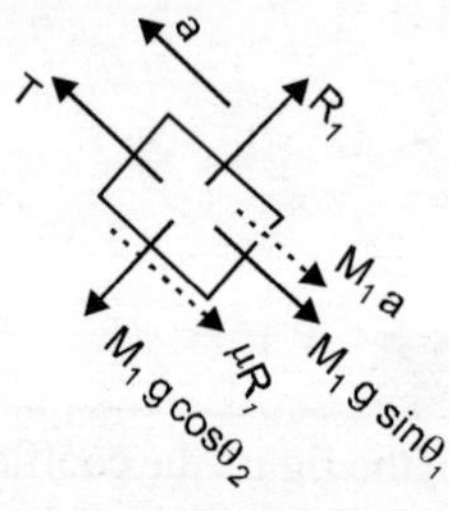

FBD of block M_1

Values of all parameters of right hand side of above equations will be known. Hence, solving the simultaneous equations we can find out two unknown T and a.

EXAMPLE 8

Find the tension in the string during motion of the system shown in figure, if

$$W_1 = 400\text{N}, W_2 = 200\text{N}, \mu = 0.2$$

SOLUTION From the body diagram of W_1, equation of force equilibrium along incline and along its normal are

$$R_1 = W_1\cos 45° \;;$$

and $\qquad T + \frac{W_1}{g}a + \mu R_1 = W_1\sin 45°$

$$\Rightarrow \qquad T = W_1 \sin 45° - \frac{W_1}{g} a - \mu W_1 \cos 45°$$

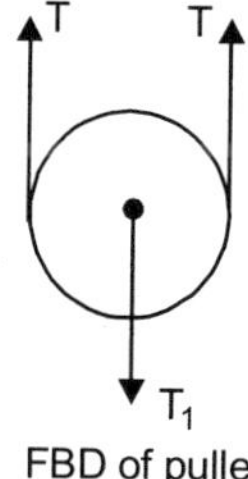

FBD of pulley

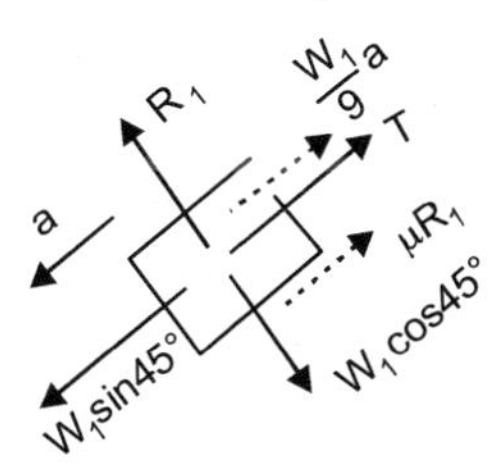

FBD of W_1

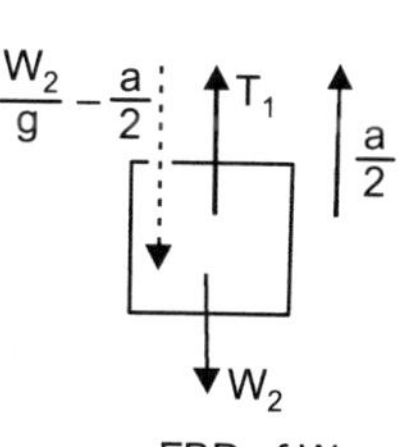

FBD of W_2

From free body diagram of moving pulley, $T_1 = 2T$.
Again, vertical force equilibrium of W_2 gives

$$T_1 = W_2 + \frac{W_2}{g} \cdot \frac{a}{2}$$

$$\Rightarrow \qquad 2T = W_2 + \frac{W_2}{g} \cdot \frac{a}{2}$$

$$\Rightarrow \qquad 2W_1 \frac{1}{\sqrt{2}} - \frac{2W_1}{g} a - 0.2W_1 \frac{2}{\sqrt{2}} = W_2 + \frac{W_2}{g} \frac{a}{2}$$

Substituting values of W_1 and W_2 we obtain, $a = 2.75 \text{m/s}^2$, $T = 114\text{N}$

Example 9

A rectangular block of weight $Q = 300$N rests on a flat car of weight $P = 150$N which may roll along the horizontal plane AB without friction. The car and the block together are to be accelerated by the weight W arranged as shown in figure. Assuming sufficient friction between the block and the car to prevent sliding find maximum value of W by which car can be accelerated. What will be this acceleration?

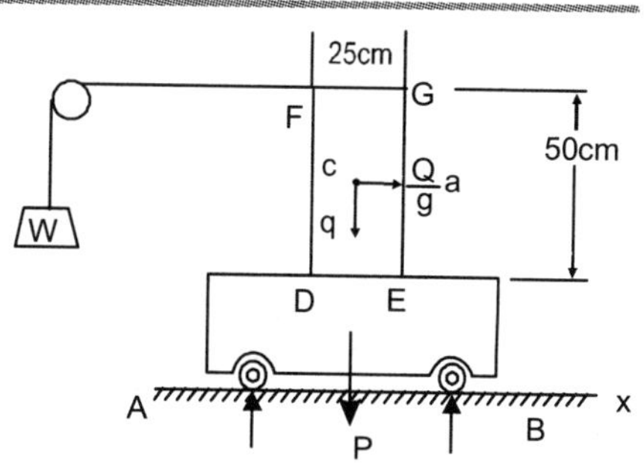

FBD of weight W

Solution Equation of dynamic equilibrium of the weight becomes,

$$T + \frac{W}{g} a = W$$

$$\Rightarrow \qquad T = W - \frac{W}{g} a$$

Equation of dynamic equilibrium of combined body of block and car comes as

$$T = \frac{P}{g} a + \frac{Q}{g} a$$

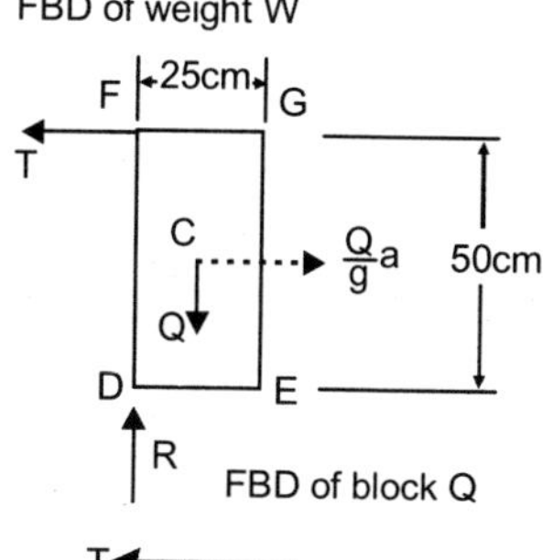

FBD of block Q

When the block is about to topple about D, moment equation of dynamic equilibrium is,

$$\frac{Q}{g} a \times 0.25 + Q \frac{0.25}{2} - T \times 0.5 = 0$$

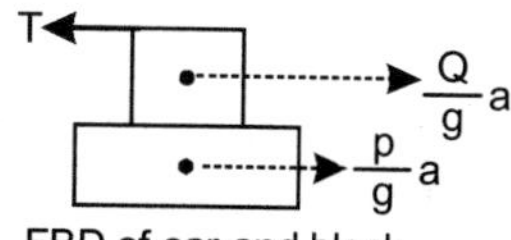

FBD of car and block

Substituting values,

$$T = \frac{450}{9.81}a = 45.87a$$

and $$\frac{150 \times 0.25}{9.81}a + \frac{150 \times 0.25}{2} = 0.5T$$

Solving above two equations, we have, $a = 0.98\text{m/s}^2$, $T = 44.96\text{N}$

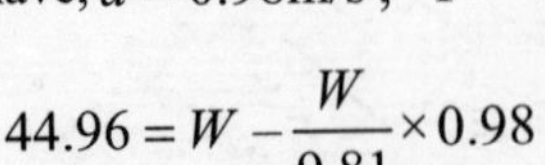

Again, $$44.96 = W - \frac{W}{9.81} \times 0.98$$

FBD of weight W

$\Rightarrow$ $$W = 49.95\text{N}$$

EXAMPLE 10

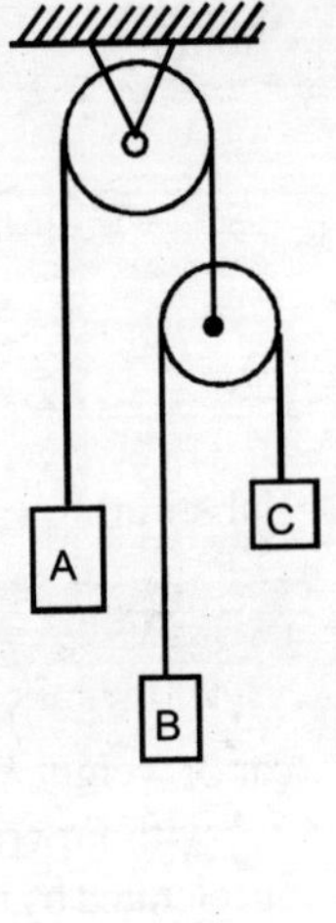

Neglecting friction and inertia of the pulleys in the system shown in figure, find the acceleration of each weight, while $W_a : W_b : W_c = 3:2:1$

SOLUTION Let $W_a = 3W$, $W_b = 2W$, $W_c = W$ and the acceleration of weights A, B, C are a_1, a_2 and a_3 respectively.

From FBD of pulley, we have

$$2T_1 = T$$

$\Rightarrow$ $$T_1 = \frac{T}{2}$$

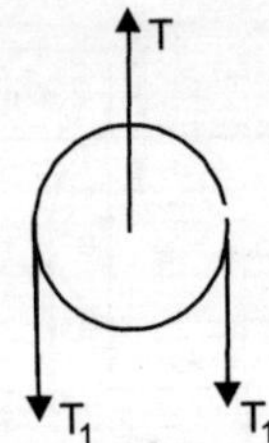

FBD of pulley

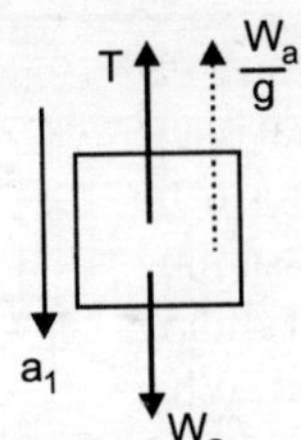

FBD of block A

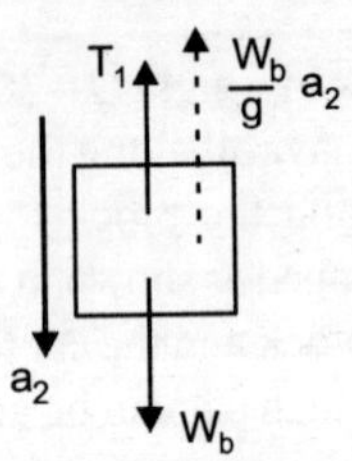

FBD of block B

From free body diagram of A, equation of dynamic equilibrium is

$\Rightarrow$ $$T + \frac{W_a}{g}a_1 = W_a$$

$\Rightarrow$ $$T = 3W - \frac{3W}{g}a_1 \quad \text{...(1)}$$

Similarly from free body diagram of B,

$\Rightarrow$ $$T_1 + \frac{W_b}{g}a_2 = W_B$$

$\Rightarrow$ $$\frac{T}{2} = 2W - \frac{2W}{g}a_2$$

$\Rightarrow$ $$T = 4W - \frac{4W}{g}a_2 \quad \text{...(2)}$$

And from free body diagram of C,

$$\Rightarrow \quad T_1 = W_c + \frac{W_c}{g} a_3$$

$$\Rightarrow \quad \frac{T}{2} = W + \frac{W}{g} a_3$$

$$\Rightarrow \quad T = 2W + \frac{2W}{g} a_3 \qquad \text{...(3)}$$

From above equations (1), (2) and (3) following simultaneous equations are obtained,

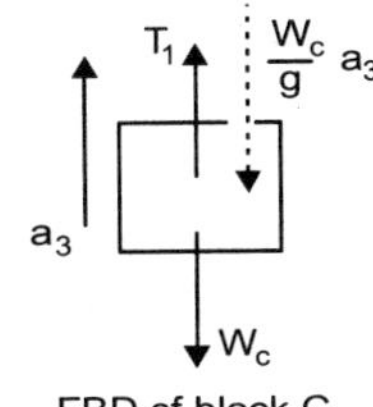

FBD of block C

$$\left.\begin{aligned} 3a_1 - 4a_2 &= -g \\ 2a_2 + a_3 &= g \\ 3a_1 + 2a_3 &= g \end{aligned}\right\}$$

Solution of these three simultaneous equations yields,

$$a_1 = \frac{g}{17}, \; a_2 = \frac{5g}{17}, \; a_3 = \frac{7g}{17}$$

EXAMPLE 11

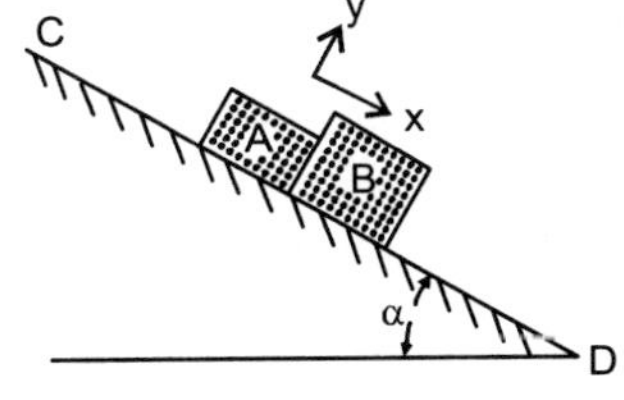

Block A and B slide down on rough incline where $\mu_A = 0.15$, $\mu_B = 0.30$, $W_A = 20$kg, and $W_B = 20$kg, $\alpha = 30°$. Find pressure P existing between A and B during motion.

SOLUTION Equations of dynamic equilibrium of block A, along the specified X and Y-axis, are

$$R_A = gW_A \cos\alpha$$

and $$P + W_A a + \mu_A R_A = gW_A \sin\alpha$$

$$\Rightarrow \quad P = g(W_A \sin\alpha - \mu_A W_A \cos\alpha) - W_A a$$

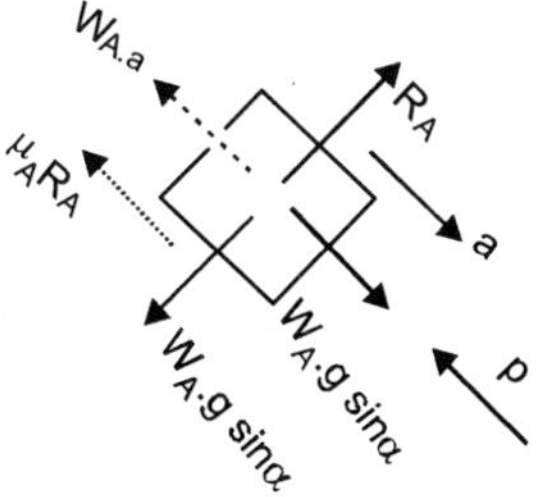

FBD of block A

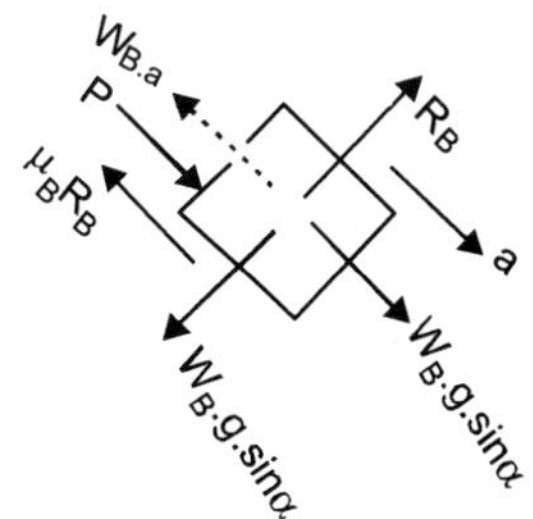

FBD of Block B

Equations of dynamic equilibrium of block B, along the specified X and Y-axis are

$$\Rightarrow \quad R_B = gW_B \cos\alpha$$

and $$P + gW_B \sin\alpha = W_B a + \mu_B R_B$$

$$\Rightarrow \quad P = W_B a + (\mu_B W_B \cos\alpha - W_B \sin\alpha)g$$

$$\Rightarrow \quad g(W_A \sin\alpha - \mu_A W_A \cos\alpha) - W_A a = W_B a + (\mu_B W_B \cos\alpha - W_B \sin\alpha)g$$

$$\Rightarrow \qquad (W_A \sin\alpha - \mu_A W_A \cos\alpha - \mu_B W_B \cos\alpha + W_B \sin\alpha)$$

$$= g\left[(W_A + W_B)\sin\alpha - (\mu_A W_A + \mu_B W_B)\cos\alpha\right]$$

$$\Rightarrow \qquad a = g\left[\sin\alpha - \left(\frac{\mu_A W_A + \mu_B W_B}{W_A + W_B}\right)\cos\alpha\right] = 2.781\text{m/s}^2$$

So, $$P = 9.81(20\times\sin 30° - 0.15\times 20\times\cos 30°) - 20\times 2.781 = 16.99\text{N}$$

EXAMPLE 12

In the adjacent figure a piston of weight W, constrained to move vertically is raised and lowered by a cam ACB, moving horizontally with constant v_0. Face of the cam has the shape of a full cosine wave. Find maximum v_0 that the cam may have without losing contact with cam follower D, through the cycle.

SOLUTION Let the displacement-time equation of motion of the piston be,

$$x = a + b\cos ky \qquad \text{...(1)}$$

From the figure it is obvious that $AB = (l + v_0 t) - v_0 t = l$

Apply following boundary conditions in equation (1)

(*a*) at A, $y = 0$ and $x = 0$

(*b*) at B, $y = l$ and $x = 0$

(*c*) at C, $y = \dfrac{l}{2}$ and $x = h$

Substitution of (*a*), (*b*) and (*c*) gives

$$b = -a, \text{ and } kl = 2\pi$$

$$\Rightarrow \qquad k = \frac{2\pi}{l}, \text{ and } a = \frac{h}{2}$$

Equation of motion finally becomes, $x = \dfrac{h}{2}\left(1 - \cos\dfrac{2\pi y}{l}\right)$

At any time instant t, path traversed $y = v_0 t$

and $$x = \frac{h}{2}\left(1 - \cos\frac{2\pi}{l}v_0 t\right)$$

Two successive differentiation yields,

$$\dot{x} = \frac{h}{2}\cdot\frac{2\pi}{l}v_0 \sin\frac{2\pi}{l}v_0 t = \frac{\pi h v_0}{l}\sin\frac{2\pi}{l}v_0 t$$

$$\ddot{x} = \frac{\pi h v_0}{l}\cdot\frac{2\pi}{l}v_0\cos\frac{2\pi}{l}v_0 t = \frac{2\pi^2 h v_0^2}{l^2}\cos\frac{2\pi}{l}v_0 t$$

If P be the vertical components of pressure between cam and follower, dynamic equation of equilibrium for piston will be

$$\frac{W}{g}\ddot{x} = P - W$$

$\Rightarrow$
$$P = W\left(1 - \frac{\ddot{x}}{g}\right)$$

From this equation, if P is to be positive, $\ddot{x} \geq -g$

The maximum (–)ve value of

$$\ddot{x} = -\frac{2\pi^2 h v_0^2}{l^2} \text{ for } v_0 t = \frac{l}{2}$$

So,
$$-\frac{2\pi^2 h v_0^2}{l^2} \geq -g$$

$\Rightarrow$
$$\frac{2\pi^2 h v_0^2}{l^2} \leq g$$

$\Rightarrow$
$$v_0 = \sqrt{\frac{g l^2}{2\pi^2 h}}$$

EXAMPLE 13

A car weighing 5000N is moving at a speed of 150m/s. The resistance to the motion of the car is largely due to air drag which is equal to $0.005v^2$. What distance will it travel before its speed is reduced to 60ms^{-1}.

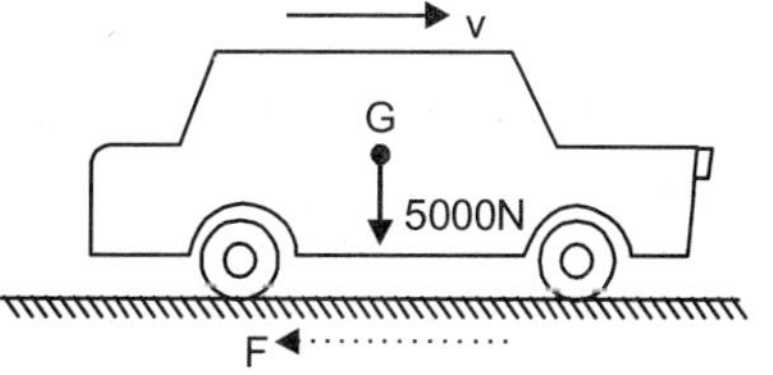

SOLUTION Here,

$$m = \frac{5000}{9.81} = 509.7\text{kg}, \; F = 0.005v^2$$

Equation of motion of the car

$$-0.005v^2 = 509.7a = 509.7\left(v \cdot \frac{dv}{dx}\right)$$

$\Rightarrow$
$$\frac{dv}{v} = -\frac{0.005}{509.7}dx = -9.81 \times 10^{-6}\, dx$$

Integrating within limits
$$\int_{150}^{60} \frac{dv}{v} = -9.81 \times 10^{-6} \int_0^x dx$$

$\Rightarrow$
$$[\ln v]_{150}^{60} = -9.81 \times 10^{-6}\, [x]_0^x$$

$\Rightarrow$
$$x = 93403.7\text{m}$$

EXAMPLE 14

An elevator of total weight 6000N starts to move upwards with a constant acceleration and acquires a velocity of 3m/s after travelling a distance of 3m. Find the tensile force in the cable during the accelerated motion. The above elevator while moving up with a velocity of 3m/s is uniformly decelerated to stop in 3 seconds. Find the pressure at the floor of the elevator under the feet of a man weighing 500N riding in the elevator.

SOLUTION (*i*) Here, $u = 0, v = 3\text{m/s}, S = 3\text{m}$

So, acceleration.

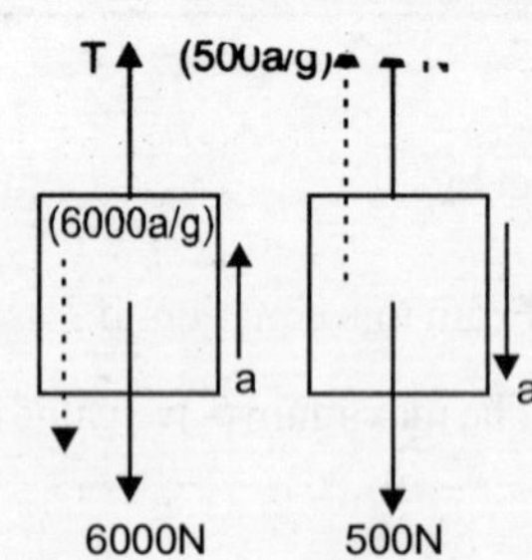

$$a = \frac{v^2 - u^2}{2S} = 1.5\text{m/s}^2$$

If T be the tension in the cable applying equation of equlibrium

$$T = W + \frac{W}{g}a = 6000\left(1 + \frac{1.5}{9.81}\right) = 6917.43\text{N}$$

(*ii*) Here, $v = 0, u = 3\text{m/s}, t = 3\text{s}$

So, deceleration

$$a = \frac{v - u}{t} = -1\text{m/s}^2$$

Equation of equilibrium $N + \frac{W_1}{g}a = W_1$

$$\Rightarrow \qquad N = W_1\left(1 - \frac{a}{g}\right) = 500\left(1 - \frac{1}{9.81}\right) = 449.03\text{N}$$

EXAMPLE 15

Two blocks A and B are held on an inclined plane 5m apart as shown in figure. The coefficients of friction between A and B and the inclined plane are 0.2 and 0.1 respectively. If the blocks begin to slide down the plane simultaneously, calculate the time and distance travelled by the each block before collision.

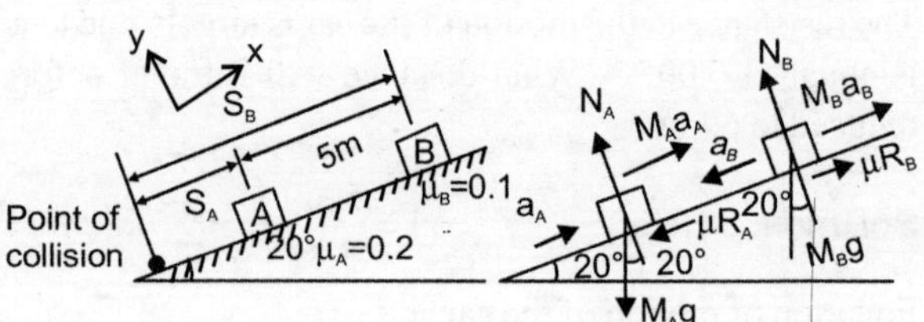

SOLUTION Let a_A and a_B be the accelerations of the block A and B. From free body diagram of block A,

$$\sum F_y = 0$$

$$\Rightarrow \qquad N_A = M_A g \cos 20°$$

$$\sum F_x = 0$$

$$\Rightarrow \qquad M_A a_A = M_A g \sin 20° - \mu_A N_A$$

$$= M_A g \sin 20° - 0.2(M_A g \cos 20°)$$

$$\Rightarrow \qquad a_A = g \sin 20° - 0.2 g \cos 20° = 1.51\text{m/s}^2$$

From free body diagram of block B,

$$\sum F_y = 0$$

$$\Rightarrow \qquad N_B = M_B g \cos 20°$$

$$\sum F_x = 0$$

$$\Rightarrow \qquad M_B a_B = M_B g \sin 20° - \mu_B N_B$$

$$= M_B g \sin 20° - 0.1(M_B g \cos 20°)$$

$\Rightarrow$

$$a_B = g\sin 20° - 0.1g\cos 20° = 2.43\text{m/s}^2$$

Let the blocks collide after a time t from release. For block B,

$$u = 0,\ a_B = 2.43\text{m/s}^2,\ S = S_B$$

So,

$$S_B = 0\times t + \frac{1}{2}\times 2.43\times t^2 = \frac{2.43}{2}t^2$$

For block A,

$$u = 0,\ a_A = 1.51\text{m/s}^2,\ S = S_A$$

So,

$$S_A = 0\times t + \frac{1}{2}\times 1.51\times t^2 = \frac{1.51}{2}t^2$$

For the blocks to collide, $S_B - S_A = 5$

$$\Rightarrow \quad \frac{2.43}{2}t^2 - \frac{1.51}{2}t^2 = 5$$

$$\Rightarrow \quad t = 3.3\text{s}$$

And,

$$S_B = \frac{2.43}{2}\times 3.3^2 = 13.23\text{m},$$

$$S_A = \frac{1.51}{2}\times 3.3^2 = 8.22\text{m}$$

EXAMPLE 16

A weight W attached to the end of a flexible rope of diameter $d = 0.5$cm is raised vertically by winding the rope on a reel as shown. The reel is turned uniformly at the rate of 2 revolution per second. Find the tension in the rope. Neglect the inertia of the rope and the lateral motion of the weight W.

SOLUTION From free body diagram of the block, $T = W + \frac{W}{g}\ddot{x}$. To find $\ddot{x}$ we have to find

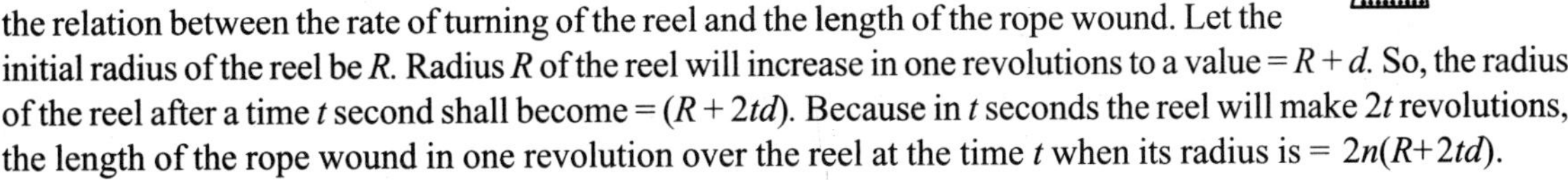

the relation between the rate of turning of the reel and the length of the rope wound. Let the initial radius of the reel be R. Radius R of the reel will increase in one revolutions to a value $= R + d$. So, the radius of the reel after a time t second shall become $= (R + 2td)$. Because in t seconds the reel will make $2t$ revolutions, the length of the rope wound in one revolution over the reel at the time t when its radius is $= 2n(R+2td)$.

Length of the rope wound per second $= 2\times 2\pi\,(R + 2td)$

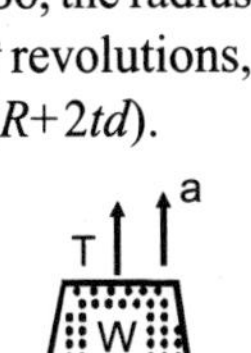

So, velocity v of the weight $\quad W = 2\times 2\pi\,(R+2td)$

So,

$$\ddot{x} = \frac{dv}{dt} = \frac{d[4\pi(R+2td)]}{dt} = 8\pi d = 8\pi\times\frac{0.5}{100} = \frac{\pi}{25}\text{m/s}^2$$

Hence,

$$T = W\left(1 + \frac{\pi}{25\times 9.81}\right) = 1.013W$$

EXAMPLE 17

Two weight $P = 40\text{N}$ and $Q = 30\text{N}$ are connected by the arrangement shown in figure. Neglecting friction and the inertia of the pulley and cord, find the acceleration a of weight Q.

SOLUTION Vertical equilibrium equation of Q,

$$S + \frac{Q}{g}a = Q$$

$$\Rightarrow \qquad S = Q\left(1 - \frac{a}{g}\right)$$

Vertical equilibrium equation of pulley, $S_1 = 2S = 2Q\left(1 - \frac{a}{g}\right)$

Vertical equilibrium equation of P, $\qquad S_1 = P\left(1 + \frac{a_1}{g}\right)$

From system if Q moves by x and corresponding movement of P is x_1 we have,

$$2x_1 + x = \text{constant}$$

$$\Rightarrow \qquad 2\frac{d^2x_1}{dt^2} + \frac{d^2x}{dt^2} = 0$$

$$\Rightarrow \qquad a_1 = -\frac{a}{2}$$

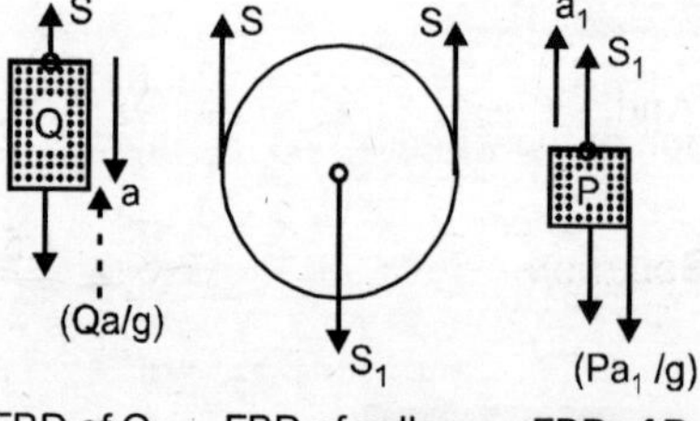

FBD of Q FBD of pulley FBD of P

Now we have,

$$\Rightarrow \qquad P\left(1 + \frac{a}{2g}\right) = 2Q\left(1 - \frac{a}{g}\right)$$

$$\Rightarrow \qquad 40\left(1 + \frac{a}{2g}\right) = 2 \times 30\left(1 - \frac{a}{g}\right)$$

$$\Rightarrow \qquad \frac{4a}{g} = 1$$

$$\Rightarrow \qquad a = 2.45\text{m/s}^2$$

EXAMPLE 18

A block of weight W, height $2h$ and width $2c$ rest on a flat car which moves horizontally with constant acceleration a. Determine (*a*) value of the acceleration a at which slipping of the block on the car will impend if the coefficient of friction is μ, and (*b*) value of the acceleration at which tipping of the block about the edge A will impend, assuming sufficient friction to prevent slipping.

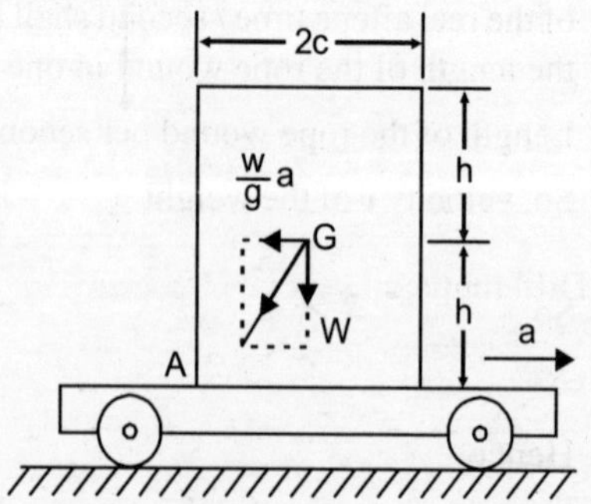

SOLUTION (*a*) Equations of equilibrium,

$$\sum F_y = 0$$

$$\Rightarrow \qquad N = W$$

$$\sum F_x = 0$$

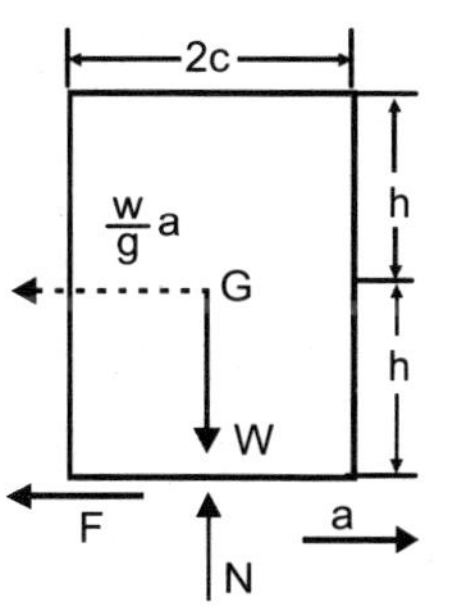

FBD for slipping of the block

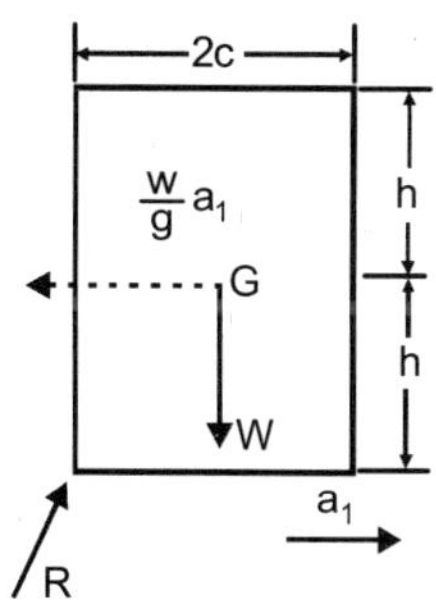

FBD for topping of the block

$\Rightarrow \qquad \dfrac{W}{g}a + F = 0$

$\Rightarrow \qquad \dfrac{W}{g}a + \mu N = 0$

$\Rightarrow \qquad \dfrac{W}{g}a + \mu W = 0$

$\Rightarrow \qquad a = -\mu g \text{ m/s}^2$

(*b*) Taking moment about *A*, $\qquad W \times c = \dfrac{W}{g}a_1 \times h$

$\Rightarrow \qquad a_1 = \dfrac{gc}{h} \text{ m/s}^2$

EXAMPLE 19

Neglecting friction and the inertia of the two pulleys, find the acceleration of the weight Q, assuming that $P = Q$.

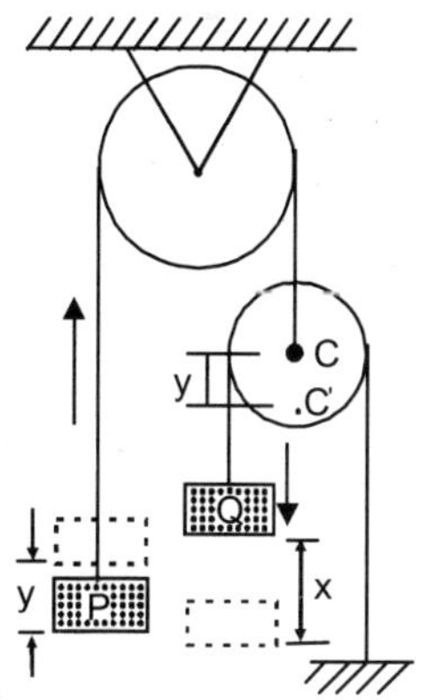

SOLUTION From FBD of block *P*, $\qquad T = P\left(1 + \dfrac{a_p}{g}\right)$

From FBD of pulley, $\qquad T = 2T_1$

From FBD of block *Q*, $\qquad T_1 = Q\left(1 - \dfrac{a_q}{g}\right)$

Let us assume, block *P* is given an upward movement by an amount *y*. So, the movable pulley *C* will come down by an amount *y* and block *Q* will come down by an amount *x*. So, we can say,

$$2y = x$$

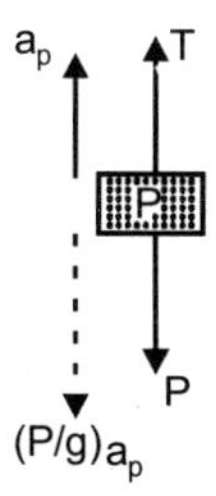

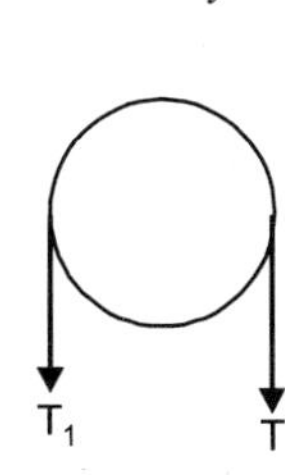

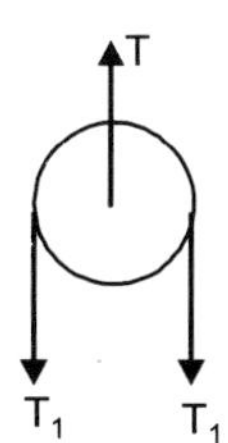

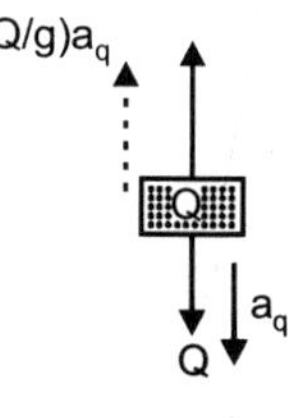

Free body diagrams

Differentiating twice, we obtain

$\Rightarrow \qquad 2\dfrac{d^2y}{dt^2} = \dfrac{d^2x}{dt^2}$

$\Rightarrow \qquad a_p = \dfrac{a_q}{2}$

Now, $$T = 2T_1$$

$$\Rightarrow \quad P\left(1+\frac{a_p}{g}\right) = 2Q\left(1-\frac{a_q}{g}\right)$$

$$\Rightarrow \quad \left(1+\frac{a_q}{2g}\right) = 2\left(1-\frac{a_q}{g}\right)$$

$$\Rightarrow \quad \frac{5a_q}{2g} = 1$$

$$\Rightarrow \quad a_q = \frac{2g}{5}$$

EXAMPLE 20

The two springs shown in the figure each has a spring constant $k = 1\text{N/cm}$ and the attached ball has the weight $W = 1\text{N}$. If the ball is initially displaced 1cm to the right, find the period of oscillation of the ball and the velocity with which it passes through its middle position. Neglect friction.

SOLUTION For each spring, restoring force, $S = kx$

Equation of equilibrium of the ball, along horizontal direction

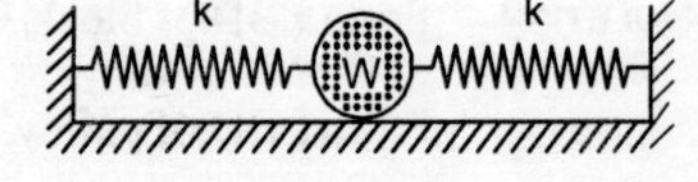

$$2S + \frac{W}{g}\ddot{x} = 0$$

$$\Rightarrow \quad 2kx + \frac{W}{g}\ddot{x} = 0$$

$$\Rightarrow \quad \ddot{x} + \frac{2k \times g}{W}x = 0$$

Comparing with the standard equation of S.H.M $\ddot{x} + \omega^2 x = 0$,

We have, $$\omega = \sqrt{\frac{2k \times g}{W}} \text{ and}$$

Time period $$\tau = \frac{2\pi}{\omega} = 2\pi\sqrt{\frac{1}{2 \times 1 \times 100 \times 9.81}} = 0.142\text{s}$$

FBD of ball

Solution of $$\ddot{x} + \frac{2k \times g}{W}x = 0 \text{ yields,}$$

$$x = C_1 \cos\sqrt{\frac{2k \times g}{W}}t + C_2 \sin\sqrt{\frac{2k \times g}{W}}t$$

So, $$\dot{x} = \sqrt{\frac{2k \times g}{W}}\left(C_1 \sin\sqrt{\frac{2k \times g}{W}}t - C_2 \cos\sqrt{\frac{2k \times g}{W}}t\right)$$

Applying boundary conditions, at $t = 0$, $x = 0.01\text{m}$ and $\dot{x} = 0$

we have, $$C_1 = 0.01, C_2 = 0.$$

Finally, $$\dot{x} = -\sqrt{\frac{2k \times g}{W}}\left(0.01\sin\sqrt{\frac{2k \times g}{W}}t\right)$$

The ball passes through the midpoint of oscillation, at $t = \frac{\tau}{2} 0.071$s.

At this instant, the velocity of the ball will be,

$$= -\sqrt{\frac{2 \times 1 \times 100 \times 9.81}{1}} \times 0.01\sin\left(\sqrt{\frac{2 \times 1 \times 100 \times 9.81}{1}} \times 0.071\right)$$

$$= -0.024\text{m/s}$$

EXAMPLE 21

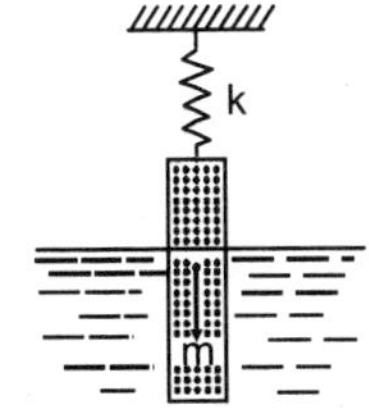

A solid cylinder of mass m and cross-sectional area A is suspended by a spring of stiffness k and hangs partially in water of density ρ as shown in figure. Find the period for small vertical oscillations of the cylinder. Neglect the inertia of water.

SOLUTION Let the cylinder be given a downward displacement x.

Increase in the tension of the spring = kx

Increase in the buoyancy force = $Ax\rho g$

Total unbalanced force acting in the direction of motion = $-(kx+Ax\rho g)$

Differential equation of motion of the cylinder

$$m\ddot{x} = -(kx + Ax\rho g)$$

$$\Rightarrow \qquad \ddot{x} + \left(\frac{k + A\rho g}{m}\right)x = 0$$

Comparing the equation S.H.M. $\ddot{x} + \omega^2 x = 0$

we can write, $$\omega^2 = \left(\frac{k + A\rho g}{m}\right)$$

and time period $$\tau = \frac{2\pi}{\omega} = 2\pi\sqrt{\frac{m}{A\rho g + k}}$$

Exercise

1. A body of mass 20kg falls on the muddy ground from a height of 39.2m. The body penetrates into the ground. Find the distance through which the body will penetrate into the ground, if the resistance by the ground to penetration is constant and equal to 980N. Take $g = 9.8\text{m/sec}^2$. **Ans.** [9.8 m]

2. A man weighing 588N dives into a swimming pool from a tower of height 9.8m. He was found to go down in water by 1.5m and then started rising. Find the average resistance of water. **Ans.** [452kgf]

3. A bullet weighing 80gmf and moving with a velocity of 350m/sec is fired into a log of wood and it penetrates to a depth of 15cm. If the bullet moving with the same velocity, were fired into a similar piece of wood 10cm thick, with what velocity would it emerge? Find also the force of resistance, assuming it to be uniform.

Ans. [202m/sec, 32666N]

4. **A car, moving on a straight level road, skidded for a total distance of 40 metres after the brakes were applied. Determine the speed of the car, just before the brakes were applied,** if the coefficient of friction between the car tyres and the road is 0.4. Take $g = 9.80 \text{m/sec}^2$. **Ans.** [63. 75km/hr]

5. The tractive force, exerted by a railway car weighing 40kN is 1600N. If the frictional resistance is 4N per kN of the railway car's weight, determine the acceleration when the railway car is moving on a level track. **Ans.** [0.353/m/s²]

6. A train of weight 2000kN starts from rest and attains a speed of 100km/hr in 4 minutes. If the frictional resistance of the track is 8N per kN of the train's weight, find the average pull required. Take $g = 9.8 \text{m/sec}^2$. **Ans.** [39.588kN]

7. A train weighing 150 tonnes moves down a slope of 1 in 200 at 20 km/hr and engine develops 40 h.p. What horse power will be required to pull the train up the slope at the same speed? **Ans.** [150.99]

8. An elevator weighs 2000N and is moving vertically downwards with a uniform acceleration. Write the equation for the elevator cable tension. Starting from rest it travels a distance of 30m during an interval of 12 seconds. Find the cable tension during this time. Neglect all other resistances to motion. What are the limits of cable tension? **Ans.** [1915.2 N, at $a = 0$, $T = 2000$ and at $a = 9.81 \text{ms}^{-2}$, $T = 0$]

9. An elevator weighing 6000N is ascending with an acceleration of 2m/sec². During this ascent its operator whose weight is 600N is standing on the scales placed on the floor. What is the scale reading? What will be the total tension in the cables of the elevator during this motion? **Ans.** [722.3 N, 7945.5 N]

10. Two bodies of weight 40N and 20N are connected to the two ends of a light inextensible string, passing over a smooth pulley. The weight of 40N is placed on a smooth horizontal surface while the weight of 20N is hanging free in air. Find (*i*) the acceleration of the system, and (*ii*) the tension in the string. Take $g = 9.81 \text{m/sec}^2$. **Ans.** [3.27m/sec², 13.33 N]

11. If in the previous problem, the horizontal surface is rough one, having a coefficient of friction between the weight 40N and the horizontal surface equal to 0.3, determine: (*i*) the acceleration of the system, and (*ii*) the tension in the string. **Ans.** [1.308 m/sec²,17.33 N]

12. Two bodies of weights 60N and 40N are connected to the two end of a light inextensible string, which passes over a smooth pulley. The weight 60N is placed on a smooth inclined plane of angle of inclination 10°, while the weight 40N is hanging free in air. Determine: (*i*) acceleration of the system, and (*ii*) tension in the string. Take $g = 9.80 \text{m/sec}^2$. **Ans.** [2.89 m/sec², 28.16 N]

13. If in previous question the inclined surface is a rough one, having coefficient of friction between the weight 60N and inclined surface equal to 0.2, determine: (*i*) the acceleration of the system, (*ii*) tension in the string, and (*iii*) the distance moved by the weight 40N in 4 seconds starting from rest. Take $g = 9.80 \text{m/sec}^2$. **Ans.** [(*i*) 1.74m/sec², (*ii*) 32.89N, (*iii*)13.92 m]

14. Find the angular acceleration of a flywheel of an engine, which weighs 1500N and has a radius of gyration 0.6 m, if the wheel is subjected to a torque of 2000N m. Take $g = 9.80 \text{ m/sec}^2$. **Ans.** [36.33 rad/sec²]

15. A flywheel, weighing 4000N and having a radius of gyration 1m, looses its speed from 500 r.p.m. to 300 rp.m. in 2 minutes. Calculate: (*i*) the retarding torque acting on it, (*ii*) change in kinetic energy during the above period, and (*iii*) change in its angular momentum during this period. **Ans.** [(*i*) 71.24N.m, (*ii*) –358074 N.m, (*iii*) 8548.77 N.m/sec]

16. A weight of 4N is suspended by a light rope wound round a pulley of 48N and radius 25cm, the other end of the rope being fixed to the periphery of the pulley. If the weight is moving downwards, determine: (*i*) acceleration of the weight 4N, and (*ii*) tension in the string. Take $g = 9.80$ m/sec^2.

Ans. [(*i*) 14 m/sec^2, (*ii*) 3.428N]

17. A cage weighing 3000 N is raised by a rope, one end of which is wrapped round a drum 100 cm in diameter and weighing 800N and having a radius of gyration of 45cm. The drum is rotated by an electric motor, which exerts constant torque of 4200Nm. If the rope is tight when drum begins to rotate, determine: (*i*) the acceleration of the cage, (*ii*) the tension in the rope, and (*iii*) the time required to raise the cage 15m from ground. Neglect the loss of energy due to friction and weight of the rope. Take $g = 9.80$ m/sec^2.

Ans. [(*i*) 16.736 m/sec^2, (*ii*) 8123.26N, (*iii*) 1.338 sec.]

18. A homogeneous solid cylinder of weight 160N whose axis is horizontal, rotates about its axis in frictionless bearings under the action of the weight of a 15N block which is carried by a rope wrapped round the cylinder. What will be the angular velocity of the cylinder three seconds after the motion starts? Assume diameter of the cylinder as 100cm. **Ans.** [9.80 rad/sec]

CHAPTER

12

PRINCIPLE OF WORK, POWER, ENERGY

12.1 WORK OF A FORCE ACTING ON A PARTICLE

Referring to Figure 12.1 (*a*), initial and final positions of the particle are defined at A and A' with position vectors r and $(r + dr)$, when it is subjected to a force F. Here dr denotes the displacement of the particle. The work of a force F corresponding to the displacement dr is defined as the quantity.

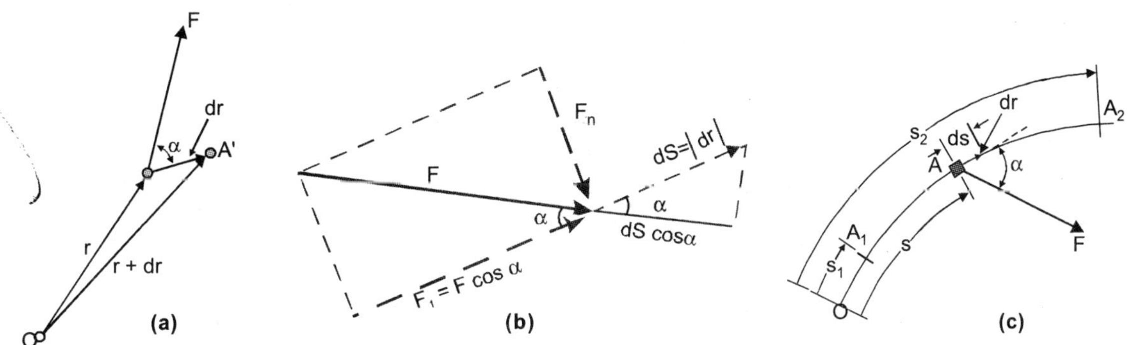

Fig. 12.1

$$dU = F \cdot dr \qquad \text{... (12.1}a\text{)}$$

$$= |F| dS \cos\alpha \qquad \text{... (12.1}b\text{)}$$

In the above equation, the magnitude of the scalar product is obtained where $|F|$ and ds are the magnitude of force vector and displacement respectively. This equation can otherwise be thought as,

$$dU = |F| dS \cos\alpha$$

$$= |F| \cos\alpha \, dS$$

$$= F_t dS \qquad \text{... (12.1}c\text{)}$$

Resolving dU in terms of component of rectangular force and displacement,

$$dU = F_x dx + F_y dy + F_z dz \qquad \text{...(12.2)}$$

Being a scalar quantity, work has a magnitude and sign, but no direction. The S.I unit of work is newton-metre (N·m) or joule (J).

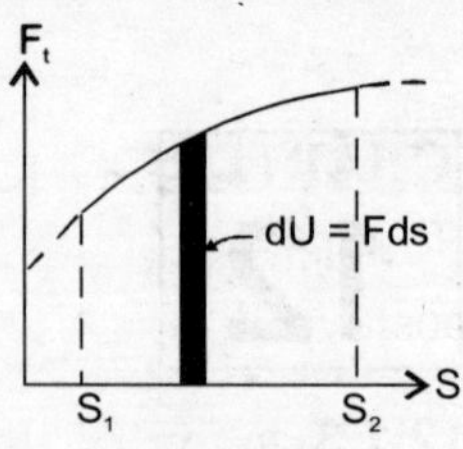

Fig. 12.1(d)

It is evident from equation [12.1(*b*)] that the work dU is positive if the angle α is acute and negative if α obtuse. So, if the force F has

(*i*) Same direction as dr, $dU = |F|\,dS$
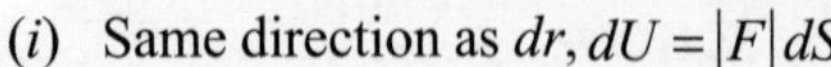

(*ii*) Opposite direction to that of dr, $dU = -|F|\,dS$

(*iii*) Perpendicular to the direction of dr, $dU = 0$.

The work of F during a finite displacement of the particle from A_1 to A_2 can be obtained by integrating equation [12.1(*a*)] along the path described by the particles as,

$$U_{12} = \int_{A_1}^{A_2} F \cdot dr = \int_{S_1}^{S_2} |F| \cos\alpha\, dS$$

$$= \int_{S_1}^{S_2} F_t\, dS \qquad \text{... (12.3)}$$

If the force F is resolved along axes of 3D-rectangular co-ordinates it can be written, similar to equation (12.2) as,

$$U_{12} = \int_{A_1}^{A_2} F_x\, dx + \int_{A_1}^{A_2} F_y\, dy + \int_{A_1}^{A_2} F_z\, dz \qquad \text{... (12.3}a\text{)}$$

Following are the description of some work in a few salient cases.

12.1.1 Work of a Constant Force in Rectilinear Motion

When a particle, moving in a straight line, is subjected to a force F of constant magnitude and constant direction as in (Fig.12.2).

$$U_{12} = \left(|F| \cos\alpha\right) \Delta x \qquad \text{... (12.4)}$$

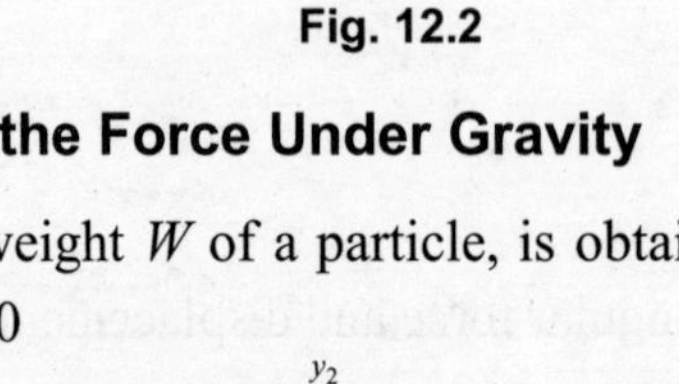

Fig. 12.2

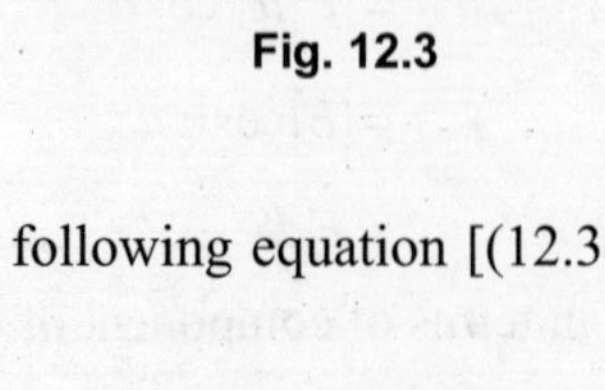

Fig. 12.3

12.1.2 Work of the Force Under Gravity

The work of the weight W of a particle, is obtained following equation [(12.3(*a*)] when, $F_x = 0$, $F_y = -W$ and $F_Z = 0$

Hence, $$U_{12} = -\int_{y_1}^{y_2} F_y\, dy$$

$$= -\int_{y_1}^{y_2} W\,dy = W(y_1 - y_2) = W \cdot \Delta y \qquad \text{... (12.5)}$$

This work is conventionally designated as *potential energy* of a particle of weight W. It is positive, if Δy is less than zero i.e., in the condition of moving the particle in downward direction.

12.1.3 Work of the Force of a Spring

Adjacent *spring-mass system* of Figure 12.4 shows the undeformed state of a spring at A_0. An amount of initial displacement of amount x_1 is applied on the spring to generate final displacement x_2 at position A_2. Here the direction of F will be opposite to the direction of movement of spring-mass system, because F is a *restoring force* or *reactive force*. If k be the spring constant,

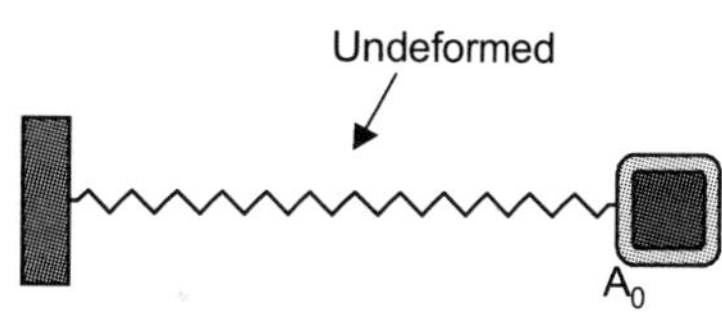

Fig. 12.4

$$dU = -F\,dx = -kx\,dx \qquad [\therefore\ F \propto x \Rightarrow F = kx]$$

$$\therefore \qquad U_{12} = -\int_{x_1}^{x_2} kx\,dx = \frac{1}{2}k\left(x_1^2 - x_2^2\right) \qquad \text{... (12.6)}$$

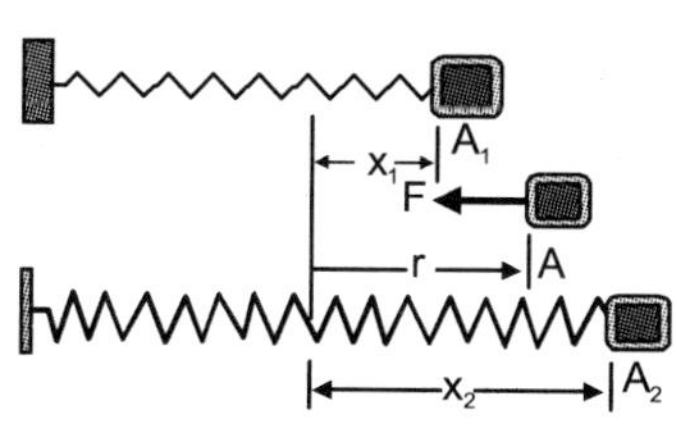

Fig. 12.5 (a)

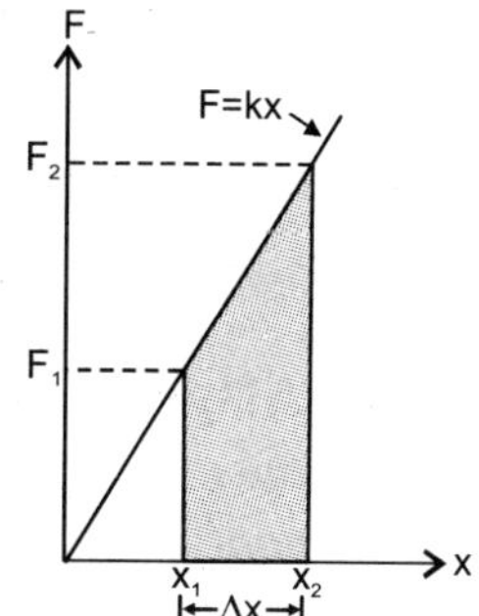

Fig. 12.5 (b)

Otherwise, $U_{12} = -\frac{1}{2}(F_1 + F_2)\Delta x$ while, $F_1 = kx_1, F_2 = kx_2$ and the work done corresponds to the shaded trapezoidal part in Figure 12.5(*b*).

12.2 KINETIC ENERGY OF A PARTICLE

Referring to Figure 12.6, a particle of mass m is subjected to a force F having components along tangential direction,

$$F_t = ma_t = m\frac{dv}{dt}$$

$$= m\frac{dv}{dS}\cdot\frac{dS}{dt}$$

$$= mv\frac{dv}{dS}$$

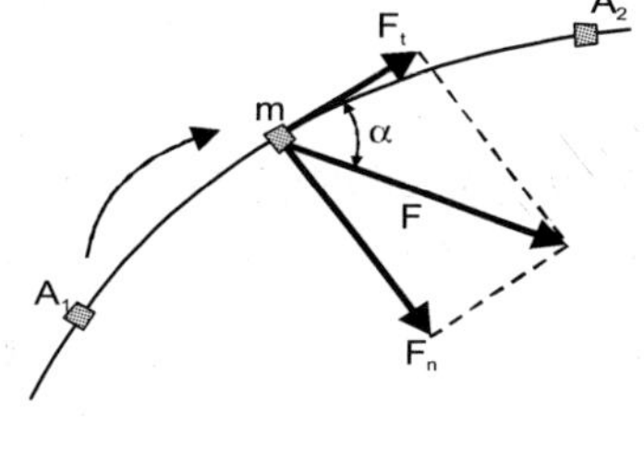

Fig. 12.6

$$\Rightarrow \qquad F_t dS = mv\, dv \qquad \text{... (12.7a)}$$

Integrating from A_1 to A_2, i.e., from S_1 to S_2 and v_1 to v_2, we find

$$\int_{S_1}^{S_2} F_t\, ds = m\int_{v_1}^{v_2} v dv = \frac{1}{2}m\left(v_2^2 - v_1^2\right)$$

$$\Rightarrow \qquad U_{12} = E_{k2} - E_{k1} \qquad \text{... (12.7b)}$$

or $$E_{k1} + U_{12} = E_{k2}$$

where, $E_{k1} = \frac{1}{2}mv_1^2$ = K.E of the particle at position 1.

$E_{k2} = \frac{1}{2}mv_2^2$ = K.E of the particle at position 2.

Thus, it is said that, when a particle moves from A_1 to A_2, under the action of force F, the work of the force F is equal to the change in the kinetic energy of the particle.

12.3 POTENTIAL ENERGY OF A PARTICLE

Work done by the force under gravity is already deduced in Article 12.1(*b*). Following equation (12.5) we can write, $U_{12} = W_{y1} - W_{y2} = E_{P1} - E_{P2}$ ... (12.8)

where, $E_{p1} = W_{y1}$ = P.E of the particle at position 1.

$E_{p2} = W_{y2}$ = P.E of the particle at position 2.

12.4 CONSERVATIVE FORCES

A force F is said to be a *conservative force* if the work U_{12} is *independent of the path*, followed by the particle, as it moves from A_1 to A_2. *Weight of a particle*, force exerted by a spring are examples of conservative forces. It is to mention here that, *frictional force is not at all a conservative force*, because the work of a frictional force can not be expressed as a change in potential energy.

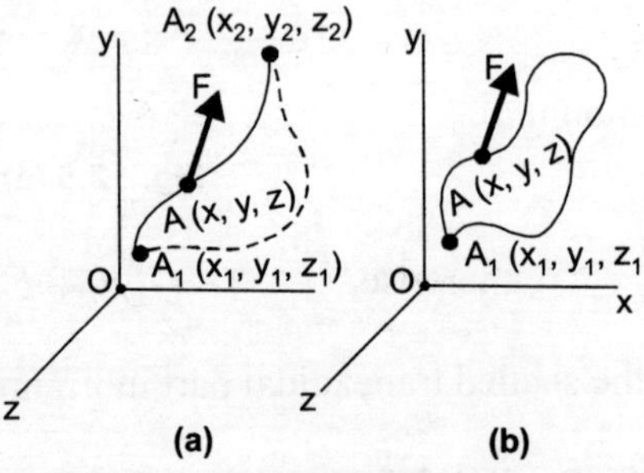

Fig. 12.7

If the particle is allowed to follow a closed path, as shown in Fig. 12.7 (*b*), the work will be zero. So, for any conservative force F,

$$\oint F \cdot dr = 0 \qquad \text{... (12.9)}$$

12.5 PRINCIPLE OF CONSERVATION OF ENERGY OF A PARTICLE

When a particle moves under the action of conservative forces, we can write from equation, 12.7(*b*).

$$U_{12} = E_{k2} - E_{k1}$$

$$\Rightarrow \qquad E_{p1} - E_{p2} = E_{k2} - E_{k1}$$

$$\Rightarrow \qquad E_{p1} + E_{k1} = E_{p2} + E_{k2}$$

$$\Rightarrow \qquad E_1 = E_2 \qquad \text{... (12.10)}$$

where, $E_1 = E_{p1} + E_{k1}$ = total energy at position 1.

$E_2 = E_{p2} + E_{k2}$ = total energy at position 2.

Equation (12.10) is the expression of principle of conservation of energy, which states that, *for any conservative force the total energy of a particle, on which it acts is constant at any position of path of movement of the particle.*

12.6 WORK OF FORCE ACTING ON A RIGID BODY

To compute work of external forces on a rigid body it is convenient to find out the *work of a couple*, instead of considering separately the work of each two forces forming the couple. Adjacent figure shows the couple M is formed by force F, acting on a rigid body. A small displacement of the rigid body brings point A to A' and B to B''. Now the total work can be conceived in two successive stages,

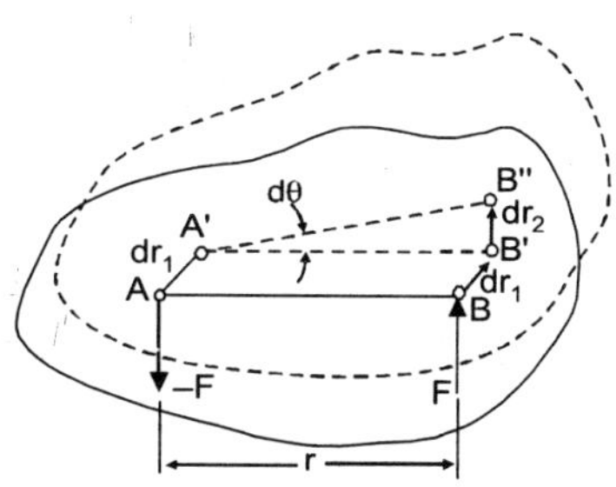

Fig. 12.8

$$dU - (\text{movement of } A \text{ to } A' \text{ and } B \text{ to } B') + (\text{movement of } B' \text{ to } B'')$$

$$= (-F \cdot dr_1 + F \cdot dr_1) + F \cdot dr_2$$

$$= |F| dr_2 = |F| r d\theta = M d\theta \qquad \text{... (12.11)}$$

Here $d\theta$ is the small angle expressed in radians through which the body rotates. For a finite amount of rotation of rigid body, from θ_1 to θ_2.

$$U_{12} = \int_{\theta_1}^{\theta_2} M d\theta \qquad \text{... (12.12}a\text{)}$$

If moment M of the couple remains constant, above equation changes its form as,

$$U_{12} = M(\theta_2 - \theta_1) \qquad \text{... (12.12}b\text{)}$$

It is to mention here that there are forces acting on rigid body, who displaces the body in a direction perpendicular to the force of action and hence perform no work. Again, when a rigid body rolls without sliding on a fixed surface, the frictional force at the point of contact does no work.

12.7 KINETIC ENERGY OF RIGID BODY IN PLANE MOTION

Let us consider a rigid body of mass m in plane motion, as shown in Figure 12.9. For any particle P_i,

Absolute velocity of P_i = velocity of mass centre G + velocity of P_i relative to a fram $Gx'y'$

i.e., $$v_i = \bar{v} + v_i' \qquad \text{... (12.13)}$$

Now kinetic energy of the system of particles forming rigid body can be written as,

$$E_{k_{RB}} = \frac{1}{2}\sum_{i=1}^{n} \Delta m_i v_i^2 = \frac{1}{2}\sum_{i=1}^{n} \Delta m_i \bar{v}^2 + \frac{1}{2}\sum_{i=1}^{n} \Delta m_i v_i'^2 \quad \text{... (12.14a)}$$

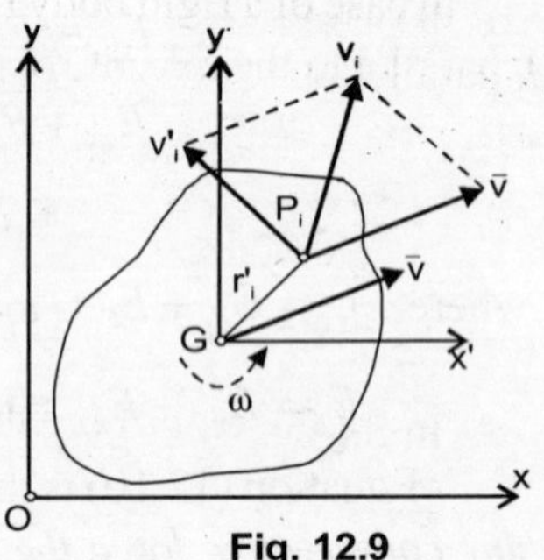

Fig. 12.9

As the velocity of the mass centre of the body $\bar{v}$ remains constant,

$$E_{k_{RB}} = \frac{1}{2} m\bar{v}^2 + \frac{1}{2}\sum_{i=1}^{n} \Delta m_i v_i'^2 \quad \text{... (12.14b)}$$

If r_i' be the distance of any particle P_i from mass centre G, $v_1' = r_1'\omega$, where ω is the angular velocity of the body at that instant above equation finally takes the shape as,

$$E_{k_{RB}} = \frac{1}{2} m\bar{v}^2 + \frac{1}{2}\sum_{i=1}^{n} \Delta m_i v_i'^2 = \frac{1}{2} m\bar{v}^2 + \frac{1}{2}\bar{I}\omega^2 \quad \text{... (12.14c)}$$

$$= \text{K.E of the body in translation} \left(\frac{1}{2} m\bar{v}^2\right)$$

$$+ \text{K.E of the body in rotation} \left(\frac{1}{2}\bar{I}\omega^2\right) \quad \text{... (12.14d)}$$

Where, moment of inertia of the rigid body (I) about the axis through G is $\frac{1}{2}\left(\sum_{i=1}^{n} r_i'^2 \, \Delta m_i\right)$.

Now we can infer that, if the body performs only translation, second term of equation 12.14 (*c*) vanishes. Similarly if the movement of body is only rotation then first term gets vanished.

12.8 PRINCIPLE OF CONSERVATION OF ENERGY OF A RIGID BODY

This principle is unique, whether for the particle dynamics or rigid body dynamics, that *the algebraic summation of potential and kinetic energy at any instant movement is constant, i.e.,* $E_1 = E_2$.

12.9 POWER AND EFFICIENCY

Power is defined as the time rate change of work done. If ΔU is the work done during time interval Δt then

$$\text{Average power } P_{\text{av}} = \frac{\Delta U}{\Delta t} \quad \text{...(12.15a)}$$

$$\text{Taking } \Delta t \text{ limiting to zero, } P_{\text{av}} = \lim_{\Delta t \to 0} \frac{\Delta U}{\Delta t} = \frac{dU}{dt} \quad \text{... (12.15b)}$$

For a particle in motion, $dU = F \cdot dr$ and

$$\text{Power} = \frac{dU}{dt} = \frac{F \cdot dr}{dt} = F \cdot v \quad \text{... (12.15c)}$$

In case of a rigid body rotating with an angular velocity ω and subjected to a couple of moment M, parallel to the axis of rotation,we can write,

$$dU = M \cdot d\theta \text{ and}$$

$$\text{Power} = \frac{dU}{dt} = \frac{Md\theta}{dt} = Mv \qquad \text{...(12.15}d\text{)}$$

In S.I the unit of power is watt (W).

$$1\,\text{W} = 1\,\text{J/s} = 1\text{N} \cdot \text{m/s} = \frac{1}{746}\text{hp}$$

Efficiency or mechanical efficiency is the ratio of output power to input power, and is symbolised by η (eta)

$$\eta = \frac{\text{power output}}{\text{power input}} \qquad \text{... (12.15}e\text{)}$$

Multiple Choice Questions

Select the best alternative (s):

1. If a mass of 20 kg falling from a height of 1.0m from rest is brought to rest by penetrating into sand by 1m, then average resistance offered by sand is
   (*a*) 100 kg (*b*) 110 kg (*c*) 150 kg
   (*d*) 200 kg (*e*) 50 kg
2. A jet engine works on the principle of conservation of
   (*a*) energy (*b*) mass (*c*) angular momentum
   (*d*) linear momentum (*e*) none of the above
3. The kinetic energy of a body rotating with an angular speed ω depends on
   (*a*) ω only (*b*) ω^2 only (*c*) its mass only
   (*d*) the distribution of mass and angular speed
   (*e*) all of the above
4. If two bodies, one light and other heavy, have equal kinetic energy, which one has a greater linear momentum?
   (*a*) heavy body (*b*) light body (*c*) both have equal momentum
   (*d*) unpredictable (*e*) none of the above

Answers

1. (*d*) 2. (*a*) 3. (*d*) 4. (*b*)

Numerical Examples

Example 1

Determine the work done by electric motor in winding up a uniform cable which hangs from a hoisting drum if its free length is 10m and weight 500N. The drum is rotated by the motor.

Solution Free length of the cable, $L = 10\text{m}$

Weight of 10m length cable, $W = 500\text{N}$

$\therefore$ Weight of cable per meter length, $w = \dfrac{500}{10} = 50\text{N}$

Consider an element of length dx at a distance x from free end.

This element is to be raised to a height of $(10 - x)$m.

Now, weight of element $= w \times dx = 50dx$N

$\therefore$ Work done on the element $dW = (50dx) \times (10 - x)$

$= 50(10 - x)dx$

$\therefore$ Total work done $W = \int_0^{10} dW = \int_0^{10} 50(10 - x)dx = 2500$ joule

Example 2

A train of weight 200kN is pulled by an engine on a level track at a constant speed of 72kmph. The resistance due to friction is 10N/kN of the train weight. Find the power of the engine. If the train is to move with an acceleration of 1m/sec² on the level track after attaining a speed of 72kmph then find the power of the engine.

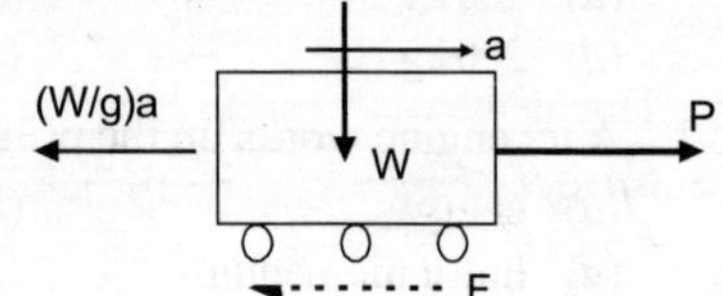

Solution Weight of train, $W = 200\text{kN}$

Speed of train, $v = 72\text{kmph} = \dfrac{72 \times 1000}{3600} = 20\text{m/s}$

Resistance due to friction, $F = 10 \times 200 = 2000$ N

Here, P is force exerted by the engine to pull the train at constant speed, $\left(\dfrac{W}{g}a\right)$ is inertia force applied to the system to keep it in dynamic equilibrium. Now applying equilibrium condition along x direction, we have

$$P - F - \frac{W}{g}a = 0 \quad \text{...(1)}$$

Case I : Here, $v = 20\text{m/s}$ and acceleration $a = 0$

$\therefore$ $P = F = 2000\text{N} = 2\text{kN}$

and power $= (2 \times 20)\text{kW} = 40\text{kW}$

Case II : Here, acceleration $a = 1\text{m/s}^2$

From equation (1),

$$P = F + \frac{W}{g}a = 2000 + \frac{200 \times 1000}{9.81} = 22387.356\text{N} = 22.387\text{kN}$$

Therefore, $\quad \text{power} = (22.387 \times 20)\text{kW} = 447.747\text{kW}$

EXAMPLE 3

A train of weight 200kN moves down a slope of 1:100 at 36kmph and engine develops a power of 40kW. If the train is pulled up at the same speed, what power will be required to pull the train.

SOLUTION Weight of the train W, $\quad = 200\text{kN}$

Slope $= 1 : 100$

$\therefore \quad \sin\theta = \frac{1}{100}$

Velocity (v) $\quad = \frac{36 \times 1000}{3600} = 10\text{m/s}$

Case I : Apply equation of equilibrium along the plane

$$F + \frac{W}{g}a - W\sin\theta - P = 0$$

As acceleration is zero, force exerted by engine $P = F - W\sin\theta$

Again $\quad 40 = P \times v = (F - W\sin\theta) \times 10 = 10\left(F - \frac{200}{100}\right)$

$\Rightarrow \quad 4 = F - 2$

$\Rightarrow \quad F = 4 + 2 = 6\text{kN}$

This is the resistance due to friction.

Case II : Apply equation of equilibrium along the plane

$$P_1 = F + W\sin\theta + \frac{W}{g}a = F + W\sin\theta \qquad [\because a = 0]$$

This is the force exerted by the engine. Power developed by the engine

$$= (F + W\sin\theta) \times v = \left(6 + \frac{200}{100}\right)10 = 80\text{kW}$$

EXAMPLE 4

A tangential force of 2000N is acting on a shaft of diameter 20mm. Find the work done by the force for one revolution of the shaft. If the shaft is rotating at 300 r.p.m, find the power of the shaft.

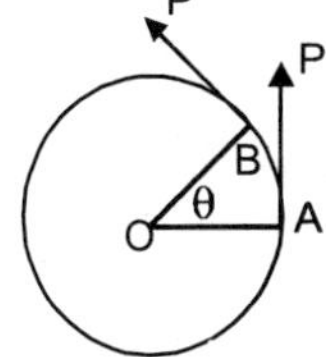

SOLUTION Here, force P, $\quad = 2000\text{N}$

Diameter of the shaft D, $\quad = 20\text{mm} = 0.02\text{m}$

Radius of the shaft R, $\quad = 0.01\text{m}$

Angle turn for one revolution $= 2\pi$ radian

$\because$ Torque $T = P \times R$

$\therefore$ Work done$(W = T\times\theta = P\times R\times\theta = 2000 \times 0.01 \times 2\pi = 125.66$ joule

$\therefore$ Power developed $= \dfrac{2\pi NT}{60} = \dfrac{2\pi \times 300 \times 2000 \times 0.01}{60} = 628.32\text{W}$

EXAMPLE 5

A bullet of mass 81gm and moving with a velocity of 300m/s is fired into a block of wood and it penetrates to a depth of 10cm. If the bullet moving with the same velocity, were fired into a similar piece of wood 5cm thick, with what velocity would it emerge? Find also the force of resistance, assuming it to be uniform.

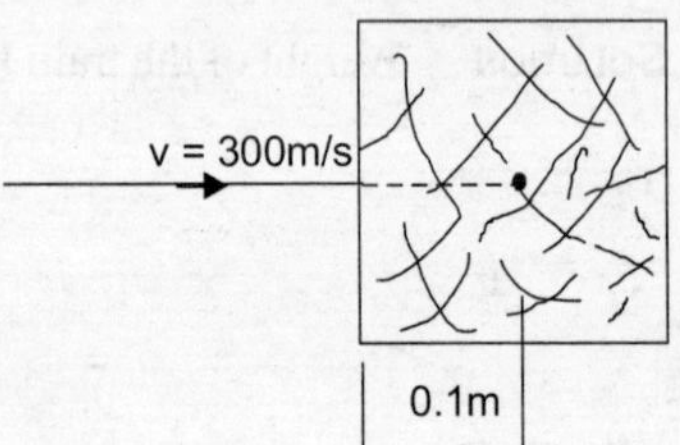

SOLUTION Mass of bullet, $m = 81\text{gm} = 0.081\text{kg}$

Velocity of the bullet, $v = 300\text{m/s}$

Depth of penetration, $x = 10\text{cm} = 0.1\text{m}$

Work done by the force of resistance $= F \times x$

K.E of the bullet $= \dfrac{1}{2}mv^2$

From work energy principle, $\dfrac{1}{2}mv^2 = F \times x$

$$\Rightarrow \quad \frac{1}{2}\times 0.081\times 300^2 = F\times 0.1$$

$$\Rightarrow \quad F = 36450\text{N}$$

If velocity of bullet after 5cm penetration be v_1 m/s,work done will be

$$= 36450 \times 0.05\text{N·m}$$

From work energy principle $= \dfrac{1}{2}mv^2 - \dfrac{1}{2}mv_1^2 = 36450 \times 0.05$

$$\Rightarrow \quad \frac{1}{2}\times 0.081\times 300^2 - \frac{1}{2}\times 0.081\times v_1^2 = 36450\times 0.05$$

$$\Rightarrow \quad v_1 = 212.132\text{m/s}$$

EXAMPLE 6

A hammer of mass 1500kg drops from a height of 60cm on a pile of mass 750kg. Find (*a*) the common velocity after impact assuming plastic impact, (*b*) the average resistance of the ground if the pile comes to rest after penetrating 5cm into the ground.

SOLUTION Mass of hammer, $M = 1500\text{kg}$.

Height of hammer fall, $h = 60\text{cm} = 0.6\text{m}$ and

Mass of pile, $m = 750\text{kg}$

Depth of penetration in one impact $(\delta x) = 5\text{cm} = 0.05\text{m}$

(*a*) If common velocity after impact be v, from law of conservation of linear momentum, we have for plastic impact,

$$Mu_1 + mu_2 = (M+m)v$$

Here, $$u_1 = \sqrt{2gh} = \sqrt{2\times 9.81\times 0.6}\ \text{m/s and } u_2 = 0$$

∴ $$1500\sqrt{2\times 9.81\times 0.6} + 750\times 0 = (1500+750)v$$

⇒ $$v = 2.29\text{m/s}$$

(*b*) If average resistance of ground to the penetration be R, from principle of work energy, we have,

$$(M+m)g\delta x + \frac{1}{2}(M+m)v^2 = R\times\delta x$$

⇒ $$(1500+750)\times 9.81\times 0.05 + \frac{1}{2}(1500+750)\times 2.29^2 = 0.05R$$

⇒ $$R = 140064.75\text{N}$$

EXAMPLE 7

From the pile and pile driver as shown in figure the following numerical data are given : $W_1 = 2000$N, $W_2 = 1000$N, $h = 10$m and coefficient of restitution $e = \frac{1}{4}$. If the resistance to penetration is constant and equals to 60000N, how many blow of the hammer will be required the drive the pile 1m?

SOLUTION Velocity of hammer before impact $u_1 = \sqrt{2gh} = \sqrt{2\times 9.81\times 10}$

Velocity of pile before impact $u_2 = 0$

Velocity of hammer after impact $= v_1$

Velocity of pile after impact $= v_2$

From conservation of linear momentum, we have,

From the equation of coefficient of restitution, we have,

$$Mu_1 + mu_2 = Mv_1 + mv_2$$

⇒ $$\frac{2000}{9.81}\sqrt{2\times 9.81\times 10} + \frac{1000}{9.81}\times 0 = \frac{2000}{9.81}v_1 + \frac{1000}{9.81}v_2$$

⇒ $$2\sqrt{2\times 98.1} = 2v_1 + v_2 \quad ...(1)$$

From the equation of coefficient of restitution, we have, $e = \dfrac{v_2 - v_1}{u_1 - u_2}$

⇒ $$\frac{1}{4} = \frac{v_2 - v_1}{\sqrt{2\times 98.1}}$$

⇒ $$v_2 - v_1 = 3.5 \quad ...(2)$$

Solving (1) and (2), we have, $v_1 = 8.17$m/s, $v_2 = 11.67$m/s. If depth of penetration per single blow is δx, from the principle of work energy, we have,

$$mg\delta x + \frac{1}{2}mv_2^2 = R \times \delta x$$

$$\Rightarrow \qquad 1000\delta x + \frac{1}{2} \times \frac{1000}{9.81} \times 11.67^2 = 60000\delta x$$

$$\Rightarrow \qquad \delta x = 0.1176\text{m}$$

$\therefore$ Number of blows of the hammer, $N = \frac{x}{\delta x} = \frac{1}{0.1176} = 8.4998 \approx 9$

Example 8

A small car of weight W starts from rest at A and rolls without friction along an inclined plane to B where it strikes a block also of weight W and initially at rest. Assuming a plastic impact at B, the car and block will move from B to C as one particle. If the coefficient of friction between the block and plane is $\mu = \frac{1}{2}$, calculate the distance x to point C where the bodies come to rest.

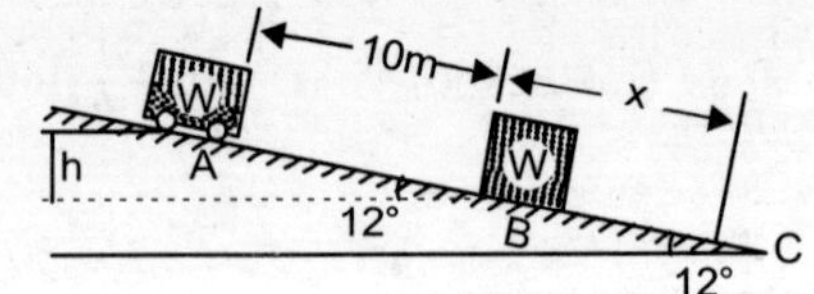

Solution Vertical height between car and block $h = 10 \sin 12°$

Velocity of car before impact $u_1 = \sqrt{2gh} = \sqrt{2 \times 9.81 \times 10 \sin 12°} = 6.387$m/s

Velocity of block before impact $u_2 = 0$

Common velocity of car and block after impact $= v$

From conservation of linear momentum, we have,

$$\frac{W}{g}u_1 + \frac{W}{g}u_2 = \frac{W+W}{g}v$$

$$\Rightarrow \qquad v = \frac{u_1}{2} = \frac{6.387}{2} = 3.193\text{m/s}$$

From principle of work energy, we have

$$\frac{1}{2} \cdot \frac{2W}{g} \cdot v^2 = (F - 2W\sin 12°)x = (\mu W \cos 12° - 2W \sin 12°)x$$

$$\Rightarrow \qquad \frac{v^2}{g} = \left(\frac{1}{2}\cos 12° - 2\sin 12°\right)x$$

$$\Rightarrow \qquad 3.193^2 = 0.072 \times 9.81x$$

$$\Rightarrow \qquad x = 14.43\text{m}$$

EXAMPLE 9

A cylinder is pulled up on an inclined plane by the tension in a rope which passes over a frictionless pulley and is attached to a 70kg mass as shown in figure. The 45kg cylinder has a radius of 600mm. The cylinder moves a distance of 5m. What will be its speed? Assume that the cylinder rolls without slip.

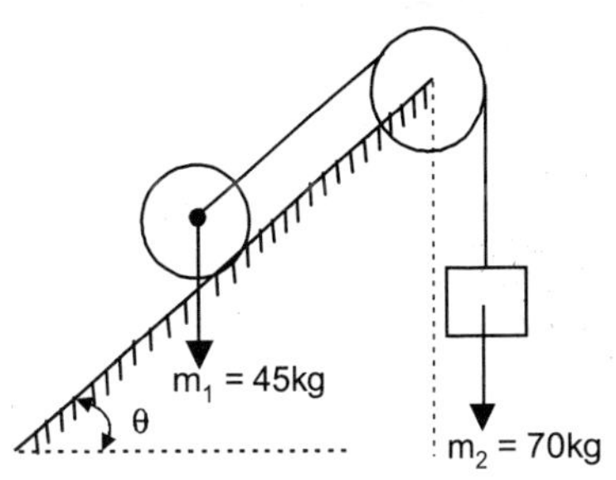

SOLUTION

Mass of cylinder, $m_1 = 45\text{kg}$

Mass of the block, $m_2 = 70\text{kg}$

Radius of cylinder, $R = 600\text{mm} = 0.6\text{m}$

Angle of inclination, $\theta = 50°$

Distance travelled, $x = 5\text{m}$

From work energy principle, for cylinder, we have,

$$W_{12} = KE_2 - KE_1$$

$$\Rightarrow \quad (T - m_1 g \sin\theta)x = \left(\frac{1}{2}m_1\dot{x}^2 + \frac{1}{2}I_1\omega^2\right) - 0$$

Substituting $I_1 = \frac{1}{2}m_1R^2$ and $\omega R = \dot{x}$

$$(T \quad m_1 g \sin\theta)x = \frac{1}{2}m_1\dot{x}^2 + \frac{1}{2}\cdot\frac{1}{2}m_1R^2\omega^2$$

$$\Rightarrow \quad = \frac{1}{2}m_1\dot{x}^2 + \frac{1}{4}m_1\dot{x}^2 = \frac{3}{4}m_1\dot{x}^2$$

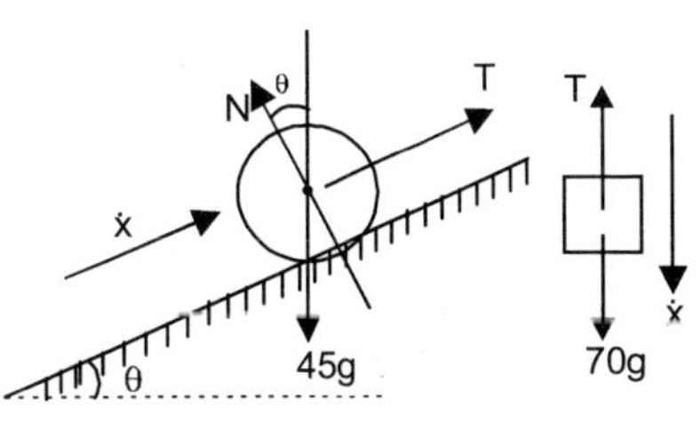

FBD of the system

From work energy principle, for block, we have, $W_{12} = KE_2 - KE_1$

$$\Rightarrow \quad (m_2 g - T)x = \left(\frac{1}{2}m_2\dot{x}^2\right) - 0$$

$$\Rightarrow \quad xT = m_2 gx - \frac{1}{2}m_2\dot{x}^2$$

On substitution we have,

$$m_2 gx - \frac{1}{2}m_2\dot{x}^2 - m_1 gx\sin\theta = \frac{3}{4}m_1\dot{x}^2$$

$$\Rightarrow \quad (m_2 g - m_1 g\sin\theta)x = \frac{3}{4}m_1\dot{x}^2 + \frac{1}{2}m_2\dot{x}^2$$

Substituting values, we obtain

$$(70 - 45\sin 50°) \times 9.81 \times 5 = \left(\frac{3}{4}\times 45 + \frac{1}{2}\times 70\right)\dot{x}^2$$

$$\Rightarrow \quad \dot{x} = 5.038\text{m/s}$$

EXAMPLE 10

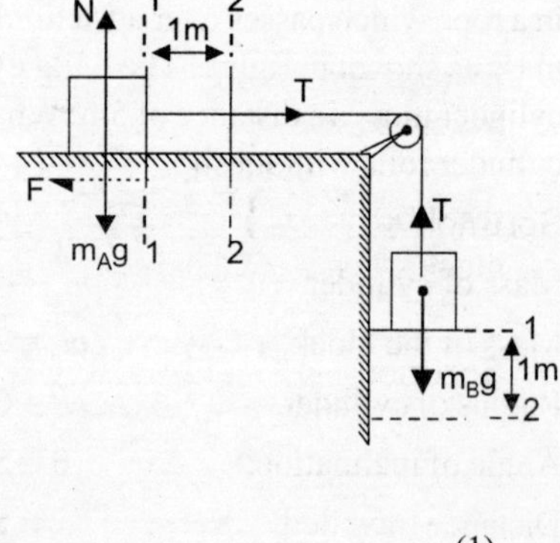

Two block A and B of mass 100kg and 150kg respectively are connected by a string as shown in figure. If the system is released from rest, find the velocity of the block A after it has moved a distance of 1m. The μ between the block and the horizontal plane is 0.20.

SOLUTION Here, $m_A = 100\text{kg}$, $m_B = 150\text{kg}$, $\mu = 0.2$, $x = 1\text{m}$

Apply work energy principle for block A, we have,

$$(T - F)x = \frac{1}{2}m_A\dot{x}^2 - 0$$

$$\Rightarrow \qquad Tx - \mu m_A gx = \frac{1}{2}\times 100\dot{x}^2$$

$$\Rightarrow \qquad Tx = 50\dot{x}^2 + \mu m_A gx \qquad \text{... (1)}$$

Apply work energy principle for block B we have,

$$m_B gx - Tx = \frac{1}{2}m_B\dot{x}^2$$

$$\Rightarrow \qquad Tx = m_B gx - \frac{1}{2}m_B\dot{x}^2 \qquad \text{... (2)}$$

So, comparing equation (1) and (2) we have,

$$m_B gx - \frac{1}{2}m_B\dot{x}^2 = 50\dot{x}^2 + \mu m_A gx$$

$$\Rightarrow \qquad (150 - 0.2\times 100)\times 9.8\times 1 = (50 + 75)\dot{x}^2$$

$$\Rightarrow \qquad \dot{x} = 3.19\text{m/s}$$

EXAMPLE 11

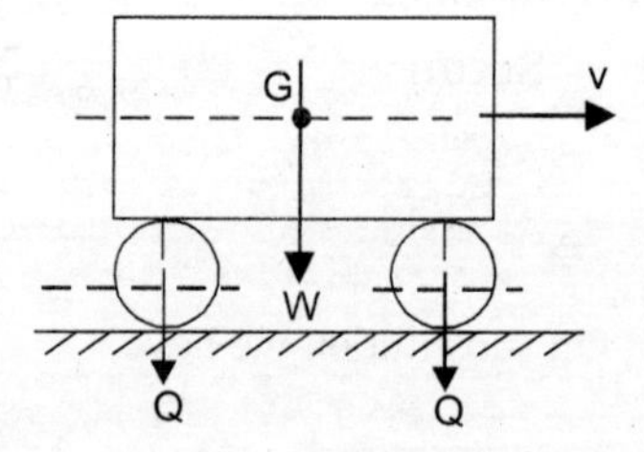

A car body of weight W has four solid disc wheels each of weight Q and rolls on a horizontal plane with a constant speed v start from rest. Find work done by the system to move a distance x from start.

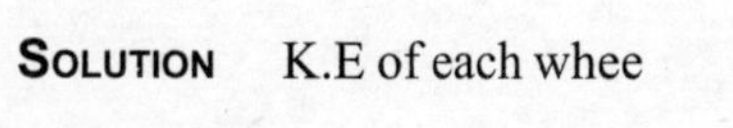

SOLUTION K.E of each whee $\qquad KE_w = \frac{1}{2}\frac{Q}{g}v_G^2 + \frac{1}{2}I_G\omega^2$

$$= \frac{1}{2}\frac{Q}{g}\omega^2 r^2 + \frac{1}{2}\cdot\frac{1}{2}\cdot\frac{Q}{g}\omega^2 r^2$$

$$= \frac{3}{4}\frac{Q}{g}\omega^2 r^2 = \frac{3}{4}\frac{Q}{g}V_G^2 = \frac{3}{4}\frac{Q}{g}V^2$$

Here, it is assumed, $\qquad v_G = v$

Total K.E of four wheels $\qquad KE_{4w} = 4\times\frac{3}{4}\frac{Q}{g}v^2 = 3\frac{Q}{g}v^2$

Total K.E of the system $\quad KE_{4w} + KE_{Car} = 3\frac{Q}{g}v^2 + \frac{1}{2}\frac{W}{g}v^2$

So, total work done $= \left(3Q + \frac{W}{2}\right)\frac{v^2}{g}$

EXAMPLE 12

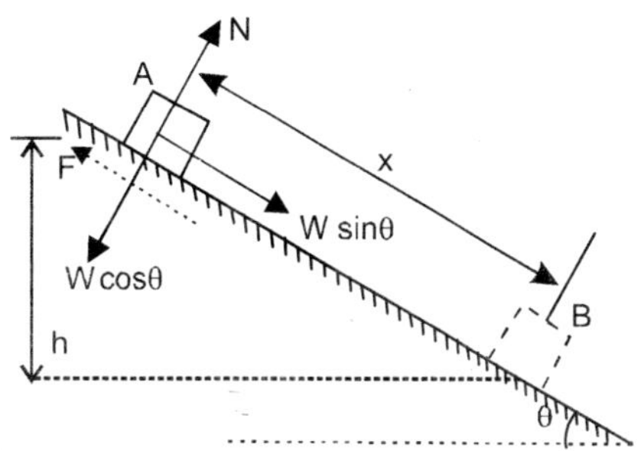

A block of weight W slides down an inclined plane from rest from an elevation h as shown in figure. Find the velocity of the block when it reaches at point B.

SOLUTION From figure, we have, $x = \frac{h}{\sin\theta}$ and

$$F = \mu N = \mu W \cos\theta$$

Applying principle of work energy,

$$W_{12} = KE_2 - KE_1$$

$$\Rightarrow \quad (W\sin\theta - \mu W\cos\theta)x = \left(\frac{1}{2}\frac{W}{g}v^2\right) - 0$$

$$\Rightarrow \quad (\sin\theta - \mu\cos\theta)\frac{Wh}{\sin\theta} = \frac{W}{g}\frac{v^2}{2}$$

$$\Rightarrow \quad v = \sqrt{2gh(1-\mu\cot\theta)}$$

EXAMPLE 13

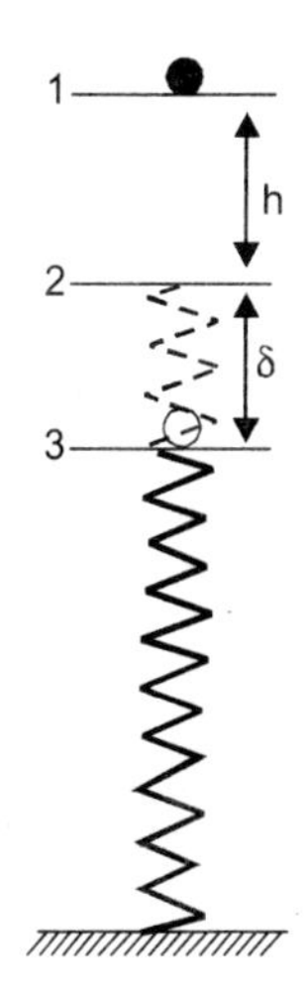

A ball of mass m is dropped on to a spring of stiffness k from a height h. Find the maximum deflection δ of the spring. Assume $m = 10\text{kg}$, $k = 500\text{N/m}$, $h = 10\text{cm}$.

SOLUTION Consider position (2) as datum. At position (1), for ball,

$$KE_1 = 0, PE_1 = mgh$$

At position (2), for ball,

$$KE_2 = \frac{1}{2}mv^2, PE_2 = 0$$

From law of conservation of energy, we have,

$$KE_1 + PE_1 = KE_2 + PE_2$$

$$\Rightarrow \quad 0 + mgh = \frac{1}{2}mv^2 + 0$$

$$\Rightarrow \quad v = \sqrt{2gh}$$

At position (3), for both ball and spring

$$KE_3 = 0, \quad PE_3 = PE_{\text{spring}} + PE_{\text{ball}} = \frac{1}{2}k\delta^2 - mg\delta$$

Total energy of ball and spring at position (3)

$$E_3 = KE_3 + PE_3 = \frac{1}{2}k\delta^2 - mg\delta$$

Again we know, $$\frac{1}{2}mv^2 = \frac{1}{2}k\delta^2 - mg\delta$$

$$\Rightarrow \quad mgh = \frac{1}{2}k\delta^2 - mg\delta$$

$$\Rightarrow \delta^2 - \frac{2\times10\times9.81}{500}\delta - \frac{2\times10\times9.81\times0.1}{500} = 0$$

$$\Rightarrow \quad \delta = 0.475\text{m} = 47.5\text{cm}$$

EXAMPLE 14

A block of mass 10kg resting on a 30º inclined plane is released. The block after travelling a distance of 0.5m along the inclined plane hits a spring of stiffness 20N/cm. Find the maximum compression of spring. Assume μ = 0.2 between block and inclined plane.

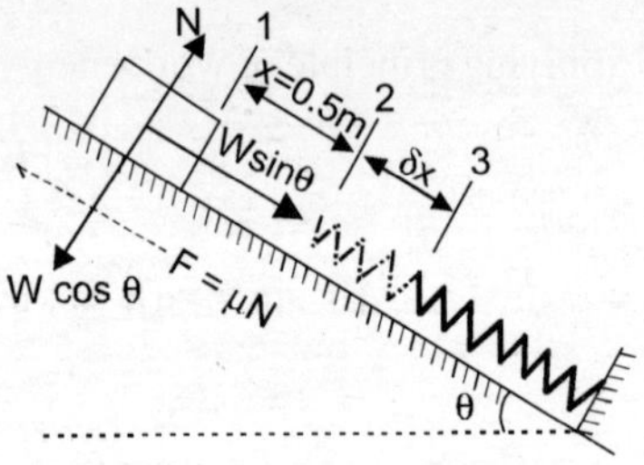

SOLUTION Here, m = 10kg, k = 20N/cm = 2000N/m, x = 0.5m, μ = 0.2, θ = 30°. Let V is the velocity after travelling a distance x and block and spring come to rest after travelling a distance δx.

Here, $$KE_1 = 0, \text{ and } KE_2 = \frac{1}{2}mv^2$$

Applying principle of work energy, we have,

$$W_{12} = KE_2 - KE_1$$

$$\Rightarrow \quad (mg\sin\theta - \mu mg\cos\theta)\,x = \frac{1}{2}mv^2$$

$$\Rightarrow \quad v^2 = 2gx(\sin\theta - \mu\cos\theta)$$

$$= 2 \times 9.81 \times 0.5$$

$$(\sin 30° - 0.2\cos 30°) = 3.206$$

$$\Rightarrow \quad v = 1.79 \text{ m/s}$$

For the position from (2) to (3)

$$KE_2 = \frac{1}{2}mv^2,\ KE_3 = 0,$$

$$W_{23} = (mg\sin\theta - \mu mg\cos\theta)\delta x - \frac{1}{2}k\delta x^2$$

Applying principle of work energy we have,

$$(mg\sin\theta - \mu mg\cos\theta)\delta x - \frac{1}{2}k\delta x^2 = -\frac{1}{2}mv^2$$

$$\Rightarrow \quad (\sin 30° - 0.2\cos 30°)\times 9.81\delta x - \frac{2000}{2\times10}\delta x^2 = -\frac{1.79^2}{2}$$

$$\Rightarrow \quad \delta x = 0.1435\text{m} = 14.35\text{cm}$$

EXAMPLE 15

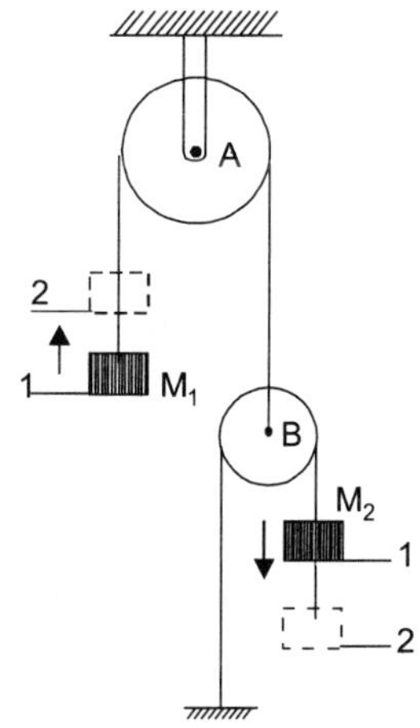

A system of two masses M_1 and M_2 arranged as shown in figure are released from rest. Find the velocity of the mass M_2 after it has fallen a vertical distance of 1m. Neglect the inertia of the pulleys. Assume $M_1 = M_2 = 20$kg.

SOLUTION At position (1), total energy $E_1 = KE_1 + PE_1 = 0 + 0 = 0$

At position (2), M_2 acquires a velocity v

after a fall of 1m, the velocity of M_1 is $v/2$.

$$PE_2 = M_1 g\left(\frac{1}{2}\right) + (-M_2 g \times 1) = \frac{20}{2}g - 20g = -10g$$

$$KE_2 = \frac{1}{2}M_1\left(\frac{v}{2}\right)^2 + \frac{1}{2}M_2 v^2 = \frac{25}{2}v^2$$

$$\therefore \quad E_2 = PE_2 + KE_2 = -10g + \frac{25}{2}v^2$$

Applying principle of conservation of energy, we have, $E_1 = E_2$

$$\Rightarrow \quad 0 = -10g + \frac{25}{2}v^2$$

$$\Rightarrow \quad v^2 = \frac{20 \times 9.81}{25}$$

$$\Rightarrow \quad v = 2.8\text{m/s}$$

EXAMPLE 16

A flexible chain of length $\frac{\pi r}{2}$ is held on smooth cylinderical surface of radius as shown in figure. Calculate the velocity v with which the chain with leave the point B if released from rest. Assume that the chain is inextensible and weight g per unit length.

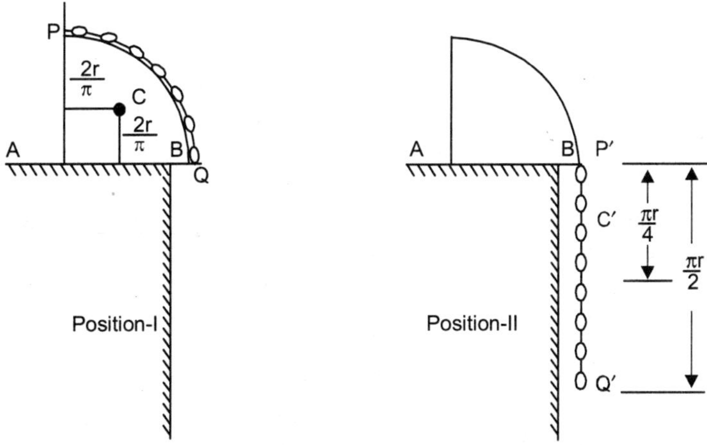

SOLUTION At position (1) centroid of chain is $\frac{2r}{\pi}$ distance with respect to AB.

$$\therefore \qquad PE_1 = mgh = \left(\frac{\gamma}{g} \times \frac{\pi r}{2}\right) g \left(\frac{2r}{\pi}\right) = \gamma r^2 \text{ and } KE_1 = 0$$

So total energy at position (1),

$$E_1 = PE_1 + KE_1 = \gamma r^2$$

At position (2) centroid of vertical chain is at $\frac{\pi r}{4}$ distance.

$$\therefore \qquad PE_2 = -\left(\frac{\gamma}{g} \times \frac{\pi r}{2}\right) g \left(\frac{\pi r}{4}\right) = -\frac{\gamma \pi^2 r^2}{8}$$

and

$$KE_2 = \frac{1}{2} \cdot \frac{\gamma}{g} \cdot \frac{\pi r}{2} \cdot V^2 = \frac{\gamma \pi r V^2}{4g}$$

So total energy at position (2),

$$\therefore \qquad E_2 = PE_2 + KE_2 = \frac{\gamma \pi r V^2}{4g} - \frac{\gamma \pi^2 r^2}{8}$$

Applying principle of conservation of energy,

$$E_1 = E_2$$

$$\gamma r^2 = \frac{\gamma \pi r v^2}{4g} - \frac{\gamma \pi^2 r^2}{8}$$

$$\Rightarrow \qquad v^2 = 4gr\left(\frac{1}{\pi} + \frac{\pi}{8}\right)$$

$$\Rightarrow \qquad v = \sqrt{gr\left(\frac{4}{\pi} + \frac{\pi}{2}\right)}$$

EXAMPLE 17

A particle of mass m resting on a smooth semicircular cylinder of radius r is prevented from moving by a string AB of length r. If the string is cut find the value of the angle θ defining the position C of the particle when it jumps off the cylindrical surface.

SOLUTION The initial position of the particle is defined by the angle α. Tangential acceleration = a_t, Normal acceleration = a_n

Let it jumps off at the point C after attaining a velocity V and is defined by the angle θ. The equations of dynamic equilibrium of the particle can be written as

$$\left.\begin{aligned} R + ma_n - mg\cos\theta = 0 \\ ma_t - mg\sin\theta = 0 \end{aligned}\right\}$$

Here,

$$a_n = \frac{v^2}{r}, \text{ and } a_t = \frac{dv}{dt}$$

When the particle jumps off it losses contact and $R = 0$.

$$\therefore \quad \frac{mv^2}{r} = mg\cos\theta$$

$$\Rightarrow \quad \cos\theta = \frac{v^2}{rg}$$

At position (1), $E_1 = PE_1 + KE_1 = 0 + 0 = 0$

At position (2), $PE_2 = -mg\,(r\cos\alpha - r\cos\theta)$ and $KE_2 = \frac{1}{2}mv^2$

$$\therefore \quad E_2 = \frac{1}{2}mv^2 - mg(r\cos\alpha - r\cos\theta)$$

Applying principle of conservation of energy, $E_1 = E_2$

$$\Rightarrow \quad 0 = \frac{1}{2}mv^2 - mgr(\cos\alpha - \cos\theta)$$

$$\Rightarrow \quad v^2 = 2gr(\cos\alpha - \cos\theta) = 2gr\left(\cos\alpha - \frac{v^2}{rg}\right)$$

$$\Rightarrow \quad \cos\theta = 2(\cos\alpha - \cos\theta)$$

$$\Rightarrow \quad \cos\theta = \frac{2}{3}\cos\alpha$$

Example 18

A roller of $m = 500$kg and radius $r = 30$cm is pushed with a constant force $P = 750$N on a rough horizontal plane. If the roller starts from rest and rolls without slipping, find the distance required to be rolled if it to acquire a velocity of 5m/s.

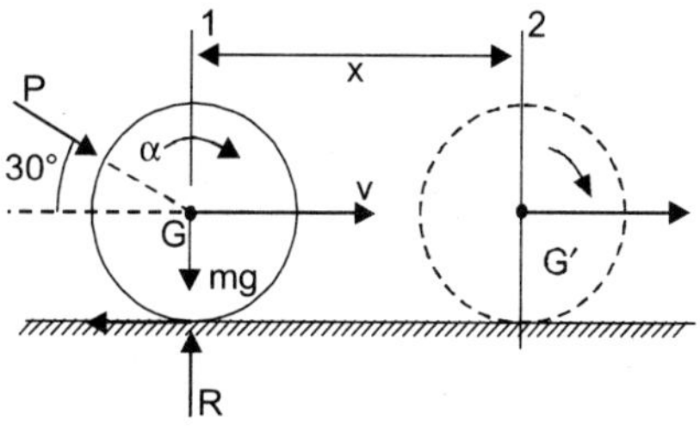

Solution Forces acting on the body and work done:

(*i*) Gravity force *mg*, acting normal to the direction of motion, does no work.

(*ii*) Normal reaction *N*, does no work, as same reason above.

(*iii*) Friction force *F*, does no work, as there is no translation of the contact point of the body.

(*iv*) Horizontal component of the external force $P\cos\theta$ does work.

(*v*) Vertical component of the external force $P\sin\theta$ does no work, as same reason above.

At position (1), $KE_1 = 0$

At position (2), $KE_2 = \frac{1}{2}mv^2 + \frac{1}{2}I\omega^2 = \frac{1}{2}mv^2 + \frac{1}{2}\cdot\frac{mr^2}{2}\cdot\frac{v^2}{r^2} = \frac{3}{4}mv^2$

Change in $KE = KE_2 - KE_1 = \frac{3}{4}mv^2 - 0 = \frac{3}{4}\times 600\times 3^2$

Work done between (1) and (2), $W_{12} = P\cos\theta\, x = 850\, x\cos 30°$

From the principle of work and energy, we have,

$$850x\cos 30° = \frac{3}{4}\times 600\times 3^2$$

$\Rightarrow$ $$x = 5.5\text{m/s}$$

EXAMPLE 19

Three bodies, a sphere, a cylinder and a hoop each the same mass m and radius r are released from rest from an inclined plane of angle θ. Determine the velocity of each of the bodies after it has rolled down the inclined plane through a distance x and also find the reaction of the velocity of the bodies.

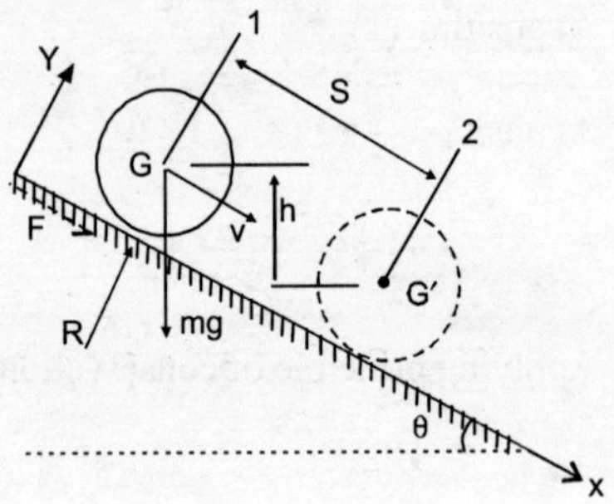

SOLUTION Forces acting on the bodies and work done:

(*i*) Component of gravitational force $mg\cos\theta$, does no work, being normal to direction of motion.

(*ii*) Component of gravitational $mg\sin\theta$, does work.

(*iii*) Frictional force F, does work as body rolls without sliding.

Let us assumed mass centre linear velocity at position (1).

$v_0 = 0$ and that at position (2) $= v$ and angular velocity $= \omega$

K.E of the body at position (1), $KE_1 = 0.$

K.E of the body at position (2), $KE_2 = \frac{1}{2}mv^2 + \frac{1}{2}I\omega^2 = \frac{1}{2}mv^2 + \frac{1}{2}I\frac{v^2}{r^2}$

Change in K.E between position (1) and (2)

$$KE_{12} = KE_2 - KE_1 = \frac{1}{2}mv^2 + \frac{1}{2}I\frac{v^2}{r^2}$$

From the principle of work and energy, we have, $\frac{1}{2}mv^2 + \frac{1}{2}I\frac{v^2}{r^2} = mgx\sin\theta$

$\Rightarrow$ $$v = \sqrt{\frac{2gx\sin\theta}{1+\frac{I}{mr^2}}}$$

Again we know values of mass moment of inertia for

$$\text{Sphere } (I_s) = \frac{2}{5}mr^2,\ \text{Cylinder}(I_c) = \frac{1}{2}mr^2,\ \text{Hoop}(I_H) = mr^2$$

So, we have the velocities as,

$$v_S = \sqrt{\frac{2gx\sin\theta}{1+\frac{I_S}{mr^2}}} = 0.845\sqrt{2gx\sin\theta}$$

$$v_C = \sqrt{\frac{2gx\sin\theta}{1+\frac{I_C}{mr^2}}} = 0.816\sqrt{2gx\sin\theta}$$

$$v_H = \sqrt{\frac{2gx\sin\theta}{1+\frac{I_H}{mr^2}}} = 0.707\sqrt{2gx\sin\theta}$$

$\therefore$ $$v_S : v_C : v_H = 845:816:707$$

EXAMPLE 20

A truck of mass 20,000kg starts from rest rolling down a gradient of 1 in 100. The truck is mounted on two pairs of axles and wheels which are1 metre in diameter, each pair having a mass of 400 kg and radius of gyration of 350mm. Find the velocity and time required to traverse 200m and also find the acceleration. Neglect axle friction and any other resistances .

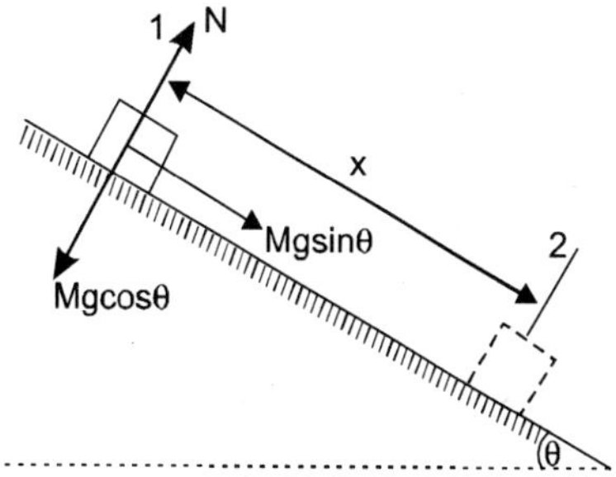

1 N x Mgsinθ 2 Mgcosθ θ

SOLUTION Mass of truck, $M = 20{,}000$ kg

Mass of wheels and axles, $m = 2 \times 400 = 800$kg

Radius of the wheels, $R = 0.5$m

Radius of gyration of wheels and axles, $r = 0.35$m

Moment of inertia of wheels and axles, $I = mr^2 = 800 \times 0.35^2$ kgm^2

Change of K.E from position (1) to (2), $KE_{12} = \frac{1}{2}Mv^2 + \frac{1}{2}\cdot mr^2 \cdot \frac{v^2}{R^2}$

From the principle of work energy, we have, $\frac{1}{2}Mv^2 + \frac{1}{2}\cdot mr^2 \cdot \frac{v^2}{R^2} = Mgx\sin\theta$

$\Rightarrow$ $$v = \sqrt{\frac{2Mgx\sin\theta}{M+\frac{mr^2}{R^2}}} = \sqrt{\frac{2\times 20000\times 9.81\times 200\times\frac{1}{100}}{20000+\frac{800\times 0.35^2}{0.5^2}}} = 6.2\text{m/s}$$

Time taken

$$t = \frac{\text{distance moved}}{\text{average velocity}} = \frac{200}{\left(\frac{6.2}{2}\right)} = 64.52\text{s} = 1\,\text{min}\,4.52\,\text{sec}.$$

If a be the acceleration of the truck down the plane, $v^2 = u^2 + 2ax = 2ax$

$\Rightarrow$ $$a = \frac{v^2}{2x} = \frac{2Mgx\sin\theta}{\left(M+\frac{mr^2}{R^2}\right)2x}$$

$$= \frac{20000 \times 9.81}{20000 + \dfrac{800 \times 0.35^2}{0.5^2}} \times \frac{1}{100} = 0.0962 \text{m/s}^2$$

EXAMPLE 21

A flywheel assembled to a shearing machine 2 × 10⁵ joules of K.E, stored in it when it runs at 200 r.p.m. If this speed falls to 150 r.p.m during the shearing operation what energy is being parted with by this flywheel? If 90% of the above energy that is parted with is imparted to the shearing during a stroke of 18mm, what is the average force due to this on the blade of the shears.

SOLUTION Original speed of the flyweel, $N_0 = 200$ r.p.m

Final speed of the flywheel $N = 150$ r.p.m

Original angular velocity, $\omega_0 = \frac{2\pi N_0}{60}$ rad/s

Final angular velocity, $\omega = \frac{2\pi N}{60}$ rad/s.

Original kinetic energy, $KE_1 = \frac{1}{2} I\omega_0^2 = \frac{1}{2} I \left(\frac{2\pi N_0}{60}\right)^2$

$\therefore$ $2 \times 10^5 = \frac{1}{2} I \left(\frac{2\pi \times 200}{60}\right)^2$

$\therefore$ $I = \frac{9 \times 10^3}{\pi^2}$

Final kinetic energy, $KE_2 = \frac{1}{2} I\omega^2 = \frac{1}{2} \cdot \frac{9 \times 10^3}{\pi^2} \left(\frac{2\pi \times 150}{60}\right)^2 = 0.1125 \times 10^5 \text{ J}$

Loss of kinetic energy, $KE_s = KE_1 - KE_2 = 1.8875 \times 10^5 \text{ J}$

Useful kinetic energy $= \frac{90}{100} \times 1.8875 \times 10^5 \text{ J}$

From principle of work energy, we have,

$$P \times 18 \times 10^{-3} = \frac{90}{100} \times 1.8875 \times 10^5$$

$\therefore$ $P = 9.4375 \times 10^6 \text{ N} = 9437.5 \text{kN}$

EXAMPLE 22

A circular cylinder of mass m and radius r rolls without slipping on a circular path of radius R. Find the period of oscillations of a cylinder when it is displaced slightly from the equilibrium position.

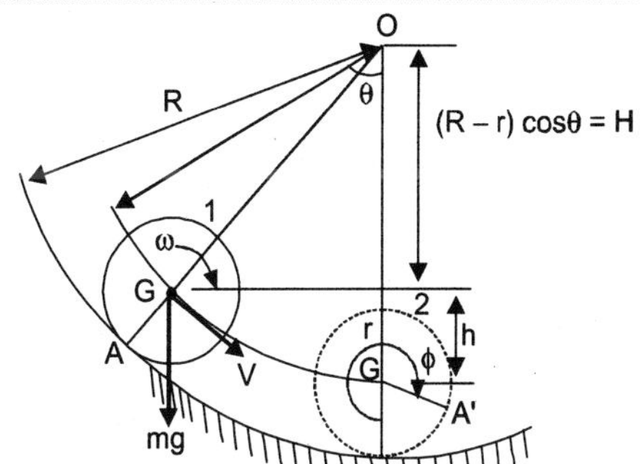

SOLUTION Let angular velocity of the line OG be $\dot{\theta}$,

then the linear velocity v of the mass centre G is $(R-r)\dot{\theta}$.

At position (1), $KE_1 = 0$ and

$$PE_1 = mg[(R-r)-(R-r)\cos\theta] = mg(R-r)\times 2\sin^2\frac{\theta}{2}$$

$$= mg(R-r)\times 2\times\frac{\theta^2}{4}$$

$$= mg(R-r)\frac{\theta^2}{2}$$

Total energy, $$E_1 = KE_1 + PE_1 = mg(R-r)\frac{\theta^2}{2}$$

At position (2), $$PE_2 = 0,\ KE_2 = \frac{1}{2}mv^2 + \frac{1}{2}I\omega^2$$

As $$v = \dot{\theta}(R-r) = \omega r \text{ and } I = \frac{mr^2}{2}$$

$\therefore$ $$KE_2 = \frac{1}{2}m\left[\dot{\theta}(R-r)\right]^2 + \frac{1}{2}\cdot\frac{mr^2}{2}\left[\frac{\dot{\theta}(R-r)}{r}\right]^2$$

$$= \frac{3}{4}m(R-r)^2\dot{\theta}^2$$

Total energy, $$E_2 = KE_2 + PE_2 = \frac{3}{4}m(R-r)^2\dot{\theta}^2$$

From principle of conservation of energy, we have, $E_1 = E_2$

$\therefore$ $$mg(R-r)\frac{\theta^2}{2} = \frac{3}{4}m(R-r)^2\dot{\theta}^2$$

$\therefore$ $$\frac{3}{2}(R-r)^2\dot{\theta}^2 - g\theta^2 = 0$$

Differentiating above with respect to time,

$$\frac{3}{2}(R-r)2\dot{\theta}\ddot{\theta} - g\times 2\dot{\theta}\theta = 0$$

$\Rightarrow$ $$\frac{3}{2}(R-r)\ddot{\theta} - g\dot{\theta} = 0$$

$\Rightarrow$ $$\ddot{\theta} - \frac{2}{3}\frac{g}{(R-r)}\theta = 0$$

[***Minus Sign appears, as when θ decrease ϕ increases.***] Comparing the above equation with the equation of S.H.M.

$\ddot{\theta} + P^2\theta = 0$, we can confirm the motion of the cylinder is S.H.M.

Here time period $\tau = \dfrac{2\pi}{P}$,

where $P^2 = \dfrac{2}{3}\left(\dfrac{g}{R-r}\right)$

So, $\tau = 2\pi\sqrt{\dfrac{3}{2}\left(\dfrac{R-r}{g}\right)}$

EXAMPLE 23

A uniform bar of mass m and length l hangs from a frictionless hinge. It is released from the horizontal position AB. Find the angular velocity and linear velocity of the centre of mass of the bar when it is in vertical positions. Also find the time period of oscillations.

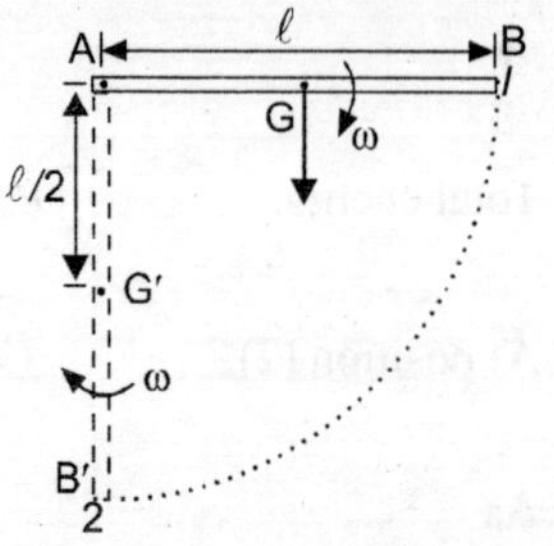

SOLUTION At position (1), $KE_1 = 0, PE_1 = 0.$

So $E_1 = 0$

At position (2), $E_2 = PE_2 + KE_2 = \left(-\dfrac{mgl}{2}\right) + \dfrac{1}{2}mv^2 + \dfrac{1}{2}I\omega^2$

$$= \left(-\frac{mgl}{2}\right) + \frac{1}{2}mv^2 + \frac{1}{2}\frac{ml^2}{12}\cdot\left(\frac{2v}{l}\right)^2$$

$$= \frac{1}{2}mv^2 + \frac{1}{6}mv^2 - \frac{1}{2}mgl$$

From principle of conservation of energy, $E_1 = E_2$

$$\Rightarrow \quad 0 = \frac{1}{2}mv^2 + \frac{1}{6}mv^2 - \frac{1}{2}mgl$$

$$\Rightarrow \quad v^2 = \frac{3}{4}gl$$

$$\Rightarrow \quad v = \sqrt{\frac{3}{4}gl}$$

Again, $\omega = \dfrac{v}{l/2} = \dfrac{\sqrt{\frac{3}{4}gl}}{l/2} = \sqrt{\dfrac{3g}{l}}$,

and $\tau = \dfrac{2\pi}{\omega} = 2\pi\sqrt{\dfrac{l}{3g}}$

EXAMPLE 24

A uniform bar AB of length L and mass m rests against two surfaces as shown in figure. If the bar starts moving from rest when in vertical position ($\theta = 90°$). Find the angular velocity of the bar as a function of its angle of rotation θ.

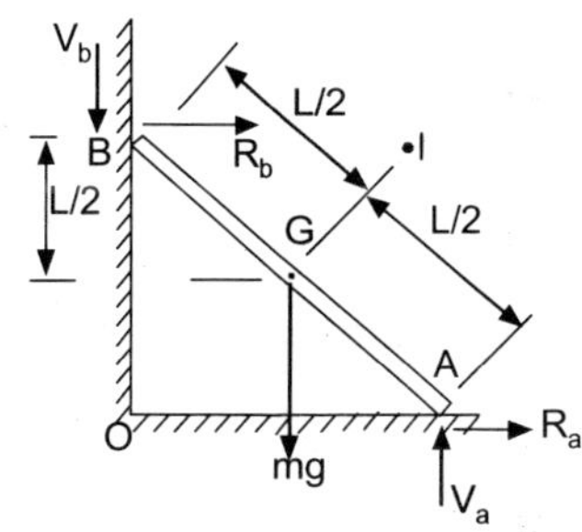

SOLUTION From figure, the end A can move along OA and the end B along OB. The instantaneous centre of rotation (I) of the bar is located. From the geometry of the figure it is clear that $BG = GA = OG = IG = L/2$. If the angular velocity of bar be ω, the centre of gravity G of the bar moves in a circular path with I as centre and its velocity $V_C = \omega L/2$.

At position (1), $$KE_1 = 0.$$

At position (2), $$KE_2 = \frac{1}{2}m \cdot V_G^2 + \frac{1}{2}I\omega^2 = \frac{1}{2}m\left(\frac{\omega L}{2}\right)^2 + \frac{1}{2}\cdot\frac{mL^2}{12}\cdot\omega^2 = \frac{1}{6}mL^2\omega^2$$

Work done between (1) and (2),

$$W_{12} = mgh = mg\frac{L}{2}(1-\sin\theta)$$

From the principle of work energy, we have, $KE_2 - KE_1 = W_{12}$

$$\Rightarrow \qquad \frac{1}{6}mL^2\omega^2 = mg\frac{L}{2}(1-\sin\theta)$$

$$\Rightarrow \qquad \omega = \sqrt{\frac{3g}{L}(1-\sin\theta)}$$

EXAMPLE 25

A smooth semicircular tube AB of radius r is fixed in a vertical plane and contains heavy flexible chain of length πr and mass $\rho\pi r$ as shown in figure. Assuming a slight disturbance to start the chain in motion, find the velocity V with which it will emerge from the open end B of the tube.

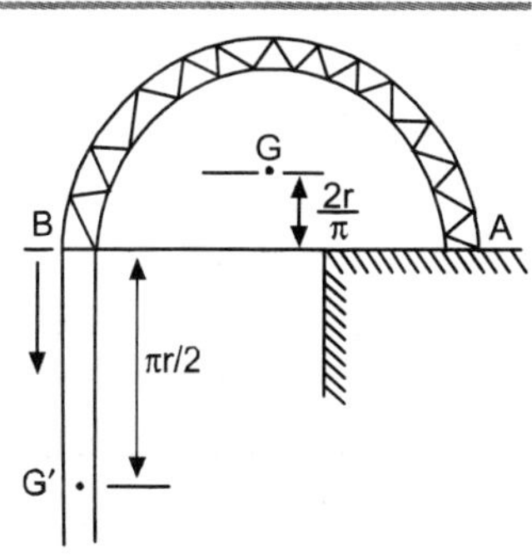

SOLUTION Length of chain $l = \pi r$, mass of the chain $m = \rho\pi r$, C.G of semicircular tube, $h = 2r/\pi$.

At position (1), $$KE_1 = 0,\ PE_1 = mgh = \rho\pi gr\left(\frac{2r}{\pi}\right) = 2\rho gr^2$$

At position (1), $$E_1 = KE_1 + PE_1 = 2\rho gr^2$$

At position (2), $$KE_2 = \frac{1}{2}\rho\pi rV^2,\ PE_2 = -\rho\pi gr\left(\frac{\pi r}{2}\right)$$

$$\Rightarrow \qquad E_2 = KE_2 + PE_2 = \frac{1}{2}\rho\pi rV^2 - \frac{1}{2}\rho\pi^2 gr^2$$

From principle of conservation of energy, $E_1 = E_2$

$$\Rightarrow \quad 2\rho g r^2 = \frac{1}{2}\rho\pi r v^2 - \frac{1}{2}\rho\pi^2 g r^2$$

$$\Rightarrow \quad v = \sqrt{\frac{4gr + \pi^2 gr}{\pi}}$$

EXAMPLE 26

Determine velocity V of the falling weight W of the system as a function of its displacement from the initial position of rest. Assume weight of the cylinder as $2W$.

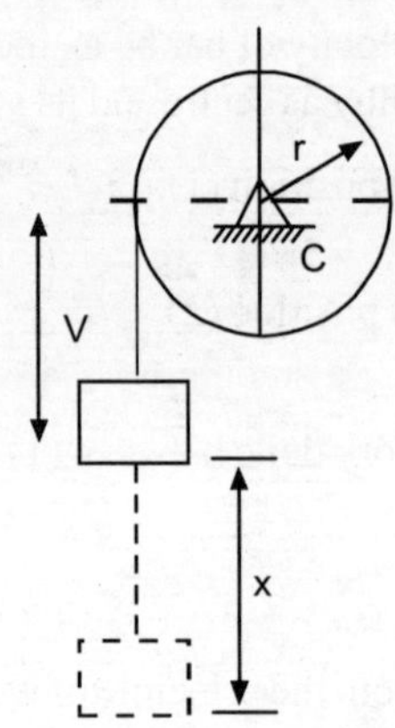

SOLUTION At position (1), $KE_1 = 0$

At position (2), $KE_2 = \frac{1}{2}I\omega^2 + \frac{1}{2}\frac{W}{g}V^2 = \frac{1}{2}\frac{W}{g}\omega^2 r^2 + \frac{1}{2}\frac{W}{g}V^2 = \frac{W}{g}V^2$

From the principle of work energy, we have,

$$W_{12} = KE_2 - KE_1$$

$$\Rightarrow \quad W \times x = \frac{W}{g}V^2$$

$$\Rightarrow \quad V = \sqrt{gx}$$

Exercise

1. A chain of length 50.5m is placed on a pulley whose radius is 16cm. One end of the chain hangs down 40m below the point it leaves the pulley. A man holds the other end of the chain and pulls it down until the length of the chain on both sides of the pulley is equal. Find the work done by the man. Given the weight of the chain as 10N per metre length and the pulley is frictionless. **Ans.** [2250N·m]
2. A train of weight 1500kN, is ascending (i.e., going upwards) a slope of 1 *in* 100 with a uniform speed of 36km/hr. Find the power exerted by the engine, if the road resistance is 5N per kN, weight of the train. **Ans.** [225kW]
3. A mass of 10kg is dropped through 2cm on a spring of stiffness k = 100N/cm. Working from the first principles. Find the maximum displacement of the spring. **Ans.** [3.19cm]
4. A wagon weighing 490kN starts from rest, runs 20 metres down a one percent grade and strikes a post. If the rolling resistance of the track is 5N per kN, find the velocity of the wagon when it strikes the post. If the impact is to be cushioned by means of one bumper spring which compresses by 1mm for 14.7kN, determine by how much this bumper spring will be compressed.
5. A pile weighing 4900N is driven into the ground by dropping freely a weight of 2940N through a height of 3 metres. If the pile is driven into ground by 150mm, calculate the average resistance of the ground and the amount of energy lost at impact. **Ans.**[29890N, 5512.5N]
6. Two railway trucks weighing 120kN and 100kN moving in the same direction at speeds of 3 metres and 5 metres per second respectively collide and subsequently move together. Calculate the common speed and the loss of kinetic energy due to impact. **Ans.** [3.91 m/sec, 11052N·m.]

7. The 125 kg concrete block A is released from rest in the position shown in the figure and pulls the 200kg log B up a 30° ramp. If the coefficient of kinetic friction between the log and the ramp is 0.5, determine the velocity of log as the block, A-hits the ground. **Ans.** [4.618m/s.]

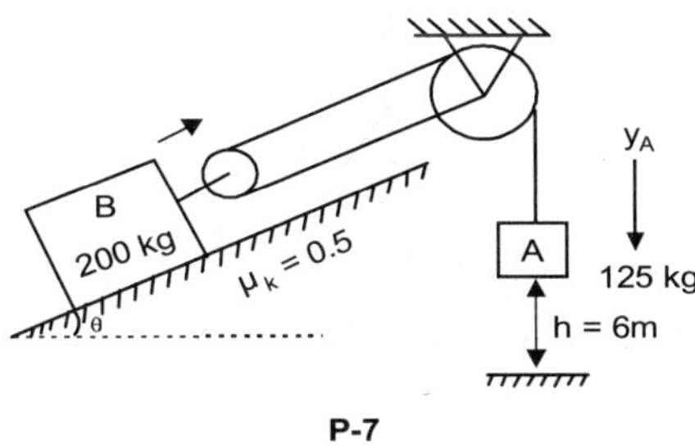

P-7

8. A spring, normally 150mm long is connected to the masses, as shown in figure. If the system is released on a smooth horizontal plane, what will be the speed of each block when the spring is again in its normal length? The spring constant is 2.1 kN/m.

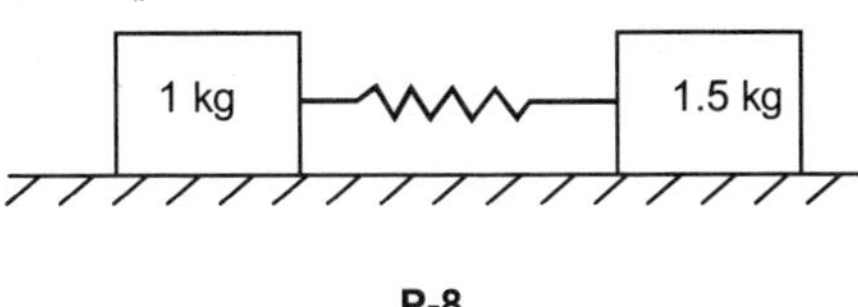

P-8

9. Block A initially rests on the spring to which it is connected by a 600mm inextensible cord which becomes tant after the system is released. What will be the stretch of the spring to bring the system to rest. The cylinder may be considered homogeneous, has a mass of 75kg, and rotates in frictionless bearings.

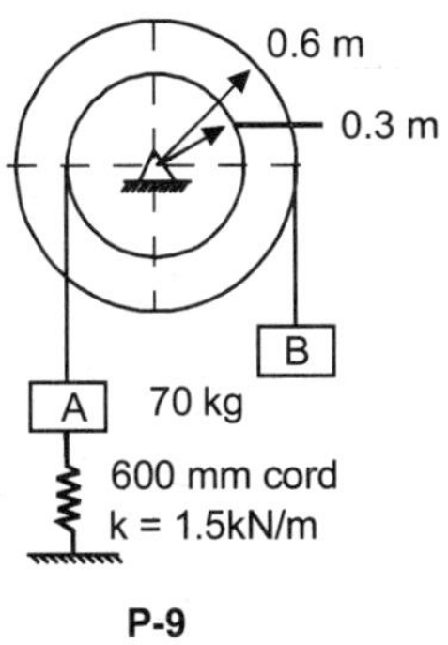

P-9

10. A weight W_1 falls through a height h onto a block of weight W_2 which is supported by spring having a spring constant K. Assuming plastic impact, determine the maximum compression δ of the spring over and above that due to the static action of W_2. The following numerical data are given:

$$W_1 = W_2 = 50\text{N},\ k = 20\text{N/cm},\ h = 7.5\text{cm}$$

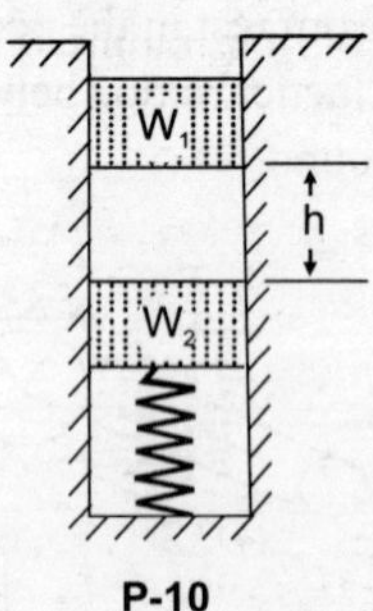

P-10

11. A collar of mass 5kg can slide along a vertical bar as shown. The spring attached to the collar is in undeformed state of length 20cms and stiffness 500N/m. If the collar is suddenly released find the velocity of the collar if it moves 15cm down.

Ans. [16.4m/s]

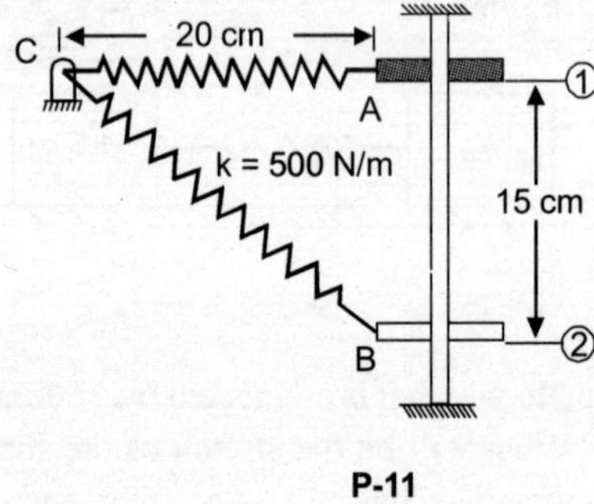

P-11

12. A block of mass *M* resting at *A* on the surface of a smooth circular cylinder of radius *r* slides in a vertical plane. At point *B* it leaves the cylinder and travels as a projectile hitting the horizontal plane at *C*. Prove that the distance *CD*= 1.46*r*.

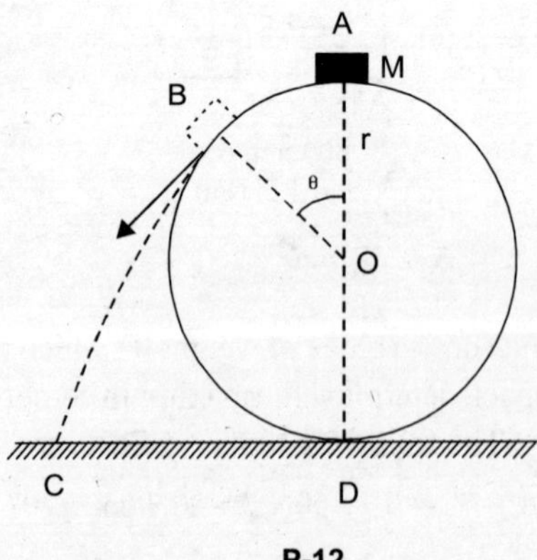

P-12

13. A flywheel rigidly attached to a shaft of 4cm radius is placed on two parallel rails fixed at an angle $\alpha = 15°$ to the horizontal. When released from rest, it attains a velocity of 16cm/s after moving a distance of 1.5m along the rails. Determine the centroidal radius of gyration of the system.

Ans. [68.9cm]

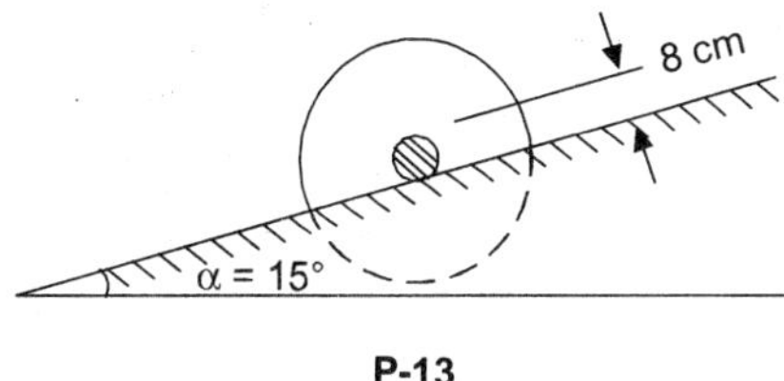

P-13

14. A right circular cylinder of mass m and radius r is suspended by a cord wound round its surface. The cylinder is allowed to fall from rest through a vertical distance h. Find the velocity of its centre of gravity.

Ans. $\left[v = \sqrt{\dfrac{4gh}{3}}\right]$

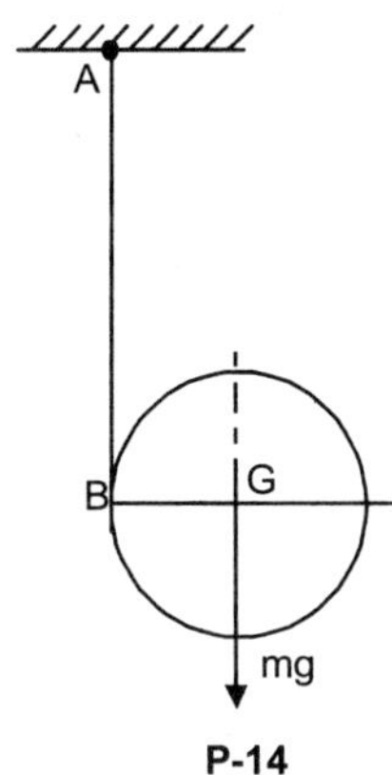

P-14

15. Bullet A of mass 0.01kg moving with a velocity of 100m/s hits a bob B of a simple pendulum horizontally. Find the maximum angle through which the pendulum swings when. (*a*) The bullet gets embedded in the bob, (*b*) The bullet rebounds from the surface of the bob with a velocity of 20m/s, (*c*) The bullet escapes from the other end of the bob with a velocity of 20m/s. Assume the mass of the bob to be 1.0kg and the length of pendulum as 1.0m.

Ans.[(*a*) 18.2°, (*b*) 22.09°, (*c*) 14.7°]

CHAPTER

13

PRINCIPLE OF LINEAR IMPULSE AND MOMENTUM

13.1 MOMENTUM AND IMPULSE OF A PARTICLE

Let us consider a mass m is subjected to a force F. If we express Newton's second law of motion in a *differential form*, it appears as

$$F = \frac{d}{dt}(mv) \qquad \dots (13.1)$$

Here mv is the *linear momentum* of the particle. Necessary manipulation of the above equation to conform into a *integrative form*, we have,

$$Fdt = d(mv)$$

$$\int_{t_1}^{t_2} F\,dt = m\int_{v_1}^{v_2} dv = mv_2 - mv_1 \qquad \dots (13.2)$$

The integral at the left hand side of the above equation, is a vector known as *linear impulse* or simply the *impulse*, of the force F during the interval of time considered. Considering 3D-rectangular co-ordinate system of axes,

$$\int_{t_1}^{t_2} F\,dt = i\int_{t_1}^{t_2} F_x\,dt + j\int_{t_1}^{t_2} F_y\,dt + k\int_{t_1}^{t_2} F_z\,dt \qquad \dots (13.3)$$

In S.I, the unit of impulse is Newton-second (N.s).

13.2 RELATION BETWEEN MOMENTUM AND IMPULSE OF PARTICLE

Now, transposing of elements in equation (13.2), we obtain,

$$mv_1 + \int_{t_1}^{t_2} F\,dt = mv_2 \qquad \dots (13.4)$$

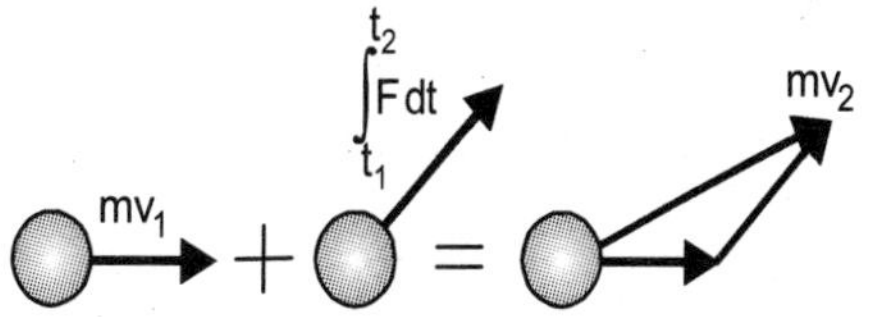

Fig. 13.1

Above equation can envisaged as *the final momentum mv_2 of the particle can be obtained by adding vectorially its initial momentum mv_1 and the impulse of the force F during the time interval considered.* A pictorial representation of this principle is shown in Figure 13.1.

In rectangular co-ordinate system of axes,

$$\left.\begin{aligned} (mv_x)_1 + \int_{t_1}^{t_2} F_x dt = (mv_x)_2 \\ (mv_y)_1 + \int_{t_1}^{t_2} F_y dt = (mv_y)_2 \\ (mv_z)_1 + \int_{t_1}^{t_2} F_z dt = (mv_z)_2 \end{aligned}\right\} \quad \ldots (13.5)$$

If several forces act on a particle, effect of impulse of forces should be taken into account as,

$$mv_1 + \sum_{i=1}^{n} \int_{t_1}^{t_2} F_i dt = mv_2 \quad \ldots (13.6)$$

If *more than a single particle* are taken into account and when subjected to several forces, the equation gets transformed as,

$$\sum_{i=1}^{n} (m_i v_i)_1 + \sum_{i=1}^{n} \int_{t_1}^{t_2} F_i dt = \sum_{i=1}^{n} (m_i v_i)_2 \quad \ldots (13.7)$$

13.3 CONSERVATION OF LINEAR MOMENTUM OF PARTICLE

If no external force is applied on a single particle or more than one particle, or more generally, if the sum of external forces is zero, the term of impulse of equation (13.7) gets vanished and the final equation gets reduced to

$$\sum_{i=1}^{n} (m_i v_i)_1 = \sum_{i=1}^{n} (m_i v_i)_2 \quad \ldots (13.8)$$

This expresses that the total momentum of the particle is conserved, and is known as *law of conservation of linear momentum of particle.*

13.4 RELATION BETWEEN MOMENTUM AND IMPULSE OF RIGID BODY

A rigid body is idealized as a system of combined effect of number of particles P_i. The system forced by the momenta of the particles at time t_1 and the system of the impulse of the external forces applied from t_1 to t_2 are together equipollent to the system formed by the momenta of the particles at time t_2.

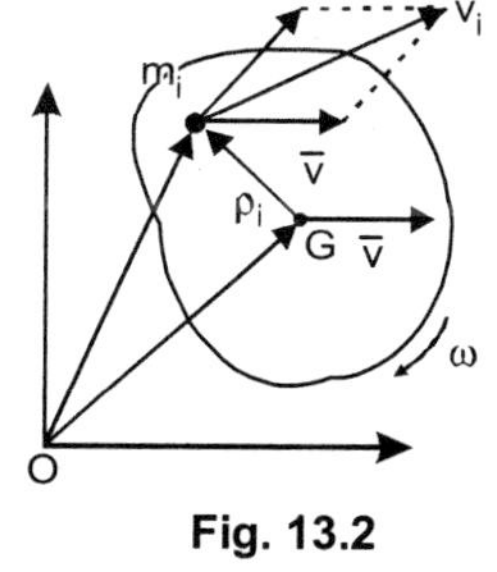

Fig. 13.2

So,

[System Momenta]$_1$ + [System Impulse]$_{1-2}$ = [System Momenta]$_2$

$$\therefore \quad \sum_{i=1}^{n}[\Delta m_i v_i]+\sum_{i=1}^{n}\int_{t_1}^{t_2} F_i\, dt=\sum_{i=1}^{n}[\Delta m_i v_i]_2 \qquad \text{... (13.9)}$$

Referring to the Figure 13.2,

$$\sum_{i=1}^{n}\Delta m_i v_i=\sum_{i=1}^{n}\Delta m_i(\bar{v}+\omega\times\rho_i)=\sum_{i=1}^{n}\Delta m_i\bar{v}+\sum_{i=1}^{n}\Delta m_i\ \omega\times\rho_i$$

$$= m\bar{v}+\omega\times m\bar{\rho}=m\bar{v}+0$$

$$= G(\text{vector}) \qquad \text{... (13.10)}$$

So from equation (13.9),
$$G_1+\sum_{i=1}^{n}\int_{t_1}^{t_2} F_1 dt=G_2 \qquad \text{... (13.11}a\text{)}$$

$$\Rightarrow \quad \sum_{i=1}^{n}\int_{t_1}^{t_2} F_i dt=G_2-G_1=\Delta G \qquad \text{... (13.11}b\text{)}$$

13.5 CONSERVATION OF LINEAR MOMENTUM OF RIGID BODY

Previously this principle has already been discussed in the context of a particle. Following the same concept, it can be said that, for a given interval of time, if no impulse of applied forces do act on the rigid body, linear momentum of the body remains constant.

For equation 13.11 (*b*),

$$\sum_{i=1}^{n}\int_{t_1}^{t_2} F_i dt=0$$

$$\Rightarrow \quad \Delta G=0$$

$$\Rightarrow \quad G_1=G_2 \qquad \text{... (13.12)}$$

This is the expression of *law of conservation of linear momentum of rigid body.*

13.6 IMPACT

A collision between two particles or rigid body occuring in a very small interval of time, during which the particles or bodies exert relatively large forces on each other, is called an *impact*.

The common normal to the surfaces in contact during the impact is called *line of impact*.

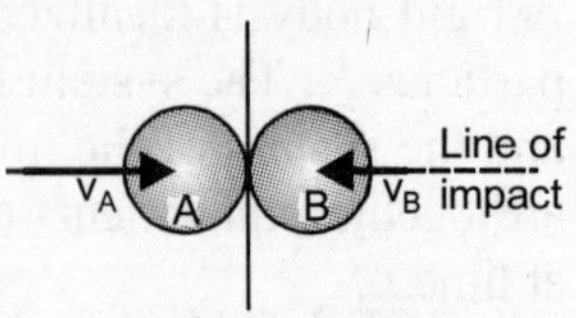

Fig. 13.3(a)

If centre of masses on two colliding particles are located on this line of impact, it is a *central impact.* Otherwise the impact is said to be *noncentral* or *eccentric impact*.

If velocities of the colliding particles are directed along line of impact, it is said to be a *direct impact* as shown in Figure 13.3(*a*). If either or both particles move along a line other than line of impact, impact is said to be *oblique impact,* as shown in Figure 13.3(*b*).

13.6.1 Direct Central Impact

Let us assume two particles *A* and *B* of mass m_A and m_B respectively move along a single straight line in velocities v_A and v_B respectively. If $v_A > v_B$, particle *A* will strike particle *B*. Different stages of impact are envisaged as

(*a*) *Just at the instant of collision, particles start deforming.*

(*b*) *Deformation continues for a certain interval of time, which is called period of deformation. So period of deformation is the time interval between initial contact for bodies and instant of maximum deformation.*

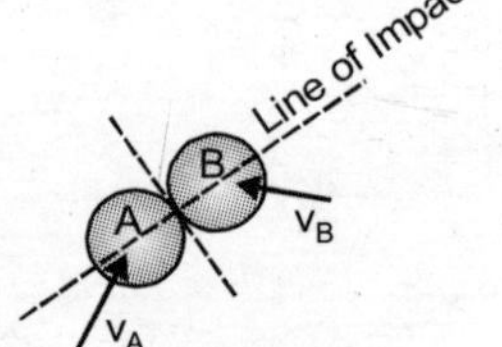

Fig.13.3b

(*c*) *At the end of period of deformation, both particles will have same velocity u. Schematically it is shown in Figure* 13.4 (*a*). *If the principle of impulse and momentum of particle A is applied at this stage, it is:*

$$m_A v_A - \int P dt = m_A u \qquad \text{... (13.13a)}$$

Here P is the force of deformation.

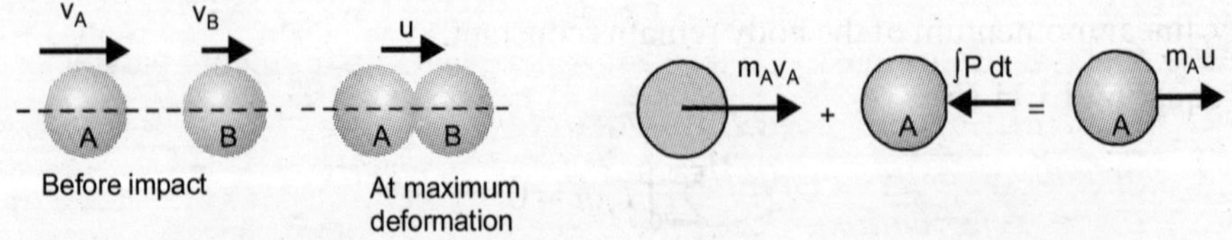

Fig. 13.4(a) **Fig. 13.4(b)**

(*d*) *After that starts period of restitution. Period of restitution is the time interval between maximum deformation condition to the instant of separation of bodies. Depending on magnitude of impact and materials of particles, the colliding particle either will get deformed permanently or regain their shapes.*

(e) *At the end of period of restitution, particles get separated with velocities v'_A and v'_B respectively. Applying principle of linear impulse and momentum of particle A, we have schematic presentation in Figures* 13.4 (*c*) *and* (*d*).

Here R is the force of restitution and

$$m_A u - \int R\, dt = m_A v'_A \qquad \text{... (13.14}b\text{)}$$

v'_A v'_B $m_A u$ + $\int R\,dt$ = $m_A v_A$

After impact — Period of restitution

Fig. 13.4(c) **Fig. 13.4(d)**

Generally force of deformation *R* and its impulse are not equal with force of restitution *P* and its impulse. The ratio of the magnitude of impulse of force of restitution to impulse of force of deformation is called *coefficient of restitution* (*e*).

So,
$$e = \frac{\int R dt}{\int P dt} = \frac{m_A u - m_A v'_A}{m_A v_A - m_A u} = \frac{u - v'_A}{v_A - u} \qquad \text{... (13.15)}$$

In the similar fashion, coefficient of restitution (*e*) can be obtained from the analysis of particle *B*, and
$$e = \frac{v'_B - u}{u - v_B}$$

So we can say from the concept of identity,

$$e = \frac{(u - v'_A) + (v'_B - u)}{(v_A - u) + (u - v_B)} = -\frac{v'_A - v'_B}{v_A - v_B} = -\frac{\text{relative velocity after impact}}{\text{relative velocity before impact}} \qquad \text{... (13.16)}$$

The value of *e* varies from 0 to1 and possibly dependent on following parameters:

(*i*) *Materials of the colliding particles, causing perfectly elastic, partially elastic or perfectly plastic deformation.*

(*ii*) *Velocity of impact.*

(*iii*) *Size of the colliding particles.*

Variation of coefficient of restitution with respect to relative velocity is shown in Figure 13.5.

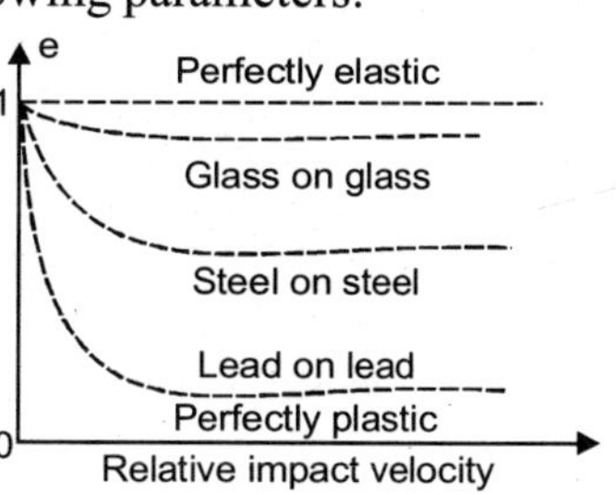

Fig. 13.5

13.6.2 Nature of Impact

If after collision, during the period of restitution, colliding particles regain their initial shapes completely, the collision is defined as *perfectly elastic impact* and the deformation occurred is defined as *perfectly elastic deformation.* In this case, period of deformation equals with period of restitution. This causes, coefficient of restitution be unity i.e., $e = 1$

So,
$$1 = -\frac{v_A' - v_B'}{v_A - v_B}$$

$$\Rightarrow \quad v_A - v_B = v_A' - v_B' \qquad ...(13.17)$$

In fully elastic impact, total energy as well as their total momentum are conserved. Following equation (13.8) we can write,

$$m_A v_A + m_B v_B = m_A v_A' + m_B v_B'$$

$$\Rightarrow \quad m_A(v_A - v_A') = m_B(v_B' - v_B) \qquad ... (13.18)$$

Rearranging (13.17),

$$(v_A + v_A') = (v_B' + v_B) \qquad ... (13.19)$$

Multiplying (13.18) and (13.19),

$$m_A(v_A^2 - v_A'^2) = -m_B(v_B^2 - v_B'^2)$$

$$\Rightarrow \quad \frac{1}{2}m_A(v_A^2 - v_A'^2) = -\frac{1}{2}m_B(v_B^2 - v_B'^2)$$

$$\Rightarrow \quad \frac{1}{2}m_A v_A^2 + \frac{1}{2}m_B v_B^2 = \frac{1}{2}m_A v_A'^2 + \frac{1}{2}m_B v_B'^2 \qquad ... (13.20)$$

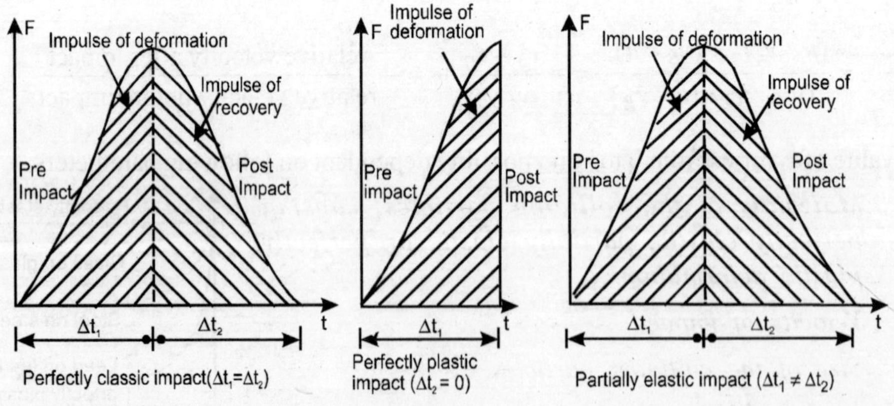

Fig. 13.6

If after collision colliding particles cannot regain at all their initial shapes, but stay together, the collision is defined as *perfectly plastic impact* and the deformation occurring is defined as *perfectly plastic deformation.* In the case $v'_A = v'_B = v'$. So, coefficient of restitution becomes zero i.e., $e = 0$.

It implies that in perfectly plastic impact, there is no period of restitution. Following equation (13.8), it can be said,

$$m_A v_A + m_B v_B = (m_A + m_B)v'_A = (m_A + m_B)v'_B = (m_A + m_B)v' \qquad \text{... (13.21)}$$

Hence total momentum of particles are conserved.

From the Figure 13.6, graphical concept of different types of impact can be visualised.

For *partially elastic impact,* where $0 < e < 1$, total kinetic energy the particles is not conserved, and the loss of kinetic energy K.E can easily be found out.

K.E of particles before impact $KE_1 = \frac{1}{2} m_A v_A^2 + \frac{1}{2} m_B v_B^2$

K.E of particles after impact $KE_2 = \frac{1}{2} m_A v'^2_A + \frac{1}{2} m_B v'^2_B$

Loss of K.E, Δ K.E $= KE_1 - KE_2 = \frac{1}{2}\left[(m_A v_A^2 + m_B v_B^2) - (m_A v'^2_A + m_B v'^2_B)\right]$

$$= \frac{1}{2(m_A + m_B)}\left[(m_A + m_B)(m_A v_A^2 + m_B v_B^2) - (m_A + m_B)(m_A v'^2_A + m_B v'^2_B)\right]$$

$$= \frac{1}{2(m_A + m_B)}\left[\left\{m_A^2 v_A^2 + m_B^2 v_B^2 + m_A m_B (v_A^2 + v_B^2)\right\} - \left\{m_A^2 v'^2_A + m_B^2 v'^2_B + m_A m_B (v'^2_A + v'^2_B)\right\}\right]$$

$$= \frac{1}{2(m_A + m_B)}\left[\begin{matrix}\left\{(m_A v_A + m_B v_B)^2 - 2m_A v_A m_B v_B + m_A m_B (v_A^2 + v_B^2)\right\} - \\ \left\{(m_A v'_B + m_B v'_B)^2 - 2m_A v'_A m_B v'_B + m_A m_B (v'^2_A + v'^2_B)\right\}\end{matrix}\right]$$

For conservation of momentum, $m_A v_A + m_B v_B = m_A v'_A + m_B v'_B$

$$\Rightarrow \qquad (m_A v_A + m_B v_B)^2 = (m_A v'_A + m_B v'_B)^2$$

Substituting in above equation,

$$\Delta \text{K.E} = \frac{1}{2(m_A + m_B)}\left[m_A m_B \left(v_A^2 + v_B^2\right) - 2m_A m_B v_A v_B - m_A m_B \left(v'^2_A + v'^2_B\right) - 2m_A m_B v'_A v'_B\right]$$

$$= \frac{1}{2(m_A + m_B)}\left[m_A m_B (v_A - v_B)^2 - m_A m_B (v'_A - v'_B)^2\right]$$

Substituting $\quad (v_A' - v_A') = -e(v_A - v_B)$,

$$\Delta \text{K.E} = \frac{1}{2(m_A + m_B)}\left[m_A m_B (v_A - v_B)^2 - m_A m_B (v_A - v_B)^2 e^2\right]$$

$$= \frac{1}{2(m_A + m_B)} m_A m_B (v_A - v_B)^2 \left(1 - e^2\right)$$

$$= \frac{m_A m_B (v_A - v_B)^2}{2(m_A + m_B)}\left(1 - e^2\right)$$

13.6.3 Oblique Central Impact

Here velocities of colliding particles are not directed along the line of impact. Referring Fig.13.7, v_A and v_B are the velocities before impact and v_A' and v_B' are those after impact. Here line of impact is assumed as x-axis and the common tangent is assumed as y-axis. While the colliding particles are perfectly smooth and frictionless, only impulse imparted on the particles during impact are due to internal forces acting along line of impact or x-axis. From this phenomenon, we can observe following situations.

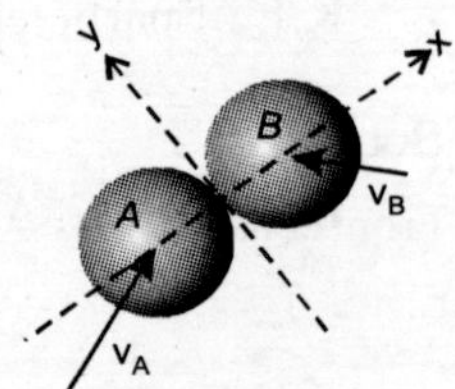

Fig. 13.7

(*i*) *The components of velocities of colliding particles are conserved along y-axis, separately.*

$$(v_A)_y = (v'_A) \qquad \text{...(13.22}a\text{)}$$

$$(v_B)_y = (v'_B)_y \qquad \text{...(13.22}b\text{)}$$

(*ii*) *The component of total momentum along x-axis of colliding particles remain conserved.*

$$m_A (v_A)_x + m_B (v_B)_x = m_A (v'_B)_x + m_B (v'_B)_x \qquad \text{...(13.22}c\text{)}$$

(*iii*) *The coefficient of restitution (e), is obtained utilizing the components of velocities along line of impact.*

$$e = \frac{(v'_A)_x - (v'_B)_x}{(v_A)_x - (v_B)_x} \qquad \text{...(13.22}d\text{)}$$

Multiple Choice Questions

Select the best alternative (s):

1. Which of the following do not have identical dimensions?
   (*a*) momentum and force of impulse
   (*b*) torque and energy
   (*c*) torque and work
   (*d*) kinetic energy and potential energy
2. If the momentum of a body is doubled, its kinetic energy will
   (*a*) increase by two times
   (*b*) increase by four times
   (*c*) remain same
   (*d*) get halved
   (*e*) reduce to one fourth

Answers

1. (*a*) 2. (*a*)

Numerical Examples

EXAMPLE 1

Prove that in a perfectly elastic impact, two colliding equal masses exchange their velocities.

SOLUTION Let us assume, $m_A = m_B = m$

From conservation of momentum,

$$mv_A + mv_B = mv'_A + m'_B \quad \text{...(1)}$$

$$\Rightarrow \quad v_A + v_B = v'_A + v'_B$$

Being perfectly elastic impact,

$$1 = -\frac{v'_A - v'_B}{v_A - v_B} \quad \text{...(2)}$$

$$\Rightarrow \quad v_A - v_B = v'_B - v'_A$$

Solving above two equations (1) and (2), we have, $v_A = v'_B$, $v_B = v'_A$.

EXAMPLE 2

If a particle strikes an immovable body of extremely large mass, prove that the particle rebounds at the velocity at which it stroke the body. Assume the impact perfectly elastic.

SOLUTION Let the particle is A and large body is B.

Here, $m_B = \alpha, v_B = 0$

From conservation of momentum, $m_A v_A = m_A v'_A + m_B v'_B$

$$\Rightarrow \quad \frac{m_A}{m_B} v_A = \frac{m_A}{m_B} v'_A + v'_B$$

$$\Rightarrow \quad v'_B = 0$$

Again, $$v_A' - v_B' = e\left(v_B - v_A\right)$$

$$\Rightarrow \quad v_A = v_A' \text{ (with opposite direction)}$$

EXAMPLE 3

A glass ball strikes a smooth horizontal floor from height H, and continues to bounce. Find an expression of velocity of the ball, after it bounces for n times.

SOLUTION If the coefficient of restitution is e, we can say, after 1st bounce,

$$v_A' = e \cdot v_A$$

After 2nd bounce, velocity of ball $$v_A'' = ev_A' = e^2 v_A$$

Similarly after 3rd bounce, $$v_A''' = e^3 v_A$$

So, after n-th bounce, $$v_A^n = e^n v_A$$

As the ball falls by the action of gravity, $$v_A = \sqrt{2gH}$$

So, $$v_A^n = e^n \sqrt{2gH}$$

EXAMPLE 4

A tennis ball is dropped on a smooth horizontal floor from which it bounces to a height of 10m. On third bounce it rises to 3m. Determine the height where from the ball is dropped and also the coefficient of restitution.

SOLUTION Let the ball strikes first time on floor at velocity v, after first impact the upward velocity v', after second impact upward velocity v'' and third impact v'''. Consider height of ball be H, for first impact.

Now, $$v = \sqrt{2gH}$$

Again, $$e = \frac{0 - v'}{v - 0} = -\frac{v'}{v}$$

$$\Rightarrow \quad v' = -ev = -e\sqrt{2gH}$$

After first bounce, second bounce and third bounce, let the rise of height be H', H'', H''' respectively.

So, after first bounce, $$v' = \sqrt{2gH'}$$

Comparing values of v', $$\sqrt{2gH'} = -e\sqrt{2gH}$$

$$\Rightarrow \quad H' = e^2 h$$

On second impact, $$e = \frac{0 - v''}{v' - 0} = -\frac{v''}{v'}$$

$\Rightarrow \qquad v'' = -ev' = -e\sqrt{2gH'}$

Again, $\qquad v'' = \sqrt{2gH''}$

So, $\qquad \sqrt{2gH''} = -e\sqrt{2gH'}$

$\Rightarrow \qquad H'' = e^2H' = e^2\left(e^2H\right) = e^4H$

Proceeding in the similar way, $\qquad H'' = e^2H'' = e^6H''$.

Here $\qquad H' = 3\text{m}$ and $H''' = 10\text{m}$

Again, $\qquad \dfrac{H'''}{H'} = \dfrac{e^6H}{e^2H} = e^4$

$\Rightarrow \qquad e^4 = \dfrac{3}{10}$

$\Rightarrow \qquad e = 0.74$

So, $\qquad H = \dfrac{H'''}{e^6} = \dfrac{3\text{m}}{(0.74)^6} = 18.269\text{m}$

EXAMPLE 5

Two identical frictionless balls A and B strike each other. Assuming $e = 0.86$, determine magnitude and direction of the velocity of the each ball after impact.

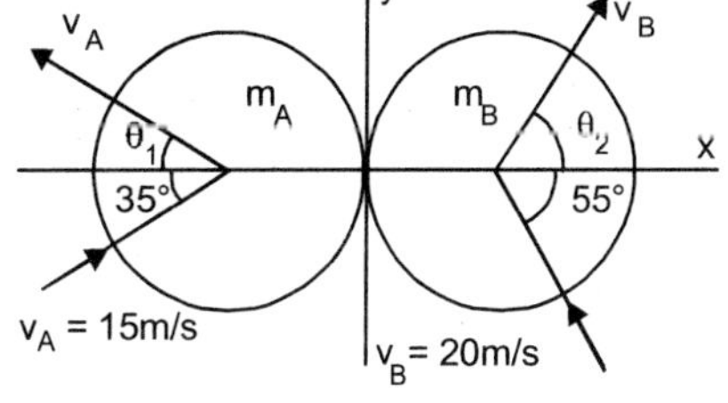

SOLUTION Here, components of striking velocities

$$(v_A)_x = 15\cos 35° = 12.29\text{m/s},$$

$$(v_A)_y = 15\sin 35° = 8.6\text{m/s}$$

$$(v_B)_x = -20\cos 55° = -11.47\text{m/s},$$

$$(v_B)_y = 20\sin 55° = 16.38\text{m/s}$$

As, velocity of each ball before and after impact, normal to line of impact remain constant,

$$(v_A')_y = (v_A)_y = 8.6\text{m/s}, (v_B')_y = (v_B)_y = 16.38\text{m/s}$$

From conservation of momentum along x-direction.

$$m_A(v_A)_x + m_B(v_B)_x = m_A(v'_A)_x + m_B(v'_B)_x$$

$\Rightarrow \qquad 12.29 - 11.47 = (v_A')_x + (v'_B)_x$

$\Rightarrow \qquad (v_A')_x + (v_B')_x = 0.82 \qquad ...(1)$

Again, $\qquad (v_B')_x - (v_A')_x = e[(v_A)_x - (v_B)_x] = 0.86\,(12.29 + 11.47) = 20.43 \qquad ...(2)$

Solving equations (1) and (2), $\qquad (v_A')_x = -9.81\text{m/s}, (v_B')_x = 10.63\text{m/s}$

After impact, resultant velocities of A and B

$$\left.\begin{aligned} v'_A &= \sqrt{(v'_A)_x^2 + (v'_A)_y^2} = \sqrt{(-9.81)^2 + 8.6^2} = 13.05\text{m/s} \\ v'_B &= \sqrt{(v'_B)_x^2 + (v'_B)_y^2} = \sqrt{10.63^2 + 16.38^2} = 19.53\text{m/s} \end{aligned}\right]$$

And their directions

$$\left.\begin{aligned} \theta_1 &= \tan^{-1}\frac{(v'_A)_y}{(v'_A)_x} = \tan^{-1}\left(-\frac{8.6}{9.81}\right) = -41.24° \\ \theta_2 &= \tan^{-1}\frac{(v'_B)_y}{(v'_B)_x} = \tan^{-1}\left(-\frac{16.38}{10.63}\right) = -57.02° \end{aligned}\right\}$$

EXAMPLE 6

Three perfectly elastic balls A, B, C of masses 3 kg, 9kg, and 18 kg are moving in the same direction with velocities 12m/s, 4m/s, 2m/s respectively. If A strikes B which in turn strikes C, prove that A and B will be brought to rest by impact.

SOLUTION Here, $m_A = 3\text{kg}, m_B = 9\text{kg}, m_C = 18\text{kg}$

$v_A = 12\text{m/s}, v_B = 4\text{m/s}, v_C = 2\text{m/s}$

The entire problem can be idealised as superposition of following two cases.

(*i*) *Considering impact of A and B*, from conservation of momentum,

$$m_A v_A + m_B v_B = m_A v'_A + m_B v'_B$$

$$\Rightarrow \quad 3 \times 12 + 9 \times 4 = 3v'_A + 9v'_B$$

$$\Rightarrow \quad v'_A + 3v'_B = 24 \qquad \text{... (1)}$$

Again, $v'_B - v'_A = e(v_A - v_B) = 1(12 - 4) = 8$ ... (2)

Solving equations (1) and (2) $v'_A = 0,\ v'_B = 8\text{m/s}$

So, after impact, A is brought to rest.

(*ii*) *Considering impact of B and C*, new velocity of B, $v_B = 8\text{m/s}$, from conservation of momentum,

$$m_B v_B + m_C v_C = m_B v'_B + m_C v'_C$$

$$\Rightarrow \quad 9v'_C + 18v'_C = 9 \times 8 + 18 \times 2$$

$$\Rightarrow \quad v'_B + 2v'_C = 12 \qquad \text{... (3)}$$

Again, $v'_C - v'_B = e(v_B - v_C) = 1(8 - 2) = 6$ ... (4)

Solving equations (3) and (4) $v'_C = 0,\ v'_C = 6\text{m/s}$

So after impact B is brought to rest. Finally superposition of cases (*i*) and (*ii*) suggests that subsequent impact of A to B to C will result A and B to stop completely.

EXAMPLE 7

Ball *A* of 2.5kg mass moving with 7.5m/s strikes ball *B* of 3.2kg mass moving at 2.75m/s. At instant of impact, velocities of *A* and *B* are parallel and inclined 35° to line of impact. If $e = 0.53$, find velocity and direction in which *A* and *B* will move after impact.

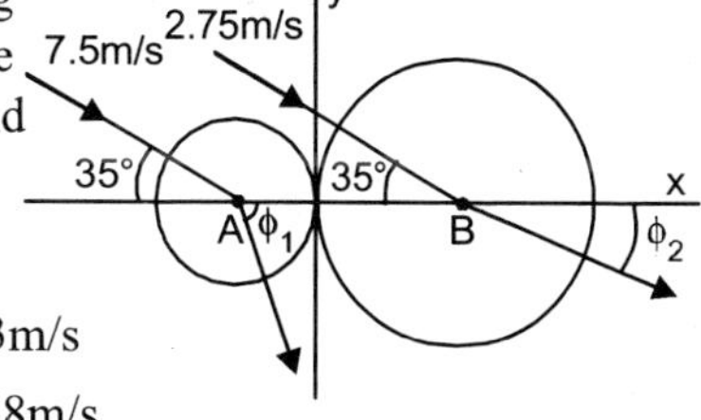

SOLUTION Here, $m_A = 2.5\text{kg}, m_B = 3.2\text{kg}, v_A = 7.5\text{m/s}, v_B = 2.75\text{m/s},$

Now, $(v_A)_x = 7.5\cos 35° = 6.14\text{m/s}, (v_A)_y = -7.5\sin 35° = -4.3\text{m/s}$

$(v_B)_x = 2.75\cos 35° = 2.25\text{m/s}, (v_B)_y = -2.75\sin 35° = -1.58\text{m/s}$

After impact,

$$(v'_A)_y = (v_A)_y = -4.3\text{m/s, and } (v'_B)_y = (v_B)_y = -1.58\text{m/s.}$$

From conservation of momentum along line of impact,

$$m_A(v'_A)_x + m_B(v'_B)_x = [m_A(v_A)_x + m_B(v_B)_x]$$

$$\Rightarrow \quad 2.5(v'_A)_x + 3.2(v'_B)_x = 2.5\times 6.14 + 3.2\times 2.25 = 22.55 \qquad ...(1)$$

Again, $$(v'_B)_x - (v'_A)_x = e\left[(v_A)_x - (v_B)_x\right] = 0.53(6.14 - 2.25) = 2.06 \qquad ...(2)$$

Solving above equations (1) and (2),

$$(v'_A)_x = 2.8\text{m/s}, (v'_B) = 4.86\text{m/s}$$

After impact final velocities of *A* and *B*,

$$\left.\begin{aligned} v'_A &= \sqrt{(v'_A)_x + (v'_A)_y} = \sqrt{2.8^2 + (-4.3)^2} = 5.13\text{m/s} \\ v'_B &= \sqrt{(v'_B)_x + (v'_B)_y} = \sqrt{4.86^2 + (-1.58)^2} = 5.11\text{m/s} \end{aligned}\right\}$$

And their direction,

$$\left.\begin{aligned} \phi_1 &= \tan^{-1}\frac{(v'_A)_y}{(v'_A)_x} = \tan^{-1}\left(-\frac{4.3}{2.8}\right) = -56.93° \\ \phi_2 &= \tan^{-1}\frac{(v'_B)_y}{(v'_B)_x} = \tan^{-1}\left(-\frac{1.58}{4.86}\right) = -18° \end{aligned}\right\}$$

EXAMPLE 8

From what height, must a heavy elastic ball be dropped on a floor, so that after rebounding thrice will reach a height of 16m. Take $e = \left(\frac{1}{2}\right)^{\frac{1}{3}}$.

SOLUTION Proceeding as is done for example (4) earlier, we can write $H''' = e^6 H$

Substituting $$H''' = 16\text{m and } e = \left(\frac{1}{2}\right)^{\frac{1}{3}},$$

we obtain $$H = 16\times\left(\frac{1}{2}\right)^{\frac{-6}{3}} = 16\times(0.5)^{-2} = 64\text{m}$$

EXAMPLE 9

A ball moving with a velocity of 7.5m/s strikes on a fixed plane making 40° with fixed plane. If coefficient of restitution is 0.5, find direction and velocity of body after impact.

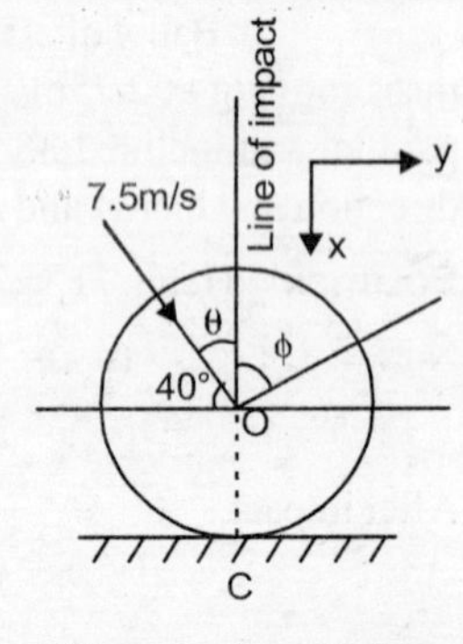

SOLUTION Here, $v_A = 7.5\text{m/s}, v_B = 0, v_B' = 0, e = 0.5$

With respect to line of impact, $\theta = 90° - 40° = 50°$

Now, $(v_A)_x = 7.5\cos 50° = 4.82\text{m/s},$

$(v_A)_y = 7.5\sin 50° = 5.75\text{m/s}$

Again, $(v_A')_y = (v_A)_y = 5.75\text{m/s}$

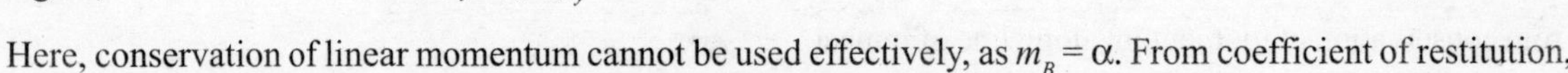

Here, conservation of linear momentum cannot be used effectively, as $m_B = \alpha$. From coefficient of restitution,

$$(v_A')_x = e(v_A)_x = 0.5 \times 4.82 = 241\text{m/s}$$

So, $$v_A' = \sqrt{(v_A')_x^2 + (v_A')_y^2} = \sqrt{2.41^2 + 5.75^2} = 6.23\text{m/s}$$

$$\phi = \tan^{-1}\frac{(v_A')_y}{(v_A')_x} = \tan^{-1}\left(\frac{5.75}{2.41}\right) = 67.26°$$

EXAMPLE 10

A bullet $m = 15$gm hits pendulum bob $M = 1.3$kg at a velocity 95m/s. If the pendulum is suspended by a string of 1.2m long, find maximum angle of swing of the pendulum when (*i*) bullet gets completely embedded in the bob, (*ii*) bullet rebounds the bob with 23m/s, (*iii*) bullet escapes the bob with 19m/s.

O
θ
1.2m
1.2m
Bullet
h
95m/s

SOLUTION Here, $m = 0.015\text{kg}, M = 1.3\text{kg}, v = 95\text{m/s}, V = 0, l = 1.2\text{m}$

Before impact,

$$mv + MV = 0.015 \times 95 + 1.3 \times 0 = 1.425\text{kgm/s}$$

(*i*) After impact, $$mV' + MV' = 1.425$$

$$\Rightarrow \quad 0.015V' + 1.3V' = 1.425$$

$$\Rightarrow \quad V' = 1.084\text{m/s}$$

Here assumed, after impact bullet and bob will move at a common velocity V'.

Applying principle of conservation of energy, the height h which the bob rises can be computed as following.

$$KE_{m+M} = PE_{m+M}$$

$$\Rightarrow \quad \frac{1}{2}(m+M)V'^2 = (m+M)gh$$

$$\Rightarrow \quad h = \frac{V'^2}{2g} = \frac{1.084^2}{2 \times 9.81} = 0.059\text{m}$$

So, angle of swing, $\theta = \cos^{-1}\left(\frac{l-h}{l}\right) = \cos^{-1}\left(\frac{1.2-0.059}{1.2}\right) = 18.04°$

(*ii*) Here, rebound velocity of bullet $v' = 23$m/s. From conservation of momentum,

$$mv' + MV' = 1.425$$

$\Rightarrow$ $$0.015 \times 23 + 1.3V' = 1.425$$

$\Rightarrow$ $$V' = 0.83\text{m/s}$$

From conservation of energy, rise of bob after impact

$$h = \frac{(V')^2}{2g} = \frac{0.83^2}{2 \times 9.81} = 0.035\text{m}$$

So, angle of swing $\theta = \cos^{-1}\left(\frac{l-h}{l}\right) = \cos^{-1}\left(\frac{1.2-0.035}{1.2}\right) = 13.87°$

(*iii*) Here escape velocity of bullet $v' = 19$m/s. From conservation of momentum,

$$mv' + MV' = 1.425$$

$\Rightarrow$ $$0.015 \times 19 + 1.3V' = 1.425$$

$\Rightarrow$ $$V' = 0.88\text{m/s}$$

From conservation of energy, rise of bob after impact

$$h = \frac{(V')^2}{2g} = \frac{0.88^2}{2 \times 9.81} = 0.039\text{m}$$

So, angle of swing $\theta = \cos^{-1}\left(\frac{l-h}{l}\right) = \cos^{-1}\left(\frac{1.2-0.039}{1.2}\right) = 14.65°$

EXAMPLE 11

A suspended sphere A of mass m is released from rest at an angle of 40° with horizontal and strikes another suspended sphere B of mass 2.8m. Determine assuming an elastic impact, the angle ϕ by which B gets deflected from its initial position.

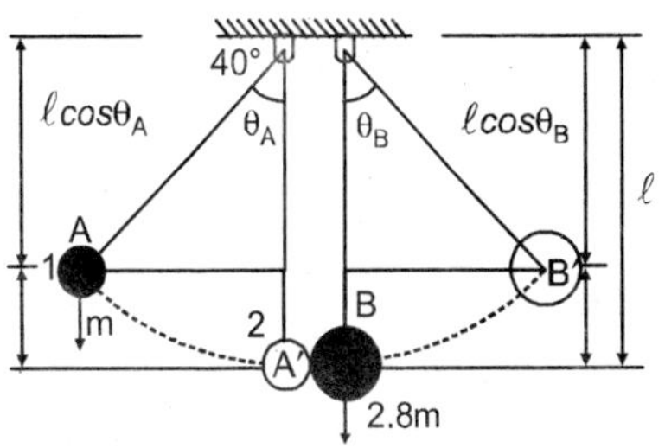

SOLUTION Sphere A moves from position (1) to (2) during impact. From the principle of conservation of energy.

$$E_{A1} = E_{A2}$$

$\Rightarrow$ $$PE_{A1} + KE_{A1} = PE_{A2} + KE_{A2}$$

$$mgl(1-\cos\theta_A) + 0 = \frac{1}{2}mv_A^2$$

$\Rightarrow$ $$v_A = \sqrt{2gl(1-\cos\theta_A)}$$

At position (2), $v_B = 0$. From conservation of momentum

$$mv_A + 2.8mv_B = mv'_A + 2.8mv'_B$$

$\Rightarrow$ $$v_A = v'_A + 2.8v'_B$$

Again, $$v'_B - v'_A = e(v_A - v_B)$$
$$= 1(v_A - 0) = v_A$$

Solving above equations, $$v'_B = 0.526v_A = 0.526\sqrt{2gl(1-\cos\theta_A)}$$

After impact B moves from position (2) to (3). Applying principle of conservation of energy

$$E_{B2} = E_{B3}$$

$\Rightarrow$ $$PE_{B2} + KE_{B2} = PE_{B3} + KE_{B3}$$

$\Rightarrow$ $$\frac{1}{2}(2.8m)\,v'^2_B = 2.8mgl\,(1-\cos\theta_B)$$

$$\frac{2.8m}{2}0.526^2 \times 2gl\left(1-\cos\theta_A\right) = 2.8mgl\left(1-\cos\theta_B\right)$$

$\Rightarrow$ $$0.277\left(1-\cos\theta_A\right) = \left(1-\cos\theta_B\right)$$

$\Rightarrow$ $$\cos\theta_B = 0.723 + 0.277\cos\theta_A$$

Substituting $\theta_A = 90° - 40° = 50°$, we obtain $\theta_B = 25.7°$

Example 12

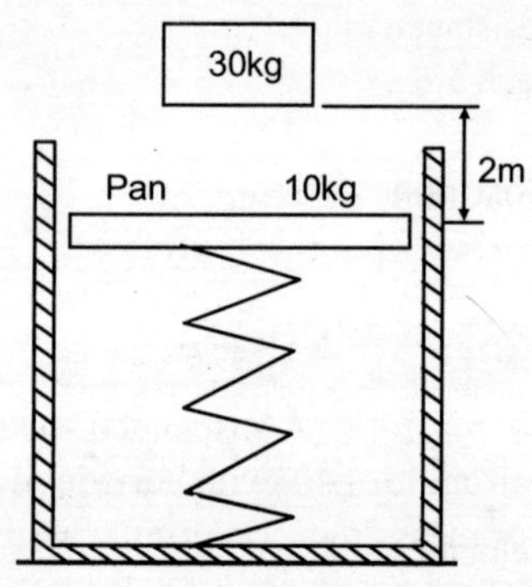

A block of mass 30kg is dropped from a height 2m onto 10kg pan of spring scale. Assuming a perfectly plastic impact, determine maximum deflection of the pan, while spring constant k=20kN/m.

Solution Here, $m_B = 30kg$, $m_P = 10kg$, $h = 2m$, $k = 20\times10^3$N/m

Velocity of block just before impact

$\Rightarrow$ $$v_B = \sqrt{2gh} = \sqrt{2\times9.81\times2} = 6.26 \text{ m/s}$$

At the instant of impact, $v_p = 0$ and after plastic impact, both block and pan attains a common velocity v'. From conservation of momentum

$$m_B v_B + m_P v_P = \left(m_B + m_P\right)v'$$

$\Rightarrow$ $$30 \times 6.26 + 10 \times 0 = (30 + 10)v'$$

$\Rightarrow$ $$v' = 4.695 \text{ m/s}$$

Static deflection or initial compression of spring due to self weight of pan

$$\delta_{St} = \frac{m_P \times g}{k} = \frac{10\times9.81}{20\times10^3} = 4.905\times10^{-3}\text{ m}$$

If both the block and spring come to rest after moving a distance δ, total compression of the spring will be,

$$\delta_T = \delta + \delta_{St} = \delta + 4.905\times10^{-3}$$

Applying conservation of energy *just after impact* and *both coming to rest*,

$$\left(KE_{\text{Spring}} + KE_B + KE_P\right)_1 = \left(KE_{\text{Spring}} + PE_B + PE_P\right)_2$$

$$\Rightarrow \quad \frac{1}{2}k\delta_{st}^2 + \frac{1}{2}m_B v'^2 + \frac{1}{2}m_P v'^2 = \frac{1}{2}k\delta_T^2 - m_B g\delta - m_P g\delta$$

$$\Rightarrow \quad \frac{1}{2}20\times10^3\times\left(4.905\times10^{-3}\right)^2 + \frac{1}{2}(30+10)4.695^2$$

$$= \frac{1}{2}20\times10^3\,\delta_T^2 - (30+10)\times9.81\delta$$

$$\Rightarrow \quad 0.24059 + 440.8605 = 10000\,\delta_T^2 - 392.4\,\delta$$

$$= 10000\delta_T^2 - 392.4\,(\delta_T - 4.905\times10^{-3})$$

$$= 10000\delta_T^2 - 392.4\,\delta_T + 1.9247$$

$$\Rightarrow \quad \delta_T^2 - 0.03924\,\delta_T - 0.043921 = 0$$

$$\Rightarrow \quad \delta_T = 0.2301$$

So, $\delta = \delta_T - \delta_{St} = 0.2301 - 4.905\times10^{-3} = 0.2252\text{m} = 22.52\text{cm}$

EXAMPLE 13

A hammer of weight 10kN strikes a pile of weight 0.3kN from a height of 102cm. During piling, the resistance of the soil was measured 83kN. Assuming plastic impact, determine depth of penetration into soil in each blow.

SOLUTION Here,

$$m_H = \frac{10\times10^3}{9.81} = 1019.37\text{kg}$$

$$m_P = \frac{0.3\times10^3}{9.81} = 30.58\text{kg}$$

$$h = 1.02m,\ R_s = 83\times10^3\ \text{N},\ v_P = 0$$

Velocity of hammer before impact $v_H = \sqrt{2gh} = \sqrt{2\times9.81\times1.02} = 4.47\text{m/s}$

If after impact, the common velocity of hammer and pile be v, from conservation of momentum,

$$m_H v_H + m_P v_P = (m_H + m_p)v$$

$$\Rightarrow \quad 1019.37\times4.47 + 30.58\times0 = (1019.37 + 30.58)v$$

$$\Rightarrow \quad v = 4.34\text{m/s}$$

If the depth of penetration in each blow is x, total energy of the system just after impact

$$E = KE_H + KE_P + PE_{\text{Penetration}}$$

$$= \frac{1}{2}m_H v^2 + \frac{1}{2}m_p v^2 + \left(m_H + m_p\right)gx$$

$$= \frac{1}{2}(1019.37 + 30.58)4.34^2 + (1019.37 + 30.58)9.81x$$

$$= 9888.22 + 10300x$$

Work done during penetration

$$W = R_s \times x = 83 \times 10^3 x$$

From principle of work energy,

$$83 \times 10^3 x = 9888.22 + 10300x$$

$$\Rightarrow \quad x = 0.136\text{m}$$

Example 14

A 8kN idealized canon with a recoil spring of $k = 4$kN/m fires a 4.6kg projectile with a muzzle velocity of 0.53km/s at 47°. Determine maximum compression of the spring.

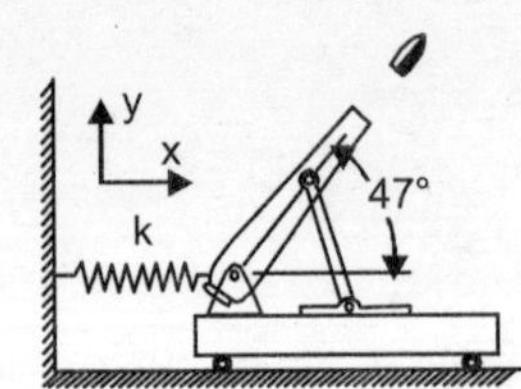

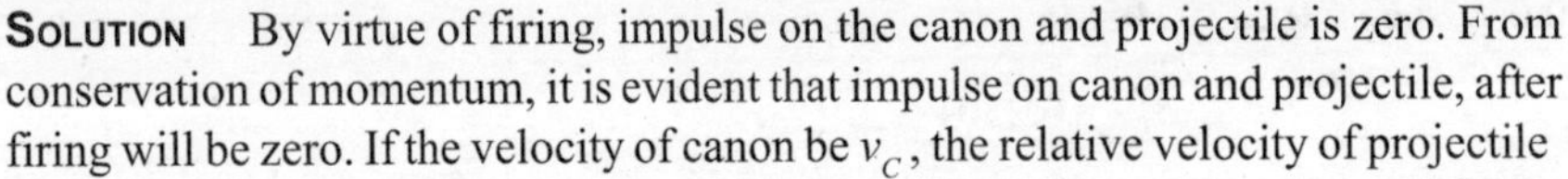

Solution By virtue of firing, impulse on the canon and projectile is zero. From conservation of momentum, it is evident that impulse on canon and projectile, after firing will be zero. If the velocity of canon be v_C, the relative velocity of projectile

$$v_P = (v_c + 530\cos 47°)$$

So,
$$m_C v_C + m_P v_P = 0$$

$$\Rightarrow \quad \frac{8000}{9.81} v_c + 4.6\,(v_c + 530\cos 47°) = 0$$

$$\Rightarrow \quad 815.49\, v_C + 4.6 v_C = -1662.71$$

$$\Rightarrow \quad v_C = -2.02 \text{ m/s}$$

The initial impulsive action an instantaneous velocity is imparted to the canon. Then motion of canon is impeded by the spring. If δ be the maximum compression of spring due to recoiling, applying *conservation of* K.E *concept,*

$$\frac{1}{2} m_c v_c^2 = \frac{1}{2} k\delta^2$$

$$\Rightarrow \quad \frac{8000}{9.81} \times (-2.02)^2 = 4000\delta^2$$

$$\Rightarrow \quad \delta = 0.912\text{m}$$

Example 15

Two identical billiard balls collide as in the figure. For a coefficient of restitution be 0.87, compute final velocities after the balls get parted. Also determine loss of kinetic energy.

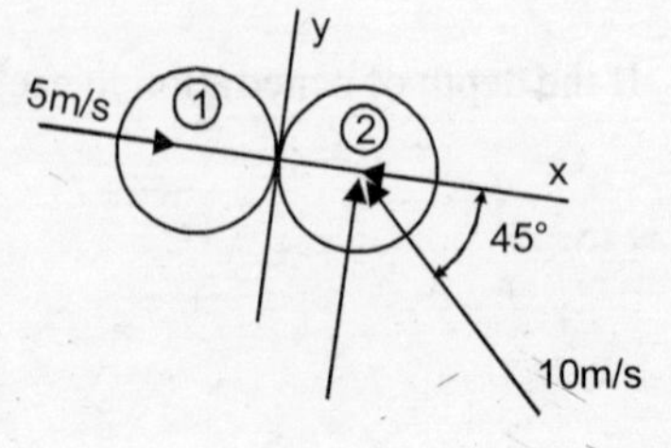

Solution Here,
$$v_1 = 5\text{m/s},\ v_2 = 10\text{m/s},$$
$$m_1 = m_2 = m,\ e = 0.87$$

Now,
$$(v_2)_x = -10\cos 45° = -7.07\text{m/s} = (-v_2)_y$$

and
$$(v_1)_x = v_1 = 5\text{m/s},\ (v_1)_y = 0$$

Applying conservation of momentum along line of impact,

$$m(v_1)_x + m(v_2)_x = m(v'_1)_x + m(v'_2)x$$

$$\Rightarrow \qquad 5 - 7.07 = (v'_1)_x + (v'_2)_x$$

$$\Rightarrow \qquad (v'_1)_x + (v'_2)_x = -2.07 \qquad \text{...(1)}$$

Again, $\qquad (v'_2)_x - (v'_1)_x = e\,[(v_1)_x - (v_2)_x]$

$$= 0.87\,(5 + 7.07) = 10.5 \qquad \text{...(2)}$$

Solving above equations (1) and (2), $(v'_1)_x = -6.285\text{m/s}$, and $(v'_2)_x = 4.215$ m/s

Final velocities, $\qquad v'_1 = (v'_1)_x = -6.285\text{m/s}$

So, $\qquad v'_2 = \sqrt{(v'_2)_x^2 + (v'_2)_y^2} = \sqrt{4.215^2 + 7.07^2} = 8.23\,\text{m/s}$

And loss of kinetic energy ΔK.E $= (KE_1 + KE_2)_I - (KE_1 + KE_2)_F$

$$= \frac{1}{2}m\left(v_1^2 + v_2^2\right) - \frac{1}{2}m\left(v'^2_1 + v'^2_2\right)$$

$$= \frac{m}{2}\left[\left(5^2 + 10^2\right) - (-6.285)^2 - 8.23^2\right] = 8.88m\ \text{N·m}$$

EXAMPLE 16

A particle of mass m hits a rigid body M at an angle α with vertical and rebounds the body at an angle β with vertical. If $M \gg m$ prove that $e = \dfrac{\tan\alpha}{\tan\beta}$.

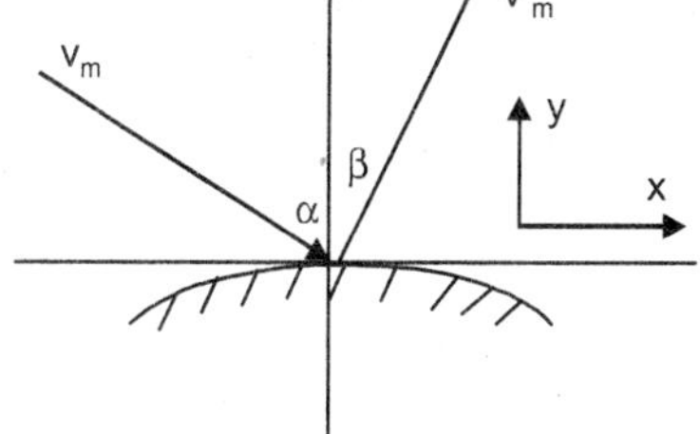

SOLUTION Here conservation of linear momentum cannot be used effectively, as $M \gg m$ and

$$v_M = 0,\ v'_M = 0$$

From coefficient of restitution, $\quad (v'_M)_y - (v'_m)_y = e[(v_m)_y - (v_M)_y]$

$$\Rightarrow \qquad 0 - (v'_m)_y = e\left[(v_m)_y - 0\right]$$

$$\Rightarrow \qquad e = -\frac{(v'_m)_y}{(v_m)_y}$$

Now, $\qquad (v_m)_y = -v_m \cos\alpha$ and $(v'_m)_y = v_m \cos\beta$

As $\qquad (v_m)_x = (v'_m)_x,\ v_m \sin\alpha = v'_m \sin\beta$

$$\Rightarrow \qquad \frac{v'_m}{v_m} = \frac{\sin\alpha}{\sin\beta}$$

So, $$e = \frac{(v'_m)_y}{(v_m)_y} = \frac{v'_m \cos\beta}{-v_m \cos\alpha} = \frac{v'_m}{v_m} \cdot \frac{\cos\beta}{\cos\alpha} = \frac{\sin\alpha}{\sin\beta} \cdot \frac{\cos\beta}{\cos\alpha} = \frac{\tan\alpha}{\tan\beta}$$

EXAMPLE 17

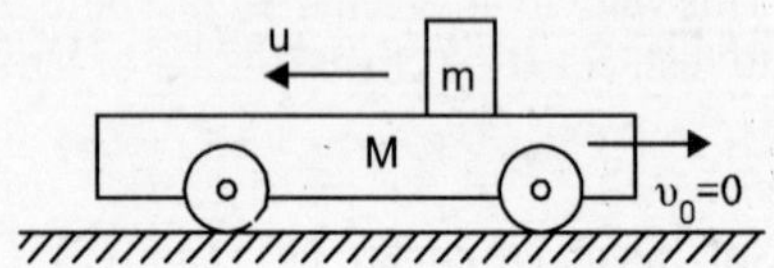

Referring to the figure, assume the car is initially at rest holds n number of people, each of mass m. If they all run with relative velocity u and jump one after another, prove that they can impart a greater velocity to the car than if they all run and jump at a time.

SOLUTION Running of all men together produces equivalent effect of running of a single man of mass nm. Assuming total velocity of the car be v_1.

$$nmu = (M + nm)v_1$$

$$\Rightarrow \qquad v_1 = \frac{nmu}{M + nm}$$

Initial momentum of the system is zero, as it is at rest.
If Δv be the increment of velocity, final momentum of the car $= M \cdot \Delta v$ and of a single man $= m(\Delta v - u)$
So, from conservation of momentum,

$$0 = M \cdot \Delta v + m(\Delta v - u)$$

$$\Rightarrow \qquad M\Delta v - mu + m\Delta v = 0$$

$$\Rightarrow \qquad \Delta v = \frac{mu}{M + m}$$

Now, when all men run and jump in succession, we can write,

$$v_2 = \frac{mu}{M + 1 \times m} + \frac{mu}{M + 2 \times m} + \frac{mu}{M + 3 \times m} + \cdots + \frac{mu}{M + n \times m}$$

As, $\frac{mu}{M+m}, \frac{mu}{M+2m}, \ldots, \frac{mu}{M+(n-1)m} > \frac{mu}{M+nm}$, we can conclude that $v_2 > v_1$

EXAMPLE 18

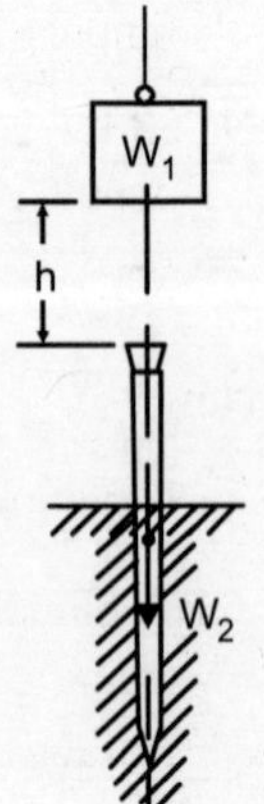

A wooden pile W_2 is driven into soil by successive blows of hammer W_1 from a height h. If the pile penetrates by δ per blow of hammer, find out total soil resistance as $R = f(\delta)$, while $W_1 = 30$kN, $W_2 = 0.5$kN, $h = 22$cm.

SOLUTION Here, the impact is perfectly plastic, as there is no rebound of hammer. Let at the instant of impact, velocity of hammer be v_1 and that of pile v_2.

So, $$v_1 = \sqrt{2gh} \text{ and } v_2 = 0$$

If the common velocity after impact is v', from conservation of momentum,

$$\frac{W_1}{g}v_1 + \frac{W_2}{g}v_2 = \frac{W_1 + W_2}{g}v'$$

$$\Rightarrow \qquad W_1\sqrt{2gh} = (W_1 + W_2)v'$$

$$\Rightarrow \qquad v' = \frac{W_1\sqrt{2gh}}{W_1 + W_2}$$

This velocity v' is acquired just before any penetration occurs. If the pile moves δ through soil, before coming to rest, equating change in KE of the system with work done by all possible forces, we obtain,

$$\frac{W_1 + W_2}{g} g\delta - R\delta = -\frac{W_1 + W_2}{2g} \frac{W_1^2 \times 2gh}{(W_1 + W_2)^2}$$

As, $(W_1 + W_2) \gg R$, we can write from above equation,

$$R = \frac{W_1^2 h}{(W_1 + W_2)\delta} = \frac{30^2 \times 0.22}{(30 + 0.5)\delta} = \frac{6.492}{\delta} \text{ kN/m}$$

Exercise

1. A shot of mass m penetrates a thickness t of a fixed plate of mass M. Show that if the plate is free to move, the thickness penetrated is $\left(\frac{M}{M+m}\right)t.$

2. Find the velocity acquired by a block of wood of mass M kg which is free to recoil, when it is struck by a bullet of m kg moving with v cm/s in a direction passing through its C.G. If the bullet embedded for a meters, show that the average resistance of the block of wood to the bullet is $\frac{Mmv^2}{2(M+m)a} N$ and that the time of penetration is $\frac{2a}{v}$ seconds during which time the block will move $\frac{ma}{M+m}$ meters.

3. A bullet of mass 25 gm leaves the barrel of a gun with a velocity of 600 metres per second. Find the impulse of the force produced by the discharge. Suppose the force lasted for 0.002 second, find the average impulsive force. **Ans.** [7500 N]

4. A bullet loses 1/20th of its velocity is passing through a plank. Find how many such uniform planks it would pass through before coming to rest, assuming the resistance of the planks to be uniform. **Ans.** [10 planks]

5. A gun is mounted on a gun carriage which can move on a smooth horizontal plane. The gun is directed at an angle 30° with the horizontal. A shot is fired which leaves the gun at an angle 40° to the horizontal. If the total mass of the gun and the carriage is m times that of the shot find the value of m. **Ans.** [2.205]

6. A bullet of mass 25 gms is shot horizontally from a gun whose barrel is 0.8 m long, the velocity of the bullet being 200 m/sec. The bullet strikes and gets buried into a wooden block of mass 60 kg floating in water, without making any splinters or causing it to rotate. Determine the velocity of the block soon after it is struck. **Ans.** [0.083 m/sec]

7. A body of mass 4 kg is thrown vertically with a velocity of 20 metres/sec. What is its kinetic energy (*i*) at the instant of propulsion, (*ii*) after one second. **Ans.** [800 N-m, 208.08 N-m]

8. A body of mass 5 kg drops freely from a height of 40 metres and penetrates into the ground by 0.75 metre. Find the average resistance to penetration and the time of penetration. **Ans.** [2662.53 N, 0.054 sec]

9. A pile of mass 550 kg is driven into the ground by dropping freely a mass of 150 kg from a height of 3 metres. If the pile is driven into the ground by 100 mm, calculate the average resistance of the soil and the amount of energy lost at impact. **Ans.** [16310 N, 3465 N-m]

10. A bullet of mass 25 gm is fired with a velocity of 1000 m/sec. It strikes a wooden block of mass 8 kg at its centre the block being suspended by a vertical string from a point 5 metres above the centre of the block. If after striking, the bullet remains embedded in the centre of the block, find the angle by which the string will swing from the vertical. **Ans.** [25°43']

11. A track weighing 200 kN and moving at 2 metres per second impacts with a buffer spring which compresses 2.5 mm/kN. Find the maximum compression of the spring. **Ans.** [451.75 mm]

12. Two spheres one of which is of mass of 30 kg with a velocity of 10 m/sec and the other of mass 15 kg with a velocity of 5 m/sec. move towards each other and impinge centrally. Find the velocity of each sphere after the impact, if the coefficient of restitution is 0.8. Find also the loss of kinetic energy at the impact.

Ans. [1 m/sec, 13 m/sec, 135 N-m]

13. A ball of mass 1 kg traverses a frictionless tube falling through a height of 1 metre and emerges horizontally out of the tube and strikes a ball of mass 2 kg hung on a rope 1.5 metres long. Determine the height to which the 2 kg ball will rise if the collision is perfectly elastic. **Ans.** [0.15 m]

14. A ball of mass 20 gm, moving with a velocity of 3 m/s strikes on a ball of mass 40 gm moving with a velocity of 1 m/s. The velocities of the two balls are parallel and inclined at 30° to the line joining their centres at the instant of impact. If the coefficient of restitution is 1/2, find: (*i*) the direction and magnitude of velocity of ball of mass 20 gm after impact; and (*ii*) the direction and magnitude of velocity of ball of mass 40 gm.

Ans. [$\phi_1 = 60°$, $v_1 = 1.732$ m/s, $\phi_2 = 16.1°$, $v_2 = 1.803$ m/s]

15. A vehicle of mass 600 kg and moving with a velocity of 10 m/s strikes another vehicle of mass 400 kg, moving at 5 m/s in the same direction. Both the vehicles get coupled together due to impact. Find the common velocity with which the two vehicles will move. Also find the loss of kinetic energy due to impact.

Ans.[8 m/s, 3000 N-m]

16. A ball of mass 1 kg is dropped on a horizontal floor from a height of 6m. The ball rebounds due to impact to a height of 25m. Find the value of coefficient of restitution between the floor and the ball. **Ans.** [0.833]

17. A ball is dropped from a height of 100 cm on a smooth floor, from which it rebounds to a height of 81 cm. Find: (*i*) coefficient of restitution; and (*ii*) the expected height of the second rebound.

Ans. [$e = 0.9$ and $h_2 = 65.61$ cm]

Question Paper – 2001

(CS/Sem-1/B.Tech/ME-101)

Time Allotted: 3 Hours **Full Marks: 70**

(*i*) The questions are of equal value.

(*ii*) The figures in the margin indicate full marks.

(*iii*) Candidates are required to give their answers in their own words as far as practicable.

(*iv*) Answer question no.1 and any six from the rest.

---

1. (*a*) Explain Varignon's theorem of resultant force. **5× 2 = 10**

(*b*) Determine the co-ordinate x_c and y_c of the centroid C of the area between the parabola $y = x^2/a$ and the straight line $y = x$. (shown in figure.)

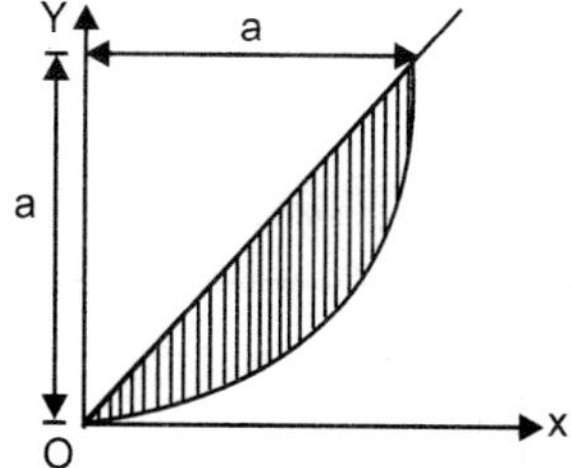

(*c*) State D' Alembert's principle of rectilinear motion.

(*d*) State the differential equation for free vibration of a particle.

(*e*) How are the yield point and ultimate strength represented in a stress-strain diagram?

2. A ball of weight $Q = 2.2$ kg rests in a right-angled trough, as shown in Figure. Determine the forces exerted on the sides of the trough at D and E if all surfaces are perfectly smooth. **10**

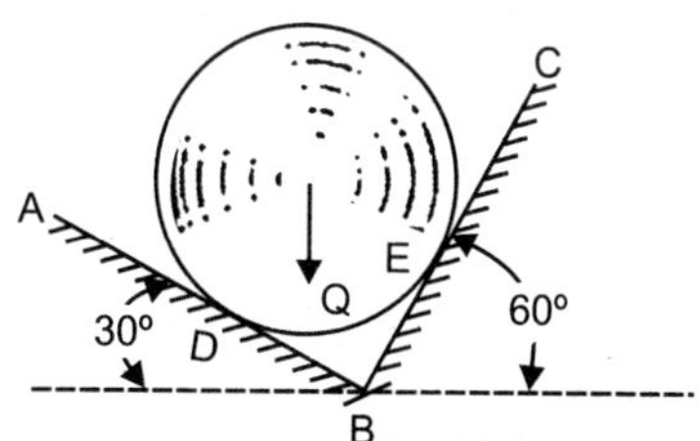

3. Two blocks of weight W_1 and W_2 rest on a rough inclined plane and are connected by a short piece of string as shown in figure. If the coefficient of friction are $\mu_1 = 0.2$ and $\mu_2 = 0.3$ respectively, find the angle of inclination of the plane for which sliding will impend. Assume $W_1 = W_2 = 5$ kg. **10**

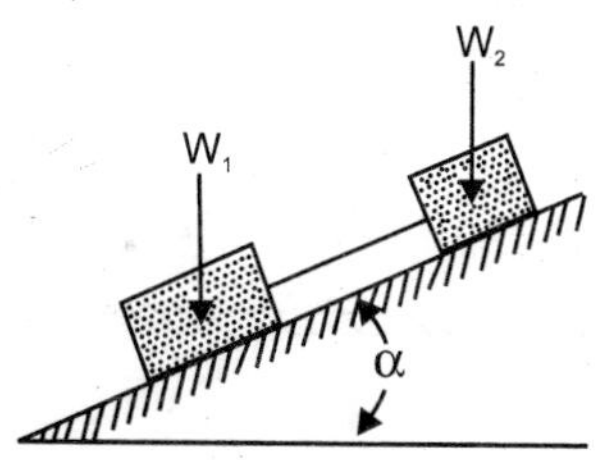

4. Determine the axial force in each bar of the plane truss loaded as shown in figure. **10**

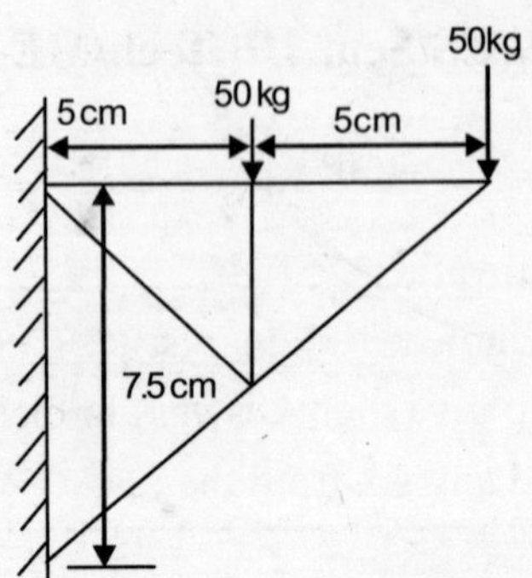

5. Using the principle of virtual work, find the reaction R_d for the system shown in figure for any position of a vertical load P on the beam AC as defined its distance x from A. **10**

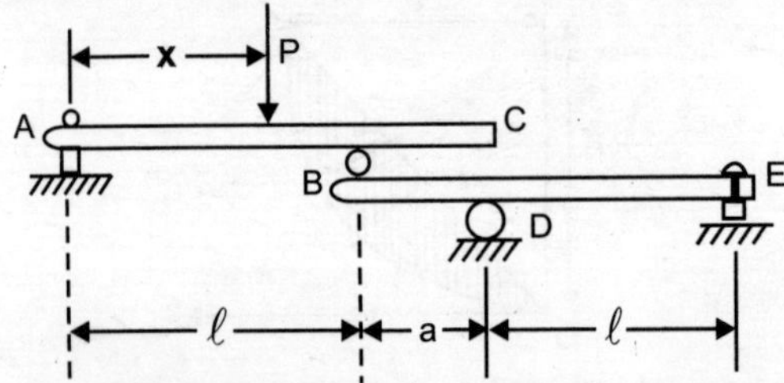

6. Weights W and $2W$ are supported in a vertical plane by a string and pulleys arranged as shown in figure. Find the magnitude of an additional weight Q applied on the left which will give a downward acceleration $a = 0.1$ g to the weight W. Neglect friction and inertia of pulleys. **10**

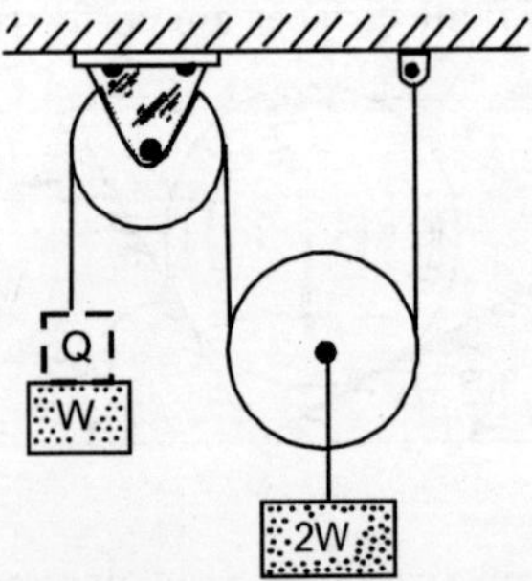

7. The two springs shown in figure each has a spring constant $k = 0.2$ kg/cm, and the attached ball has the weight $W = 0.5$kg. If the ball is initially displaced 2 cm to the right, find the period of oscillation of the ball and the velocity with which it passes through its middle position. Neglect friction. **10**

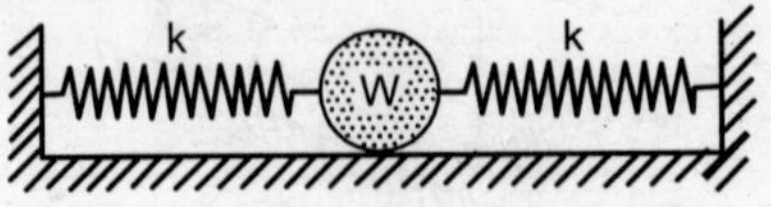

8. Construct the complete shear force and bending moment diagrams for the beam in figure. **10**

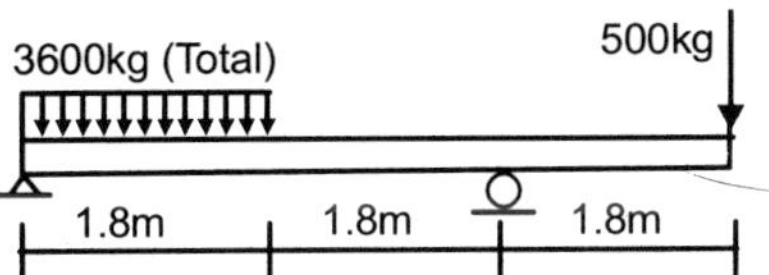

9. Construct Mohr's circle for the element as shown in figure, if $\sigma_x = \sigma_y = 0$ and $\tau_{xy} = 350$ kg/ cm^2. From this construction, find principal stresses σ_1 and σ_2 and also the angle ϕ_a between the x-axis and the direction of the principal stress σ_1. **10**

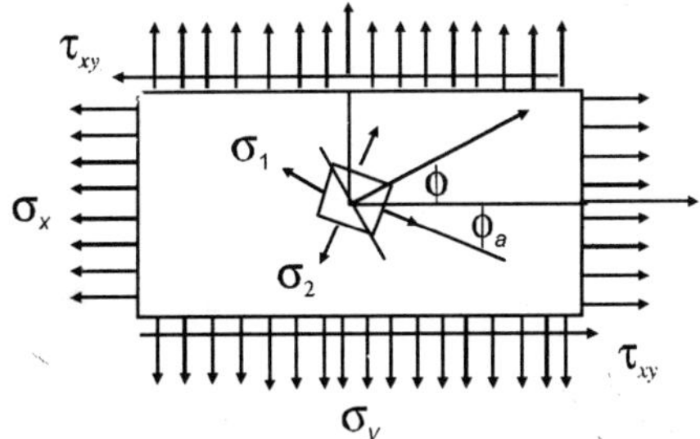

10. A truncated conical tank having the dimensions shown in figure is filled with water (w = 1gm/ cm^3). Calculate the membrane stress σ_1 and σ_2 for an element of the wall situated as shown in the figure if wall thickness $t = 0.03$cm. **10**

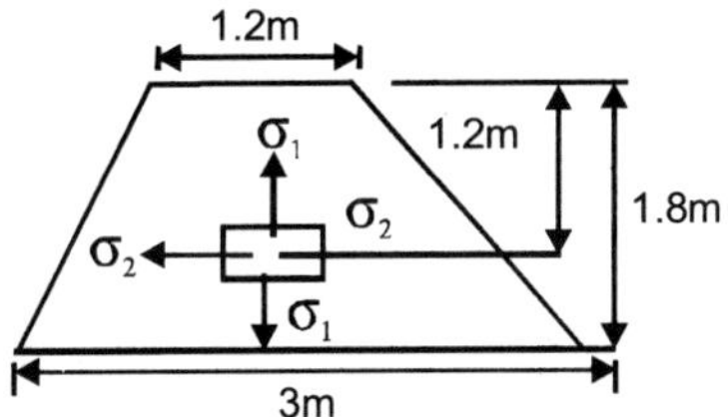

Question Paper – 2002

(CS/ME–101/Sem-1/2002-2003)

Time Allotted: 3 Hours **Full Marks: 70**

(*i*) The questions are of equal value.
(*ii*) The figures in the margin indicate full marks.
(*iii*) Candidates are required to give their answers in their own words as far as practicable.
(*iv*) Answer question no.1 and any six from the rest.

---

1. Write the correct alternative.

(*a*) A just rigid truss with 6 joints must contain following number of members. (*i*) 7, (*ii*) 6, (*iii*) 9, (*iv*) 11

(*b*) Efficiency of a screw jack with helix angle and angle of friction is given by

(*i*) $\frac{\tan\alpha}{\tan(\alpha+\phi)}$ (*ii*) $\frac{\tan\alpha}{\tan\alpha+\tan\phi}$ (*iii*) $\frac{\tan\alpha}{\tan(\alpha-\phi)}$ (*iv*) $\frac{\tan(\alpha+\phi)}{\tan\alpha}$

(*c*) Energy absorbed in a body when it is strained within elastic limit is known as

(*i*) strain energy (*ii*) resilience
(*iii*) modulus of resilience (*iv*) toughness.

(*d*) Moment of inertia of a rectangular area of base b and height h about an axis through centroid and parallel to the base b is given by

(*i*) $\frac{bh^3}{3}$ (*ii*) $\frac{bh^3}{4}$ (*iii*) $\frac{bh^3}{8}$ (*iv*) $\frac{bh^3}{3}$

(*e*) Which statement is wrong ?

(*i*) tangent of angle of friction is equal to the coefficient of friction.
(*ii*) friction force is independent of area of contact.
(*iii*) sine of angle of repose gives the coefficient of friction.
(*iv*) none of the above.

(*f*) Equation of motion of a particle is $s = 2t^3 - t^2 - 2$, where s is displacement in meters and t is time in seconds. Acceleration of the particle after 1 second will be

(*i*) 8m/sec^2 (*ii*) 9m/sec^2 (*iii*) 10m/sec^2 (*iv*) 5m/sec^2

(*g*) If u_1, u_2 are the velocities of approach of two moving bodies in same direction and the velocities of separation V_1 and V_2 then as per Newton's law of collision between elastic bodies, coefficient of restitution e is given by

(*i*) $e = \frac{v_1 - v_2}{u_1 - u_2}$ (*ii*) $e = \frac{u_2 - u_1}{v_1 - v_2}$ (*iii*) $e = \frac{v_2 - v_1}{u_1 - u_2}$ (*iv*) $e = \frac{v_1 - v_2}{u_2 - u_1}$

(*h*) Hoop stress for a thin cylinder of diameter *d* and thickness *t*.

(*i*) $\frac{pd}{4t}$ (*ii*) $\frac{pd}{t}$ (*iii*) $\frac{pd}{2t}$ (*iv*) $\frac{2pd}{t}$

(*i*) In case of Simple Harmonic Motion period of oscillation is given by

(*i*) $T = \frac{2\omega}{\pi^2}$ (*ii*) $T = \frac{2\pi}{\omega}$ (*iii*) $T = \frac{\omega}{2\pi}$ (*iv*) $T = \frac{\pi}{\omega}$

(*j*) Proof stress is

(*i*) stress corresponding to proportional limit
(*ii*) stress causing material to break
(*iii*) stress causing a specific permanent deformation usually 0.01 % or less
(*iv*) not related with engineering **1 × 10**

2. A vertical load *P* is supported by a triangular bracket as shown in figure. Find the forces transmitted to the bolts *A* and *B*. Assume bolt *B* fits loosely in a vertical slot in the plate. **10**

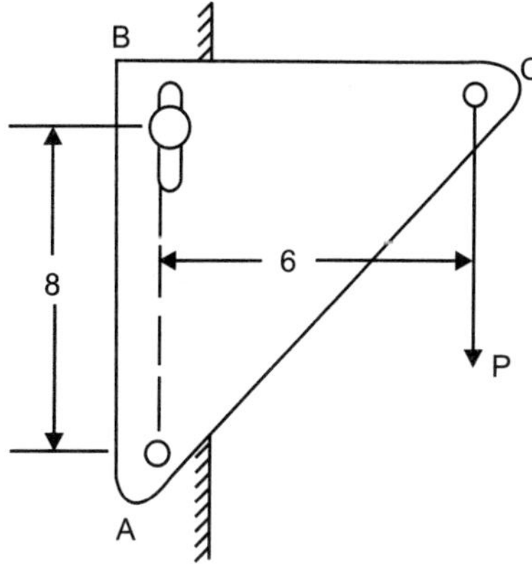

3. (*a*) State Pappus theorem. **4**

(*b*) Using it find the centroid of a semicircular area of radius *r*. **6**

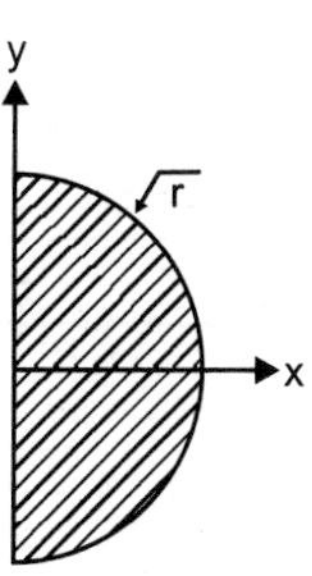

4. Determine the axial force in the bar *CD* of the following figure. **10**

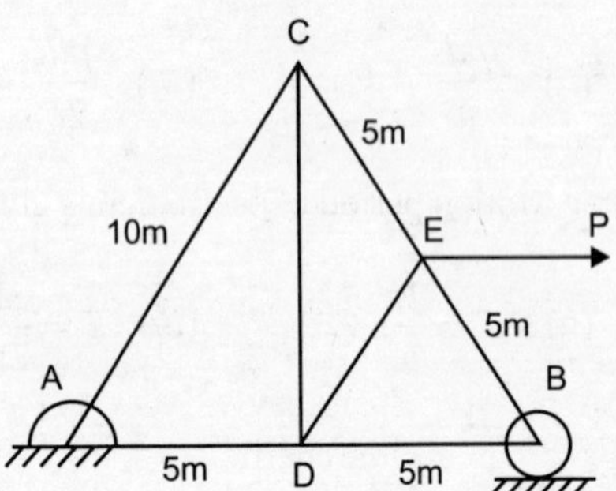

5. (*a*) State the principle of virtual work. **2**

(*b*) Using it determine the magnitude of the horizontal force *P* to be applied at point *A* of the slender prismatic bar *AB* of length 2m and weight 1000N, to hold it in a vertical plane and supported by smooth surfaces at *A* and *B* as shown in figure. **8**

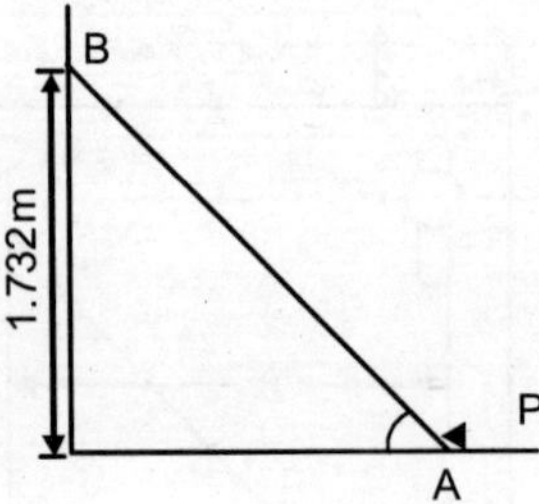

6. (*a*) Explain principal stress. **2**

(*b*) At a certain point within a stressed body principal stresses are σ_x = 80MPa and σ_y = –40MPa. Determine σ and τ on a plane whose normal is at 60° with the *x*-axis. Construct the Mohr's circle and show result on a sketch of different element. **8**

7. Writing the moment and force equations, draw the Shear force and Bending Moment diagram for the beam loaded as shown in figure. **10**

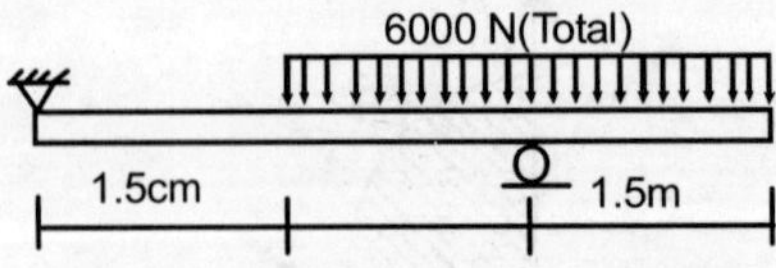

8. In figure a small car of weight *W* starts from rest at *A* and rolls without friction along an inclined plane to *B* where it strikes a block also of weight *W* and initially at rest. Assuming a plastic impact at *B*, the car and the block will move from *B* to as one particle if the coefficient of friction between the block and the plane is μ = 0.5, calculate the distance *x* to point *C* where the bodies come to rest. **10**

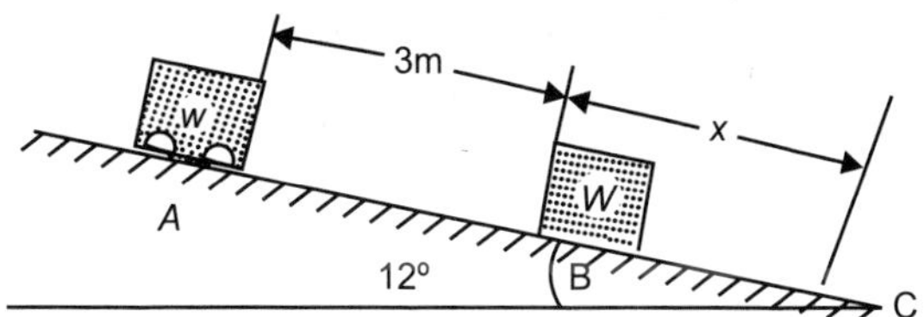

9. A steel rod 750 mm long has a cross-sectional area of 1200mm^2 over 300 mm of its length and 1800 mm^2 over the remaining 450mm. At a temperature $T = 10$ °C, the rod fits exactly between unyielding walls at its two ends. Compute the maximum compressive stress that will exist in the rod at a temperature $T = 40$ °C. The modulus of elasticity is $E = 2\times10^5$ N/mm^2 and the coefficient of thermal expansion is $\alpha = 12\times10^{-6}$mm/°C. **10**

10. The overhanging beam is loaded by the uniformly distributed load as well as concentrated force as shown in figure. Determine the deflection of point *A* of the beam. **10**

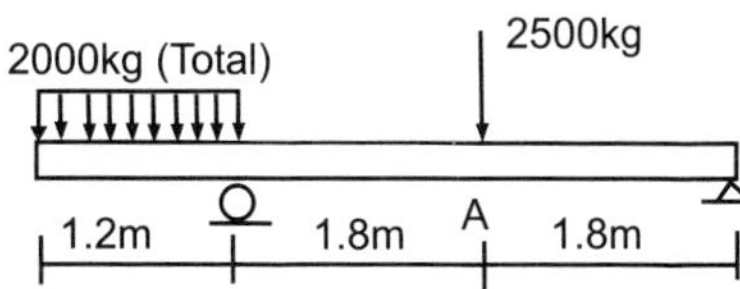

11. A rigid bar *AB*, 9m long, is suspended by two vertical rods at its ends and hangs in horizontal position under its own weight as shown in figure. The rod at *A* is brass; length 3m, cross-sectional area 1000mm^2, modulus of elasticity, 10×10^4 N/mm^2. The rod at *B* is steel; length 5m, cross-sectional area 500mm^2, modulus of elasticity 2×10^5 N/mm^2. At what distance x from *A* may a vertical load *P* be applied if the bar is to remain horizontal after the load is applied? **10**

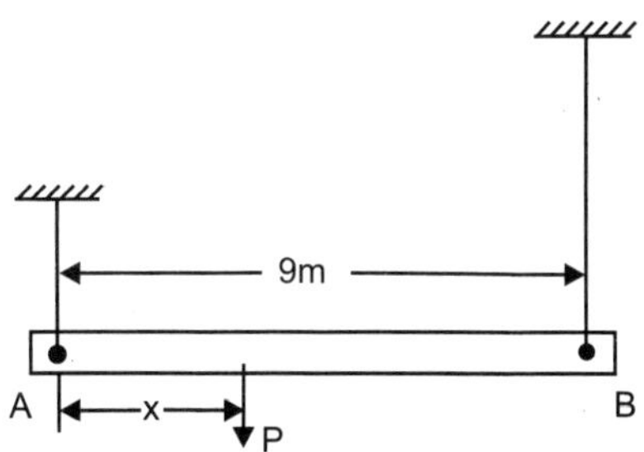

12. Draw the shear force and bending moment diagrams for the beam loaded as shown in figure. **10**

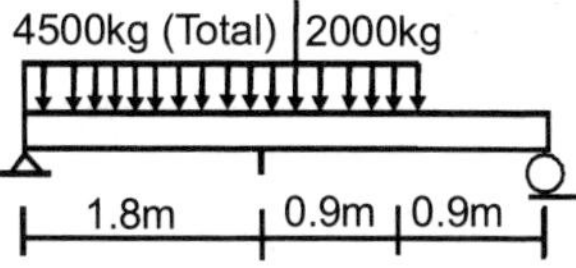

Question Paper – 2003

(CS/Sem-1/ME-101/2003)

Time Allotted: 3 Hours **Full Marks: 70**

(*i*) The questions are of equal value.
(*ii*) The figures in the margin indicate full marks.
(*iii*) Candidates are required to give their answers in their own words as far as practicable.
(*iv*) Answer question no.1 and any six from the rest.

---

1. Choose the correct answer: **1×10**

(*a*) Which one of the following statements is correct? A rigid body in translation

(*i*) can move along a straight or curved path (*ii*) can move only in a straight line
(*iii*) cannot move on a circular path (*iv*) must undergo curvilinear motion

(*b*) Moment of inertia of a quadrant about its $X-X$ axis is given by

(*i*) $0.055r^4$ (*ii*) $0.04r^4$ (*iii*) $0.06r^4$ (*iv*) r^4

(*c*) In case of a screwjack the condition for maximum efficiency is given by

(*i*) $\alpha = 45° + \dfrac{\phi}{2}$ (*ii*) $\alpha = 45° - \dfrac{\phi}{2}$ (*iii*) $\alpha = 60° - \dfrac{\phi}{2}$ (*iv*) $\alpha = 60° + \dfrac{\phi}{2}$

(*d*) The period of simple pendulum will be double if

(*i*) its length is doubled (*ii*) its length is quadrupled
(*iii*) the mass of its bob is doubled (*iv*) its length is halved.

(*e*) Which one of the following does not have identical dimensions?

(*i*) momentum and impulse (*ii*) torque and energy.
(*iii*) torque and work (*iv*) moment of a force and angular momentum

(*f*) For a truss to be over-stable and statically indeterminate, the condition for *m* (number of members) and *j* (number of joints) is given by

(*i*) $m \geq 2j - 3$ (*ii*) $m > 2j - 3$ (*iii*) $m = 2j - 3$ (*iv*) $m = 2j$

(*g*) Longitudinal stress in a thin cylindrical shell as compared to hoop stress is

(*i*) half (*ii*) two times (*iii*) equal (*iv*) none of the above

(h) Thermal strain caused in the material of a composite body due to change in temperature will be of

(*i*) same nature (*ii*) opposite nature (*iii*) same magnitude (*iv*) none of the above

(*i*) Maximum deflection in cantilever beam with U.D.L. over the entire span is given by

(*i*) $\dfrac{WL^3}{48EI}$ (*ii*) $\dfrac{5WL^3}{384EI}$ (*iii*) $\dfrac{WL^3}{3EI}$ (*iv*) $\dfrac{WL^3}{8EI}$

(*j*) True stress ($\bar{\sigma}$) is related to simple stress (σ) and strain (ε) by

(*i*) $\bar{\sigma} = \sigma(1-\varepsilon)$ (*ii*) $\bar{\sigma} = \sigma(1+\varepsilon)$ (*iii*) $\bar{\sigma} = \sigma\varepsilon$ (*iv*) $\bar{\sigma} = \dfrac{\sigma}{\varepsilon}$

2. (*a*) State and prove Lami's Theorem. **5**

(*b*) A uniform wheel of 60cm diameter rests against a rigid rectangular block of 15cm height as shown in figure below. Find the magnitude and direction of the least pull through the centre of the wheel that will just turn the wheel over the corner of the block. All surfaces are smooth. Find the reaction of the block at point *C*. The wheel weight 5kN. **5**

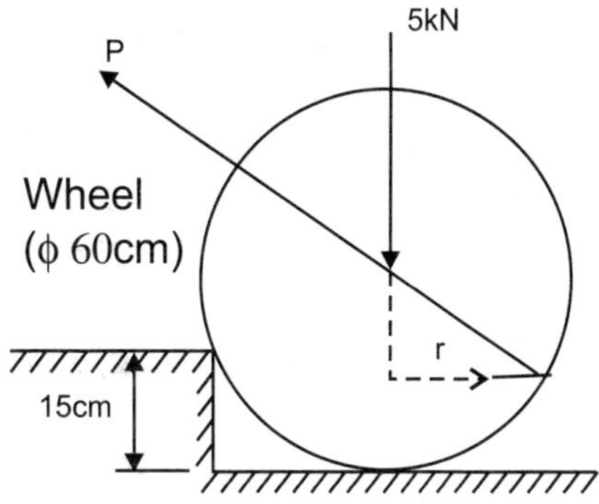

3. (*a*) State the parallel axes theorem of moment of inertia (MI) of lamina. **2**

(*b*) Determine MI of the inverted *T* section as shown in figure about its centroidal axis parallel to the base. **8**

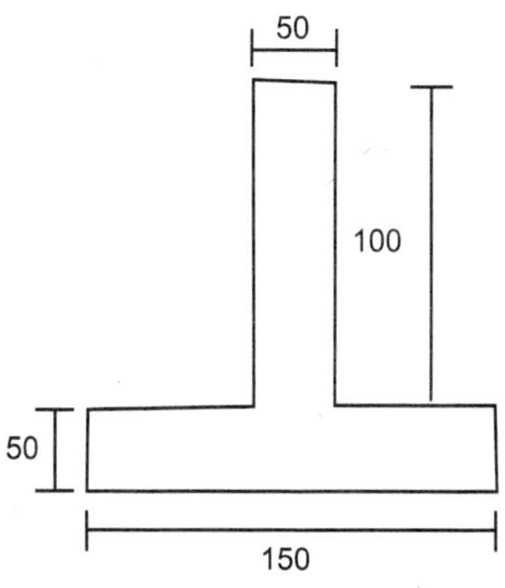

4. (*a*) State the laws of static friction. **2**

(*b*) As shown in the figure, find the minimum value of horizontal force *P* applied to the lower block that will keep the system in equilibrium, given the coefficient of friction between the lower block and floor = 0.25, between the upper block and vertical wall = 0.3 and between the blocks = 0.20. **8**

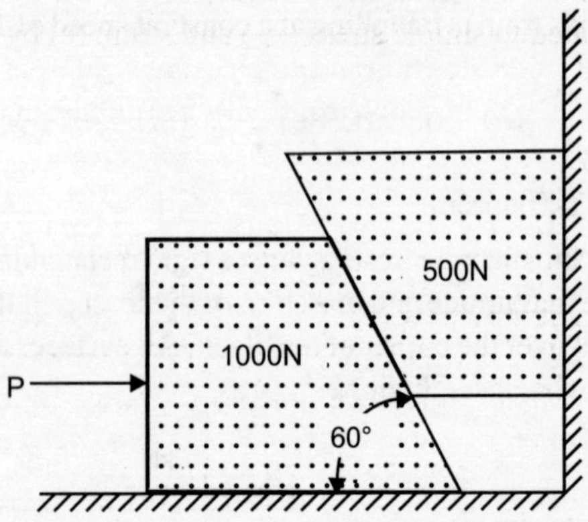

5. Write the shear force and bending moment equation for the beam shown in figure. Sketch the shear force and bending moment diagram and locate the position of occurrence of maximum bending moment. **10**

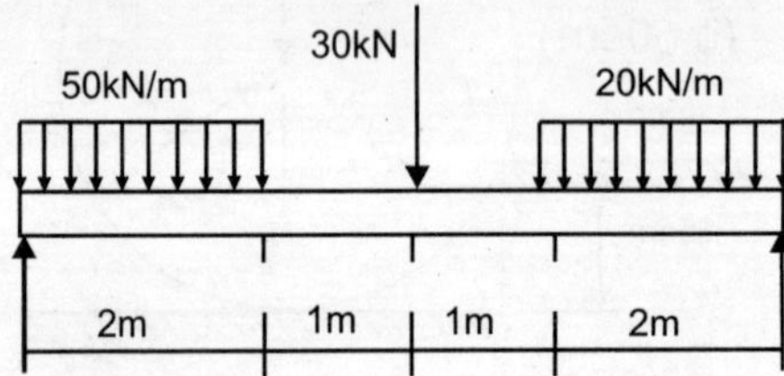

6. Determine the deflections at the free end and middle of the span of the cantilever beam loaded as shown in the figure. The flexural rigidity (i.e., EI) of the beam is uniform throughout its length. **10**

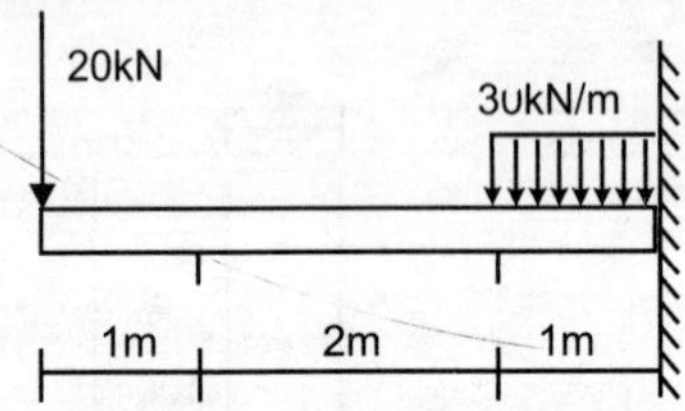

7. A state of stress is specified in the figure. Determine the normal and shearing stress on.
   (*a*) principal planes
   (*b*) the planes of maximum-shearing stress. Show the results of part (*a*) and (*b*) on complete sketches of differential element. **10**

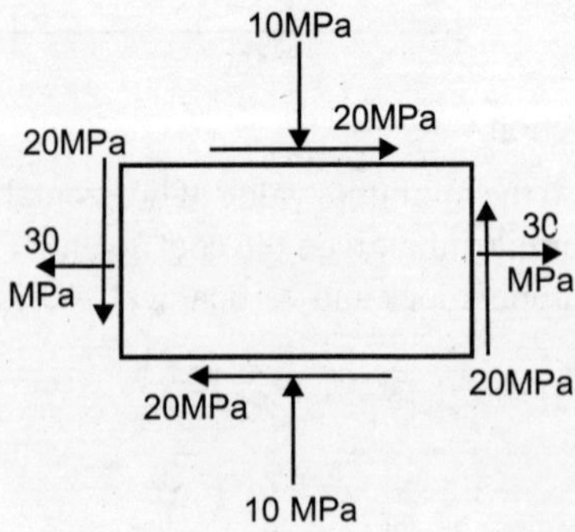

8. (*a*) A 120 tonnes goods train is travelling at a constant speed of 100 km/hr while total resistance against the motion due to ground friction and air pressure is 50N per tonne weight. Suddenly the last wagon weighing 20 tonnes gets decoupled and falls behind the main train.

Determine—

(*i*) the acceleration and deceleration of the main train and decoupled wagon respectively.

(*ii*) the distance between the two after 20 seconds. **7**

(*b*) The acceleration a of a body starting from rest is given by $a = 4 - 0.002s$ where a is acceleration in m/s^2 and s is the distance travelled in metres. Find the velocity of the body when it has travelled 100m and the distance travelled when the body is again at rest. **3**

9. (*a*) State d' Alembert's principle. **2**

(*b*) Two bodies A and B are connected by an inextensible string as shown in figure. Find acceleration of the bodies and tension in the string by using d' Alembert's principle. **8**

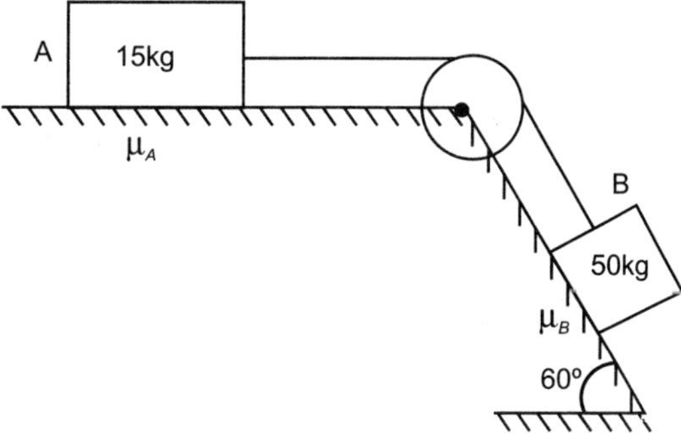

Question Paper – 2004

(CS/B.Tech/Sem-1/ME-101/2004)

Time Allotted: 3 Hours **Full Marks: 70**

(*i*) The figures in the margin indicate full marks

(*ii*) Assume necessary data, if any.

(*iii*) Answer question no.1 and any six from the rest.

---

1. Choose the correct answer: **1 × 10**

(*a*) Two non-collinear parallel equal forces acting in opposite direction.

(*i*) balance each other (*ii*) constitute a moment
(*iii*) constitute a couple (*iv*) constitute a moment of a couple
(*v*) constitute a resultant couple

(*b*) The coefficient of friction depends on

(*i*) area of contact (*ii*) shape of surfaces
(*iii*) strength of surfaces (*iv*) nature of surfaces
(*v*) all of the above

(*c*) Moment of Inertia of a rectangular area of base *b* and height *d* about *x*-axis is given by

(*i*) $\dfrac{bd^3}{3}$ (*ii*) $\dfrac{bd^3}{4}$ (*iii*) $\dfrac{bd^3}{6}$

(*iv*) $\dfrac{bd^3}{12}$ (*v*) $\dfrac{bd^3}{8}$

(*d*) According to principle of transmissibility of forces, the effect of a force upon a body is

(*i*) maximum when it acts at the center of gravity of a body.
(*ii*) different at different points in its line of action.
(*iii*) the same at every point in its line of action.
(*iv*) minimum when it acts at the C.G of the body.
(*v*) none of the above.

(*e*) d'Alembert's principle is used for

(*i*) reducing the problem of kinetics to equivalent statics problem
(*ii*) determining stresses in the truss
(*iii*) stability of floating bodies
(*iv*) designing safe structures
(*v*) solving kinematic problems

(*f*) When a body slides down an inclined surface of inclination θ, the acceleration f of the body is given by

(*i*) $f = g$ (*ii*) $f = g \sin\theta$ (*iii*) $f = g \cos\theta$ (*iv*) $f = g \tan\theta$ (*v*) $f = g/\sin\theta$

(g) The kinetic energy of a body rotating with an angular speed ω depends on

(i) ω only (ii) ω^2 only (iii) its mass only

(iv) the distribution of mass and angular speed

(v) all of the above

(h) The material having same elastic properties in all directions are called

(i) idea material (ii) uniform material (iii) elastic material

(iv) isotropic material (v) anisotropic material

(i) Poisson's ratio is defined as

(i) longitudinal stress and longitudinal strain (ii) longitudinal stress and lateral stress

(iii) lateral stress and longitudinal stress (iv) lateral stress and lateral strain

(v) none of the above

(j) The maximum strain energy that can be stored in a body is known as

(i) impact energy (ii) resilience (iii) proof resilience

(iv) modulus of resilience (v) toughness

2. (a) State Varignon's Theorem of moments. **2**

(b) A sphere hangs by a chord and rests on a smooth vertical wall. Determine the tension in the chord and the reaction exerted by the wall against the roller. Given: Mass of the sphere = 100kg, Diameter of the sphere = 200mm, Length of the chord = AC = 100mm **8**

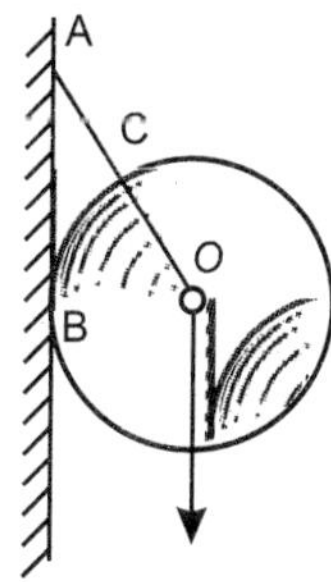

3. (a) State the Laws of Friction. **3**

(b) Explain clearly the term 'Angle of Repose'. **2**

(c) A block of weight 1600N is in contact with a plane inclined at 30° to the horizontal. A force P parallel to the plane is applied on the body as shown in figure. The coefficient of friction between the contact surface is 0.20. Find the value of P to just cause the motion to impend up the plane. **7**

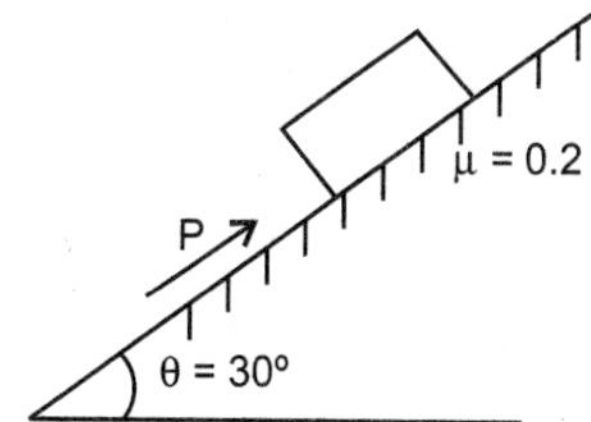

4. (*a*) Calculate the location of the centroid of the *L*-section shown in figure. **6**

(*b*) Calculate the centroidal moment of inertia (of the same figure) parallel to the base. **6**

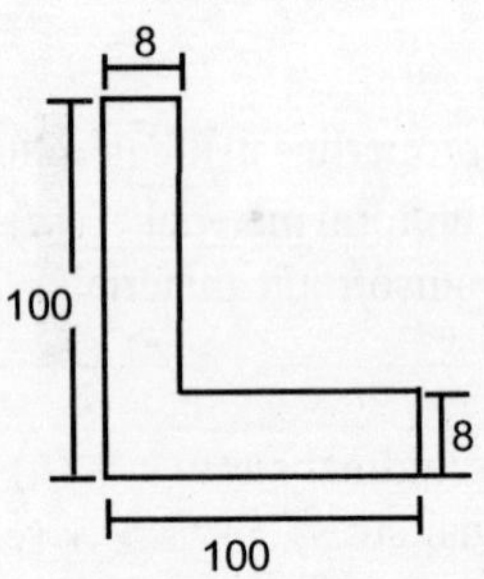

5. A simply supported plane truss of span 7.5 m is supported at *A* and *B* joint *D*, as shown in the figure. Find (*i*) the reactions at *A* and *B* and (*ii*) forces in the members of the truss. Identify the members that are subjected to maximum tensile and compressive forces. **4+6+2**

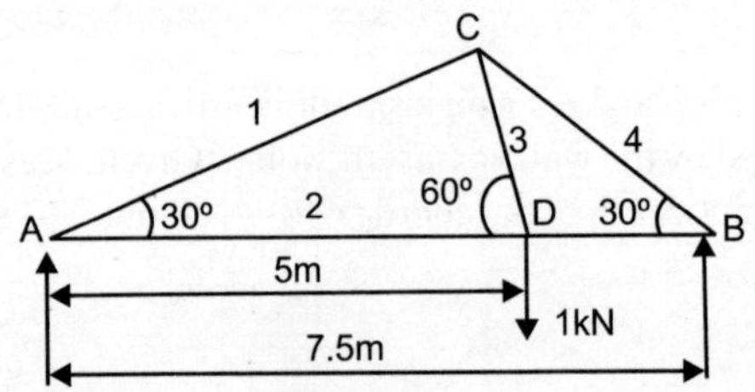

6. The bar shown in figure is subjected to a tensile load of 152kN. Find the diameter of the middle portion if the stress there is to be limited to 140N/mm². Find also the length of the middle portion if the total elongation of the bar is to be 0.16 mm. Take $E = 2\times 10^5$N/mm². **12**

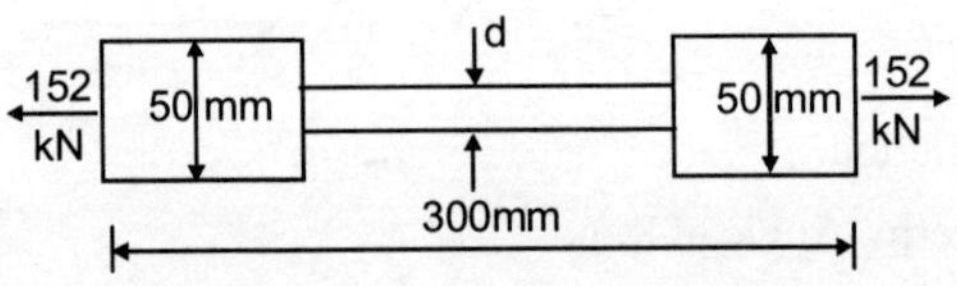

7. Write of the shear force and bending moment equations for the beam shown in figure. Sketch the shear force and bending moment diagrams and locate the position of maximum bending moment. **12**

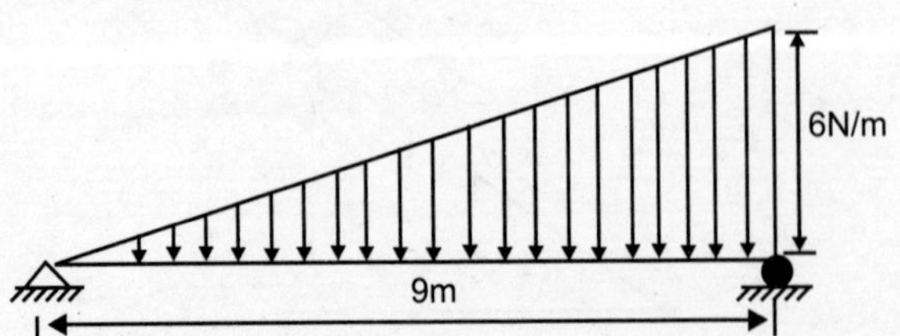

8. A shot is fired with a bullet with an initial velocity of 40m/s from a point. 20m in front of a vertical wall 10m high shown in figure. Find the angle of projection with horizontal to enable the shot to just clear the wall. **12**

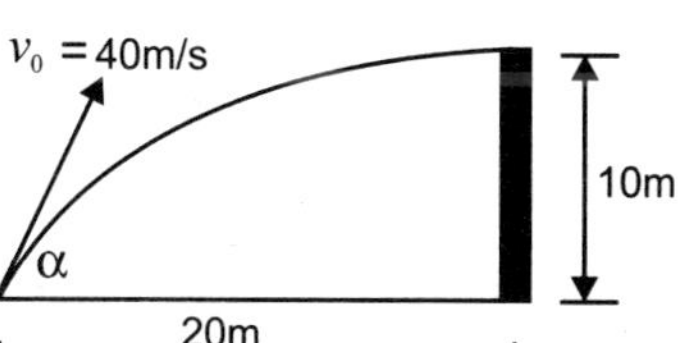

9. (*a*) State and explain d' Alembert's Principle. **4**

(*b*) A system of weights and pulleys is arranged in a vertical plane as shown in figure. Neglecting friction and the inertia of the pulleys, find the acceleration of each weight if their magnitudes are in the ratio $W_a : W_b : W_c = 3 : 2 : 1$ **8**

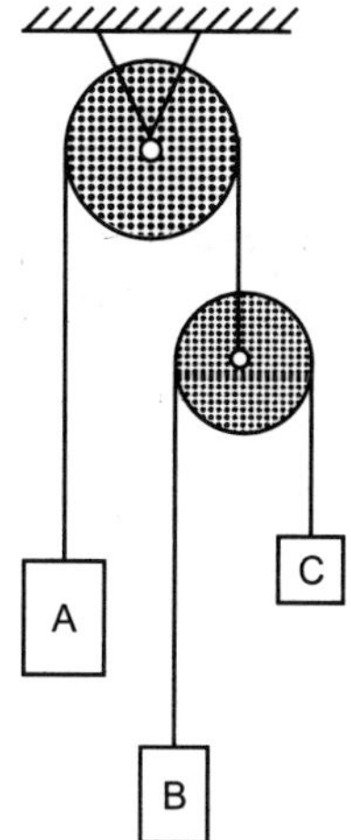

Question Paper – 2005

(CS/B-TECH/SEM-I/ME-I01/03)

Time Allotted: 3 Hours **Full Marks: 70**

(*i*) The questions are of equal value.

(*ii*) The figures in the margin indicate full marks.

(*iii*) Candidates are required to give their answers in their own words as far as practicable.

(*iv*) Answer question no. 1 which is compulsory and any six from the remaining.

---

1. Choose the correct answers with proper justification: **5 × 2=10**

(*a*) First area moments (of a plane surface area) about centroidal axes is equal to

(*i*) zero (*ii*) non-zero

(*b*) For a two-dimensional equilibrium (static) problem, the maximum number of unknowns that can be evaluated using equilibrium equations are.

(*i*) one (*ii*) three (*iii*) six

(*c*) Thermal stress is induced within a material due to

(*i*) free expansion (*ii*) free contraction

(*iii*) free expansion or contraction (*iv*) restricted expansion or contraction

(*v*) none of these

(*d*) When a body slides down an inclined surface (of inclination θ) the acceleration f of the body is

(*i*) $f = g$ (*ii*) $f = g \sin\theta$ (*iii*) $f = g \cos\theta$ (*iv*) $f = g \tan\theta$

(*e*) The kinetic energy of a body rotating with an angular speed ω depends on

(*i*) ω only (*ii*) ω^2 only (*iii*) mass only

(*iv*) the distribution of mass and angular speed (*v*) all of these

2. (*a*) State and prove perpendicular axis theorem of area moment of inertia. **4**

(*b*) Locate the centroid of the quadrant of a circle of radius *r*. Shown in figure. **6**

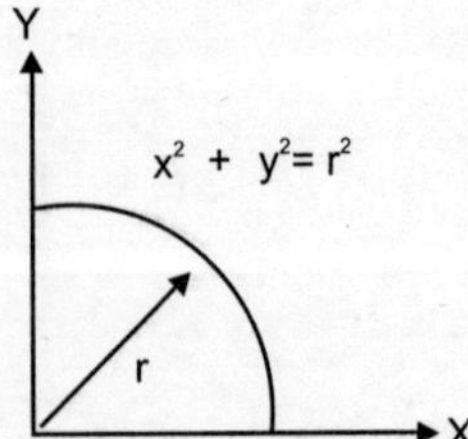

3. (*a*) State the laws of static friction. **3**

(*b*) A block of weight $W_1 = 1290$N rests on a horizontal surface and supports another block of weight $W_2 = 570$N on top of it as shown in figure. Block of weight W_2 is attached to a vertical wall by an inclined string *AB*. Find the force *P* applied to the lower block, that will be necessary to cause the slipping to impend, Given:

Coefficient of friction between blocks (1) and (2) = 0.25. Coefficient of friction between (1) and horizontal surface = 0.40 **7**

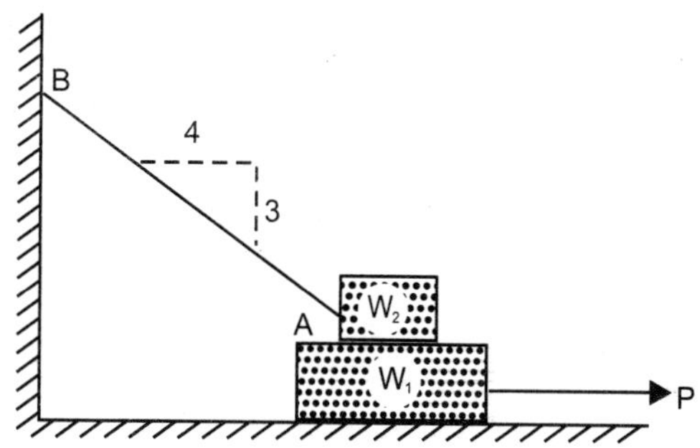

4. (*a*) State and prove Lami's theorem. **3**

(*b*) Define free body diagram. **2**

(*c*) Two cylinders of diameters 60mm and 30mm weighing 160N and 40N respectively are placed as shown. Assuming all the contact surfaces to be smooth, find the reactions at *A*, *B*, and *C*, shown in figure. **5**

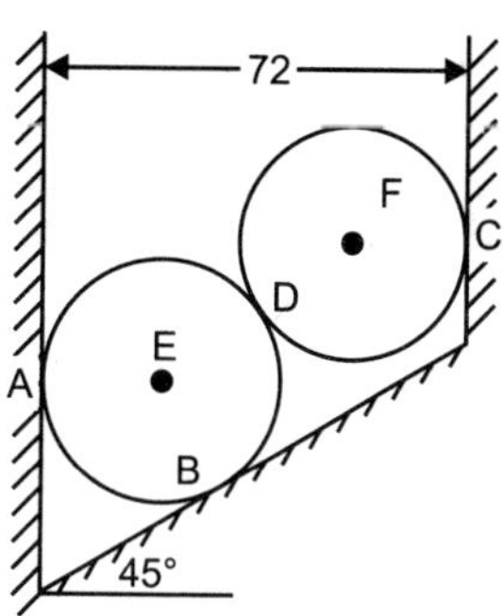

5. (*a*) State the principle of virtual work. **3**

(*b*) Using the principle of virtual work, find the value of the angle θ defining the configuration of equilibrium of the system as shown in figure. The balls *D* and *E* can slide freely along the bars *AC* and *BC* but the string *DE* connecting them is inextensible. **7**

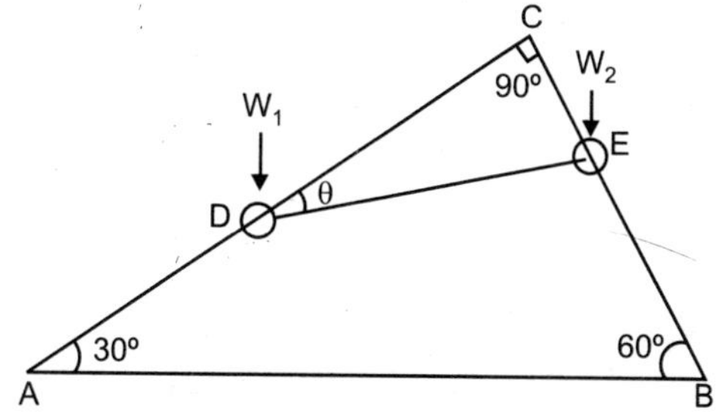

6. (*a*) Define Hooke's law. **3**

(*b*) A bronze bar 3m long with a cross-sectional area of $320mm^2$ is placed between two rigid walls as shown in figure. At a temperature of –20°C, the gap $\Delta = 2.5$mm. Find the temperature at which the compressive strees in the bar will be $\sigma = 35$MPa. Use $\alpha = 18 \times 10^{-6}$ m/m°C and $E = 80$ GPa. **7**

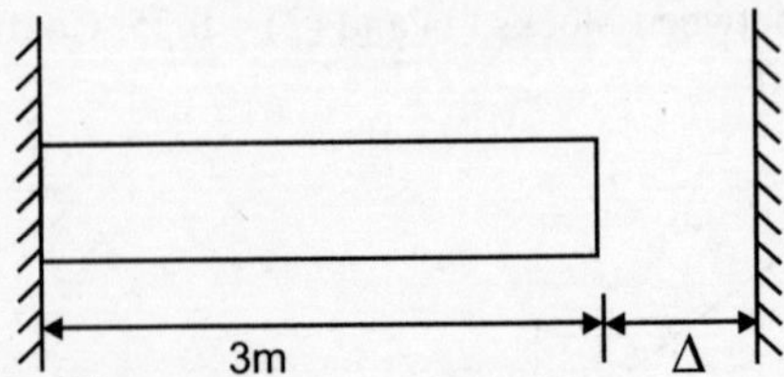

7. (*a*) State Coulomb's Law of friction. **2**

(*b*) A block of weight $W_1 = 500$N rests on a horizontal surface and supports on top of it another block of weight $W_2 = 100$N. The block W_2 is attached to a vertical wall by the inclined string *AB*. Find the magnitude of the horizontal force *P* applied to the lower block as shown in figure that will be necessary to cause slipping to impend. The coefficient of static friction for all contiguous surface is $\mu = 0.3$. **8**

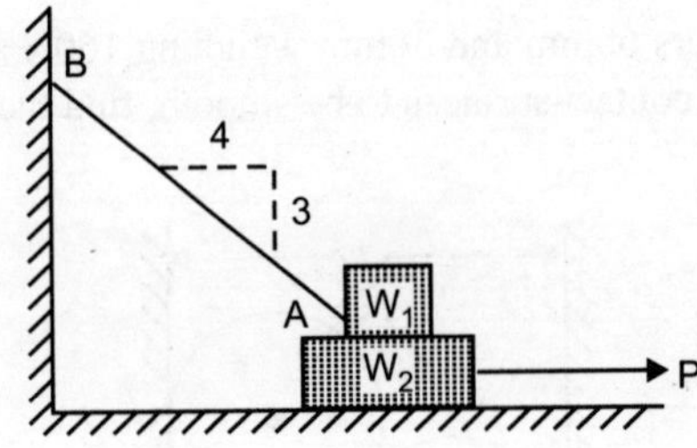

8. (*a*) A particle moving in the $x-y$ plane undergoes a displacement $\overline{S} = (4\overline{i} + 6\overline{j})$m with a constant force $\overline{F} = (-5\overline{i} + 10\overline{j})\,N$ acting on it. Calculate the work done, magnitude of force and magnitude of displacement. **5**

(*b*) A 5kg block slides from rest at point *A* along a frictionless inclined plane making an angle 25° with horizontal. Determine the speed of the block at *B* at a distance of 3m from *A*. **5**

9. (*a*) Distinguish between particle and rigid body. **2**

(*b*) A ball is dropped vertically on to a 20° inclined plane at *A*. The direction of rebound forms an angle of 35° with strikes the inclined plane at *B*, determine.

(*i*) the velocity of rebond at *A*.

(*ii*) the time required for the ball to travel from *A* to *B*. **8**

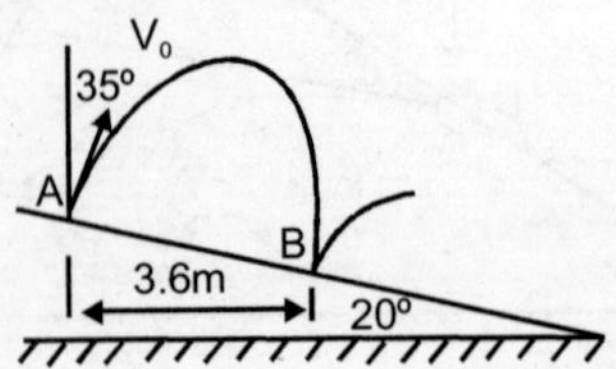

Question Paper – 2006

(CS/Sem-1/B.Tech/ME-101/06)

Time Allotted: 3 Hours **Full Marks: 70**

Group – A
(Multiple Choice Questions)

(*i*) The questions are of equal value.

(*ii*) The figures in the margin indicate full marks.

(*iii*) Candidates are required to give their answers in their own words as far as practicable.

(*iv*) Answer question no.1 and any six from the rest.

---

1. Choose the correct answer from the given alternatives in each of the following questions: **10 × 1 = 10**

(*a*) Centre of gravity of a solid cone lies on the axis at the height,

(*i*) one-fourth of the total height above base (*ii*) one-third of the total height above base
(*iii*) one-half of the total height above base (*iv*) three-eighth of the total height above base
(*v*) none of these.

(*b*) In the equestion of virtual work, which of the following forces is (are) neglected?

(*i*) Reaction of any smooth surface with which the body is in contact.
(*ii*) Reaction of a rough surface of a body which rolls on it without slipping
(*iii*) Reaction at a point or an axis, fixed in space, around which a body is constrained to turn
(*iv*) All of these (*v*) None of these

(c) M.I. of rectangular area of base b and height d about z-axis is given by

(*i*) $bd^3/3$ (*ii*) $bd^3/4$ (*iii*) $bd^3/6$ (*iv*) $bd^3/12$ (*v*) $bd^3/8$

(*d*) M.I. of circular area whose diameter is d about an axis perpendicular to the area passing through its centre is given by

(*i*) $\pi d^4/64$ (*ii*) $\pi d^4 / 32$ (*iii*) $\pi d^4/12$ (*iv*) $\pi d^4/16$ (*v*) $\pi d^4/24$

(*e*) The acceleration due to gravity of a particle falling towards earth is

$$a = -g\frac{R^2}{r^2}$$

where r is the distance from centre of the earth to the particle, g = acceleration due to gravity. If the particle is projected from a tower of height h. its escape velocity (V_e) will be

(*i*) $V_e = \left(\frac{2gR}{R+h}\right)^{1/2}$ (*ii*) $V_e = 2g\left(\frac{R}{R+h}\right)^2$ (*iii*) $V_e = \left(\frac{2g}{R+h}\right)^{1/2} R.$

(*f*) In order to avoid overturning for a vehicle moving on a level curved path, the maximum permissible velocity must be

(*i*) $V_{max} \leq [gra/h]^{1/2}$ (*ii*) $V_{max} \geq [gra/h]^{1/2}$
(*iii*) $V_{max} \leq [hra/g]^{1/2}$ (*iv*) $V_{max} \geq [hg/(ra)]^{1/2}$

where V = velocity of the vehicle moving along a circular path
r = radius of the path
h = height of c.g. of the vehicle from the track
$2a$ = wheel to wheel distance.

(*g*) A body is resting on a plane inclined at an angle of 30° to horizontal. What force would be required to slide it down, if the co-efficient of friction between body and plane is 0.3?

(*i*) zero (*ii*) 1 kg (*iii*) 5 kg
(*iv*) would depend on weight of the body (*v*) none of these.

(*h*) On a plane resultant stress is inclined at an angle of 30° to the plane. If the normal stress on the plane is 50 N/mm², the shear stress on the plane will be

(*i*) 43.3 N/mm² (*ii*) 86.6 N/mm² (*iii*) 100 N/mm² (*iv*) None of these.

(*i*) Equation of motion of a particle is $s = 2t^3 - t^2 - 2$, where s is displacement in metres and t is time in seconds. Acceleration of the particle after 1 second will be

(*i*) 8 m/sec² (*ii*) 9 m/sec² (*iii*) 10 m/sec² (*iv*) 5 m/sec²

(*j*) The ratio of lateral strain to the linear strain within elastic limit is known as

(*i*) Young's modulus (*ii*) Bulk modulus
(*iii*) Modulus of elasticity (*iv*) Poisson's ratio

Group – B
(Short Answer Questions)

Answer any three questions. **3×5 = 15**

2. A ball is dropped onto a fixed horizontal surface from height h_0. The co-efficient of restitution is e. Show that the total vertical distance D, travelled by the ball before it comes to rest is

$$D = h_0\left(\frac{1+e^2}{1-e^2}\right)$$

3. A bullet of mass m, moving with a horizontal velocity v, hits a stationary block of mass M, suspended by a massless string of length L. The bullet gets embedded in the block after impact and the two together swings up., Show that the maximum angle of swing (*i.e.*, angle made by the string with the vertical) is

$$\theta = \cos^{-1}\left(1 - \frac{M^2v^2}{2gL(M+m)^2}\right)$$

4. State and prove Varignon's theorem. What is meant by a free-body diagram?
5. With a neat sketch, explain stress-strain diagram for a ductile material.

6. What is D'Alembert's principle? What is the advantage of using the principle? How does it differ from Newton's second law of motion?

Group – C
(Long Answer Questions)

Answer any three questions. **3× 15 = 45**

7. (*a*) Determine the axial moment of inertia of the T-section shown in Fig. 1 about the centroidal axis parallel to base.

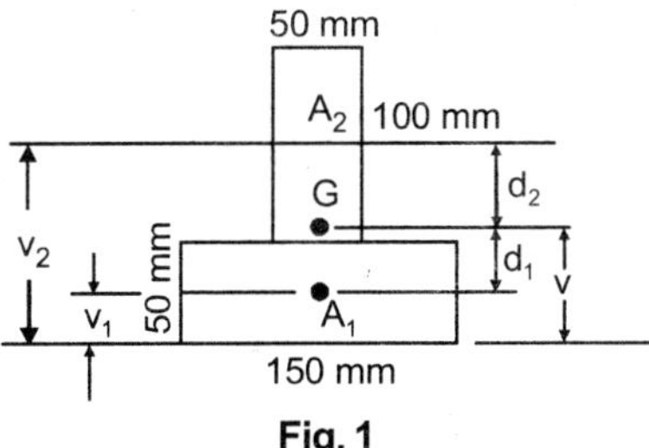

Fig. 1

(*b*) A steel tube 4.5 cm external diameter and 3 mm thick encloses centrally a solid copper bar of 3 cm dia. The bar and the tube are rigidly connected together at the ends at a temperature of 30°C. Find the stress in each metal when heated to 180°C. Also find the increase in length if original length of assembly is 30 cm. Given, $\alpha_{st} = 1.08 \times 10^{-5}$ per °C, $E_{copper} = 1.1 \times 10^{-6}$ kg/cm^2.

$$\alpha_{Copper} - \text{per } ^\circ C, E_{Steel} - 2.1 \times 10^{-6} \text{kg/cm}^2,$$

8. (*a*) A ball of weight *W* rests upon a smooth horizontal plane and has attached to its centre two strings *AB* and *AC* which pass over frictionless pulleys at *B* and *C* and carry loads P and *Q* respectively, as shown in Fig. 2. If the string *AB* is horizontal, find the angle α that the string *AC* makes with the horizontal when the ball is in a position of equilibrium. Also find the pressure *R* between the ball and the plane.

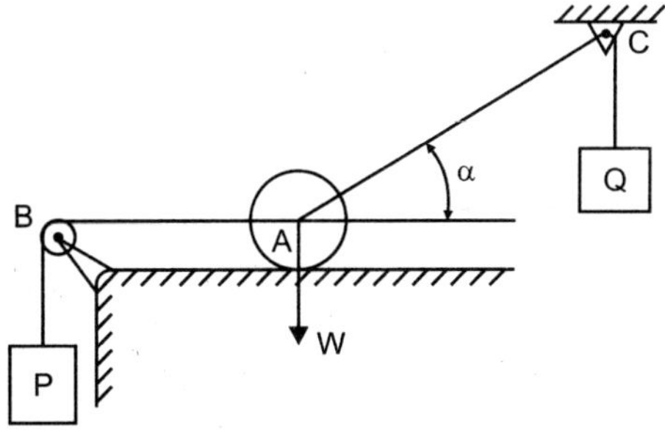

Fig. 2

(*b*) Find the acceleration of a falling weight *P* hanging over a pulley by a string connecting a block *Q* as shown in the Fig. 3, the co-efficient of friction between block *Q* and the horizontal plane if slides is *u*. Neglect inertia of the pulley and friction on its axle. Given, $P = 10$ kgf, $Q = 12$kgf, $u = \frac{1}{2}$.

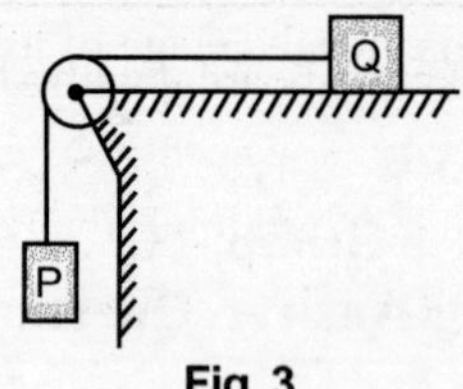

Fig. 3

9. (*a*) Determine the maximum ratio h/b for which the homogeneous block will slide without tipping under the action of force P as shown in the Fig. 4. The co-efficient of static friction between the block and the incline is u_s.

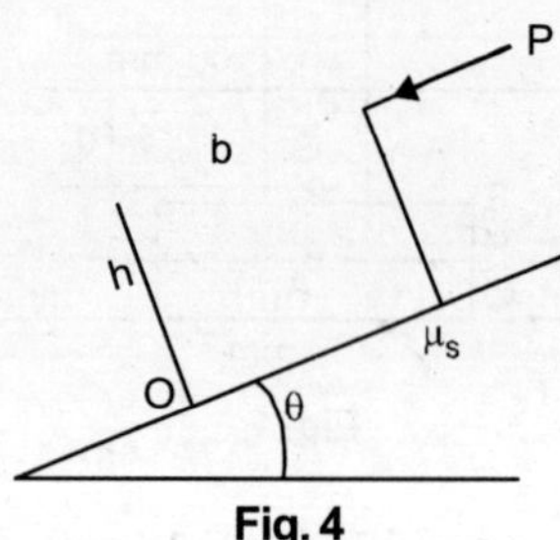

Fig. 4

(*b*) Two blocks of weight W_1 and W_2 are located on two incline planes as shown in Fig. 5. Assuming the contact surfaces to be frictionless, $W_1 = 200$ N, $W_2 = 300$ N, determine the angle θ for equilibrium.

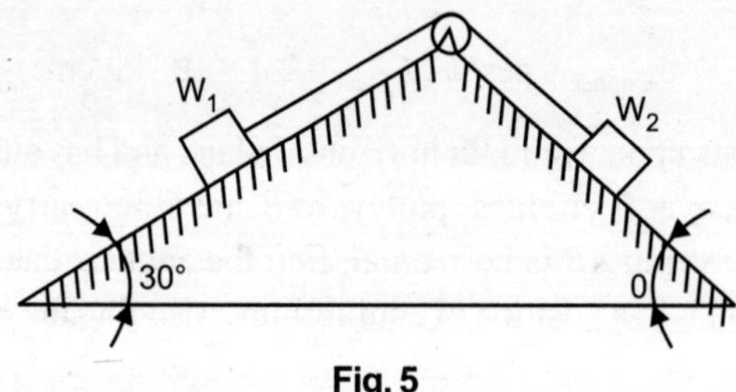

Fig. 5

10. (*a*) To anticipate the dip and hump in the road, the driver of a car applies his brakes to produce a uniform deceleration. His speed is 100 km/hr at the bottom A of the dip and 50 km/hr at the top C of the hump, which is 120 m along the road from A. If the passengers experience a total acceleration of 3 m/s^2 at A and if the radius of curvature of the hump at C is 150 m, calculate

(*i*) the radius of curvature π at A

(*ii*) the acceleration at the inflexion point B, and

(*iii*) the total acceleration at C.

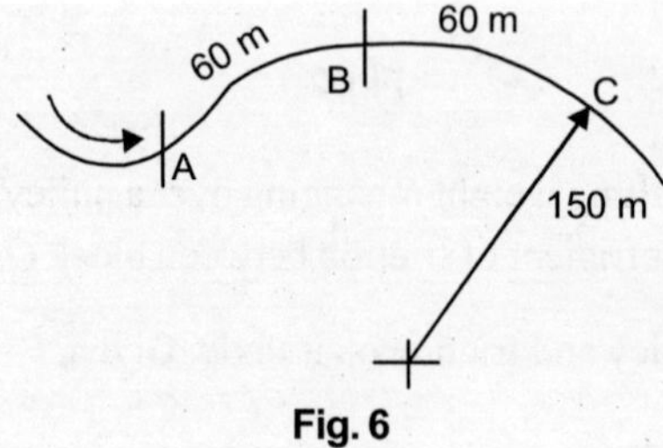

Fig. 6

(*b*) A broad jumper approaches his take-off board *A* with a horizontal velocity of 10 m/s. Determine the vertical component v_y of the velocity of his centre of gravity at take-off for him to make the jump shown in Fig. 7. What is the vertical rise *h* of his centre of gravity?

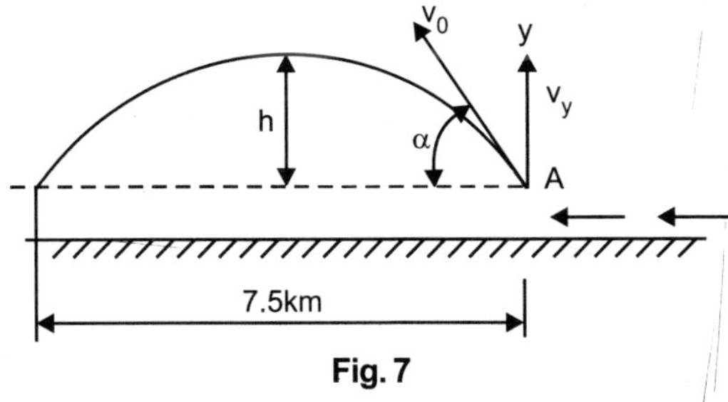

Fig. 7

11. Two equal uniform rods *AC* and *CB* are freely jointed at *C* and rest in a vertical plane with the ends *A* and *B* in contact with a rough horizontal plane. If the equilibrium be limiting and μ is the co-efficient of friction, show that

$$\sin \angle ACB = \frac{4\mu}{1 + 4\mu^2}.$$

Question Paper – 2007

Time Allotted: 3 Hours **Full Marks: 70**

GROUP – A

(Multiple Choice Type Questions) **10 × 1 = 10**

1. Choose the correct alternatives for the following:

(*i*) Two non-collinear parallel equal forces in opposite direction

(*a*) balance each other (*b*) constitute a moment
(*c*) constitute a couple (*d*) constitute a moment of couple
(*e*) constitute a resultant couple.

(*ii*) The centre of gravity of a uniform lamina lies at

(*a*) the centre of heavy portion (*b*) the bottom surface
(*c*) the mid-point of its axes (*d*) all of these
(*e*) none of these.

(*iii*) The ratio of limiting friction and reaction is known as

(*a*) co-efficient friction (*b*) angle of friction
(*c*) angle of repose (*d*) sliding friction
(*e*) friction resistance.

(*iv*) d' Alembert's principle is applied to solve problems related to

(*a*) statics (*b*) stress of a structure
(*c*) dynamics (*d*) none of these.

(*v*) Materials having same elastic properties in all directions are called

(*a*) ideal material (*b*) isotropic material
(*c*) elastic material (*d*) uniform material.

(*vi*) The energy absorbed in the body when it is strained within the elastic limit is

(*a*) strain energy (*b*) resilience
(*c*) toughness (*d*) modulus of resilience.

(*vii*) For stable equilibrium the potential energy will be

(*a*) maximum (*b*) minimum
(*c*) zero (*d*) none of these.

(*viii*) If the velocity of projection is u m/sec and the angle of projection is $\alpha°$, the maximum height of the projectile on a horizontal plane is

(a) $\frac{u^2 \cos^2 \alpha}{(2g)}$ (b) $\frac{u^2 \sin^2 \alpha}{(2g)}$ (c) $\frac{u^2 \tan^2 \alpha}{(2g)}$ (d) $\frac{u^2 \sin^2 \alpha}{g}$

(e) $\frac{u \sin \alpha}{g}$

(ix) A jet engine works on the principle of conservation of

(a) energy (b) mass (c) angular momentum
(d) linear momentum (e) none of these.

(x) When a body slides down an inclined surface (angle of inclination = θ), the acceleration f of the body is given by

(a) $f = g$ (b) $f = g \sin \theta$
(c) $f = g\cos \theta$ (d) $f = \tan \theta$ (e) $f = g/\sin \theta$.

GROUP – B
(Short Answer Type Questions)

Answer any three of the following. **3 × 5 = 15**

2. (a) State Varignon's Principle.

(b) Determine the tension in the tie rod $AC = 300$ mm when a circular roller of weight $Q = 450$ N and radius $r = 150$ mm is rest against a vertical wall at B as shown in Fig. 1.

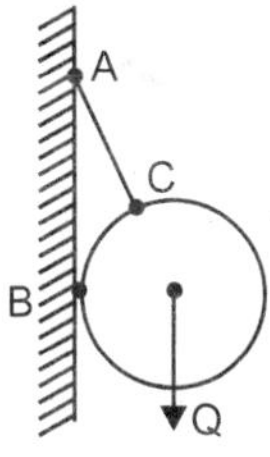

Fig. 1

3. (a) State and prove Lami's theorem. (b) State the principle of Transmissibility of forces.

4. (a) Malleability (b) Resilience (c) Toughness and (d) Poisson's ratio.

5. (a) State d' Alembert principle.

(b) The position of a particle is given by, $S = 4t^3 + 3t^2 - 18t + 5$, when S is in metre, t in second. Determine the velocity and acceleration at $t = 3$ seconds.

6. A force given by $F = 3i + 2j - 4\,k$ is applied at the point $P(1, -1, 2)$. Find the moment of the force F about the point $O(2, -1, 3)$.

GROUP – C
(Long Answer Type Questions)

Answer any three of the following. **3 × 15 = 45**

7. (*a*) The tension in the supporting cable *AB* (Fig. 2) is 10 kN. Write the force which the cable exerts on the boom *BC* as a vector T. Determine the angle θ_x, θ_y and θ_z which the line of action of *T* forms with the positive *x*-, *y*-and z-axes.

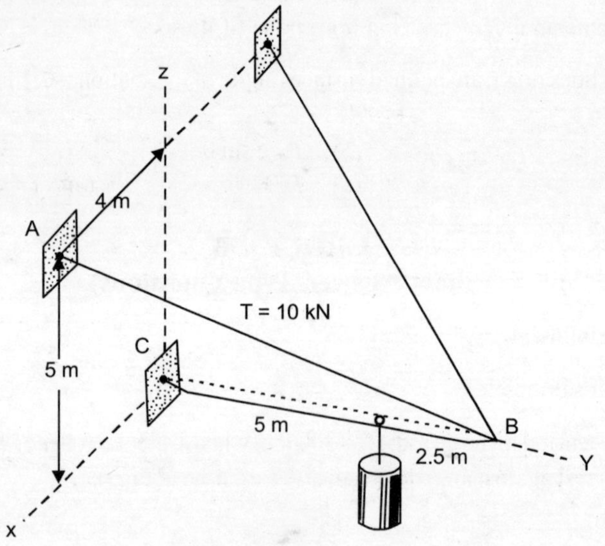

Fig. 2

(*b*) A roller of radius $r = 12$ cm and weight $Q = 5$ kN is to be rolled over a curb of height $h = 6$ cm by a horizontal force *P* applied to the end of a string wound around the circumference of the roller as shown in Fig. 3. Find the magnitude of *P* required to start the roller over the curb. There is sufficient friction between the roller surface and the edge of the curb to prevent slip at *A*.

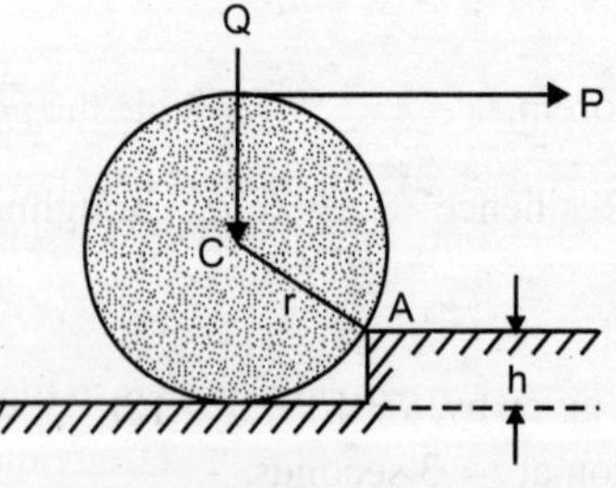

Fig. 3

8. (*a*) A slender prismatic bar *AB* of length *l* and weight *Q* stands in a vertical plane and is supported by smooth surfaces at *A* and *B* as shown in Fig. 4. Using the principle of virtual work, find the magnitude of the horizontal force *P* applied at *A* if the bar is in equilibrium.

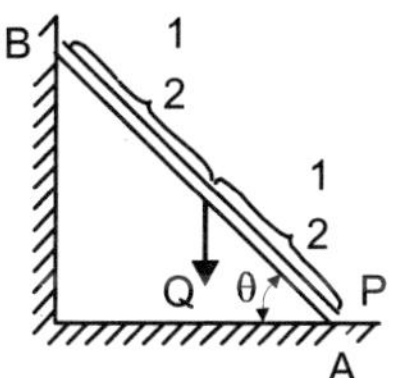

Fig. 4

(*b*) A projectile is launched with an initial speed of 200 m/s at an angle of 60° (Fig. 5) with respect to the horizontal. Compute the range *R* as measured up the incline.

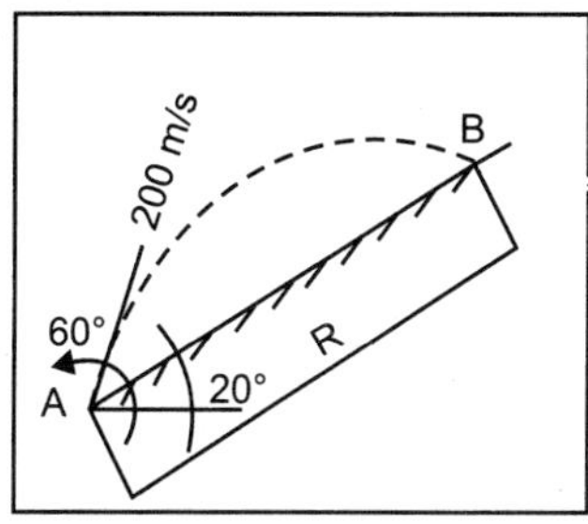

Fig. 5

9. (*a*) Two rectangular blocks of weights W_1 and W_2 are connected by a flexible cord and rest upon a horizontal and an inclined plane, respectively, with the cord passing over a pulley as shown in Fig. 6. In the particular case where $W_1 = W_2$ and the coefficient of static friction μ is the same for all contiguous surfaces, then find the angle of α of inclination of the inclined plane at which motion of the system will impend. Neglect friction in the pulley.

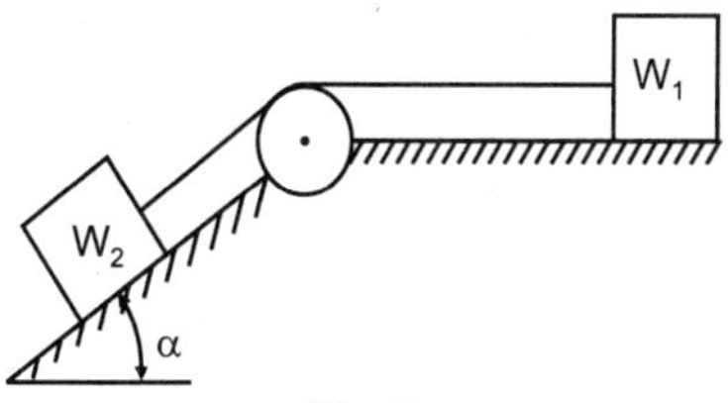

Fig. 6

(*b*) A particle is moving along a circular path having a radius of 4 m such that its position as a function of time is given by θ = cos 2*t*, where θ is in radians and *t* is in seconds. Determine the magnitude of the velocity of the particle when θ = 30°.

10. (*a*) A slender bar *AB* of length *l* which remains always in the same vertical plane has its ends *A* and *B* constrained to remain in contact with a horizontal floor and a vertical wall, respectively as shown in Fig. 7. The bar starts from a vertical position and the end *A* is moved along the flor with constant velocity v_0 so that its displacement OA = v_0 t. Find the displacement time and acceleration time equations for the vertical motion of the end *B* of the bar.

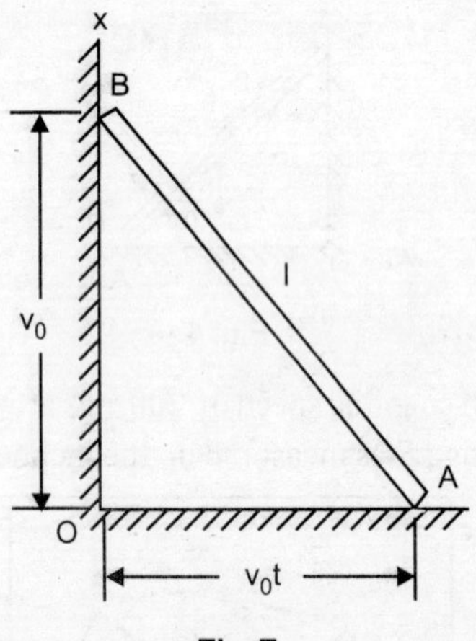

Fig. 7.

(*b*) A 20 tonnes goods train is travelling at a constant speed of 100 km/ hr while total resistance against the motion due to ground friction and air pressure is 50 N per tonne weight. Suddenly the last wagon weighing 20 tonnes gets decoupled and falls behind the main train. Determine:

(i) the acceleration and deceleration of the main train and decoupled wagon respectively

(ii) the distance between the two after 20 seconds.

11. (*a*) Referring to Fig. 8 determine the coordinates of the centre of the circular hole cut in a thin plate so that this point will be the centre of gravity of the remaining shaded area.

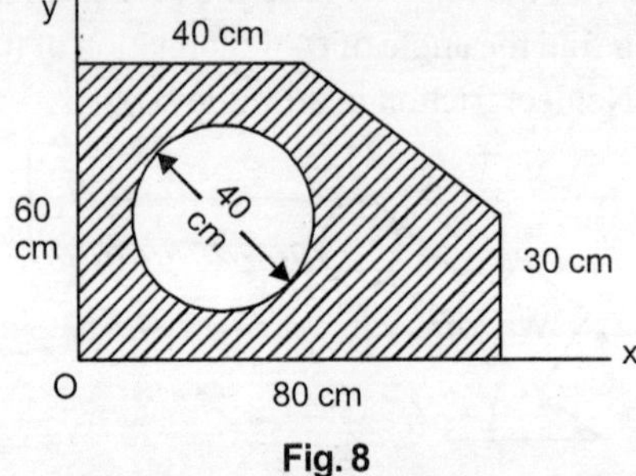

Fig. 8

(*b*) In Fig. 9 a lever is attached to a spindle by means of a square key 6 mm × 6 mnı by 2.5 cm long. If the average shear stress in the key not to exceed 700 kg/cm². What is the safe value of the load P applied to the end of the lever?

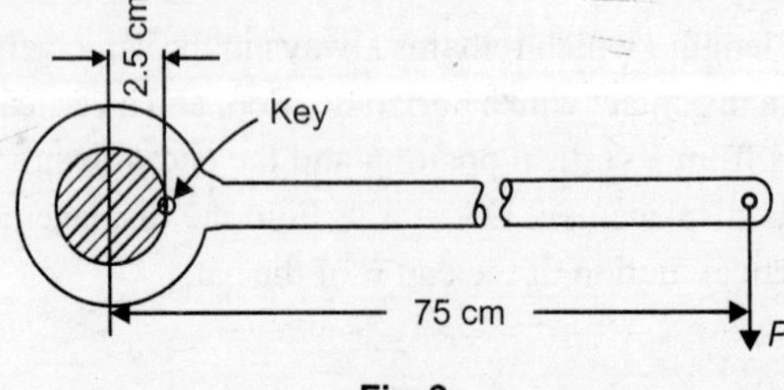

Fig. 9

Question Paper – 2008

Time: 3 Hours **Full Marks: 70**

GROUP–A

(Multiple Choice Type Questions)

1. Choose the correct alternatives for the following: **10 × 1 = 10**

(*i*) Free body diagram can be applied only in

(*a*) dynamic equilibrium problem (*b*) static equilibrium problem

(*c*) both dynamic and static equilibrium problems (*d*) nont of these

(*ii*) The conditions of equilibrium for coplanar non-concurrent forces are

(*a*) $\Sigma F_X = 0$; $\Sigma F_Y = 0$ (*b*) $\Sigma F_X = 0$; $\Sigma M = 0$

(*c*) $\Sigma F_Y = 0$; $\Sigma M = 0$ (*d*) $\Sigma F_X = 0$; $\Sigma F_Y = 0$; $\Sigma M = 0$

(*iii*) The centre of gravity of solid hemisphere of radius R is

(*a*) $3R/8$ (*b*) $R/2$

(*c*) $3R/4$ (*d*) none of these

(*iv*) Equation of motion of a particle is $s = 2t^3 - t^2 - 2$ where s is displacement in metre and t is time in second. Acceleration of the particle after 1 second will be

(*a*) 8 m/sec^2 (*b*) 9 m/sec^2

(*c*) 10 m/sec^2 (*d*) 5 m/sec^2

(*v*) When a body slides down an inclined surface of inclination θ, the acceleration f of the body is given by

(*a*) $f = g$ (*b*) $f = g \sin \theta$

(*c*) $f = g \cos \theta$ (*d*) $f = g/\sin \theta$.

(*vi*) The maximum strain energy that can be stored in a body is known as

(*a*) impact energy (*b*) resilience

(*c*) proof resilience (*d*) modulus of resilience.

(*vii*) When two ships are moving along inclined directions, then the time when the two ships will be closest together depends upon

(*a*) velocity of one of the ships (*b*) velocity of both the ships

(*c*) angle between the two directions (*d*) all of these.

(*viii*) The maximum height of a projectile on a horizontal range is

(*a*) $\frac{(u^2 \sin 2\alpha)}{2g}$ (*b*) $\frac{(u^2 \sin \alpha)}{2g}$

(*c*) $\frac{(u^2 \sin^2 2\alpha)}{2g}$ (*d*) $\frac{(u^2 \sin^2 \alpha)}{2g}$

(*ix*) The differential equation of falling body under gravity is

(*a*) $\ddot{x} = 0, \quad \ddot{y} = 0$ (*b*) $\ddot{x} = 0, \quad \ddot{y} = -g$

(*c*) $\ddot{x} = c, \quad \ddot{y} = -g$ (*d*) $\dddot{x} = 0, \quad \dddot{y} = 0$

(*e*) none of these

(*x*) If a momentum of a body is doubled, its kinetic energy will

(*a*) increase by two times (*b*) increase by four times

(*c*) remain same (*d*) get halved

(*e*) reduce to four times.

GROUP–B

(Short Answer Type Questions)

Answer any three questions. **3×5 = 15**

2. What do you mean by a Free Body Diagram? Draw the Free Body Diagrams of the following as shown below: **1 + 2 + 2**

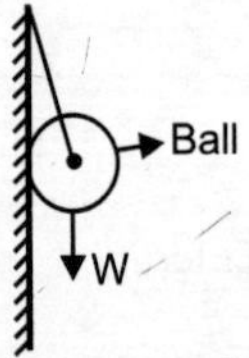

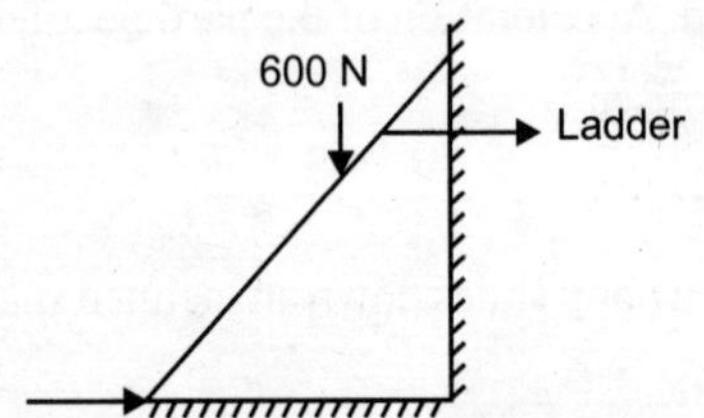

3. Determine the force P required to intend the motion of the block B shown in the figure below. Take $\mu = 0.3$ for all surfaces of contact, where μ = coefficient of friction. **5**

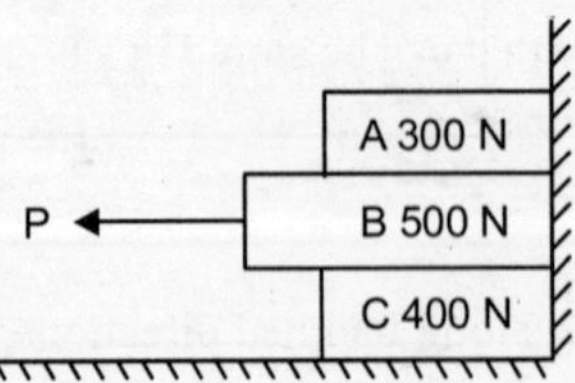

4. A force $\vec{F} = 3i - 4j + 12k$ acts at a point A whose coordinates are (1, –2, 3) m. Compute.

(*a*) moment of force about origin **3**

(*b*) moment of force about the point B (2, 1, 2) m. **2**

5. If the string AB is horizontal, find the angle that the string AC makes with the horizontal when the ball is in a position of equilibrium. Also find the pressure R between the ball and the plane.

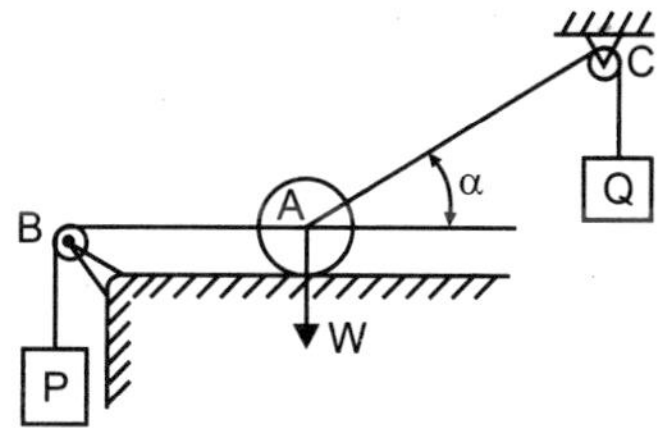

GROUP–C

(Long Answer Type Questions)

Answer any three questions. **3 × 15 = 45**

6. (*a*) A short semi-circular right cylinder of radius r and weight W rests on a horizontal surface and is pulled at right angles to its geometric axis by a horizontal force P applied at the middle B of the front edge as shown. Find the angle α that the flat face will make with the horizontal plane just before sliding begins if the coefficient of friction at the line of contact A is μ. The gravity force W must be considered as acting at the centre of gravity C as shown in the figure.

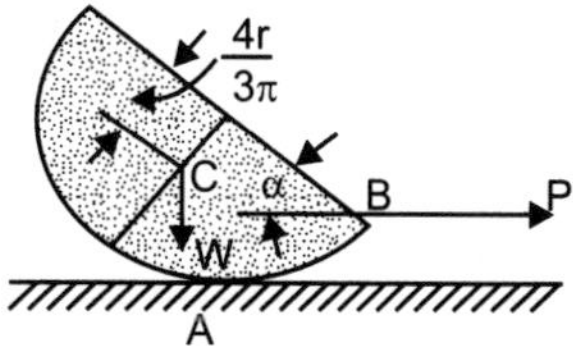

(*b*) Determine the coordinates of the centroid C of the area of the circular sector OBD of radius r and central angle α. **8 + 7**

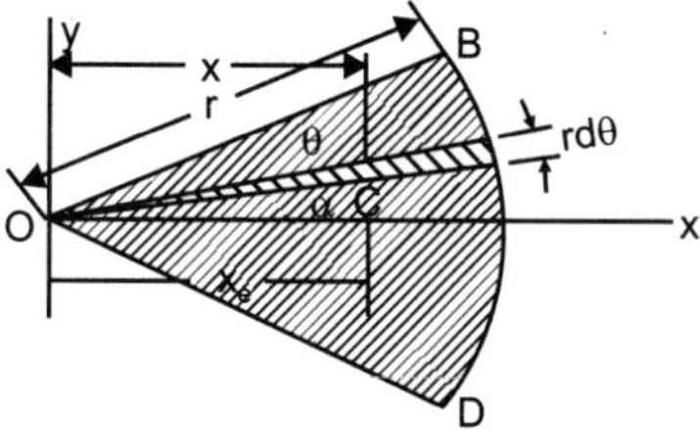

7. (*a*) The following details refer to the bar as shown:

Portion	Length	Cross-Section
AB	600 mm	40 × 40 mm
BC	800 mm	30 × 30 mm
CD	1000 mm	20 × 20 mm

If the load P_4 = 80 kN, P_2 = 60 kN and P_3 = 40 kN, find the extension of the bar, where $E = 2 \times 10^5$ N/mm^2. **7**

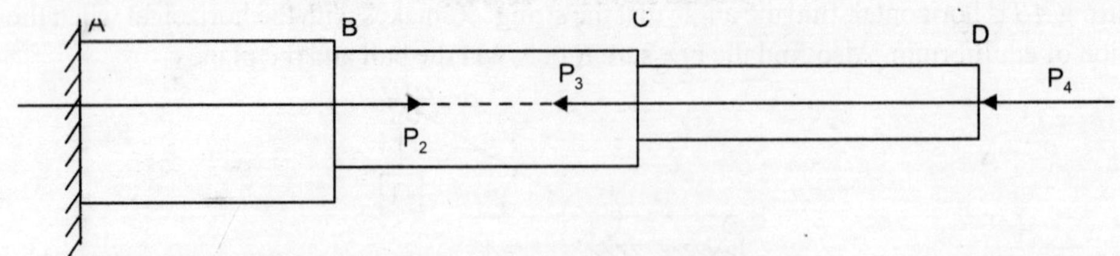

(*b*) Calculate the increase in stress for each segment of the compound bar shown in figure, if the temperature increases by 100°F. Assume that the supports are unyielding and that the bar is suitably braced against buckling. **8**

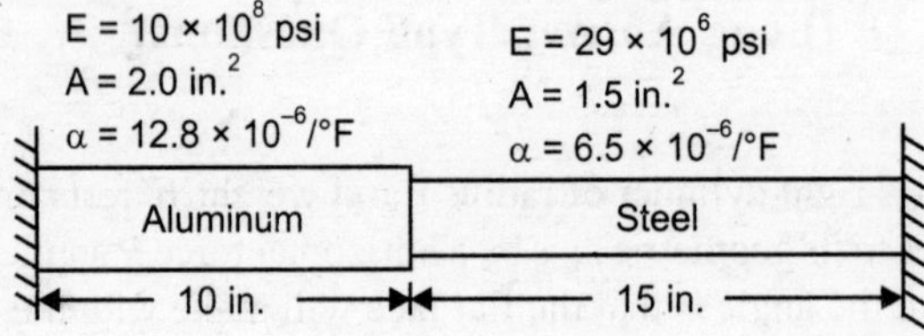

8. (*a*) A spring normally 150 mm long is connected to the two masses as shown in figure and compressed 50 mm. If the system is released on a smooth horizontal plane, what will be the speed of each block when the spring is again in is normal length? The spring constant is 2100 N/m. **8**

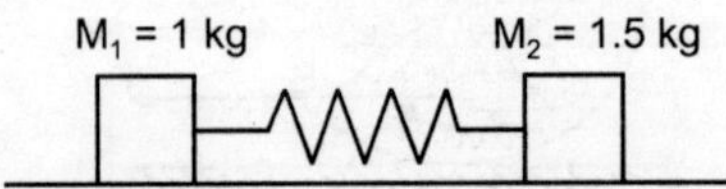

(*b*) A projectile is aimed at a mark on the horizontal plane through the point of projection and falls 12 m short when the angle of projection is 15°, while it overshoots the mark by 24 m when the same angle is 45°. Find the angle of projection to hit the mark. Assume no air resistance. Consider the velocities of projections are constant in all cases. **7**

9. (*a*) Two inclined rollers, each of weight Q = 100 kgf are supported by aninclined plane and a vertical wall as shown below. Assuming smooth surfaces, find the reactions induced at the points *A*, *B* and *C*. *7*

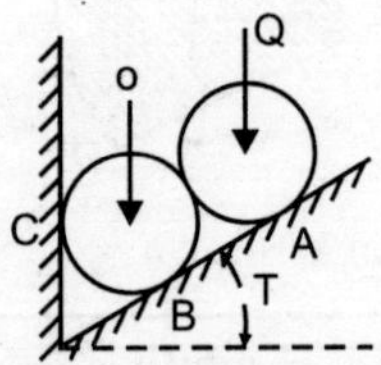

(*b*) Two blocks of weight W_1 and W_2 rest as shown. If the angle of friction of each block is φ, find the magnitude and direction of the least force *P* applied to the upper block that will induce sliding. **8**

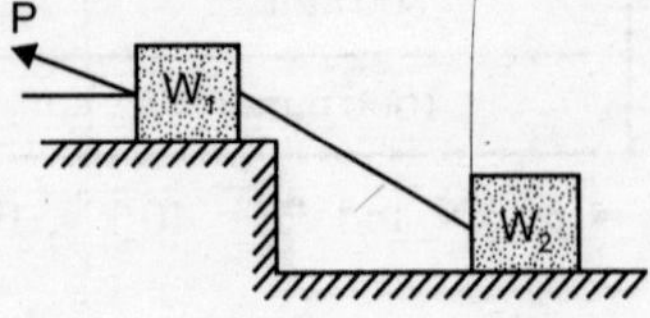

Question Paper – 2009

CS/B.Tech/SEM-1/ME-101/2009-10

Time Allotted: 3 Hours **Full Marks: 70**

The figures in the margin indicate full marks.
Candidates are required to give their answers in their own words as far as practicable.

GROUP–A
(Multiple Choice Type Questions)

1. Choose the correct alternatives for the following: **10 × 1 = 10**

(*i*) Lami's theorem is applicable to

(*a*) equilibrium of two co-planar, concurrent forces
(*b*) equilibrium of three co-planar, concurrent forces
(*c*) equilibrium of three co-planar, non-concurrent forces
(*d*) none of these

(*ii*) Strain energy is the

(*a*) maximum energy which can be stored in a body
(*b*) energy stored in a body when stressed to the elastic limit
(*c*) energy stored in a body when stressed to the breaking point
(*d*) none of these.

(*iii*) Poisson's ratio is defined as

(*a*) longitudinal stress by lateral stress (*b*) lateral stress by longitudinal stress
(*c*) longitudinal strain by lateral strain (*d*) lateral strain by longitudinal strain.

(*iv*) Free body diagram of a body is drawn

(*a*) by isolating the body its surrounding (*b*) by indicating the forces acting on it
(*c*) both of these (*d*) none of these.

(*v*) If a momentum of a body is doubled, its kinetic energy will

(*a*) increase by two times (*b*) increase by four times
(*c*) remain same (*d*) get halved
(*e*) reduced to four times.

(*vi*) A body falling freely from a height of 10 m rebounds from the floor. If it losses 20% of its energy ir the impact how high will it rebound?

(*a*) 10 m (*b*) 8 m
(*c*) 12 m (*d*) none of these.

vii) The dot product of two orthogonal vector is

(*a*) one (*b*) zero
(*c*) no definite value (*d*) none of these.

(*viii*) The centre of gravity of a uniform lamina lies at

(*a*) the centre of heavy portion (*b*) the bottom surface

(*c*) the mid-point of its axis (*d*) none of these.

(*ix*) If the velocity of projectile is u and the angle of projection is a, the maximum height attained by a projectile above the horizontal plane is

(*a*) $u^2 \cos^2 \alpha/2g$ (*b*) $u^2 \sin^2 \alpha/2g$

(*c*) $u^2 \tan^2 \alpha/2g$ (*d*) $u^2 \sin^2 \alpha/g$.

(*x*) Three forces $\sqrt{3}p$, p and $2p$ acting on a particle are in equilibrium. If the angle between first and second be 90°, the angle between second and third will be

(*a*) 30° (*b*) 60°

(*c*) 120° (*d*) 150°.

GROUP–B

(Short Answer Type Questions)

nswer any *three* of the following. **3 × 5 = 15**

(*a*) State D'Alembert's principles.

(*b*) A smooth circular cylinder of radius 1.5 is lying in a rectangular groove is shown in Fig. 1. Find the reactions at the surfaces of contact, if there is no friction and the cylinder weighs 1000 N. **1 + 4**

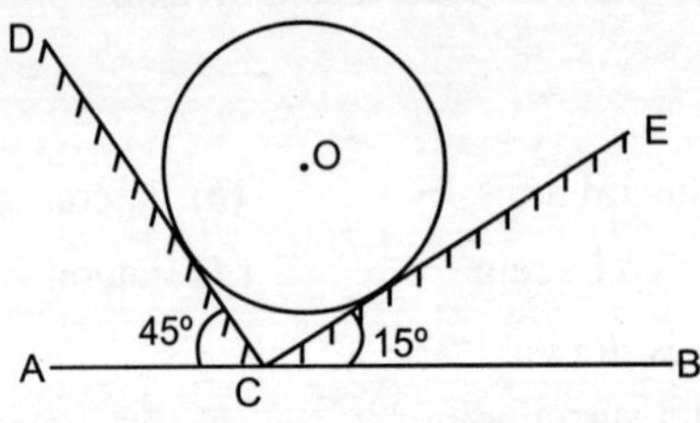

Fig. 1

Refer to the Fig. 2, determine the range of values of mass m_0 so that the 100 kg block will neither move up nor slip down the inclined plane. The coefficient of static friction for the surfaces in contact is 0.3. **5**

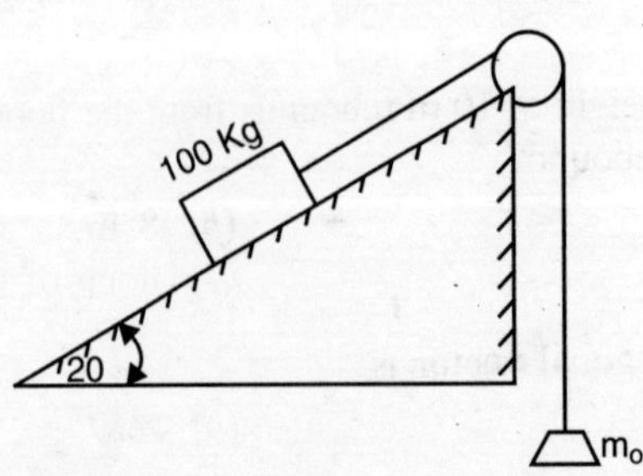

Fig. 2

4. (*a*) State Varignon's principle.

   (*b*) A circular roller of weight 100 N and radius 10 cm hangs by a ties rod AB = 20 cm and rests against a smooth vertical wall at C as shown in Fig. 3. Determine the force F in the rod.

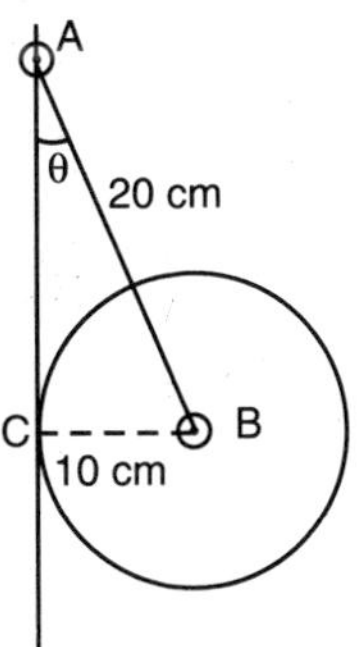

Fig. 3

5. Referring to Fig. 4, r = 12 cm. Q = 500 N and h = 6 cm. Find magnitude of P required to start the roller over curb.

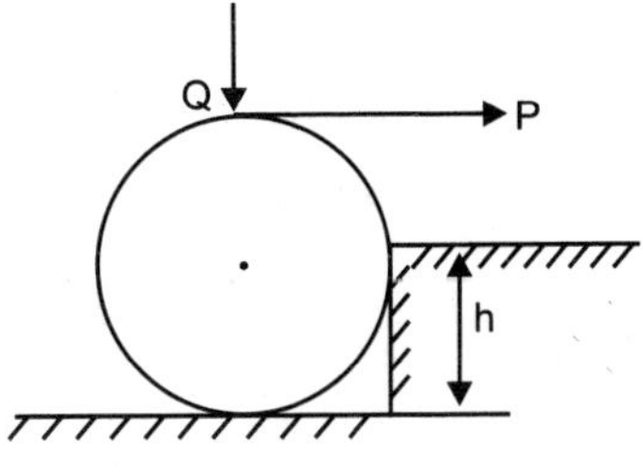

Fig. 4

6. Two smooth circular cylinders of Fig. 5, each of weight W = 100 N and radius r = 6 cm are connected by a string AB of length l = 16 cm and rest upon a horizontal plane, supporting a third cylinder of weight Q = 200 N and radius r = 6 cm above them. Find the tension S in the string AB and the pressure produced by the floor at points of contact D and E.

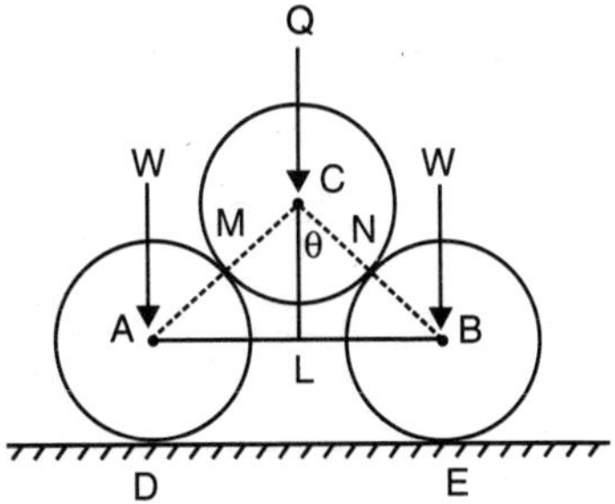

Fig. 5

GROUP–C

(Long Answer Type Questions)

Answer any *three* of the following. **3 × 15 = 45**

7. (*a*) A 150 kg man stands on the mid-point of a 50 kg ladder as shown in Fig. 6. Assuming that floor and the wall are perfectly smooth, find the reactions at points *A* and *B*.

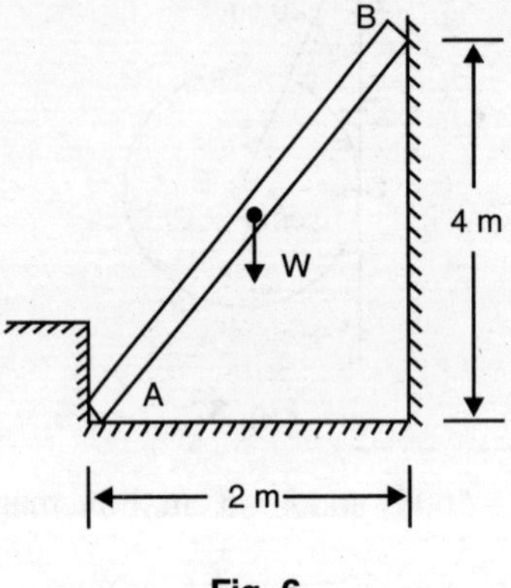

Fig. 6

(*b*) Determine the moment of inertia for the *T* section (as shown in Fig. 7) with respect to a centroidal axis parallel to *x*-axis. All dimensions are in mm. **8 + 7**

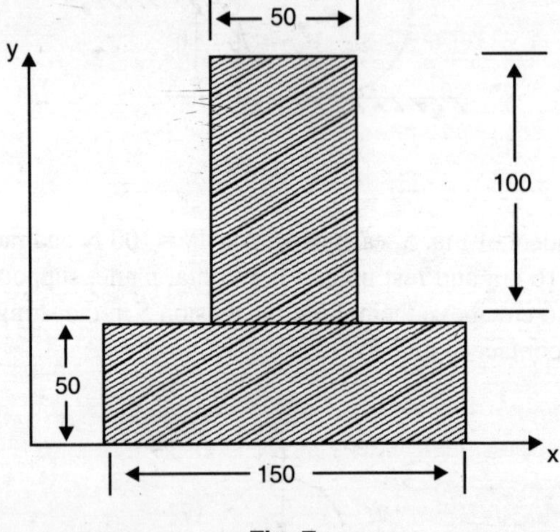

Fig. 7

8. (*a*) Prove that the volumetric strain of a rectangular bar is the algebraic sum of strains of length, width and height.

(*b*) Show that elongation of a conical bar under its own weight is independent of its base diameter but on length only.

(*c*) Determine the strain energy stored within a bar of length *l*, cross-sectional area *A*, density ρ and modulus of elasticity *E*, hanging vertically due to its own height. **6 + 4 + 5**

9. (*a*) Two spheres *P* and *Q* rests in the channel as shown in Fig. 8. The sphere *P* has a diameter 400 mm and weight of 200 N, whereas the sphere *Q* has diameter 500 mm and weight 500 N. If bottom width of the channel is 500 mm and with one side vertical and other side inclined at 60°, determine the reaction induced in the contacts. **8**

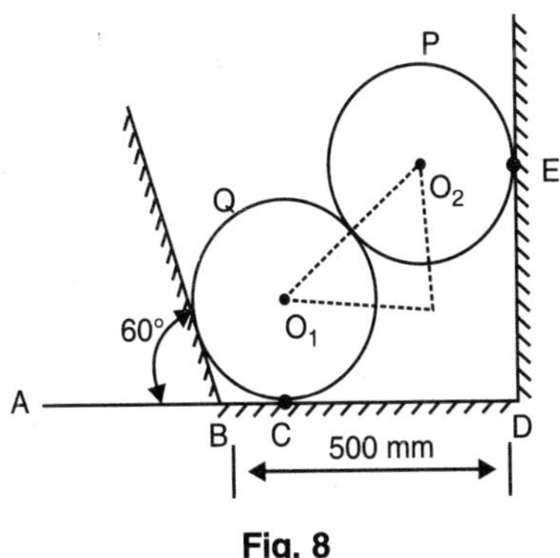

Fig. 8

(*b*) In the Fig. 9 shown, find the minimum value of horizontal force *P* applied to the lower block that will keep the system in equilibrium. Given, coefficients of friction between lower block and floor = 0.25, between the upper block and the vertical wall = 0.30, between the two blocks = 0.20. **8 + 7**

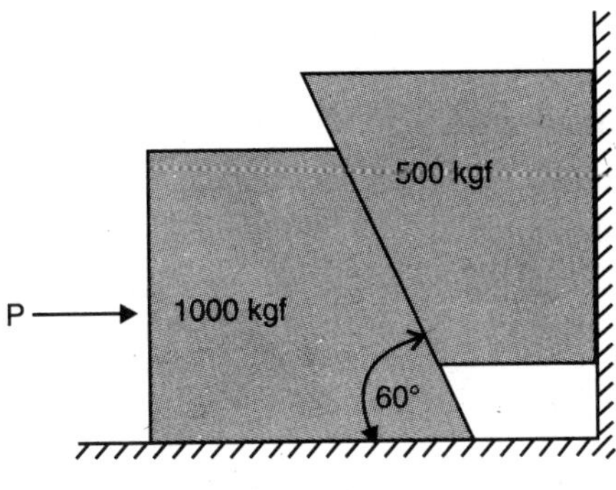

Fig. 9

10. (*a*) State the principle of virtual work. **3**

(*b*) Two blocks weighing W_1 and W_2 resting on smooth inclined planes are connected by an inextensible string passing over a smooth pulley as shown in Figure 10. Find the value of W_2 when $W_1 = 500$ N and $\alpha = 30°$, $\beta = 60°$. **7**

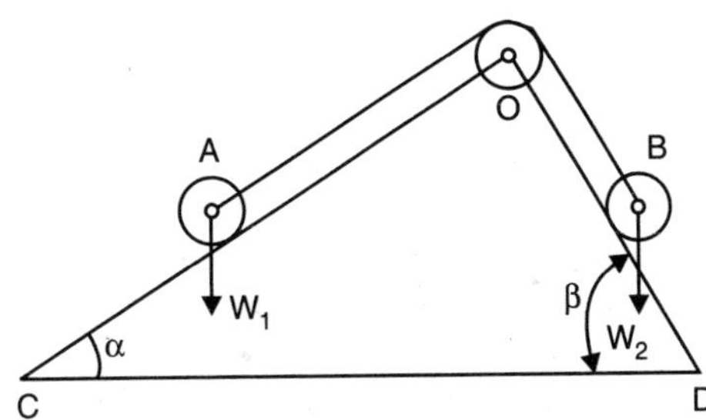

Fig. 10

(*c*) Determine velocity V of the falling weight W of the system as shown in Fig. 11 as a function of its displacement from the initial position of rest. Assume weight of the cylinder as $2W$. **5**

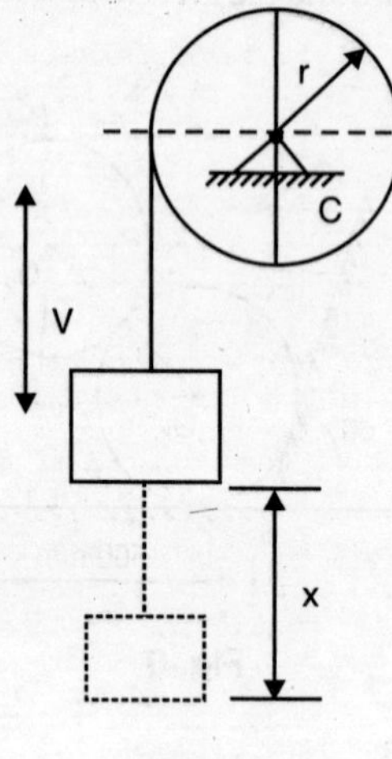

Fig. 11

11. (*a*) From top of a tower, 60 m high a bullet is fired at an angle of 20° up the horizontal with velocity 120 m/s. Determine

(*i*) time of flight

(*ii*) horizontal range of ground

(*iii*) maximum height of the bullet from ground

(*iv*) velocity of the bullet after 8 seconds.

Assume horizontal ground at the foot of the tower.

(*b*) Determine the tension in the strings and accelerations of two blocks of masses 150 kg and 50 kg connected by a string and a frictionless, weightless pulley as shown in Fig. 12. **10 + 5**

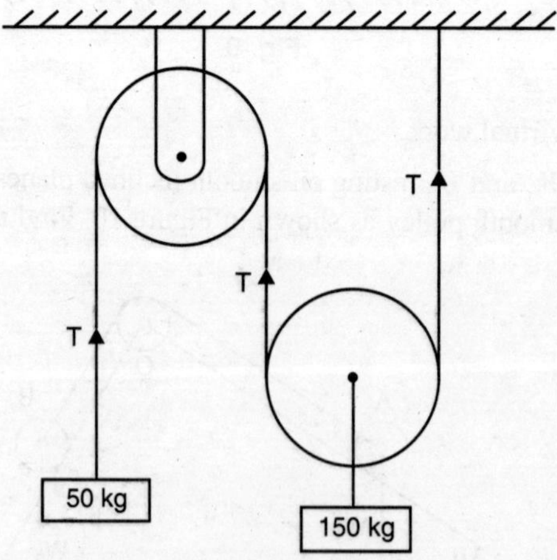

Fig. 12

CS/B.TECH (NEW)/SEM-1/ME 101/2010-11

ENGINEERING MECHANICS

Time Allotted : 3 Hours *Full Marks : 70*

The figures in the margin indicate full marks
Candidates are required to give their answers in their own words as far as practicable

GROUP A
(Multiple Choice Type Questions)

1. Choose the correct alternatives for the following: **10 × 1 = 10**

(*i*) Mass moment of inertia of body is

(a) moments of its inertia

(b) rotational analogue of mass

(c) Inertial moment about the centroidal axis

(d) none of the above

(*ii*) The centre of percussion of a rigid body is a point

(a) through which resultant of all forces acts

(b) where minimum external force acts

(c) where impact is made

(d) all of the above

(*iii*) Two coplanar couples having equal and opposite moments

(a) balance each other

(b) produce a couple and an unbalance force

(c) an equivalent

(d) all of the above

(*iv*) The moment of inertia of a rectangle having base b and height h with respect to its base is

(a) $\frac{1}{3}h^2h^2$ (b) $\frac{1}{3}b^3h$

(c) $\frac{1}{3}bh^3$ (d) $\frac{1}{3}b^2h^2$

(*v*) The D'Alembert's principle

(a) is based upon the presence of inertia force

(b) provides advantage over Newton's law

(c) is purely a hypothetical principle

(d) allows a dynamic problems to be treated as a static one

(*vi*) Centroid of a line segment

(a) must lie on the line

(b) may not lie on the line

(c) must be same as the centre of gravity

(d) none of the above

(*vii*) Volumetric strain of a rectangular body subjected to an axial force, in terms on terms of linear strain ε and Poisson's ratio √ is given by

(a) □[1+2√] (b) □[1−2√]

(c) □[1+√] (d) □[1−√]

(*viii*) Poisson's ratio is defined as

(a) lateral stress and lateral strain

(b) longitudinal stress and longitudinal strain

(c) lateral stress and longitudinal stress

(d) none of the above

(*ix*) Relative velocity of $\vec{A}$ with respect to $\vec{B}$ is defined as

(a) $\vec{V}_{A/B} = \vec{V}_B - \vec{V}_A$ (b) $\vec{V}_{A/B} = \vec{V}_A - \vec{V}_B$

(c) $\vec{V}_{A/B} = \vec{V}_B + \vec{V}_A$ (d) none of these

(*x*) When a change in length takes place, the strain is known as

(a) linear strain (b) lateral strain

(c) shear strain (d) volumetric strain

GROUP B

(Short Answer Type Questions)

Answer any *three* of the following:

3 × 5 = 15

2. The position vector of a particle moving in the *x*-*y* plane at time t = 3.60 s is 2.76 jm. At t = 3.62 s its position vector has become 2.791–3.33 jm. Determine the magnitude v of its average velocity during this interval and the angle θ made by v with x -axis.

3. What do you mean by a free body diagram? Draw the FBD from the given Fig. 1. **1 + 4**

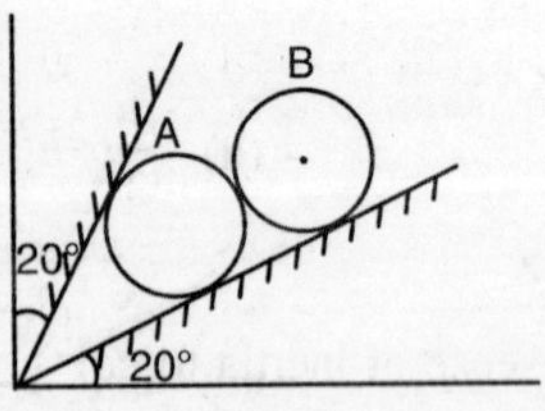

Fig. 1

4. If two equal tensions T in the pulley cable are 400 N, cable as shown in Fig. 2 express in vector notation the force R exerted on the pulley by the two tensions. Determine the magnitude of R.

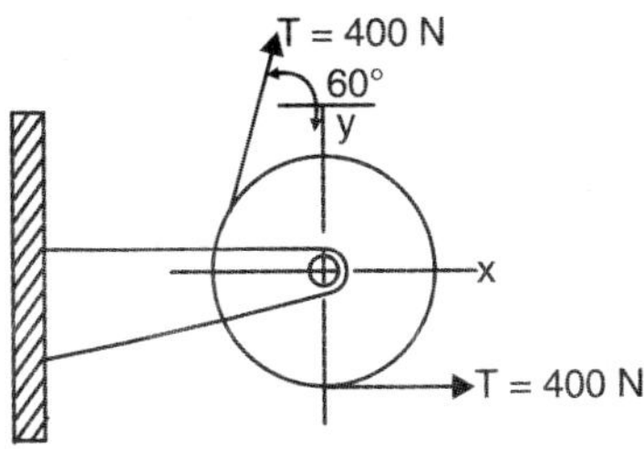

Fig. 2

5. (*i*) State and prove Lami's theorem.

(*ii*) Define free-body diagram.

3 + 2

6. A bullet of mass *m* moving with a horizontal velocity *v*, hits a stationary block of mass *M*, suspended by a massless string of length *L*. The bullet gets embedded in the block after impact and the two together swings up. Show that the maximum angle of swing (*i.e.* angle made by the string with the vertical) is

$$\theta = \cos^{-1}\left(1 - \frac{m^2 v^2}{2gL(M+m)^2}\right)$$

GROUP C

(Long Answer Type Questions)

Answer any *three* of the following:

3 × 15 = 45

7. (a) A 50 N block is released from rest on an inclined plane which is making an angle of 35° to the horizontal (Fig. 3). The block starts from 'A' slides down a distance of 1–2 m and strikes a spring with a stiffness of 8 kN/m. The coefficient of friction between the inclined plane and the block is 0.25. Determine (*i*) the amount the spring gets compressed and (*ii*) distance of the block will rebound up the plane from the compressed position.

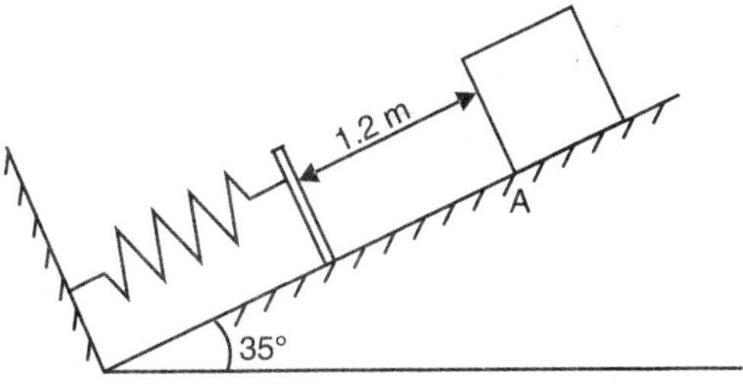

Fig. 3 **8**

(b) A reinforced concrete column having cross-section 300 mm × 300 mm is provided with 9 bars of 20 mm diameter (Fig. 4). The column carried a load of 300 KM. Find the stress developed in the steel bars and concrete. Take Es = 2.1×10^5 N/mm^2 and Ec = 0.14×10^5 N/mm^2.

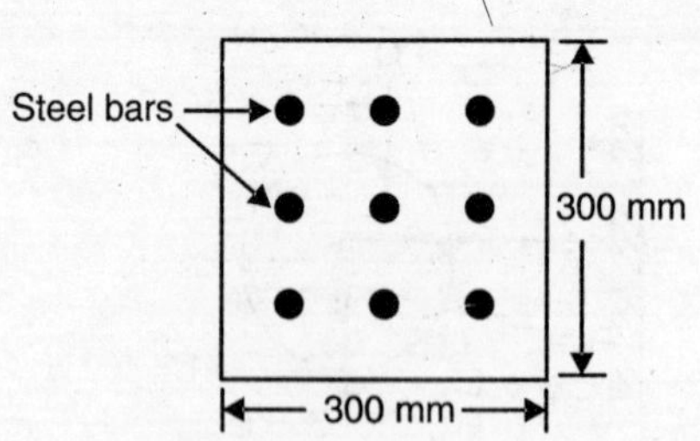

Fig. 4 **7**

8. A rigid bar AB is hinged to a vertical wall and supported horizontally by a tie rod CD as shown in Fig. 5. The cross-sectional area of CD is A = 0.5 sq. cm and its allowable stress in tension is σ_w = 1, 500 kg/sq. cm. Find the safe value of P and the corresponding vertical displacement Δ_B of B. Modulus of elasticity of tie rod E = 2×10^6 kg/sq. cm.

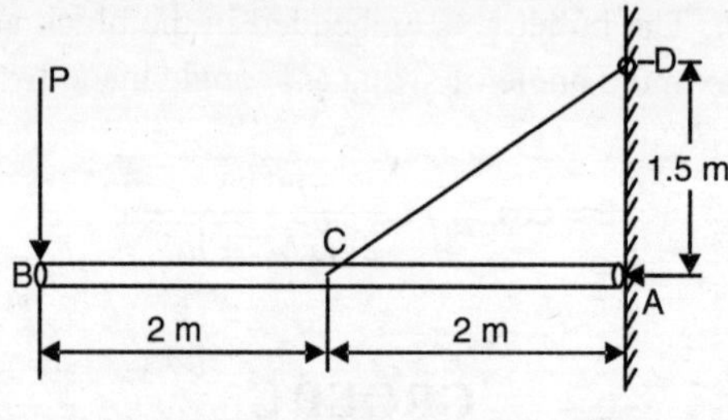

Fig. 5

9. (a) A block of weight W1 = 200 N rests on a horizontal surface and supports on top of it, another block of weight W2 = 50 N. The block W2 is attached to a vertical wall by the inclined string AB. Find the magnitude of the horizontal force P applied to the lower block as shown in Fig. 6, that will be necessary to cause slipping to impend. The coefficient of static friction for all contiguous surfaces is μ = 0.3.

7

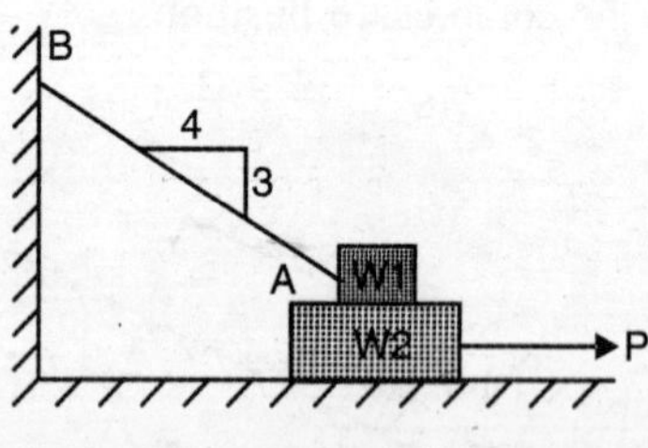

Fig. 6

(b) Two smooth circular cylinders, each of weight W = 100 N and radius r = 6 cm are connected at the centres by a string AB of length l = 16 cm and rest upon a horizontal plane, supporting above them a third cylinder of weight Q = 200 N and radius l = 6 cm as shown in Fig. 7. Find the force S in the string AB and pressure produced on the floor at D and E.

8

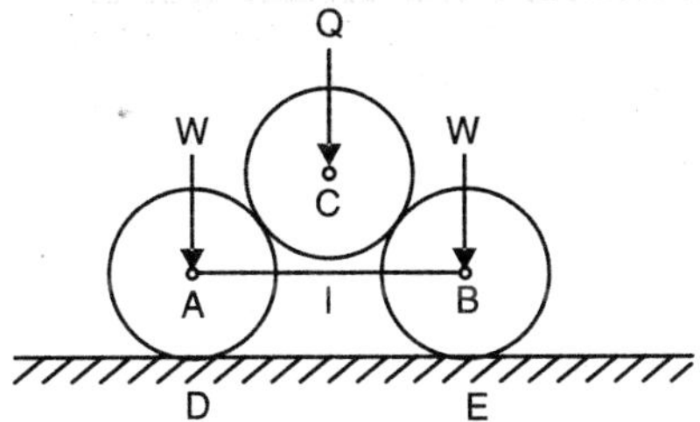

Fig. 7

10. (a) Two rollers of diameters 60 mm and 30 mm weighing 200 N and 150 N respectively. They are supported by an inclined plane and vertical walls as shown in Fig. 8. Assuming smooth surfaces, determine the reactions at the contact surfaces.

(b) Find out the moment of inertia about centroidal axes of an area as shown in Fig. 9.

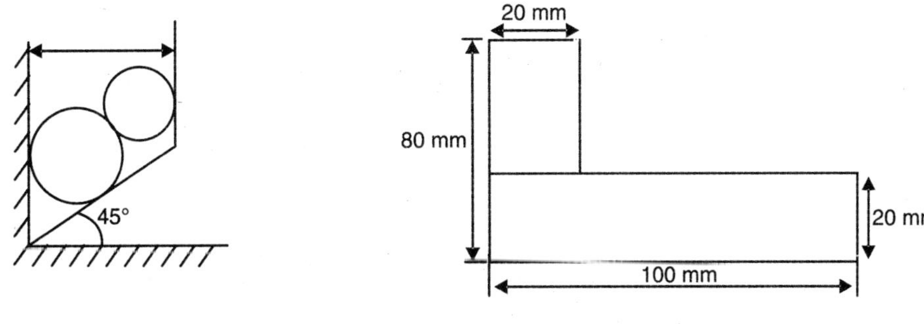

Fig. 8 Fig. 9

11. (a) Calculate the increase in stress for each segment of the compound bar shown in Fig. 10. If the temperature increases by 100°F, assume that the supports are unyielding and that the bar is suitably braced against buckling.

8

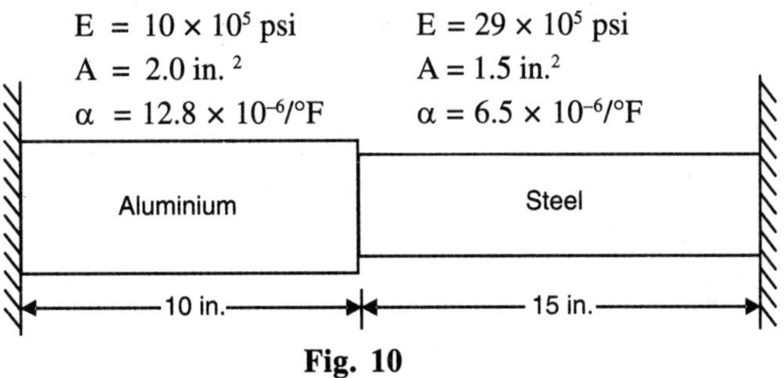

Fig. 10

(b) In Fig. 11, a lever is attached to a spindle 2.5 cm in diametre by means of a square key 6 mm × 6 mm if the average shear stress in the key not to exceed 700 N/cm². What is the safe value of the load P applied to the end of the lever?

7

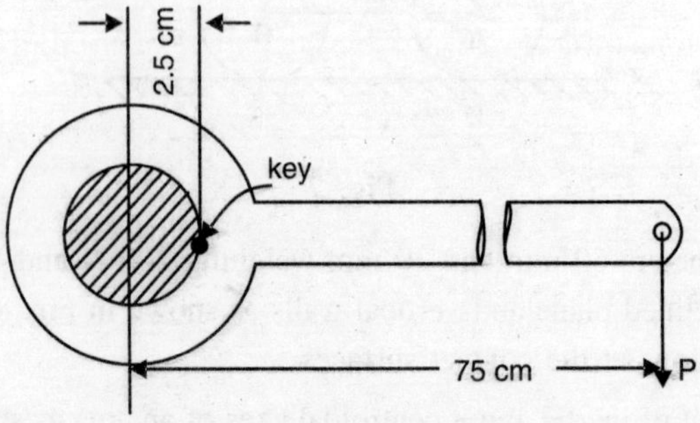

2.5 cm
key
75 cm
P

Fig. 11

Bibliography

1. Timoshenko, S. and Young, D. H., *Engineering Mechanics*, McGraw-Hill International Book Company, Singapore.
2. Tayal, A. K., *Engineering Mechanics: Statics and Dynamics*, Umesh Publications, Delhi.
3. Bansal, R. K., *A Text Book of Engineering Mechanics*, Laxmi Publication (P) Ltd., New Delhi.
4. Chakraborty, S., *Fundamental Concepts in Engineering Mechanics*, Everest Publishing House, Pune.
5. Bhavikatti, S.S. and Hegde, A.V., *Problems and Solutions in Enginering Mechanics*, New Age International Publishers, New Delhi.
6. Meriam, J. L. and Kraige, L.G., *Engineering Mechanics: Statics and Dynamics*, John Wiley & Sons, Inc., New Delhi.
7. Beer-Johnston, *Vector Mechanics for Engineers: Statics & Dynamics*, TMH Publishing Company Ltd., New Delhi.
8. Shames, I. H., *Engineering Mechanics: Statics & Dynamics*, PHI Pvt. Ltd., New Delhi.
9. Timoshenko, S.P. and Young, D. H., *Element of Strength of Materials*, EWP Pvt. Ltd., New Delhi.
10. Bansal, R. K., *A Text Book of Strength of Materials*, Laxmi Publications (P) Ltd., New Delhi.
11. Prakash Rao, D. S., *Strength of Materials,* University Press (I) Ltd., Hyderabad.

INDEX